Lecture Notes in Physics

Editorial Board

R. Beig, Wien, Austria
W. Beiglböck, Heidelberg, Germany
W. Domcke, Garching, Germany
B.-G. Englert, Singapore
U. Frisch, Nice, France
P. Hänggi, Augsburg, Germany
G. Hasinger, Garching, Germany
K. Hepp, Zürich, Switzerland
W. Hillebrandt, Garching, Germany
D. Imboden, Zürich, Switzerland
R. L. Jaffe, Cambridge, MA, USA
R. Lipowsky, Golm, Germany
H. v. Löhneysen, Karlsruhe, Germany
I. Ojima, Kyoto, Japan
D. Sornette, Zürich, Switzerland
S. Theisen, Golm, Germany
W. Weise, Garching, Germany
J. Wess, München, Germany
J. Zittartz, Köln, Germany

The Lecture Notes in Physics

The series Lecture Notes in Physics (LNP), founded in 1969, reports new developments in physics research and teaching – quickly and informally, but with a high quality and the explicit aim to summarize and communicate current knowledge in an accessible way. Books published in this series are conceived as bridging material between advanced graduate textbooks and the forefront of research to serve the following purposes:

• to be a compact and modern up-to-date source of reference on a well-defined topic;

• to serve as an accessible introduction to the field to postgraduate students and nonspecialist researchers from related areas;

• to be a source of advanced teaching material for specialized seminars, courses and schools.

Both monographs and multi-author volumes will be considered for publication. Edited volumes should, however, consist of a very limited number of contributions only. Proceedings will not be considered for LNP.

Volumes published in LNP are disseminated both in print and in electronic formats, the electronic archive is available at springerlink.com. The series content is indexed, abstracted and referenced by many abstracting and information services, bibliographic networks, subscription agencies, library networks, and consortia.

Proposals should be sent to a member of the Editorial Board, or directly to the managing editor at Springer:

Dr. Christian Caron
Springer Heidelberg
Physics Editorial Department I
Tiergartenstrasse 17
69121 Heidelberg/Germany
christian.caron@springer.com

Günter Reiter Gert R. Strobl (Eds.)

Progress in Understanding of Polymer Crystallization

Editors

Günter Reiter
CNRS-UHA
Institut de Chimie
des Surfaces et Interfaces (ICSI)
15 rue Jean Starcky
68057 Mulhouse Cedex, France
E-mail: G.Reiter@uha.fr

Gert R. Strobl
Universität Freiburg
Fakultät für Physik und Mathematik
Hermann-Herder-Str. 3
79104 Freiburg, Germany
E-mail: gert.strobl@physik.uni-freiburg.de

Günter Reiter and Gert R. Strobl, *Progress in Understanding of Polymer Crystallization*, Lect. Notes Phys. 714 (Springer, Berlin Heidelberg 2007), DOI 10.1007/b11903420

Ccost

COST is an intergovernmental European framework for international cooperation between nationally funded research activities. COST creates scientific networks and enables scientists to collaborate in a wide spectrum of activities in research and technology. COST activities are administered by the COST Office.

European
Science
Foundation

ESF provides the COST Office through an EC contract.

COST is supported by the EU RTD Framework Programme.

Library of Congress Control Number: 2006936104

ISSN 0075-8450
ISBN-10   3-540-47305-X Springer Berlin Heidelberg New York
ISBN-13   978-3-540-47305-3 Springer Berlin Heidelberg New York

This work is subject to copyright. All rights are reserved, whether the whole or part of the material is concerned, specifically the rights of translation, reprinting, reuse of illustrations, recitation, broadcasting, reproduction on microfilm or in any other way, and storage in data banks. Duplication of this publication or parts thereof is permitted only under the provisions of the German Copyright Law of September 9, 1965, in its current version, and permission for use must always be obtained from Springer. Violations are liable for prosecution under the German Copyright Law.

Springer is a part of Springer Science+Business Media
springer.com
© Springer-Verlag Berlin Heidelberg 2007

The use of general descriptive names, registered names, trademarks, etc. in this publication does not imply, even in the absence of a specific statement, that such names are exempt from the relevant protective laws and regulations and therefore free for general use.

Typesetting: by the author and techbooks using a Springer LATEX macro package
Cover design: WMXDesign GmbH, Heidelberg

Printed on acid-free paper   SPIN: 11903420   54/techbooks   5 4 3 2 1 0

Preface

The present volume of the Lecture Notes in Physics, entitled ***Progress in Understanding of Polymer Crystallization***, originated from the series of biennial Discussion Meetings on Polymer Crystallization in the Black Forest initiated in 1999. The meetings were triggered by new experimental observations that cannot be treated within the framework of conventional wisdom. They were debated together with results of computer simulations and new theoretical approaches. Various new ideas are now emerging from these discussions that deserve to be presented in a connected manner and in written form. Based on a positive response from the participants, we decided to collect these findings and views in a volume of the *Lecture Notes in Physics*.

The compilation of contributions presented here covers all aspects of polymer crystallization, and includes general views, specific observations, and also comparisons with previously proposed concepts in order to set them into perspective to the currently discussed ideas. Thus, this compiled collection reflects the state of the art in polymer crystallization and deals with various topics:

- Novel general views and concepts that help to advance our understanding of polymer crystallisation.
- Nucleation phenomena.
- Long living melt structures affecting crystallization.
- Confinement effects on crystallization.
- Crystallization in flowing melts.
- Fluid mobility restrictions caused by crystallites.
- The role of mesophases in the crystal formation.

In the context of the above-listed topics, there are several questions still open and often controversially debated. They concern nucleation – *Which conditions are necessary to nucleate the polymer crystallization process?*. The applicability of thermodynamic concepts – *Do concepts of equilibrium thermodynamics such as "melting" and "crystallization" describe correctly the nonequilibrium metastable nature of polymer crystals?* Morphological aspects – *How can morphological changes, which are observed during crystallization and melting of polymers, be interpreted?* And theoretical approaches – *How can general theories, developed for growth and relaxation phenomena, be applied or extended to crystallization of complex macromolecular systems?*

We sincerely hope that these Lecture Notes contribute to the progress of understanding of polymer crystallization, by providing a summary of the present state of the art to all active workers in the field and by raising the interest of newcomers.

Mulhouse, Freiburg, *Günter Reiter*
June 2006 *Gert Strobl*

Contents

1 Shifting Paradigms in Polymer Crystallization
Murugappan Muthukumar 1
1.1 Introduction 1
1.2 Classical View 3
    1.2.1 Lamellar Thickness 3
    1.2.2 Lamellar Growth 3
1.3 Results 6
    1.3.1 Nucleation of a Lamella 7
    1.3.2 Free Energy Landscape 8
    1.3.3 Spontaneous Selection of Lamellar Thickness and Shape .. 9
    1.3.4 Growth Front 11
    1.3.5 Kinetics at the Growth Front 11
1.4 Conclusions 15
References 16

2 Theoretical Aspects of the Equilibrium State of Chain Crystals
Jens-Uwe Sommer 19
2.1 Introduction 19
2.2 Thermodynamic Considerations about the Equilibrium Shape of a Polymer Single Crystal 20
2.3 The Brush State of the Amorphous Fraction is Thermodynamically Suppressed 23
2.4 Extended Chain Crystals and Sliding Entropy 24
2.5 The Slip-Loop Model for the Entropy of the Amorphous Fraction of a Single Chain Crystal 26
2.6 Tight Loops and Effective Fold Surface Tension for Single Chain Crystals 29
2.7 Many Chain Crystals 32
2.8 The Role of Bending Rigidity for the Formation of Small Loops .. 34
2.9 Tilting in Extended Chain Crystals 36

2.10 Summary and Conclusion 38
References 43

3 Intramolecular Crystal Nucleation
Wenbing Hu 47
3.1 Nucleation Mechanism of Polymer Crystallization 47
3.2 Concept of Molecular Nucleation 51
3.3 Intramolecular Nucleation Model 52
    3.3.1 Primary Crystal Nucleation in a Single Chain 52
    3.3.2 Secondary Crystal Nucleation in a Single Chain 55
    3.3.3 From Molecular Nucleation to Intramolecular Nucleation . 59
3.4 Concluding Remarks 60
References 61

4 Kinetic Theory of Crystal Nucleation Under Transient Molecular Orientation
Leszek Jarecki 65
4.1 Introduction 65
4.2 Time Evolution of the Chain Distribution Function 67
4.3 Free Energy and Orientation Distribution of the Chain Segments . 72
4.4 Crystal Nucleation Rate 77
4.5 Conclusions 84
References 85

5 Precursor of Primary Nucleation in Isotactic Polystyrene Induced by Shear Flow
Toshiji Kanaya, Yoshiyuki Takayama, Yoshiko Ogino, Go Matsuba and Koji Nishida 87
5.1 Introduction 87
5.2 Experimental 88
5.3 Results and Discussion 89
5.4 Conclusion 95
References 95

6 Structure Formation and Glass Transition in Oriented Poly(Ethylene Terephthalate)
Koji Fukao, Satoshi Fujii, Yasuo Saruyama, Naoki Tsurutani 97
6.1 Introduction 97
6.2 Experiments 99
6.3 Structural Change at Isothermal Annealing Process 100
    6.3.1 X-ray Diffraction Patterns 100
    6.3.2 Integrated Intensity as a Function of Annealing Time .. 103
    6.3.3 Kinetic Model Analysis 104
6.4 Structural and Thermal Change During the Heating Process 109
    6.4.1 X-ray Scattering Patterns During the Heating Process . 109
    6.4.2 Thermal Properties of Oriented PET 110

6.5	Concluding Remarks		115
References			116

7 How Do Orientation Fluctuations Evolve to Crystals?
Zhicheng Xiao, Jan Ilavsky, Gabrielle G. Long, Yvonne A. Akpalu 117

7.1	Introduction		117
7.2	Materials and Methods		119
	7.2.1	Sample Preparation	119
	7.2.2	Differential Scanning Calorimetery (DSC)	120
	7.2.3	Simultaneous WAXS and SAXS	120
	7.2.4	USAXS Measurements	121
	7.2.5	Small Angle Light Scattering (SALS)	122
7.3	Results and Discussion		123
References			130

8 Role of Chain Entanglement Network on Formation of Flow-Induced Crystallization Precursor Structure
Benjamin S. Hsiao 133

8.1	Introduction	133
8.2	Current Opinions on Flow-Induced Crystallization Precursor Structures	134
8.3	Role of High Molecular Weight Species in Flow-Induced Crystallization	138
8.4	New Insights on the Molecular Mechanism of Shish-Kebab Formation	142
	8.4.1 Kebab Growth Follows the Diffusion-Controlled Like Process	142
	8.4.2 Thermal Stability of Flow-Induced Shish-Kebab Scaffold	143
8.5	Relationship between Micro-rheology and Precursor Morphology	146
8.6	Concluding Remarks	147
References		147

9 Full Dissolution and Crystallization of Polyamide 6 and Polyamide 4.6 in Water and Ethanol
Marjoleine G.M. Wevers, Vincent B.F. Mathot, Thijs F.J. Pijpers, Bart Goderis, Gabriel Groeninckx 151

9.1	Introduction		151
9.2	Experimental Section		153
	9.2.1	Materials	153
	9.2.2	Preparation and Characterization of the Samples	154
9.3	Results and Discussion		156
	9.3.1	Dissolution and Crystallization of PA6 in Water by DSC	156
	9.3.2	Influence of Dissolution of PA6 on Molar Mass Distribution by SEC	158
	9.3.3	Influence of Dissolution of PA6 on the Crystallinity Found by DSC	160

	9.3.4	Dissolution and Crystallization of PA6 by WAXD 161
	9.3.5	Dissolution of Other Polyamides and in Various Solvents . 165

9.4 Conclusions..166
References 167

10 Small Angle Scattering Study of Polyethylene Crystallization from Solutions
Howard Wang 169
10.1 Introduction 169
10.2 Experiment 171
10.3 Results and Discussion 171
10.4 Conclusion 177
References 177

11 Morphologies of Polymer Crystals in Thin Films
*Günter Reiter, Ioan Botiz, Laetitia Graveleau, Nikolay Grozev,
Krystyna Albrecht, Ahmed Mourran, Martin Möller* 179
11.1 Introduction 179
11.2 Experimental Section 182
11.3 Results and Discussion 183
    11.3.1 Changes in Morphology with Crystallization Temperature 183
    11.3.2 Dependence of Morphology on Initial Film Thickness 186
    11.3.3 The Kinetics of Crystal Growth and the Effect
            of Changing Temperature 190
    11.3.4 "Decoration" of Flat-On Lamellar Crystals
            by Ripples and Spirals 193
    11.3.5 Orientation of the Crystalline Lamellae
            with Respect to the Substrate 195
11.4 Conclusions...197
References 198

12 Crystallization of Frustrated Alkyl Groups in Polymeric Systems Containing Octadecylmethacrylate
Elke Hempel, Hendrik Budde, Siegfried Höring, Mario Beiner..........201
12.1 Introduction 201
12.2 Side-chain Crystallization in Poly(n-octadecylmethacrylate)......203
12.3 Confined Crystallization in Microphase-separated
      Poly(styrene–*block*–octadecylmethacrylate) Copolymers 212
12.4 Conclusions...224
References 226

13 Crystallization in Block Copolymers with More than One Crystallizable Block
Alejandro J. Müller, María Luisa Arnal, Vittoria Balsamo 229
13.1 Introduction 229
13.2 Double Crystalline AB and ABA Copolymers 230

13.3	ABC Triblock Linear and Star Shaped Terpolymers	251
13.4	Conclusions	256
References		257

14 Monte Carlo Simulations of Semicrystalline Polyethylene: Interlamellar Domain and Crystal-Melt Interface
Markus Hütter, Pieter J. in 't Veld, Gregory C. Rutledge 261

14.1	Introduction, Motivation		261
14.2	Methodology		263
	14.2.1	Force Field, Virial Calculation of Stress	263
	14.2.2	Simulation Setup	264
	14.2.3	Thermal and Elastic Properties of Interlamellar Domain	267
	14.2.4	Energy and Stresses in the Crystal-Melt Interface	268
14.3	Results and Discussion		270
	14.3.1	Conformational Properties	270
	14.3.2	Thermal and Elastic Properties of Interlamellar Domain	271
	14.3.3	Properties of the Crystal-Melt Interface	275
	14.3.4	Internal Energy of the Interface	278
	14.3.5	Interface Stresses	278
14.4	Summary and Discussion		279
	14.4.1	Entire Interlamellar Domain	280
	14.4.2	Sharp Crystal-Melt Interface	280
	14.4.3	Perspectives	281
References			282

15 The Role of the Interphase on the Chain Mobility and Melting of Semi-crystalline Polymers; A Study on Polyethylenes
Sanjay Rastogi, Dirk R. Lippits, Ann E. Terry, Piet J. Lemstra 285

15.1	Introduction		286
15.2	Control of Entanglement Density Upon Crystallization		289
	15.2.1	Crystallization via Dilute Solution	289
	15.2.2	Exploitation of the Hexagonal Phase in Polyethylene	292
	15.2.3	Via Synthesis	293
15.3	Influence of the Interphase on Molecular Mobility in Crystalline Domains		295
15.4	From the Interphase to the Interface: The Welding of Semi-crystalline Polymers		296
15.5	Influence of Chain Folding on the Unit Cell		297
	15.5.1	Monodisperse Ultra-long Linear Alkanes	298
	15.5.2	Monodisperse Ultra-long Branched Alkanes	302
	15.5.3	Homogeneous Copolymers of Ethylene-1-Octene	308
15.6	Beyond Flexible Polymers: Rigid Amorphous Fraction		313
15.7	Influence of the Interphase on the Polymer Melt		315

21.6 Conclusions.. 454
References 455

22 Atomistic Simulation of Polymer Melt Crystallization by Molecular Dynamics
Numan Waheed, Min Jae Ko, Gregory C. Rutledge 457
22.1 Introduction 457
22.2 Methods 460
22.3 Results and Discussion 461
    22.3.1 Nucleation 461
    22.3.2 Growth 468
22.4 Conclusions... 476
References 478

23 A Multiphase Model Describing Polymer Crystallization and Melting
Gert Strobl 481
23.1 Introduction 481
23.2 Experimental Findings 483
    23.2.1 Crystallization Line and Melting Line 483
    23.2.2 Effects of Counits and Diluents 485
    23.2.3 Recrystallization Processes 488
23.3 A Multiphase Model of Polymer Crystallization and Melting 492
    23.3.1 Thermodynamic Scheme 494
23.4 Examples of Application 496
23.5 Conclusion 500
References 502

Index 503

List of Contributors

Yvonne A. Akpalu
Department of Chemistry and
Chemical Biology
Rensselaer Polytechnic Institute
Troy, NY 12180, USA
akpaly@rpi.edu

Krystyna Albrecht
DWI an der RWTH Aachen
Pauwelsstr. 8
52056 Aachen, Germany
albrecht@dwi.rwth-aachen.de

María Luisa Arnal
Materials Science Department
Universidad Simón Bolívar
Aptdo. 89000, Caracas 1080-A
Venezuela
marnal@usb.ve

Finizia Auriemma
Dipartimento di Chimica
Università di Napoli "Federico II"
Complesso Monte S.Angelo
Via Cintia
80126 Napoli, Italy
finizia.auriemma@unina.it

Vittoria Balsamo
Materials Science Department
Universidad Simón Bolívar
Aptdo. 89000, Caracas 1080-A
Venezuela
vbalsamo@usb.ve

Mario Beiner
FB Physik
Martin-Luther-Universität
Halle-Wittenberg
D-06099 Halle, Germany
beiner@physik.uni-halle.de

Ioan Botiz
Institut de Chimie des Surfaces
et Interfaces
ICSI-CNRS
15, rue Jean Starcky, B.P. 2488
68057 Mulhouse, France
ioan.botiz@uha.fr

Hendrik Budde
FB Chemie
Martin-Luther-Universität
Halle-Wittenberg
D-06099 Halle
Germany and Fraunhofer
Pilotanlagenzentrum für
Polymersynthese und -verarbeitung
Value Park, Bau A74
D-06258 Schkopau, Germany
hendrik.budde@iap.fraunhofer.de

Claudio De Rosa
Dipartimento di Chimica
Università di Napoli "Federico II"
Complesso Monte S.Angelo
Via Cintia
80126 Napoli, Italy
claudio.derosa@unina.it

Felice De Santis
Department of Chemical and Food
Engineering
University of Salerno
Via Ponte don Melillo
I-84084 Fisciano (SA), Italy
fedesantis@unisa.it

Tiberio A. Ezquerra
Instituto de Estructura de la Materia
C.S.I.C. Serrano 119
Madrid 28006, Spain
imte155@iem.cfmac.csic.es

Satoshi Fujii
Department of Polymer Science
Kyoto Institute of Technology
Matsugasaki
Sakyo-ku, Kyoto 606-8585, Japan
a9330712@edu.kit.ac.jp

Koji Fukao
Department of Polymer Science
Kyoto Institute of Technology
Matsugasaki
Sakyo-ku, Kyoto 606-8585, Japan
fukao@kit.ac.jp

Mari-Cruz García-Gutiérrez
Instituto de Estructura de la Materia
C.S.I.C. Serrano 119
Madrid 28006, Spain
Imtc304@iem.cfmac.csic.es

Bart Goderis
Laboratory of Macromolecular
Structural Chemistry
Division of Molecular
and Nanomaterials
Department of Chemistry
Katholieke Universiteit Leuven
Celestijnenlaan 200F
3001 Heverlee, Belgium
bart.goderis@chem.kuleuven.be

Laetitia Graveleau
Institut de Chimie des Surfaces
et Interfaces
ICSI-CNRS
15, rue Jean Starcky, B.P. 2488
68057 Mulhouse, France

Gabriel Groeninckx
Laboratory of Macromolecular
Structural Chemistry
Division of Molecular
and Nanomaterials
Department of Chemistry
Katholieke Universiteit Leuven
Celestijnenlaan 200F
3001 Heverlee, Belgium
gabriel.groeninckx@chem.kuleuven.be

Nikolay Grozev
Institut de Chimie des Surfaces
et Interfaces
ICSI-CNRS
15, rue Jean Starcky, B.P. 2488
68057 Mulhouse, France
nikolay.grozev@uha.fr

Elke Hempel
FB Physik
Martin-Luther-Universität
Halle-Wittenberg
D-06099 Halle, Germany
elke.hempel@physik.uni-halle.de

Jamie K. Hobbs
Department of Chemistry
University of Sheffield
Dainton Building, Brook Hill
Sheffield S3 7HF UK
Jamie.hobbs@sheffield.ac.uk

Siegfried Höring
FB Chemie
Martin-Luther-Universität
Halle-Wittenberg
D-06099 Halle, Germany
hoering@chemie.uni-halle.de

Benjamin S. Hsiao
Department of Chemistry
Stony Brook University
Stony Brook, NY 11794-3400, USA
bhsiao@notes.cc.sunysb.edu

Wenbing Hu
Department of Polymer Science and Engineering
State Key Laboratory of Coordination Chemistry
School of Chemistry and Chemical Engineering
Nanjing University
210093 Nanjing, China
wbhu@nju.edu.cn

Markus Hütter
Department of Materials
ETH Zürich
CH-8093 Zürich, Switzerland
markus.huetter@mat.ethz.ch

Jan Ilavsky
X-Ray Science Division
Argonne National Laboratory
9700 S. Cass Avenue
Argonne, IL 60439, USA
ilavsky@aps.anl.gov

Leszek Jarecki
Institute of Fundamental Technological Research
Polish Academy of Sciences
Swietokrzyska 21
00-049 Warsaw, Poland
ljarecki@ippt.gov.pl

Toshiji Kanaya
Institute for Chemical Research
Kyoto University
Uji Kyoto-fu 611-0011, Japan
kanaya@scl.kyoto-u.ac.jp

Ryuichiro Kawano
Tokyo Institute of Technology
Department of Organic and Polymeric Materials
Ookayama 2-12-1-S8-37
Meguroku, Tokyo, Japan
kawano-ryuichiro@jpo.go.jp

Min Jae Ko
Department of Chemical Engineering
Massachusetts Institute
of Technology
Cambridge, MA 02139, USA
mjko@mit.edu
Current address: HD Display Center
LCD Business Samsung
Electronics Co. Asan, Korea
minjae.ko@samsung.com

Piet J. Lemstra
Department of Chemical Engineering
Dutch Polymer Institute
Eindhoven University of Technology
Den Dolech 2, P.O. Box 513
5600MB Eindhoven
The Netherlands
p.j.lemstra@tue.nl

Dirk R. Lippits
Department of Chemical Engineering
Dutch Polymer Institute

Eindhoven University of Technology
Den Dolech 2, P.O. Box 513
5600MB Eindhoven
The Netherlands
d.lippits@tue.nl

Gabrielle G. Long
X-Ray Science Division
Argonne National Laboratory
9700 S. Cass Avenue
Argonne, IL 60439, USA
gglong@aps.anl.gov

Al Mamun
Tokyo Institute of Technology
Department of Organic and
Polymeric Materials
Ookayama 2-12-1-S8-37
Meguroku, Tokyo, Japan
mamun@mbox.op.titech.ac.jp

Vincent B.F. Mathot
Laboratory of Macromolecular
Structural Chemistry
Division of Molecular
and Nanomaterials
Department of Chemistry
Katholieke Universiteit Leuven
Celestijnenlaan 200F
3001 Heverlee, Belgium
SciTe, Ridder Vosstraat 6
6162 AX Geleen, The Netherlands
vincent.mathot@scite.nl

Go Matsuba
Institute for Chemical Research
Kyoto University
Uji Kyoto-fu 611-0011, Japan
gmatsuba@scl.kyoto-u.ac.jp

Martin Möller
DWI an der RWTH Aachen
Pauwelsstr. 8
52056 Aachen, Germany
moeller@dwi.rwth-aachen.de

Ahmed Mourran
DWI an der RWTH Aachen
Pauwelsstr. 8
52056 Aachen, Germany
mourran@dwi.rwth-aachen.de

Alejandro J. Müller
Materials Science Department
Universidad Simón Bolívar
Aptdo. 89000, Caracas 1080-A
Venezuela
amuller@usb.ve

Murugappan Muthukumar
Polymer Science and Engineering
Department
Materials Research Science and
Engineering Center
University of Massachusetts
Amherst, MA 01003, USA
muthu@polysci.umass.edu

Koji Nishida
Institute for Chemical Research
Kyoto University
Uji Kyoto-fu 611-0011, Japan
knishida@scl.kyoto-u.ac.jp

Aurora Nogales
Instituto de Estructura de la Materia
C.S.I.C. Serrano 119
Madrid 28006, Spain
emnogales@iem.cfmac.csic.es

Yoshiko Ogino
Institute for Chemical Research
Kyoto University
Uji Kyoto-fu 611-0011, Japan
yoshiko@pmsci.kuicr.kyoto-u.ac.jp

Norimasa Okui
Tokyo Institute of Technology
Department of Organic and
Polymeric Materials
Ookayama 2-12-1-S8-37
Meguroku, Tokyo, Japan
nokui@o.cc.titech.ac.jp

Kinga Pielichowska
Department of Chemistry and
Technology of Polymers
Cracow University of Technology
ul. Warszawska 24,
31-155 Kraków, Poland
kingafle@chemia.pk.edu.pl

Krzysztof Pielichowski
Department of Chemistry and
Technology of Polymers
Cracow University of Technology
ul. Warszawska 24,
31-155 Kraków, Poland
kpielich@usk.pk.edu.pl

Thijs F. J. Pijpers
Laboratory of Macromolecular
Structural Chemistry
Division of Molecular
and Nanomaterials
Department of Chemistry
Katholieke Universiteit Leuven
Celestijnenlaan 200F
3001 Heverlee, Belgium
thijs.pijpers@tiscali.nl

Sanjay Rastogi
Department of Chemical Engineering
Dutch Polymer Institute
Eindhoven University of Technology
Den Dolech 2, P.O. Box 513
5600MB Eindhoven
The Netherlands
s.rastogi@tue.nl

Günter Reiter
Institut de Chimie des Surfaces
et Interfaces
ICSI-CNRS
15, rue Jean Starcky, B.P. 2488
68057 Mulhouse, France
g.reiter@uha.fr

Gregory C. Rutledge
Department of Chemical Engineering
Massachusetts Institute of
Technology
Cambridge, MA 02139, USA
rutledge@mit.edu

Alejandro Sanz
Instituto de Estructura de la Materia
C.S.I.C. Serrano 119
Madrid 28006, Spain
emsanz@iem.cfmac.csic.es

Yasuo Saruyama
Department of Polymer Science
Kyoto Institute of Technology
Matsugasaki
Sakyo-ku, Kyoto 606-8585
Japan
saruyama@kit.ac.jp

Igors Šics
Department of Chemistry
State University of New York at
Stony Brook
Stony Brook, NY 11794-3400,USA
isics@bnl.gov

Jens-Uwe Sommer
Institut de Chimie des Surfaces
et Interfaces
ICSI-CNRS
15, rue Jean Starcky, P.B. 2488
F-68057 Mulhouse, France
Present address: Leibniz-Institut of
Polymer Research
Hohe Strasse 6
D-01069 Dresden, Germany
sommer@ipfdd.de

Andrea Sorrentino
Department of Chemical and Food
Engineering
University of Salerno
Via Ponte don Melillo
I-84084 Fisciano (SA), Italy
asorrentino@unisa.it

List of Contributors

Gert Strobl
Physikalisches Institut
Albert-Ludwigs-Universität Freiburg
79104 Freiburg, Germany
strobl@uni-freiburg.de

Yoshiyuki Takayama
Institute for Chemical Research
Kyoto University
Uji Kyoto-fu 611-0011, Japan
takayama@pmsci.kuicr.kyoto-u.ac.jp

Ann E. Terry
Department of Chemical Engineering
Dutch Polymer Institute
Eindhoven University of Technology
Den Dolech 2, P.O. Box 513
5600MB Eindhoven
The Netherlands
ISIS Facility
Rutherford Appleton Laboratory
Chilton, Didcot, Oxfordshire
OX11 0QX, England, UK
a.e.terry@tue.nl

Giuseppe Titomanlio
Department of Chemical and Food Engineering
University of Salerno
Via Ponte don Melillo
I-84084 Fisciano (SA), Italy
gtitomanlio@unisa.it

Naoki Tsurutani
Department of Physics
Kyoto University
Kyoto 606-8502, Japan
turutani@scphys.kyoto-u.ac.jp

Susumu Umemoto
Tokyo Institute of Technology
Department of Organic and Polymeric Materials
Ookayama 2-12-1-S8-37, Meguroku
Tokyo, Japan
sumemoto@o.cc.titech.ac.jp

Pieter J. in 't Veld
Sandia National Laboratories
Albuquerque, NM 87185, USA
pjintve@sandia.gov

Numan Waheed
Maurice Morton Institute of Polymer Science
University of Akron
Akron, OH 44325-3909, USA
nw11@uakron.edu

Howard Wang
Department of Materials Science and Engineering
Michigan Technological University
Houghton, MI 49931, USA
wangh@mtu.edu

Marjoleine G.M. Wevers
DSM Research
P.O. Box 18
6160 MD Geleen, The Netherlands
Laboratory of Macromolecular Structural Chemistry
Division of Molecular and Nanomaterials
Department of Chemistry
Katholieke Universiteit Leuven
Celestijnenlaan 200 F
3001 Heverlee, Belgium
SciTe, Ridder Vosstraat 6
6162 AX Geleen, The Netherlands
marjoleine.wevers@chem.kuleuven.be

Zhicheng Xiao
Department of Chemistry and Chemical Biology
Rensselaer Polytechnic Institute
Troy, NY 12180, USA
xiaoz@rpi.edu

1

Shifting Paradigms in Polymer Crystallization

Murugappan Muthukumar

Polymer Science and Engineering Department, Materials Research Science and
Engineering Center, University of Massachusetts, Amherst, MA 01003, USA
muthu@polysci.umass.edu

Abstract. Classical concepts on polymer crystallization are under revision, due to the demands made by a wealth of new information acquired from experiments and computer simulations that explore the time-resolved molecular details of the crystallization process. A brief summary of the classical ideas and the contrasting new results is presented here.

1.1 Introduction

The crystallization of flexible polymers from solutions and melts is a challenging fundamental problem in polymer physics. The challenge is obvious when we consider how a highly entangled collection of interpenetrating polymer chains would order into a crystal. We might think that such a process will never be complete, due to the topological connectivity of the polymer molecules. Yet, polymer crystals exhibit a myriad of morphologies with rich hierarchies of molecular organization, and in distinct contrast to those from non-polymeric systems.

From a conceptual point of view, polymer crystallization is frustrated by relatively large free energy barriers, which arise from the necessity to reorganize polymer conformations into ordered states. For the case of small molecules, the crystallization process proceeds by the mechanism of nucleation and growth. Here, the first step is the formation of a nucleus of the crystalline phase. The free energy landscape F associated with the sizes of the nuclei is sketched in Fig. 1.1a, where η is a measure of the size of the ordered state, which in turn is proportional to the degree of crystallinity. If the size of the freshly formed nucleus is greater than a critical size, then nucleation occurs by crossing the nucleation barrier in free energy. The next step is the growth of such nuclei. During the growth stage for small molecules, there are no significant barriers, and the process essentially completes with an eventual degree of crystallinity of unity.

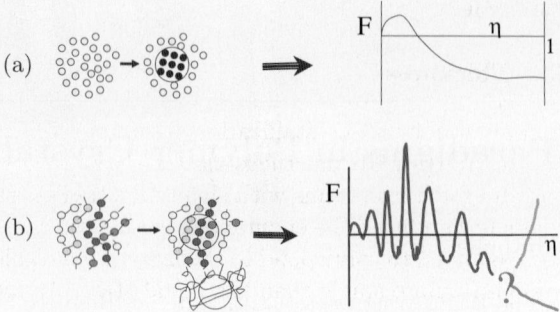

Fig. 1.1. Sketches of free energy landscapes. (**a**) small molecules; (**b**) polymer chains

In contrast, the free energy landscape for polymer crystallization is very different, as sketched in Fig. 1.1b. As in the small molecule case, during the nucleation stage, there are nuclei constituted by many monomers. However these monomers can come from different chains or from different locations along the chain contour of the same chain. This results in a competition by several nuclei for the acquisition of monomers from the strands not yet incorporated into the crystalline phase. Therefore, during the growth stage, the free energy landscape has many metastable states, which are frustrated by their immediate barriers. These barriers are due to the free energy cost involved in the rearrangement of chain conformations originally distributed among many nuclei into fewer nuclei with greater crystalline order.

One of the key questions is whether the degree of crystallinity ever reaches the maximum value of unity at temperatures below the melting temperature. In other words, in the sketch of Fig. 1.1b, the issue is whether the free energy minima continue to decrease as η increases or the global free energy minimum is at a smaller value of η. Is semicrystallinity a thermodynamic state or merely a kinetic manifestation? Further, among the many metastable states that are separated by free energy barriers, are there some significant mesomorphic states that can be taken as long-lived precursors to the final state?

Aided by heroic efforts [1–35] by the polymer community with exquisite structural elucidations and elegant theoretical attempts during the past seven decades, the answers to the above questions are still evolving. Availability of sensitive synchrotron radiation, atomic-force-microscopy techniques, and molecular modeling have spurred recent intense interest in following the mechanism of polymer crystallization. Very exciting phenomenology is currently being developed worldwide, by exploring the details at the molecular level in every stage of crystallization, both from solutions and melts. In this chapter, we restrict ourselves to the salient features of new concepts emerging from theory and modeling, in contrast to the classical ideas on polymer crystallization.

1.2 Classical View

1.2.1 Lamellar Thickness

The central premise of polymer crystallization is that the single crystals form lamellae, which are roughly 10 nm thick platelets with regular facets, and chains fold back and forth into stems with chain direction essentially perpendicular to the lamellar surface. The spontaneously selected lamellar thickness of about 10 nm is assumed to correspond to the first viable stable crystal, with its free energy slightly more stable than the liquid state. As indicated in the sketch of Fig. 1.2, where F is plotted against the ratio λ of lamellar thickness to the extended chain length, this initial thin lamella is kinetically stabilized by a huge free energy barrier sketched by the dashed curve. However, if sufficient time is allowed for the system to cross the free energy barrier, the free energy of the system will evolve to the global minimum corresponding to the limit of lamellar thickness being the extended chain length. The experimentally observed lamellar thickness of 10 nm is about two orders of magnitude smaller than thermodynamic estimates, and this situation is attributed to the large free energy barrier.

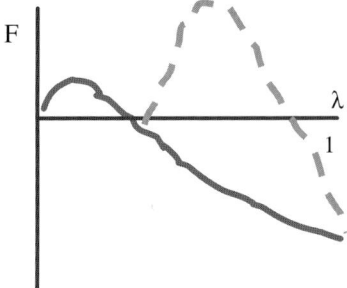

Fig. 1.2. Free energy barrier for nucleation. Dashed curve indicates the huge barrier for further evolution of initially formed lamellae

1.2.2 Lamellar Growth

The major theory [3–7] of polymer crystallization, due primarily to Lauritzen and Hoffman (LH), is a generalization of small-molecule crystallization theory of surface nucleation and growth to incorporate chain folding. In the model of LH theory (Fig. 1.3a), polymer molecules are assumed to attach at the growth front in terms of stems, each of length comparable to the lamellar thickness L. For each polymer molecule, the first step is to place its first stem at the growth surface, whose lateral dimension is taken as L_p. This step is assumed to be associated with a nucleation. The barrier for this step was assumed

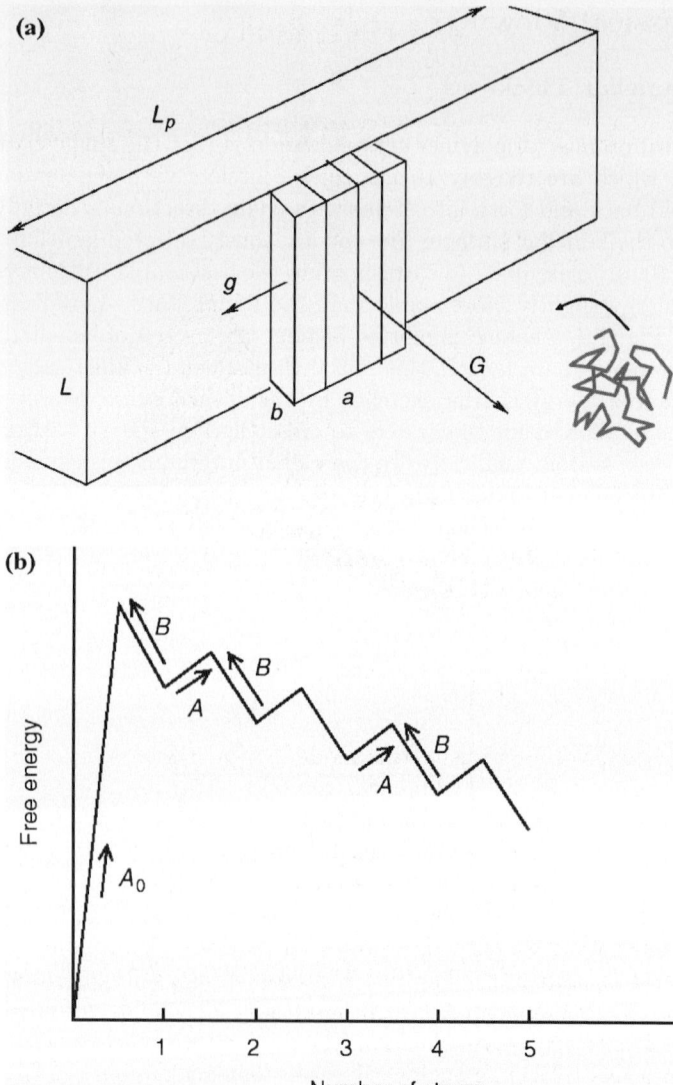

Fig. 1.3. Lauritzen-Hoffman theory. (a) model; (b) free energy barrier

to arise from a combination of gain in bulk free energy in the formation of the parallelepiped stem and the cost in free energy to make the additional surfaces. This barrier is the first peak in Fig. 1.3b. In the estimation of the barrier, macroscopic thermodynamics is taken to be valid even at the stem level. After the first step with a nucleation rate of i, the secondary nucleus spreads out laterally with the rate g. The thickness of the stem is a along the lateral direction and b along the growth direction with growth rate G.

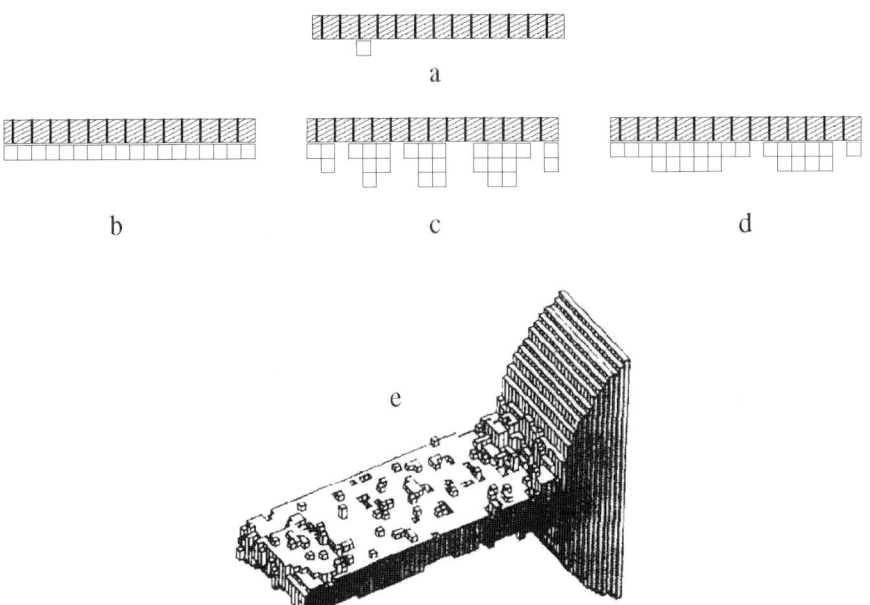

Fig. 1.4. (a)–(d) Regimes in LH model. (e) Sadler-Gilmer model of roughening

One of the main conclusions of this model is that the minimum thickness of a stable lamella is $2\sigma/\Delta F$, where σ is the fold surface energy and ΔF is the gain in free energy density. Also, in the LH theory, three regimes have been identified. Let the dark area in Figs. 1.4a–d represent the growth front and each square correspond to the cross-section of a stem. In regime I (Fig. 1.4b), secondary nucleation controls $G(g \gg i)$. In regime III (Fig. 1.4d), prolific multiple nucleation controls the growth. In between these limiting regimes, Sanchez and DiMarzio [8] identified a crossover regime II (Fig. 1.4c), where nucleation rate is more rapid than in I and less than in III.

In the LH theory, crystallization kinetics in these three regimes are expressed in terms of phenomenological parameters such as surface free energies and crystal thickness, and experimental parameters such as degree of undercooling, molecular weight and concentration. There are two key conclusions from the LH theory. The first is that the minimum thickness of a stable lamella is $2\sigma/\Delta F$, where σ is the fold surface energy and ΔF is the gain in free energy density (which is proportional to the quench depth ΔT). The second conclusion is that $G \sim \exp(-K/T\Delta T)$, where the parameter K is independent of temperature. Molecular details of g and i are not available. From an empirical point of view, the values of parameters of LH theory to fit experimental results have been argued [6, 11] to be unrealistic and there have been several extensions [6].

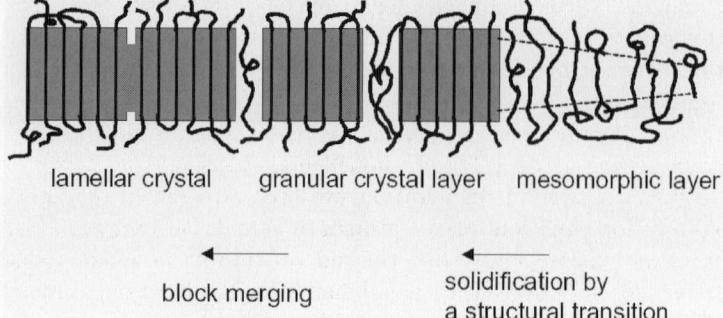

Fig. 1.5. Strobl's model

Another key concept of polymer crystallization is due to Sadler and Gilmer (SG). According to the SG model, inspired by the thermal roughening phenomenon observed in small molecular systems, even shorter stems than the stable stem length can attach at the growth front (Fig. 1.4e). The pinning of such short stems interrupts the crystal growth. The pinning must be removed to resume crystallization. Instead of the nucleation barrier of the LH theory, it is the removal of shorter stems that is the controlling factor for growth kinetics. Relying on computer simulations, SG argued that the major conclusions of the LH theory can be reproduced with the roughening model without invoking stem-nucleation.

More recently, Strobl [18] has stimulated a discussion by arguing that in all processes of polymer crystallization, a mesomorphic precursor phase is first formed before the crystalline phase. Blocks of this mesomorphic state then attach to the growth front, as sketched in Fig. 1.5. This model, inspired by the observation of the hexatic phase in short n-alkanes [19, 20], is actively contested by the polymer community. Whether such mesomorphic phases are stable intermediates before the formation of the crystalline phase and whether a critical chain stiffness is required for such mesomorphic phases are being discussed. Several laboratories worldwide are pursuing experiments to explore this aspect of polymer crystallization.

1.3 Results

During the recent years, we have attempted to evaluate the underlying assumptions and ideas of the current growth theories mentioned above under different conditions and to provide molecular interpretation of the various phenomenological parameters appearing in the LH theory. We have approached the various issues by a combination of tools. First, we have performed Langevin dynamics simulations of many chains in dilute solutions crystallizing into lamellae. Our simulations [29, 32] are based on the united-atom model, and folded-chain-lamellae form due to a competition, mediated by chain connectivity, between chain stiffness (arising from torsional energy) and attraction between non-bonded segments. This exercise has yielded tremendous insight into several major issues. Next, using the input from Langevin dynamics simulations of tens of chains, we [36] have performed coarse-grained simulations to follow the growth of lamellae of thousands of chains, by using the Monte Carlo method. In addition, we [37] have solved numerically the reaction-diffusion equation for the growth of cylindrical tablets in a medium of diffusing polymer chains, with the aid of suitable boundary conditions. Further, we [26] have derived thermodynamic results by using statistical mechanics of polymer chains. The main results are summarized below.

1.3.1 Nucleation of a Lamella

The very early stage of lamellar formation is nucleation as in the small molecular systems, except that the polymer now is long enough to participate in several nuclei. As shown in Fig. 1.6, several 'baby nuclei' are formed, connected by the same single chain. The strands connecting these baby nuclei are flexible with considerable configurational entropy. As time progresses, the monomers in the flexible strands are reeled into the baby nuclei while orientational order in each nucleus increases, making them 'smectic pearls'. Simultaneously, the competition between nuclei for further growth dissolves some nuclei. Eventually, folded-chain structure emerges. During this process, the density fluctuations are seen to be exponentially growing with time, and the growth rate $\Omega(q)$ is found to be $\Omega(q) \propto q^2(1 - \kappa q^2)$, where κ is a positive constant and q is the scattering wave vector. The time-evolution of the density fluctuations calculated using our simulations is in qualitative agreement with time-resolved X-ray scattering measurements on crystallizing polymers [12], and is different from expectations based on a spinodal mode of polymer crystallization [14].

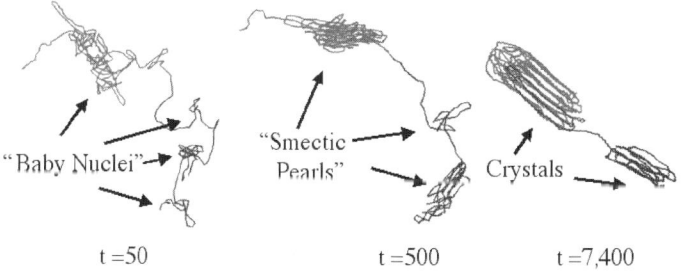

Fig. 1.6. Birth of a lamella

1.3.2 Free Energy Landscape

The typical free energy landscape for a folding chain is given in Fig. 1.7. In Fig. 1.7a, the free energy is plotted against the lamellar thickness L and the orientational order parameter S for a chain of $N = 200$ segments. The landscape is highly corrugated, consisting of several metastable states separated by barriers and a global minimum. Equivalent result is shown in Fig. 1.7b for $N = 300$, where the free energy is plotted against L^2. Accompanying the free energy landscape, the simulations show that the lamellar thickness is quantized, as seen in experiments. First, thinner lamellae form, which then thicken over a period of time. The lamellar thickening proceeds through several metastable states, each metastable state corresponding to a particular number of folds per chain. Eventually, the lamellae settle into an equilibrium thickness, which is much smaller than the extended chain dimension.

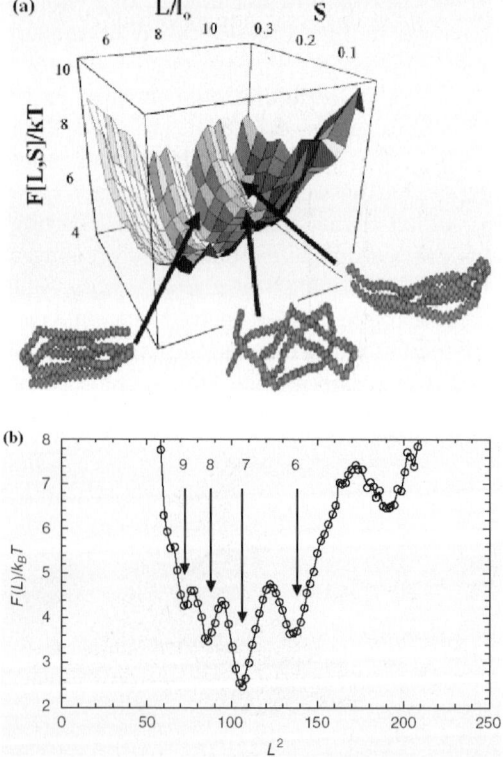

Fig. 1.7. (a) Simulated free energy landscape. (b) Global minimum corresponds to small lamellar thickness

1.3.3 Spontaneous Selection of Lamellar Thickness and Shape

Experimentally observed ratio of thickness along the chain axis to width of a solution-grown polyethylene crystal is about two orders of magnitude smaller than values allowed by existing equilibrium considerations. It has been repeatedly argued in the literature that the lamellar thickness is kinetically selected and that, if enough time is granted, the lamella would thicken to the extended chain dimension. Our simulations clearly show that the global free energy minimum corresponds to a finite lamellar thickness, which is much smaller than the extended chain thickness. Motivated by these simulation results, we [26] have formulated an exactly solvable statistical mechanics model. We considered a very long chain of N segments assembled into a lamellar tablet (Fig. 1.8a) with μ stems in full registry, each of m segments (lamellar thickness is m in units of segment length), and the rest of the segments distributed outside the tablet as correlated loops and tails. Remarkably, the free energy minimum corresponds to a finite thickness, as shown in Fig. 1.8b. This can be attributed to the entropic stability at non-zero temperatures arising from the

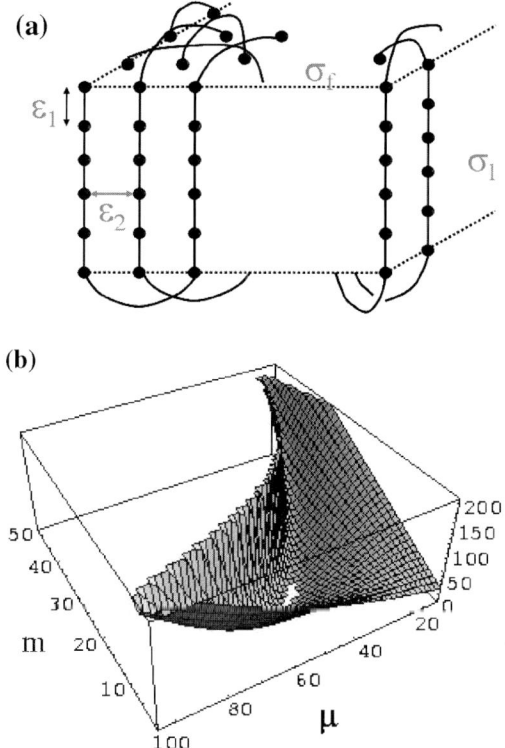

Fig. 1.8. (a) Model for a calculation of entropic stabilization of finite lamellar thickness. (b) Calculated global minimum for finite thickness

numerous ways of distributing a given length into many loops. The value of the equilibrium thickness depends on chain stiffness energy (ϵ_1), nearest-neighbor interaction energy (ϵ_2), lateral surface energy (σ_l), and the fold surface energy (σ_f). For this model, the free energy is derived by considering bulk term, interface term, and loop free energy. By calculating the free energy landscape, we have determined the initial lamellar thickness (smallest stable thickness), critical lamellar thickness, and the equilibrium thickness. The temperature dependence of these three lengths is given in Fig. 1.9 for the chain length of 10,000 in units of bond length, and typical values of the various parameters. At the melting temperature T_m, the equilibrium lamellar thickness is much smaller than the extended chain value. It must be stressed that the equilibrium lamellar thickness is about three orders of magnitude smaller than the extended chain dimension, as illustrated in Fig. 1.10.

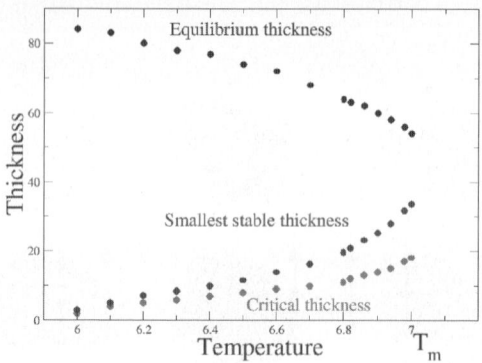

Fig. 1.9. Temperature dependence of equilibrium thickness, smallest stable thickness, and the critical thickness

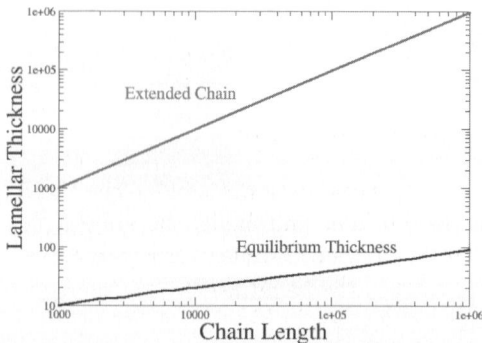

Fig. 1.10. Equilibrium thickness is orders of magnitude smaller than the extended chain dimension

1.3.4 Growth Front

In order to get insight into the nature of stems of the LH model and how stems attach at the growth front, we have monitored polymer chains at the growth front as they attach. Figure 1.11 shows the simultaneous adsorption and epitaxial registry of a diffusing chain at the growth front of a 'live' lamella (where all chains behind the growth front are allowed dynamics) as the simulation time progresses. More importantly, for situations corresponding to crystallization in dilute solutions, there are no barriers for the attachment of stems, in disagreement with the assumptions of the LH theory. Furthermore, our simulations (Fig. 1.12) of many chains near a fixed growth front shows a cooperative growth involving simultaneous homogeneous and heterogeneous nucleation (frame f is the end-view as in Fig. 1.12), in marked contrast to the LH mechanism of stem-wise addition of polymer molecules.

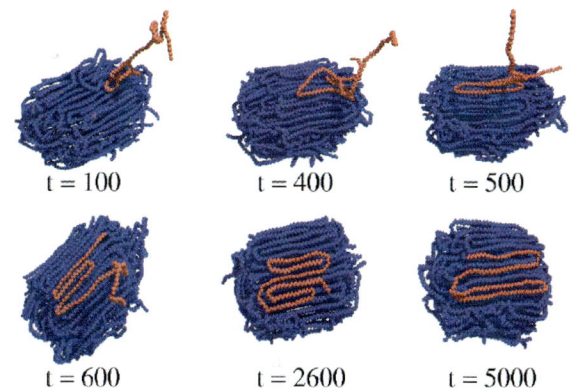

Fig. 1.11. Chains attach at the growth front by adsorption

1.3.5 Kinetics at the Growth Front

Based on the vast experimental data available in the literature on the growth kinetics of lamellae in solutions and melts, the systems can be classified into two groups. In Group A, valid for solution-grown crystals with relatively low molar mass of the polymer, the linear growth rate G depends on the crystallization temperature as sketched in Fig. 1.13a. Near the melting temperature, $G \propto \Delta T$. In addition, in this Group, G depends on molar mass M and polymer concentration C, according to $G \propto M^{-\mu}C^{\gamma}$, where the effective exponent γ is in the wide range of 0.2–2.0, and the effective exponent μ is a complicated function of the experimental conditions [27]. On the other hand, in Group B, valid for melt-grown crystals and solution-grown crystals with very high molar mass, the linear growth rate obeys the exponential law, $G \propto \exp(-K/T\Delta T)$,

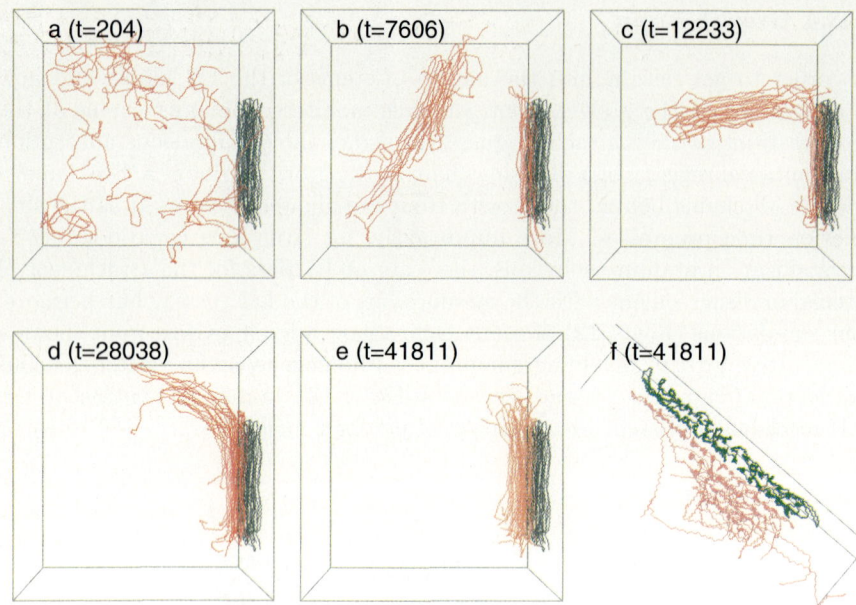

Fig. 1.12. Cooperative growth

as sketched in Fig. 1.13b, where K is a parameter. Furthermore, the molecular weight dependence of the growth rate is non-monotonic and depends on the crystallization temperature, as illustrated by the empirical Okui plot of Fig. 1.13c.

We [37] have developed a model to unify the Groups A and B, and to calculate the dependence of the growth rate on polymer concentration and molecular weight. The main idea is to focus on the molecular details at the growth front at higher concentrations and longer chains. Our Langevin dynamics simulations of lamellar growth from a solution where multiple chains are competing at the growth front, show that the growth is considerably slowed by the congestion of interpenetrating un-adsorbed chains at the growth front. Typical trajectories are shown in Fig. 1.14. The slowness, in comparison with the case of isolated chains getting adsorbed individually, arises from the reduction in the number of configurations of chains due to entanglements. This results in an entropic barrier at the growth front. In view of this observation, we have considered a new model of lamellar growth, as sketched in Fig. 1.15. We identify a boundary layer of thickness Λ in front of the growth front at $R(t)$. Inside this boundary layer, the local monomer concentration C_{in} is higher than that (C_o) in the bulk and the boundary layer is associated with a free energy barrier. The boundary values of polymer concentration C_s and C_b, respectively at $R(t)$ and the outer edge $B(t)$ of the boundary layer can

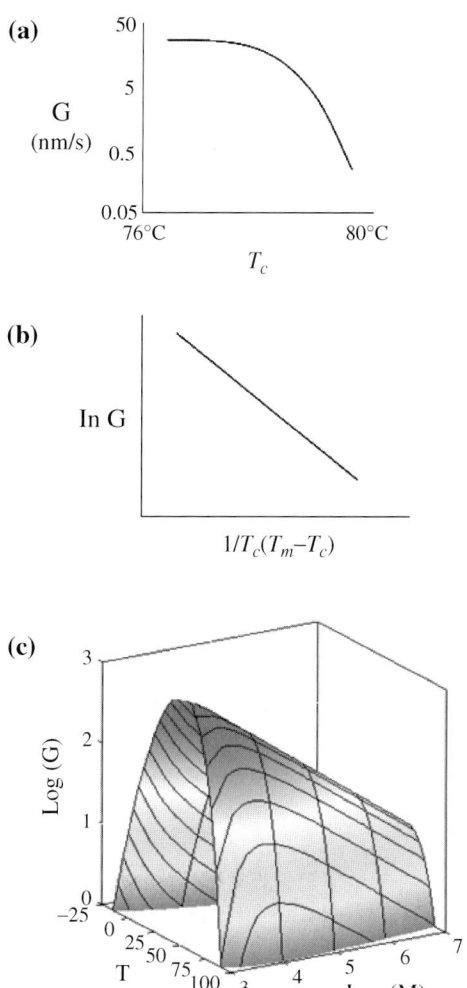

Fig. 1.13. (a) Group A behavior. (b) Group B behavior. (c) Okui plot

be different, depending on the nature of the barrier. We then calculate the growth rate in the steady state for this model with a barrier.

This model allows us to capture the limiting behaviors of the Groups A and B discussed above. The Group A behavior corresponds to the situation of insignificant barrier. Now the growth law is analogous to that of small molecules. The rate G is determined by the relative kinetics of adsorption and desorption, and $G \propto C(1 - \exp(-\Delta H \Delta T/k_B T_m T))$, where ΔH is the latent heat of fusion, ΔT is the quench depth, k_B is the Boltzmann constant. It turns out that $G \propto \Delta T$ near the melting temperature. Let G_0 be the rate in this limit of absence of any entropic barrier. When the barrier in the boundary layer is

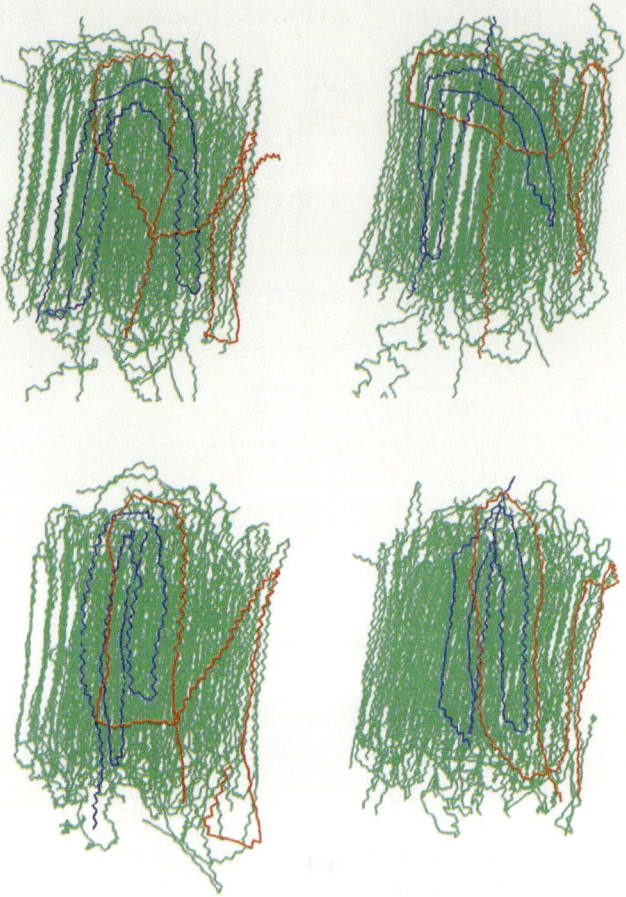

Fig. 1.14. Jamming of chains at the growth front, resulting in entropic barriers

significant, as in the case of large molecular weight polymers and melts, then Group B is expected to arise. If the barrier height is inversely proportional to ΔT, as is the case with entropic barriers [26], the saddle point approximation gives the form of $G \propto G_0(D_{in}/D_{bulk})\exp(-\alpha/T\Delta T)$, where D_{in} and D_{bulk} are the diffusion coefficients of the polymer inside and outside, respectively, of the boundary layer, and α is a temperature-independent factor. Furthermore, the values of the apparent exponents γ and μ for the dependence of the growth rate on polymer concentration and molecular weight, $G \propto C^\gamma M^{-\mu}$ follow from the dependencies of D_{in}, D_{bulk}, and T_m on concentration and molecular weight. As an example, the value of γ can be shown [37] to be 0.5 for crystallization from good solutions.

Thus the new model offers an opportunity to unify the diverse behaviors of kinetics of polymer crystallization observed experimentally.

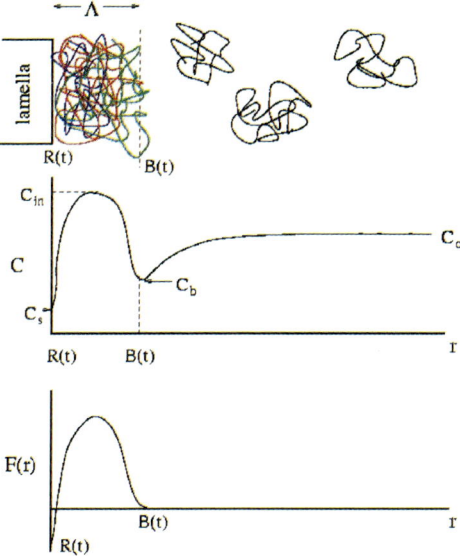

Fig. 1.15. New entropic barrier model for growth with molecular congestion at the growth front

1.4 Conclusions

The major result arising from our theoretical considerations is that chain entropy is the most dominant controlling factor that distinguishes polymer crystallization from ordering of small molecules. The substantial reduction of conformational entropy of the chains during the ordering process dictates how the ordering process proceeds. In addition, the energy considerations used in the crystallization of small molecules must naturally be accounted for. It is the free energy of the system $E - TS$ (E and S being the energy and entropy, respectively) that determines the course of polymer crystallization and the nature of the ultimate crystalline states. The LH theory and its modifications focus on energy considerations. In contrast, our work has included the entropic contributions as well.

The manifestations of chain entropy are present at all temperatures, except at $T = 0$. Two major conclusions that emerge from considerations of free energy by including chain entropy are the following:

(1) The free energy landscape for a single lamella exhibits a set of barriers, many metastable states (separated by free energy barriers), and a globally stable state. Each of these metastable states has a thickness that is much smaller than the extended chain length. Among the metastable states, even the first viable state with its free energy just below that of the melt is long-lived, due to the barrier for thickening. The thickness of this long-lived metastable

state increases with temperature, in a qualitatively similar manner to the Gibbs-Thompson law. However, if enough time is granted for this metastable state to evolve, then the equilibrium thickness would be reached for each temperature. The equilibrium thickness decreases with temperature, until the approach of the equilibrium melting temperature. The equilibrium melting temperature does not correspond to that of extended chain dimensions.

(2) The lateral growth faces a free energy barrier, due to temporal crowding of entangled chains at the growth front. The linear growth rate G assumes the form,

$$G \sim (D_{in}/D_{bulk})\exp(-1/T\Delta T)[1 - \exp(\Delta H \Delta T/kT_m T)] \qquad (1.1)$$

Where ΔT is the quench depth, T_m is the melting temperature, ΔH is the latent heat of fusion, T is the temperature, D_{in} is the diffusion coefficient inside the growth zone with the barrier, and D_{bulk} is the diffusion coefficient away from the zone. The first two terms on the right hand side become unimportant for small molecules and for dilute solutions of the polymer.

These results are qualitatively different from the classical views on polymer crystallization. Although the new entropic model seems to capture the general trends of phenomenology, much more work is required to make quantitative comparisons with experimental facts. However, there is a promise of unification of ideas on crystallization from small molecules and from polymer chains.

Acknowledgements

The author is deeply indebted to his students and postdoctoral associates who have collaborated with him in the past. In particular, acknowledgment is made to Dr. A. Kundagrami and Mr. J. Zhang whose work is briefly mentioned here. Financial support for the present work was provided by the National Science Foundation Grant DMR-0209256.

References

[1] K.H. Storks, An electron diffraction examination of some linear high polymers, J. Amer. Chem. Soc. **60**, 1753–1761 (1938).
[2] A. Keller, A note on single crystals in polymers: evidence for a folded chain configuration, Phil Mag. **2**, 1171–1175 (1957); P.H. Till, The growth of single crystals of linear polyethylene J. Polym. Sci. **24**, 301–306 (1957); E. W. Fischer, Stufen- und spiralfoermiges Kristalwachstum beit Hochpolymeren, Z. Naturforsch. **12a**, 753–754 (1957).
[3] J.D. Hoffman, G.T. Davis and J.I. Lauritzen, in *Treatise on Solid State Chemistry*, Edited by N.P. Hannay, Plenum, N.Y. 1976; Vol. 3, Chapter 7, pp 497–614.
[4] Organization of macromolecules in the condensed phase, Disc. Faraday Soc. **68** (1979).

[5] P.J. Phillips, Polymer crystals, Rep. Prog. Phys. **53**, 549–604 (1990).
[6] K. Armistead and G. Goldbeck-Wood, Polymer crystallization theories, Adv. Polym. Sci. **100**, 219–312 (1992).
[7] J.I. Lauritzen and J.D. Hoffman, Theory of formation of polymer crystals with folded chains in dilute solution, J. Res. Nat. Bur. Std. **64A**, 73–102 (1960).
[8] I.C. Sanchez and E.A. DiMarzio, Dilute solution theory of polymer crystal growth: a kinetic theory of chain folding, J. Chem. Phys. **55**, 893–908 (1971).
[9] G. Allegra, Chain folding and polymer crystallization: A statistical-mechanical approach, J. Chem. Phys. **66**, 5453–5463 (1977).
[10] D.M. Sadler and G.H. Gilmer, A model for chain folding in polymer crystals: rough growth faces are consistent with the observed growth rates, Polymer **25**, 1446–1452 (1984).
[11] J.J. Point and M. Dosière, Crystal growth rate as a function of molecular weight in polyethylene crystallized from the melt: an evaluation of the kinetic theory of polymer crystallization, Polymer **30**, 2292–2296 (1989).
[12] M. Imai, K. Mori, T. Mizukami, K. Kaji and T. Kanaya, Structural formation of poly(ethylene terephthalate) during the induction period of crystallization: 1. Ordered structure appearing before crystal nucleation, Polymer **33**, 4451–4456 (1992).
[13] J.D. Hoffman and R.L. Miller, Kinetics of crystallization from the melt and chain folding in polyethylene fractions revisited: theory and experiment, Polymer **38**, 3151–3212 (1997).
[14] P.D. Olmsted, W.C.K. Poon, T.C.B. McLeish, N.J. Terrill and A.J. Ryan, Spinodal-assisted crystallization in polymer melts, Phys. Rev. Lett. **81**, 373–376 (1998).
[15] G. Reiter and J-U. Sommer, Crystallization of adsorbed polymer monolayers, Phys. Rev. Lett. **80**, 3771–3774 (1998).
[16] N.V. Pogodina and H.H. Winter, Polypropylene crystallization as a physical gelation process, Macromolecules **31**, 8164–8172 (1998).
[17] Y. Akpalu, L. Kielhorn, B.S. Hsiao, R.S. Stein, T.P. Russell, J. van Egmond and M. Muthukumar, Structure development during crystallization of homogeneous copolymers of ethene and 1-octene: Time-resolved synchrotron X-ray and SALS measurements, Macromolecules **32**, 765–770 (1999).
[18] G. Strobl, From the melt via mesomorphic and granular crystalline layers to lamellar crystallites: A major route followed in polymer crystallization?, Eur. Phys. J. E **3**, 165–183 (2000).
[19] B. Lotz, What can polymer structure tell about polymer crystallization processes?, Eur. Phys. J. E **3**, 185–194 (2000).
[20] S.Z.D. Cheng, C.Y. Li and L. Zhu, Commentary on polymer crystallization: Selection rules in different length scales of a nucleation process, Eur. Phys. J. E **3**, 195–197 (2000).
[21] J.K. Hobbs, T.J. McMaster, M.J. Miles and P.J. Barham, Direct observations of the growth of spherulites of poly(hydroxybutyrate-co-valerate) using atomic force microscopy, Polymer **39**, 2437–2446 (1998).
[22] Y.K. Godovsky and S.N. Magonov, Atomic force microscopy visualization of morphology and nanostructure of an ultrathin layer of polyethylene during melting and crystallization, Langmuir **16**, 3549–3552 (2000).
[23] M. Al-Hussein, and G. Strobl, The melting line, the crystallization line, and the equilibrium melting temperature of isotatic polystyrene, Macromolecules **35**, 3895–3913 (2002).

[24] A. Wurm, R. Soliman, and C. Schick, Early stages of polymer crystallization - a dielectric study, Polymer **44**, 7467–7476 (2003).
[25] S.Z.D. Cheng, and B. Lotz, Nucleation control in polymer crystallization: structural and morphological probes in different length- and time-scales for selection processes, Phil. Trans. Ro. Soc. London, A-Mathemat. Phys. and Engr. Sci **361**, 517–536 (2003).
[26] M. Muthukumar, Molecular modelling of nucleation in polymers, Phil. Trans. R. Soc. Lond. A **361**, 539–556 (2003).
[27] S. Umemoto, and N. Okui, Power law and scaling for molecular weight dependence of crystal growth rate in polymeric materials, Polymer **46**, 8790–8795 (2005).
[28] S. Rastogi, D.R. Lippits, G.W.M. Peters, R. Graf, Y.F. Yao, H.W. Spiess, Heterogeneity in polymer melts from melting of polymer crystals, Nature Materials **4**, 635–641 (2005).
[29] C. Liu and M. Muthukumar, Langevin dynamics simulations of early-stage polymer nucleation and crystallization, J. Chem. Phys. **109**, 2536–2542 (1998).
[30] J.P.K. Doye and D. Frenkel, Kinetic Monte Carlo simulations of the growth of polymer crystals, J. Chem. Phys. **110**, 2692–2702 (1999).
[31] J-U. Sommer and G. Reiter, Polymer crystallization in quasi-two dimensions. II. Kinetic models and computer simulations, J. Chem. Phys. **112**, 4384–4393 (2000).
[32] P. Welch and M. Muthukumar, Molecular mechanism of polymer crystallization from solution, Phys. Rev. Lett. **87**, 218302-1-218302-4 (2001).
[33] T. Yamamoto, Molecular dynamics simultion of polymer ordering. II. Crystallization from the melt, J. Chem. Phys. **115**, 8675–8680 (2001).
[34] A. Toda, Kinetic barrier of pinning in polymer crystallization: Rate equation approach, J. Chem. Phys. **118**, 8446–8455 (2003).
[35] T. Yamamoto, Molecular dynamics modeling of polymer crystallization from the melt, **45** 1357–1364 (2004).
[36] J. Zhang and M. Muthukumar, preprint.
[37] A. Kundagrami and M. Muthukumar, foreprint.

2

Theoretical Aspects of the Equilibrium State of Chain Crystals

Jens-Uwe Sommer

Institut de Chimie des Surfaces et Interfaces (ICSI-CNRS), 15 rue Jean Starcky, P.B. 2488, F-68057 Mulhouse, France. Present address: Leibniz-Institut of Polymer Research, Hohe Strasse 6, Dresden, Germany

Abstract. The equilibrium state of polymer single crystals is considered by explicitly taking into account the amorphous fraction formed by loops and tails of the chains. Using ideal chain statistics, a general expression for the free energy excess of the amorphous part is derived. I show that tight loops and close reentries are favored under experimental conditions for under-cooling of polymer single crystals. For many chain crystals, I show that the lamellar thickness increases with the number of chains in the crystal, and that extended chain conformations are thermodynamically favored when the number of chains in the crystal is sufficiently large. The role of finite bending rigidity of chains is discussed for folded chain crystals, as well as tilt effects in extended chain crystals.

2.1 Introduction

Since the discovery of folded chain crystals [1–3] polymer crystals are considered as meta-stable systems which properties are controlled by kinetic effects [4,5]. This point of view is supported by many observations, such as the spontaneous thickening of lamellae [6], the dependence of the melting behavior on the thermal history of bulk samples [7], and spontaneous morphological transformations as observed in thin films [8,9]. Moreover, true thermodynamic coexistence is not observed in polymers, the crystallization temperature being generally lower than the melting temperature. The under-cooling necessary to obtain polymer crystals under laboratory conditions can be as large as 100 K. Furthermore, it is commonly believed that equilibrium forms of polymer crystals consist of extended chains and that such (usually extraordinary thick lamellae) are usually not observed under experimental conditions. Exceptions are short chains such as n-Alkanes [10] and polyethylene (PE) under high external pressure [11].

In contrast to crystals formed by small molecules, the positions of the individual monomers in polymer crystals are restricted due to their connectivity, and the polymer chain as a whole has to undergo a transition from the random coil (high entropy) state to a partially folded or extendend (low entropy) state. Thus, viewed on the scale of individual chains, the crystallization transition involves an *internal transition of the molecule* itself. This causes a kinetic barrier as the chain has to be rearranged into the ordered conformation, a process which bears some similarity to the folding transition of protein molecules [12]. However, in contrast to proteins which are supposed to attain their stable ground state within a short time, polymer crystals get trapped in meta-stable states. Using this paradigm, attention has focused on the understanding of the kinetical effects during the formation of chain crystals far away from equilibrium states [13].

On the other hand, not much attention has been paid to a thorough mathematical description of the equilibrium state of polymer single crystals. This involves the calculation of the free energy excess of the amorphous part formed by loops and tails of the chains. In the past there were attempts to explain properties of crystalline polymers with equilibrium concepts addressing the coexistence of crystalline and amorphous phases in the semi-crystalline state [14], in particular aimed to explain their broad melting behavior, see [15], as well as the phenomena of partial reversible melting [16,17]. Recently, this issue has been raised again by Muthukumar [18] who emphasized the possibility of folded chain states as the equilibrium form of polymer single crystals. His approach has been originally addressed to crystals formed by single chains as they can be studied in computer simulations [19,20]. Here, the extended chain form can be trivially excluded.

In this work, I consider several aspects of the equilibrium state of polymer single crystals using the model proposed by Muthukumar which will be extended to multi-chain crystals. In particular I will show that extended chain crystals are the equilibrium form for many chain crystals if sufficiently many chains are accessible and I will give a simple argument for their thermodynamic stability with reference to folded chain crystals. Furthermore, the role of finite flexibility of chains is discussed as well as the tilt of stems in extended chain crystals.

2.2 Thermodynamic Considerations about the Equilibrium Shape of a Polymer Single Crystal

Throughout this work the free energy of chain segments will be defined with respect to their value in an amorphous unrestricted chain in the melt phase, i.e. the free energy is expressed as the difference to that of the liquid phase. Using the approximation of Gaussian statistics for individual chains in the melt phase, the free energy per segment can be written as $-kT \ln c$, where c denotes the number of states available for the segment in a free chain. In the following, we take the statistical segment of length b as the basic unit of the chain.

For an infinitely extended crystal, we denote the latent heat of fusion per statistical segment by ϵ_0. Let us now consider the free energy of a segment, ϵ, at a temperature T below the melting temperature T_0. In a first order thermodynamic approximation, we obtain

$$\epsilon = \frac{\epsilon_0 \Delta T}{T_0}, \qquad (2.1)$$

with

$$\Delta T = T_0 - T . \qquad (2.2)$$

This approximation can be improved for larger under-cooling by the following expression [16, 21]

$$\epsilon = \frac{\epsilon_0 \Delta T T}{T_0^2} . \qquad (2.3)$$

As an example, we consider polyethylene (PE) where the heat of fusion per mol of CH2-units is about 4.11 kJ. Taking a statistical segment formed by 6 chemical units, we obtain $\epsilon_0 = 4.1 \cdot 10^{-20}$ J. Using $T_0 = 414$ K and $T = 300$ K in Eq. (2.3) yields $\epsilon \simeq 0.8 \cdot 10^{-20}$ J which corresponds to about 2 kT. This gives us an orientation for the values of ϵ in the experimental relevant range of under-cooling.

In a next step, we consider the finite size of a single crystal formed by μ crystalline stems (oriented orthogonal to the cylinder cross-section) each comprising m statistical segments, as sketched in Fig. 2.1. The excess free surface energy of the amorphous fraction (loops and tails) is denoted by σ_f. For simplicity, we use the term "surface tension" instead of the term "excess surface free energy" in the following. Furthermore, we assume a spherical shape of the cylinder cross-section. The latter property, however, agrees rather nicely with recent simulations of single chain crystals [20]. The free energy can be written as

$$F = -\mu m \epsilon + 2\mu \sigma_f + \sigma \sqrt{\mu} m , \qquad (2.4)$$

where $\sigma = 2\sqrt{\pi} \sigma_e$ represents lateral surface tension of the lamella and σ_f denotes the surface tension of the fold surface.

With the condition

$$N \simeq \mu m = \text{const} , \qquad (2.5)$$

we obtain

$$F = -N\epsilon + N\sigma \frac{1}{\sqrt{\mu}} + 2\mu \sigma_f . \qquad (2.6)$$

The equilibrium solution is readily obtained:

$$\mu^* = N^{2/3} \alpha \qquad (2.7)$$
$$m^* = N^{1/3}/\alpha . \qquad (2.8)$$

Here, I have introduced the *shape factor* α given by

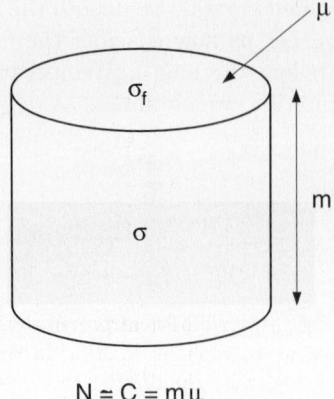

Fig. 2.1. Cylinder model for the single crystal. The cross-section contains μ crystalline stems of length m. The (tight) folds and ends are comprised in an excess free energy σ_f. Almost all monomers are considered to be contained in the crystalline fraction

$$\alpha = \left(\frac{\mu}{m^2}\right)^{1/3} = \left(\frac{\sigma}{4\sigma_f}\right)^{2/3}. \qquad (2.9)$$

Note that this solution corresponds to Wulff's construction for the equilibrium shape of a cylindrical crystal [22].

In the above consideration, the surface tension, σ_f, has been introduced *ad hoc*. Its measurement is non-trivial since equilibrium crystals are usually extended chain crystals with a large lateral extension, i.e. $\mu \gg \mu^*$, so that the shape factor cannot be directly obtained. Usually, the value for σ_f is inferred from the melting line of the non-equilibrium crystal according to a Gibbs-Thompson approach, see [14]. I note that in this case neither the surface tension can be truly assumed to be an equilibrium property, nor can the validity of the Gibbs-Thompson extrapolation be tested independently. For a criticism of the Gibbs-Thompson approach for non-equilibrium polymer crystals, see [23].

It is therefore desirable to calculate the contribution of σ_f from equilibrium models which will provide more insight into the nature of the amorphous fraction. Clearly, we are restricted here to simplified models for the chain and the crystal part. As a first step, a two-phase model for the single crystal has to be introduced, which is illustrated in (Fig. 2.2). Segments can be exchanged freely between the crystalline (C) and the amorphous fraction (A) by conserving the total number of segments:

$$N = C + A = \text{const}, \qquad (2.10)$$

Than, Eq. (2.4) can be generalized to

2 Theoretical Aspects of the Equilibrium State of Chain Crystals

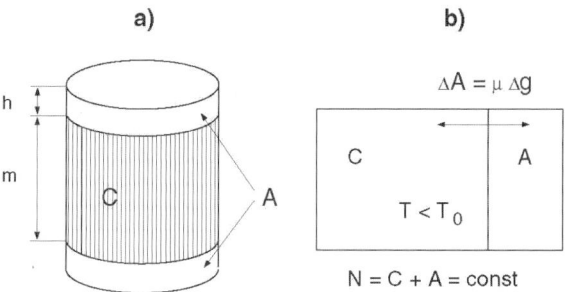

Fig. 2.2. Two-phase model for a cylindrical polymer crystal. (**a**) Loops and tails are explicitly considered as an amorphous fraction in thermodynamic equilibrium with the crystalline fraction. The height of the amorphous layers is denotes by h, keeping the notation for m as the length of the crystalline stems. Both length scales are considered in units of statistical segments. (**b**) Illustration of the thermodynamic equilibrium system. Segments can be exchanged between two phases and the temperature is considered to be lower than equilibrium melting temperature

$$F = -\mu m \epsilon + \sigma\sqrt{\mu m} + F_a = F_c + F_a , \qquad (2.11)$$

where F_a denotes the free energy of the amorphous fraction with respect to the state of free chains and F_c represents the free energy of the crystalline fraction as discussed above. In the following, I will outline tractable statistical mechanical models for the amorphous fraction to understand the origin of the fold surface tension in equilibrium crystals.

2.3 The Brush State of the Amorphous Fraction is Thermodynamically Suppressed

Let us assume that the amorphous fraction forms a dense layer with an average loop length of $g \gg 1$ segments on either side of the crystal, and that the surface is sufficiently extended to obtain a homogeneous density of segments c_A in the amorphous fraction. Then, the height of the amorphous layers, see Fig. (2.2) is given by

$$h = \frac{g}{2}\left(\frac{1}{c_A \xi^2}\right) \sim g , \qquad (2.12)$$

where the distance between the crystalline stems is denoted by ξ which corresponds to a crystallographic value of a few Å. The relation $h \sim g$ corresponds to a brush-like state where the loops and tails are extended in the direction perpendicular to the surface. Using a scaling approach [24], the free energy per loop can be written as

$$F_g \sim kT\left(\frac{h}{bg^{1/2}}\right)^2 \sim g , \qquad (2.13)$$

where we have used Eq. (2.12), and b denotes again the length of a statistical segment. Thus, we can write $F_g = rkTg$, with a numerical constant r and the free energy of the brush-decorated crystal can be written as

$$F = \mu rkTg + \mu\epsilon g - \epsilon N + \sigma\sqrt{\mu}(N/\mu - g) \, . \tag{2.14}$$

The third term in this expression is the free energy of a crystal without an amorphous fraction. The free energy excess of the amorphous part is dominated by the first two terms which are both strictly positive. The first term corresponds the effort for stretching the chains in the brush state while the second term corresponds to the increase of free energy by pulling g segments out of the crystalline phase. Thus, under equilibrium conditions, where g is a variational parameter, the stable solution corresponds to the absolute minium of g which is possible to form a loop conformation. I note that the correction due to lateral surface tension (last term in Eq. (2.14)) is also positive for $\sqrt{\mu} > \sigma/(rkT + \epsilon)$ which corresponds to a small number of stems. This calculations clearly demonstrate that a dense layer of long loops (and tails) does not correspond to a stable equilibrium state of the polymer crystal. In particular the brush-like state merely adds a free energy of several kT to each loop or tail which is transformed into the amorphous phase.

It is interesting to add that also individual chain tails are not favored thermodynamically. Here, we simply obtain

$$F = \gamma\mu\epsilon g - \epsilon N + \sigma\sqrt{\mu}(N/\mu - \gamma g) \, , \tag{2.15}$$

where faction of long loops/tails is given by $\gamma \ll 1$. Again, there is no stable solution for finite value of g, if the lateral extension of the crystal is not too small ($\sqrt{\mu} > \sigma/\epsilon$). This result is easy to understand: An isolated loop/tail with $g \gg 1$ just increases the free energy by a value of $g\epsilon$ without any compensation as referred to the equilibrium amorphous state.

2.4 Extended Chain Crystals and Sliding Entropy

In the section above we have tacitly assumed that the anchor points of the loops and tails are fixed. However, the possibility to distribute the amorphous segments in all possible ways along a given chain will give rise an an addition entropy as compared to the liquid state.

Let us consider a laterally infinitely extended polymer crystal. Each (extended) chain of length N_{ch} is composed of a (central) crystalline part made of m segments enclosed by $g = N_{ch} - m$ amorphous segments, which is illustrated in (Fig. 2.3). Since the crystalline part can be located anywhere along the chain this corresponds to a *sliding entropy* of

$$S_{slide} = k \ln g \, , \tag{2.16}$$

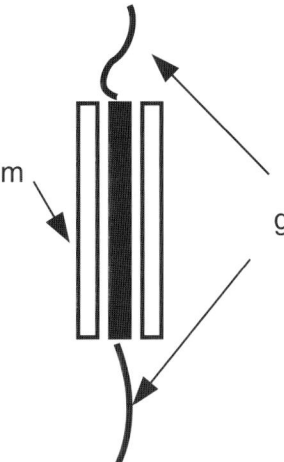

Fig. 2.3. Single chain within an extended chain crystal. Sliding of the chain trough the crystal phase (comprising m monomers per chain) is possible if g monomers are placed in the amorphous phase

where a constant S_0 can be suppressed. Thus, the free energy of a single chain in the extended chain crystal can be written as

$$F_{ext} = -kT \ln g + \epsilon g - \epsilon N_{ch} \, . \tag{2.17}$$

Minimization of Eq. (2.17) yields the equilibrium fraction of amorphous monomers per chain:

$$g_e = \frac{kT}{\epsilon} \, . \tag{2.18}$$

If we remember our example of PE given in section (2.2), we would obtain a small value of g_e. However, extended chain crystals can be observed rather close to the equilibrium melting temperature, where ϵ can become only fractions of kT. Using Eq. (2.3), we obtain

$$g_e = \frac{kT_0^2}{\epsilon_0 \Delta T} \, . \tag{2.19}$$

The nominator leads to a divergency of g_e when approaching T_0 [1]. A similar effect has been already discussed by Fischer [16] and Zachmann [15] in the context of equilibrium pre-melting in semi-crystaline polymers.

---

[1] For short chain crystals such as obtained for n-Alkanes, the equilibrium melting temperature T_0 must be replaced be maximum equilibrium melting temperature corresponding to the finite thickness of the crystals for $m = N_{ch}$. This takes into account a certain melting point depression due to the bare surface tension of the top and bottom surface.

However, Eqs. (2.18) and (2.19) are only valid for moderate values of g_e. Large values lead to the brush state in the amorphous fraction where the logarithmic entropy gain due to sliding is quickly compensated by the linear penalty term due to chain stretching, see Eq. (2.13). Taking into account Eqs. (2.13) and using the same symbols as in Eq. (2.14), we obtain

$$g_e = \frac{kT}{\epsilon + rkT}, \qquad (2.20)$$

which regulates the divergency for $\epsilon \to 0$. On the other hand, the crystal can avoid part of the stretching free energy by tilting the stems thus increasing the distance between stems projected onto the top and bottom surfaces. This issue will be discussed in Sect. 2.9.

In the above consideration I have neglected the surface tension of the lateral surfaces by assuming an infinitely extended crystal. In many experimental situations where extended chain crystals are studied, this approximation is justified since the lateral extension can be orders of magnitude larger than the height of the crystal. The free energy for finite crystal with the shape factor α, see Eq. (2.9) is given by

$$F = \alpha^3 m^2 F_{ext} + \sigma \alpha^{3/2} m^2 . \qquad (2.21)$$

For $\alpha \ll 1$, the second term dominates the free energy of the crystal, and the chain must obtain a folded conformation. An extreme case being a crystal formed by a single chain only, where the extended conformation can not be stable at all, since no crystalline bonds can be formed.

An interesting question arrises of how many chains are necessary to make the extended chain form the stable solution. The above considerations suggest $\alpha \simeq 1$. For $m \gg 1$ and $g \ll m$ (the latter is again related to the avoidance of the brush state) we have $m \simeq N_{ch}$ and the number of chains necessary for the extended chain form is given by $n_{ext} \simeq N_{ch}^2$. Using a more rigorous approach, I will show further below that this results is qualitatively correct.

2.5 The Slip-Loop Model for the Entropy of the Amorphous Fraction of a Single Chain Crystal

In the last section I have shown that sliding of chains yields to an additional entropy which favors a finite fraction of amorphous tails. This idea can be generailzed to folded chain conformations as sketched in (Fig. 2.4). Here, I will consider a crystal made of a single chain.

The essential idea is to assume that *all segments of the amorphous part can be distributed in all possible ways among the various loops and tails* for a given stem length m and for a given number of stems μ. The equilibrium solution is than obtained by minimizing the resulting free energy with respect to both variables. This shall be denoted as the *slip-loop model*.

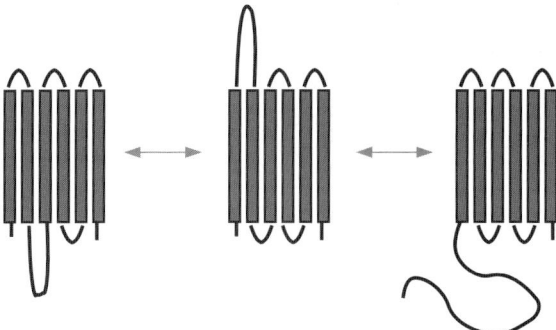

Fig. 2.4. Sketch of the slip-loop model for the amorphous part. The segments in the amorphous fraction can be arbitrarily distributed among the $\mu - 1$ loops and the both tails

In order to proceed, I have to make some assumptions about the chain statistics and about the form of the crystal. The latter should be given again by the model sketched in (Fig. 2.2). In particular, all stems should have the same length. To start with a tractable model, I will further ignore excluded volume interactions between the segments of the amorphous fraction as well as the conformational constraints due to the impenetrable crystalline surface. Furthermore, I treat the chain statistics as Gaussian and ignore effects of finite flexibility of the chain. These relaxed conditions overestimate the entropy of the amorphous fraction. I will reconsider these approximations in the context of the exact solution for the idealized model.

The number of conformations available for a Gaussian chain with g segments starting at r_0 and ending at r with respect to the free unconstrained chain is given by

$$G(r_0, r; g) = \left(\frac{1}{4\pi l^2 g}\right)^{3/2} \exp\left(-\frac{(r-r_0)^2}{4l^2 g}\right) \Delta v, \quad (2.22)$$

with $l^2 = b^2/6$. The factor Δv compensates for the formally infinitely sharp localization of the end-segment of the chain in a continuous space and denotes the uncertainty of the localization of the end-segment. The physical meaning of this factor will be discussed further below. We call $G(x, x'; g)$ the Green function. The free energy difference of the restricted chain with respect to a free chain is than given by $F = -kT \ln G$ which corresponds to my notation of the free energy in this work. The mathematical task is completed if the Greensfunction of the amorphous part G_a has been calculated.

The contribution of the tails can be explicitly taken into account, since each tail just provides $G_t = 1$ (integration of Eq. (2.22) over $r/\Delta v$). This yields

$$G_a(A) = \int_0^A dn(A-n) G_L(n, \mu - 1), \quad (2.23)$$

where G_L denotes the contribution from loops only. Note that the contribution to Eq. (2.23) is only due to the various positions the loop part, made of n segments, can take within the amorphous part made of A segments. The details of the calculation of the loop part is more technical and can be found in Appendix A. The result is given by

$$G_L(A, \mu - 1) = \frac{1}{\kappa^{\mu-1}} \sqrt{\frac{(\mu-1)\xi^2}{4\pi A l^2} \frac{1}{A}} \exp\left\{-\frac{(\mu-1)^2 \xi^2}{4A l^2}\right\}. \quad (2.24)$$

Here, I have introduced the dimensionless *localization parameter*

$$\kappa = 4\pi l^2 \xi / \Delta v, \quad (2.25)$$

where ξ characterizes the minimal distance between the loop ends, see also Sect. 2.3. Note the similarity between Eq. (2.24) and the single chain result of Eq. (2.22). Using Eq. (2.23) the final solution reads

$$G_a(A, \mu) = \frac{4A}{\kappa^\mu} \cdot \left[\left(\frac{1}{4} + \frac{1}{2}y\right) \text{erfc}(\sqrt{y}) - \frac{\sqrt{y}}{2\sqrt{\pi}} e^{-y}\right] = \frac{4A}{\kappa^\mu} \cdot f(y), \quad (2.26)$$

where I have introduced the *scaling variable* y, defined by

$$y = \frac{(\mu-1)^2 \xi^2}{4A l^2} \quad (2.27)$$

and erfc(y) denotes the complimentary error function (erfc(y) = $\frac{2}{\sqrt{\pi}} \int_y^\infty dx e^{-x^2}$).

In the following I consider only the case $\mu \gg 1$, which is the physical relevant solution for single chain crystals. The scaling variable can be related to the *average loop length* in the amorphous fraction

$$g = \frac{A}{\mu}. \quad (2.28)$$

by

$$y = \frac{a}{kT} \frac{\mu}{g}, \quad (2.29)$$

where

$$a = \frac{3}{2} kT (\xi/b)^2. \quad (2.30)$$

denotes the maximal energy of the Gaussian spring which is formed by a single loop. The scaling variable y thus denotes the spring energy in units of kT related to μ loops, containing g segments each. Assuming an average free energy per loop of the order of kT, we can conclude that the physical relevant case is given by

$$y \gg 1. \quad (2.31)$$

The opposite case of $y \ll 1$ can only be realized if the average loop length is very large ($g \gg \mu$). The latter must be excluded in order to avoid the brush regime. However, Eq. (2.31) can also be justified without referring the brush regime. Generally, we obtain from Eq. (2.27)

$$A \sim \mu^2 \quad \text{for} \quad y \sim 1 \ . \tag{2.32}$$

The free energy effort to transfer A segments into the amorphous state is given by $\epsilon A \sim \mu^2$. I will show below that the solution for the case $y \gg 1$ leads to a surface excess which scales proportional to μ only.

Using Eqs. (2.26) in the limiting case (2.31), the free energy of the amorphous fraction can be written as

$$F_a = -kT \ln G_a = \mu \left(2\sigma_{f0} + \frac{a}{g}\right) \quad \text{for} \quad y \gg 1 \text{ and } \mu \gg 1 \ , \tag{2.33}$$

with

$$2\sigma_{f0} = kT \ln \kappa \ . \tag{2.34}$$

For details, see Appendix B.

The localization parameter κ, see Eq. (2.25), can be related to the entropic restriction of an anchoring segment compared to a segment in a free chain. I will therefore consider κ as the ratio of the number of states of the end segments in the free chain compared to the anchored state. In a rough approximation the segments which directly anchor to the crystalline stem loose about half of the degrees of freedom being restricted to the half space. Therefore, the anchoring contribution might be estimated as $\kappa \simeq 2$ for each anchored segment. The corresponding free energy contribution per stem is thus comparable to kT. This free energy excess gives rise to an *entropic* surface tension, σ_{f0}, which increases with temperature.

I note that the solution in Eq. (2.33) is equally obtained using the loop part only, see Eq. (2.24). This indicates that tails do not play an singular role.

2.6 Tight Loops and Effective Fold Surface Tension for Single Chain Crystals

Using Eq. (2.11) we get for the free energy of the single chain crystal

$$F = F_c + F_a = -\mu m \epsilon + \sigma \sqrt{\mu} m + \mu \left(2\sigma_{f0} + \frac{a}{g}\right) \ . \tag{2.35}$$

The state of thermodynamic equilibrium is given by the minimum of F with respect to μ, g and m under the constraint of Eq. (2.10). A solution can be obtained analytically for $N \gg 1$, which is the physically relevant case. The direct solution of the minimization problem is given in Appendix C.

However, the solution presented in Appendix C can be rederived using a simple argument which reveals the essential physics most clearly. For $N \gg 1$, we disregard the lateral surface tension and assume that the optimal value

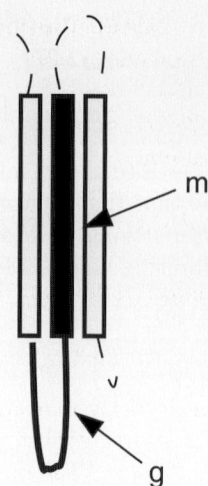

Fig. 2.5. Sketch of a single loop-stem element in the crystal

of g can be obtained by minimization for a single stem-loop element which is sketched in Fig. 2.5. The corresponding free energy reads

$$F_{sl} = 2\sigma_{f0} + g\epsilon + \frac{a}{g} = 2\sigma_{f0} + g\epsilon + \frac{3}{2}kT\frac{(\xi/b)^2}{g} \ . \qquad (2.36)$$

Here, the second term is attributed to the transition of g monomers into the disordered phase and the third term represents the g-dependent part of the free energy from Eq. (2.33) related to a single loop. This latter part, however, agrees exactly with the free energy stored in a loop of g segments with the end-to-end separation of ξ. Thus, the essential free energy balance is between melting a segment and the corresponding decrease of the free energy of a Gaussian spring which is prolongated by one segment. Minimization of F_{sl} with respect to g gives

$$g_0^2 = \frac{a}{\epsilon} = \frac{3}{2}\frac{\xi^2}{l^2}\frac{kT}{\epsilon} \ , \qquad (2.37)$$

which agrees with the solution for the full minimization problem given in Appendix C. According to our discussion in Sect. 2.2, the value of ϵ is not expected to become very small under usual experimental conditions. Therefore, the solution above indicates the formation of tight loops. Physically speaking, g_0^2 represents the ratio between the maximum free energy of the Gaussian spring to the free energy loss by pulling a segment out of the crystaline phase.

Being at the limit of validity, the Gaussian statistics used so far has to be scrutinized. This concerns in the first place the effect of finite bending rigidity which involves a fine-graining of the model towards a length scale smaller than the statistical segment length. I will come back to this issue in Sect. 2.8. On the other hand, for equilibrium crystals it should be possible (at least theoretically) to consider also small values of ϵ thus approaching close to the equilibrium melting point. In this case, g_0 can become sufficiently large to justify the Gaussian statistics.

Using the result of Eq. (2.37), we obtain for the minimal free energy excess per loop of the amorphous fraction

$$F_{sl} = F_{sl} = 2\sigma_{f0} + g_0\epsilon + \frac{a}{g_0} = 2\sigma_{f0} + 2\sqrt{\epsilon a} = 2\sigma_f , \qquad (2.38)$$

where I have introduced the *effective fold surface tension* σ_f defined as

$$\sigma_f = \sigma_{f0} + \sqrt{a\epsilon} . \qquad (2.39)$$

The optimal shape is now easily derived from the free energy of the single crystal taking into account the lateral surface tension

$$F = -\epsilon N + 2\mu\sigma_f + \sigma\sqrt{\mu}m . \qquad (2.40)$$

The relation between m and μ is given by $m + g_0 = N/\mu$. For $m \gg 1$, we can disregard the difference between m and N/μ, and we are let to the effective one-phase approach of Eq. (2.4). The shape factor is given by

$$\alpha = \left(\frac{\sigma}{4\sigma_f}\right)^{2/3} = \left(\frac{1}{4}\frac{\sigma}{\sigma_{f0} + \sqrt{a\epsilon}}\right)^{2/3} . \qquad (2.41)$$

Thus, we obtain the equilibrium values of the extension of the single crystal:

$$\mu^* = N^{2/3}\left(\frac{1}{4}\frac{\sigma}{\sigma_{f0} + \sqrt{a\epsilon}}\right)^{2/3} \qquad (2.42)$$

$$m^* = \frac{N}{\mu} - g_0 \simeq \frac{N}{\mu} = N^{1/3}\left(\frac{1}{4}\frac{\sigma}{\sigma_{f0} + \sqrt{a\epsilon}}\right)^{-1/3} \qquad (2.43)$$

As I have shown, the origin for the finite amorphous fraction formed by the (prevailing) loops is due to the balance between the entropic spring force created by the finite separation of the anchoring segments on the one hand side and the effort to remove the loop segments from the thermodynamically preferred crystalline phase on the other side.

There is another interesting conclusion from our free energy argument concerning the value of ξ. In the calculation it was introduced as the smallest possible separation between the end points of the loops. This mathematical argument can now be supported by a physical argument: Since the entropy of a loop increases quadratically with the distance ξ, see Eq. (2.36), in thermodynamic equilibrium the smallest possible distance is favored. Thus, loops have the tendency to close, i.e. tight folds are preferentially formed by thermodynamic reasons.

2.7 Many Chain Crystals

The results obtained in the last section can be readily extended to the case of many chain crystals. As I have shown, the essential argument which leads to the equilibrium state of the amorphous fraction can be reduced to the free energy balance of a single loop. As far as the number of folds per chain is large, i.e. the role of tails is only minor, the result of Eq. (2.37) holds true also within many chain crystals. A formal mathematical analysis of this problem can be found in [25]. Given a crystal thickness of m (not yet optimized), the free energy for a single chain within the many chain crystal is given by

$$F_{ch} = -\epsilon N_{ch} + 2\sigma_f \frac{N_{ch}}{g_0 + m} \, , \qquad (2.44)$$

where $\mu_{ch} = N_{ch}/(g_0+m)$ can be replaced again by N_{ch}/m for $m \gg 1$. Then, the free energy for the overall crystal formed by n chains is given by

$$F = nF_{ch} + \sigma m\sqrt{\mu n} \, . \qquad (2.45)$$

Introducing the total number of segments $N = nN_{ch}$, we obtain

$$F = -\epsilon N + 2\sigma_f \frac{N}{m} + \sigma\sqrt{Nm} \, , \qquad (2.46)$$

an expression which is again fully equivalent to Eq. (2.4). The solution for the equilibrium thickness m^* then reads

$$m^* = (nN_{ch})^{1/3} \left(\frac{4\sigma_f}{\sigma} \right)^{2/3} \sim n^{1/3} \, . \qquad (2.47)$$

This result tells us that the equilibrium thickness of the crystal is growing with the number of chains. Thus, at a certain point the thickness can become larger than the extension of the individual chains, and the extended chain crystal becomes the equilibrium form. With $m^* = N_{ch}$, I obtain

$$n_{ext} = N_{ch}^2 \alpha^3 \, . \qquad (2.48)$$

Further thickening is hampered by additional surface tension which is created by stacking several chains in one stem. This result corroborates the conclusion obtained at the end of Sect. 2.4, where I have approached the problem from the opposite limit of extended chain crystals.

In order to appreciate the values calculated above, I consider the example of an extended chain crystals formed by PE under high external pressure [11]. A polymer chain of about $100,000$ g/mol of molecular weight for PE corresponds to a value of $N_{ch} = 1000$. Using Eq. (2.48), the estimated number of chains necessary to reach the stretched state amounts to about $1,000,000$. This corresponds to a lateral size of the crystal of a few hundred nanometers.

On the other hand, given the usual thickness of *non-equilibrium* PE crystals of the order of 10 nm, the same amount of chains require a lateral size of the order of a few micrometers which is within the experimentally observed range. Thus, non-equilibrium polymer single crystals can have the *potential* to form extended chain crystals in equilibrium.

The phase diagram for the equilibrium crystal is sketched in Fig. 2.6. If the number of chains, n, is increased, the thickness of the equilibrium crystal grows as the third root of n until it reaches the extended chain state at $n = n_{ext}$.

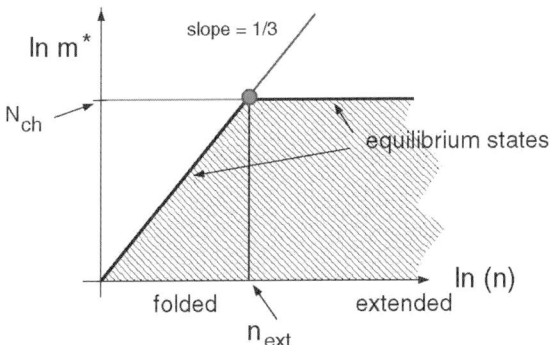

Fig. 2.6. Phase diagram of a polymer single crystal. The equilibrium states are indicated by the thick line. Non-equilibrium states (hatched area) are usually observed below the equilibrium line

When we approach the extended chain crystal, the approximation for the free energy of the amorphous fraction of the folded chain, Eq. (2.44), must be corrected to account for the dominating role of tails. This can be done using the full result for the free energy for the amorphous fraction, but the essential physics can be obtained in a much simpler way. In Sect. 2.4, the exact expression the free energy of an extended chain crystal has been derived. Here, only the tails contribute to the free energy of the amorphous fraction. In this case, only sliding of the chain (positioning of the crystal stem within the chain) is responsible for a finite amount of amorphous material per chain. By contrast, in case of folded chain crystals, the Gaussian spring energy of loops competes with the crystallization energy and the sliding term is reduced to a small contribution when many folds are formed. At the cross-over between both regimes, the amount of sliding entropy becomes increasingly important and eventually prevails the contribution from the loops. The transition is finally discontinuous because an integer number of folds have to be formed.

The only part which is left to prove is the stability of the extended chain form with respect to the folded chain form. This can be easily inferred from the following Gedankenexperiment as sketched in Fig. 2.7. Let us consider a single chain crystal formed by a huge chain of length N in thermodynamic

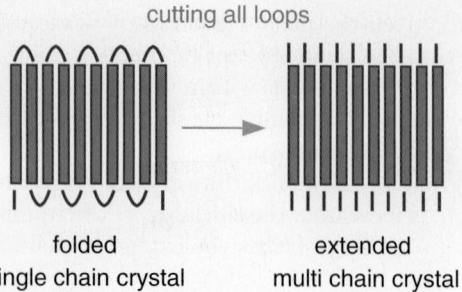

Fig. 2.7. An extended chain crystal is obtained by cutting all loops of a single chain, folded crystal

equilibrium with $\mu \gg 1$ and $m \gg 1$, see left part of Fig. 2.7. Then, we cut all loops and obtain a multi chain crystal formed by N/μ extended chains having the length of $N_{ch} = m + g_0$ monomers, see right part of Fig. 2.7. Note that N_{ch} can be arbitrary large, since $m \sim N^{1/3}$, see Eq. (2.43). The free energy change for the transition from the folded to the extended chain crystal is given by

$$\Delta F = \mu \left(-\frac{a}{g} - kT \ln g \right) , \qquad (2.49)$$

which is *strictly negative*. The first term corresponds to the opening of the loops (release of Gaussian stretching free energy) and the second term corresponds to the fee energy gain due to the *independent* sliding motions of the individual stems as is has been derived in Sect. 2.4. Thus, the extended chain form is thermodynamically preferred, if the freedom to open chain loops is given.

The essential conclusion from this paragraph is that folded chain crystals are equilibrium forms only if the number of chains contained in the crystal is limited. Here, the extended chain form can violate the optimal crystal shape according to the Wulff construction.

2.8 The Role of Bending Rigidity for the Formation of Small Loops

So far, I have considered the Gaussian chain model based on coarse-graining on the scale of a statistical segment length. For folded chain crystals, the equilibrium loop length according to Eq. (2.37) turns out to be close to unity if experimental values for the under-cooling are considered. Such small loops, however, have to bear a considerable amount of bending energy and the Gaussian approximation is limited. In this section, I will discuss the effect of finite bending rigidity for a continuous chain model. This will address the situation of tight loops only where the Gaussian approach fails. The chain is described by the path $\boldsymbol{r}(s)$ parameterized by the arc length along the chain's contour.

The bending rigidity B of the homogeneous chain (*worm-like chain model*) is related to the statistical segment by [26]

$$B = 2kTb .\qquad(2.50)$$

Then, the energy of a loop of length gb can be written as

$$E = kTb \int_0^{gb} ds \left(\frac{d\mathbf{t}(s)}{ds}\right)^2 ,\qquad(2.51)$$

where $\mathbf{t}(s) = d\mathbf{r}(s)/ds$ denotes the normalized tangent vector of the chain at position s. Here, g, denotes a real number which can be smaller than unity. For a loop-like conformation, one obtains

$$E = \zeta 4\pi^2 kT \frac{1}{g} ,\qquad(2.52)$$

where the constant ζ accounts for a non-trivial form of the loop. The value $\zeta = 1$ gives the result for a circle. The statistical weight to be taken into account for each loop can be thus written as

$$G_w(s) \sim \exp\left\{-\frac{1}{4}\frac{a_0^2}{g}\right\} ,\qquad(2.53)$$

with

$$a_0^2 = 16\zeta\pi^2 .\qquad(2.54)$$

Here, the index "w" reminds to the worm-like chain model. Note that this approach is only valid if fluctuations of the chain's contour are not dominating. Thus, it represents the complementary case to the Gaussian approach, were only fluctuation are taken into account.

Nevertheless, there is a strong similarity between the exponentials of Eqs. (2.53) and (2.22). To proceed, it is worth noting, that the essential part of the Laplace-transform which allows the calculation of the multiple integral for G_L, see Appendix A, is determined by a stationary point of the Laplace-integral only and hence (within this approach) the same result is obtained using G_{wl} instead of G. Thus, following the same steps as presented in Appendix A, the essential part of the free energy of the amorphous fraction can be written as

$$F_a = \mu \left(2\upsilon_{f0} + \frac{a}{g}\right) ,\qquad(2.55)$$

where the constant a is now defined as

$$a = 4\zeta\pi^2 kT .\qquad(2.56)$$

I note that the bare surface tension, σ_{f0}, has an empirical meaning only, although it must be still related to the localization of the end-points of the loop. In fact, being constant, σ_{f0} is not important for the physical most significant conclusion about the average loop length. Mapping the redefinition of a onto the results obtained in Sect. 2.6, I get

$$g_0^2 = \frac{a}{\epsilon} = 4\zeta \pi^2 \frac{kT}{\epsilon} \ . \tag{2.57}$$

Comparing this result to Eq. (2.37), one can conclude that the equilibrium loop length is larger (in units of the statistical segment length), although a physical interpretation is now possible for g smaller than unity. Using reasonable estimates for PE at room temperature, see Sect. 2.2, a value for g_0 of the order of a few statistical segments (depending on the value of ζ) is predicted. This, however, means that even under such conditions, the *optimal loop length does not correspond to the absolute minimum* (given by the tightest fold which can be formed [4]), but may contain a few persistence lengths.

To conclude, the calculation using a worm-like chain model gives further evidence for the formation of small loops, but suggests that these loops are not necessarily limited by the chemical structure. This is easy to understand, since the free energy loss for pulling one segment out of the crystalline phase is only of the order of kT (or less) at experimental temperatures.

2.9 Tilting in Extended Chain Crystals

As I have shown in the previous sections, the equilibrium state of a polymer single crystal contains extended chains only, if sufficiently many chains are available. An interesting aspect in the calculation of Sect. 2.4 is the existence of a *positive* entropy related to the amorphous fraction due to the sliding of chains, see Eq. (2.16). This, however, is quickly balanced by the excluded volume interactions between the tails, as has been discussed in the context of Eq. (2.20).

Now, there is a possibility to reduce the effect of excluded volume interactions by tilting the crystal stems with respect to the top and bottom surface. This idea is illustrated in Fig. 2.8.a). The increase of surface area per chain is given by

$$\xi'^2 = \xi^2 / \cos \alpha \ , \tag{2.58}$$

where the tilting angle α is defined between the stem orientation and the normal to the interface between the crystaline and the amorphous phase. Using the scaling approach to the brush limit of Sect. 2.3, see Eqs. (2.12) and (2.13), the free energy contribution for a single chain with respect to the brush state can be written as

$$F_{brush} = kTg \frac{r}{(c_A v_0)^2} \cos^2 \alpha \ , \tag{2.59}$$

where r denotes a constant which cannot be obtained from scaling. The symbol $v_0 = \xi^2 b$ denotes the segment volume as used for the derivation assuming a

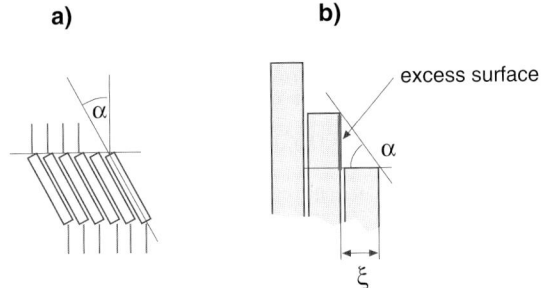

Fig. 2.8. Tilted extended chain crystal. (**a**) Tilted stems lead to a decrease of the grafting density of the tails. The angle α is defined between the stem orientation and the normal of the interface between the crystaline and the amorphous phase. (**b**) Tilting of stems increases the free energy by creating excess surface

dry brush state. Thus the product $c_A v_0$ should be close to unity. To abbreviate the notation, I introduce the constant $r' = r/(c_A v_0)^2$.

On the other hand, tilting gives rise to the formation of an excess surface per chain as illustrated in Fig. 2.8.b). The corresponding free energy excess per chain is given by

$$F_{exe} = 2s'\xi \tan \alpha \frac{\epsilon}{b} = 2s\epsilon \tan \alpha \ . \tag{2.60}$$

Note that the excess free energy (as well as the brush free energy) has to be taken on both sides of the crystaline fraction which gives rise to the factor of two. Here, I have assumed that the excess free energy is related due to a missing neighbor effect to the free energy difference ϵ, and s (s') denotes again a constant. In a more general approach, a surface tension σ_{exe} could be introduced instead of $s\epsilon$. By using $s\epsilon$ in Eq. (2.60), it is tacitly assumed that the excess surface tension vanishes if the system approaches the equilibrium melting point.

Using Eqs. (2.16) and (2.17), the total free energy per chain can be written as

$$F = F_{ext} + F_{brush} + F_{exe} = -kT \ln g + \epsilon g - \epsilon N_{ch} + kTgr' \frac{1}{1+q^2} + 2s\epsilon q \ , \tag{2.61}$$

with

$$q = \tan \alpha \ . \tag{2.62}$$

The minimization problem for F with respect to q and g can be solved in the limit of small values of ϵ and yields to

$$q^3 \simeq \frac{r'}{s} \left(\frac{kT}{\epsilon}\right)^2 \ . \tag{2.63}$$

Details can be found in Appendix D.

[7] M. Al-Hussein and G. Strobl. The melting line, the crystallization line, and the equilibrium melting temperature of isotactic polystyrene. *Macromolecules*, 35(5):1672–1676, February 2002.

[8] G. Reiter, G. Castelein, and J.-U. Sommer. Liquidlike morphological transformations in monolamellar polymer crystals. *Phys. Rev. Lett.*, 86(26):1918–5921, June 2001.

[9] J.-U. Sommer and G. Reiter. Morphogegesis of lamellar polymer crystals. *Europhys. Lett.*, 56(5):755–761, December 2001.

[10] G. Ungar, J. Stejny, A. Keller, and M. C. Whiting. The crystallization of ultralong normal paraffines – the onset of chain folding. *Science*, 229(4711):386–389, 1985.

[11] S. Rastogi, M. Hikosaka, H. Kawabata, and A. Keller. Role of mobile phase in the crystallization of polyetylene. 1. matastability and lateral growth. *Macromolecules*, 24:6384–6391, 1991.

[12] J. A. Subirana. Elucidation of chain folding in polymer crystals: Comparison with proteins. *Trends in Pol. Sci.*, 5(10):321–326, October 1997.

[13] K. Armistead and G. Goldbeck-Wood. Polymer crystallization theories. *Adv. Polym. Sci.*, 100:219–312, 1992.

[14] G. Strobl. *The Physics of Polymers*. Springer, Berlin, Heidelberg, N.Y., 2 edition, 1997.

[15] H. Zachmann. Der Eeinfluss der Konfigurationsentropie auf das Kristallisations- und Schmelzverhalten von hochpolymeren Stoffen. *Kolloid Z. Z. Polym.*, 216–217:180–191, 1967.

[16] E. W. Fischer. Das grenzflächenschmelzen der kristallite in teilkristallisierten hochpolymeren. teil i: Theoretische grundlagen. *Kolloid Z. Z. Polym.*, 218(2):97–114, June 1967.

[17] W. Hu, T. Albrecht, and G. Strobl. Reversible surface melting of pe and peo crystallites indicated by tmdsc. *Macromolecules*, 32(22):7548–7554, 1999.

[18] M. Muthukumar. Molecular modelling of nucleation in polymers. *Phil. Trans. R. Soc. Lond. A*, 361:539–556, 2003.

[19] P. Welch and M. Muthukumar. Molecular mechanisms of polymer crystallization from solution. *Phys. Rev. Lett.*, 87(21):218302, November 2001.

[20] L. Larini, A. Barbieri, P. A. Rolla, and D. Leporini. Equilibrated polyethylene single-molecule crystals: molecular-dynamics simulations and analytic model of the global minimum of the free-energy landscape. *J. Phys: Condens. Matter*, 17(19):L199–L208, 2005.

[21] J. D. Hoffman. Thermodynamic driving force in nucleation and growth processes. *J. Chem. Phys.*, 29(5):1192–1193, November 1958.

[22] G. Wulff. On the question of speed of growth and dissolution of crystal surfaces. *Zeitschrift für Kristallographie und Mineralogie*, 34:449, 1901.

[23] J.-U. Sommer and G. Reiter. Crystallization in ultra-thin polymer films morphogenesis and thermodynamical aspects. *Thermochimica Acta*, 432(2):135–147, June 2005.

[24] P. de Gennes. *Scaling Concepts in Polymer Physics*. Cornell University Press, Ithaca and London, 1979.

[25] J.-U. Sommer. The role of the amorphous fraction for the equilibrium shape of polymer single crystals. *Eur. Phys. J. E*, 19, 413–422, 2006.

[26] M. Doi and S. Edwards. *The Theory of Polymer Dynamics*. Clarendon Press, Oxford, 1986.

[27] D. S. M. de Silva, X. B. Zeng, G. Ungar, and S. J. Spells. Chain tilt and surface disorder in lamellar crystals. a ftir and saxs study of labeled long alkanes. *Macromolecules*, 35(20):7730–7741, September 2002.

[28] D. S. M. de Silva, X. B. Zeng, G. Ungar, and S. J. Spells. On perpendicular and tilted chains in lamellar crystals. *J. Macromol. Sci.*, B42(3,4):915–927, May 2003.

[29] J.-P. Gorce and S. J. Spells. Ftir studies of conformational disorder: crystal perfecting in long chain n-alkanes. *Polymer*, 45(10):3297–3303, May 2004.

[30] A. Silberberg. Distribution of conformations and chain ends near the surface of a melt of linear flexible macromolecules. *J. Coll. Interface Sci.*, 90(1):86–91, November 1981.

3

Intramolecular Crystal Nucleation

Wenbing Hu

Department of Polymer Science and Engineering, State Key Laboratory of
Coordination Chemistry, School of Chemistry and Chemical Engineering, Nanjing
University, 210093 Nanjing, China
wbhu@nju.edu.cn

Abstract. We review how the nucleation mechanism of polymer crystallization could be assigned to intramolecular processes and what are the preliminary benefits for understanding some fundamental crystallization behaviors. The speculative concept of molecular nucleation and the theoretical model of intramolecular nucleation have been elucidated in a broad context of classical nucleation theory. The focus is on explaining the phenomenon of molecular segregation caused by polymer crystal growth.

3.1 Nucleation Mechanism of Polymer Crystallization

Polymer crystallization follows a typical nucleation-growth mechanism. According to the classical nucleation theory [1–3], nucleation implies a size threshold for the growth of the crystalline phase, which is a consequence of rate competition between the body free-energy gain and the surface free-energy penalty. Thus, in the free-energy landscape, crystallization can be described as

$$\Delta F = -n\Delta f + A\sigma \quad (3.1)$$

where n is the number of particles participating the crystalline phase, Δf is the body free-energy gain for each particle entering into the new phase, A is the total interfacial area of the new phase, and σ the surface free-energy density absorbing all prefactors.

Homogeneous primary crystal nucleation is a key process that the crystalline phase spontaneously emerges from a homogeneous bulk polymer melt. It is well known that polymers are anisotropic chain-like molecules exhibiting a significant difference between the properties parallel and perpendicular to the chain. Therefore, one may assume polymer crystallites as cylindrical bundles of chain stems with a radius r and a length l, in a simple estimation to the free-energy change of primary nucleation. Accordingly, Eq. (3.1) can be rewritten in the following way:

$$\Delta F = -\pi r^2 l \Delta f + 2\pi r l \sigma + 2\pi r^2 \sigma_e \quad (3.2)$$

where σ and σ_e denote the lateral and bundle-end surface free-energy densities respectively. Minimizing the free-energy change with respect to r and l separately, the critical free-energy barrier for primary nucleation can be estimated as $\Delta F_c = 8\pi\sigma^2\sigma_e/\Delta f^2$, with $r_c = 2\sigma/\Delta f$ and $l_c = 4\sigma_e/\Delta f$. At high crystallization temperatures, $\Delta f = \Delta h - T_m \Delta s \approx \Delta h(1 - T_m/T_m^0) = \Delta h \Delta T/T_m^0$, where Δh and Δs are the heat and entropy of fusion, respectively, T_m^0 is the equilibrium melting point for the infinite chain length, and ΔT is the supercooling, hence $\Delta F_c \sim \Delta T^{-2}$. This result has been well identified by the droplet nucleation experiments on homogeneous primary crystal nucleation of bulk polymers [4]. The same experiments allow us to determine the values of surface free-energy densities.

In his famous book [5], Wunderlich has summarized two extreme paths of polymer crystal nucleation. The first one is the fringed-micelle nucleation with all the crystalline stems stretching out of the bundle-end surface, as the so-called intermolecular crystal nucleation. The second one is the folded-chain nucleation with most of crystalline stems truncated with adjacent chain-foldings at the bundle-end surface, as the so-called intramolecular crystal nucleation. For polyethylene (PE), the fold-end surface free energy σ_e has been estimated to be about 90 erg/cm^2 [6], while with respect to the fringed-micelle crystallites, the bundle-end surface free energy is much higher on account of the conformational entropy loss of those stretched chains, as estimated by Zachman [7], $\sigma_e \approx 280$ erg/cm^2. Therefore, the critical free-energy barriers will prefer to choose the path of intramolecular nucleation rather than that of intermolecular nucleation. This kinetic preference gives rise to the so-called chain-folding principle of polymer crystallization, as illustrated in Fig. 3.1. Since the crystal thickening is quite slow, the chain-folding crystallites can

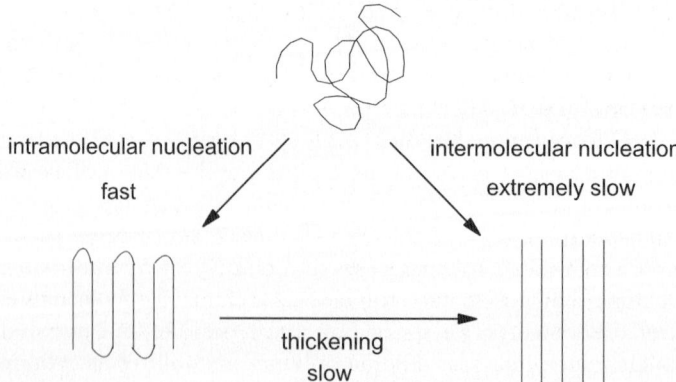

Fig. 3.1. Illustration to the chain-folding principle of polymer crystallization. The metastable chain-folding is a favorite pathway in the kinetic selections of polymer crystal nucleation

generally be regarded as in a metastable state, which has been identified by the experimental observations on single lamellar crystals dating back to 1950s [8].

The surviving crystallites in the nucleation process will continue to undergo crystal growth. The least thermodynamic condition for the lateral crystal growth of lamellar crystallites is $\partial \Delta F/\partial r \leq 0$. In experiments, such as the polarized-light microscopy or the small-angle light scattering, the linear growth rates of lamellar crystals and spherulites are usually found to be constant in the homogeneous melt. Such a constant growth rate implies an interface-controlled mechanism for polymer crystal growth rather than a diffusion-controlled mechanism. A good candidate for the interface-controlled mechanism is the surface crystal nucleation, also known as the so-called secondary crystal nucleation [5]. This mechanism can be described as two-dimensional crystal nucleation on a smooth crystal-growth front, to initiate the formation of a new layer on the growth front.

The free-energy change for secondary crystal nucleation can be estimated in a way similar to the above primary nucleation. Assuming a rectangular bundle of crystalline stems with the lateral size a and the stem length l,

$$\Delta F = -al\Delta f + 2l\sigma + 2a\sigma_e \quad (3.3)$$

The critical free-energy barrier for secondary nucleation is thus $\Delta F_c = 4\sigma\sigma_e/\Delta f$, with $a_c = 2\sigma/\Delta f$ and $l_c = 2\sigma_e/\Delta f$. One can see that the chain-folding principle is still applicable to the secondary nucleation. The temperature dependence of the free energy barrier $\Delta F_c \sim \Delta T^{-1}$ has also been well identified in experimental observations [5].

The critical stem length of secondary nucleation exhibits the temperature dependence of ΔT^{-1}, which is quite coincident with the experimental observations on the lamellar thickness of polymer crystallites, although the matured lamellar thickness under experimental observations may not be necessarily equal to the thickness during growth on account of the potential isothermal thickening right after crystal growth [5]. On the other hand, since the lateral sizes of lamellar crystallites have been well developed, $r \gg l$, and the contribution of the lateral surface free energy to the stability of the whole crystallite can be omitted from Eq. (3.2). We then have

$$\Delta F = -\pi r^2 l \Delta f + 2\pi r^2 \sigma_e \quad (3.4)$$

According to Eq. (3.4), both the least stability condition ($\Delta F = 0$) and the least lateral-growth condition ($\partial \Delta F/\partial r = 0$) give the same result $l = 2\sigma_e/\Delta f$. This expression is consistent with the critical stem length of secondary nucleation, which can actually be regarded as a microscopic interpretation to the least lateral-growth condition. Therefore, there mainly exist two candidates to explain the observed temperature dependence of lamellar thickness, the least stability condition and the least lateral-growth condition. The concentric strips on the PE single lamellar crystals grown under oscillating temperatures have been clearly observed, demonstrating that the growth thickness of lamellar crystallites is highly sensitive to changes in the temperature [9, 10]. This experiment justifies the least lateral-growth condition, since the growth thickness, if determined by the least stability of the whole crystallite, could not exhibit such a high sensitivity to the temperature.

The classical theory to elucidate the secondary nucleation of polymer crystal growth is based on the Lauritzen-Hoffman model [11–16]. This model assumes that on a smooth crystal-growth front, the first stem forming a new layer takes the responsibility for the free-energy barrier. In the following lateral spreading, each stem reels in with the inherent adjacent chain-folding and compensates the free-energy barrier in $a_0 b_0 (l \Delta f - 2\sigma_e)$, where a_0 and b_0 are the sectional linear sizes of each stem. Compared to the barrier height, the compensation of each stem is supposed to be very small, so $l \sim l_{min} \equiv 2\sigma_e/\Delta f$, another consistence under the framework of the least lateral-growth condition.

The rate of crystal nucleation has an exponential dependence on the critical free-energy barrier as suggested first by Becker [17].

$$I = I_0 exp(-\frac{\Delta U + \Delta F_c}{kT}) \quad (3.5)$$

where I_0 is a prefactor, whose value has been specified for condensed systems of small atomic particles by Turnbull and Fisher [18], ΔU is an activation-energy barrier for diffusion across the phase boundary, and k the Boltzmann constant. In Eq. (3.5), ΔU can be associated with the sluggishness of molecules that becomes significant at low temperatures, while ΔF_c becomes larger at high temperatures. Therefore, the nucleation rate gets depressed at both high and low temperatures, leading to a bell-shape dependence on the temperature. In addition, the maximum nucleation rate I_{max} and the corresponding temperature T_{max} change with chain length [5].

However recently, a master curve for the rates of primary crystal nucleation of poly(ethylene succinate) with variable chain lengths has been found in the reduced forms I/I_{max} vs T/T_{max}, where the chain-length effect appears to be subtracted [19]. For the secondary crystal nucleation, the same story holds true for the linear crystal growth rates of poly(ethylene succinate) [20] and several other polymers [21]. According to Eq. (3.5), such a subtraction of the chain-length effect can only be attributed to the prefactor I_0 rather than the exponential term, unless the chain-length dependence has a linear relation with temperature (inconceivable with our current experience). This rational implies that the barriers in the exponential term, especially the critical free-energy barrier for crystal nucleation, should be independent of chain length. This conclusion is quite reasonable since it also has been drawn from the experimental observations on both the primary crystal nucleation rates of PE extended-chain and folded-chain crystals [22, 23]. Note that according to the chain-folding principle of polymer crystallization, the extended-chain crystals may still be produced along the route of chain folding followed with a slow crystal thickening, as observed for PE mesomorphic crystals under high pressures [24] as well as for paraffin crystals exhibiting the self-poisoning phenomenon [25]. In short, experimentalists have found that the critical free-energy barrier for the intramolecular crystal nucleation should be independent of the chain length.

3.2 Concept of Molecular Nucleation

At low temperatures, the basic morphology of polymer crystals is the spherulite, with bundles of lamellar crystals grown from a nucleus center and followed with continuous branching to make a radial structural equivalence. According to the Keith-Padden phenomenological theory, the occurrence of branching is related with the fractionation of polymers as well as the segregation of impurities during crystal growth [26, 27]. The molar-mass fractionation on polymer crystal growth is one of the unique crystallization behaviors of polymers distinguished from those small molecules [5]. This molecular-segregation phenomenon appears on crystal growth primarily for long-chain fractions of polydisperse polymers. It has been observed that in spherulites the long-chain fractions are enriched in the early-grown thick crystals called dominant lamellae, while the short-chain fractions are rich in the later-grown thin crystals called subsidiary lamellae [28]. In experiments, three scenarios of such a molecular segregation have been observed. The first one originates from the thermodynamic driving force, when the melting point of the segregated short-chain fractions is lower than the crystallization temperature, like the performance of small solvents in the conventional monotectic polymer solutions [29]. The second one still refers to complete segregation but under small supercoolings for the folded-chain crystallization of short-chain fractions [30–32]. The third one is a partial segregation of the short-chain fractions under large supercoolings. Under even larger supercoolings, co-crystallization of long-chain and short-chain fractions occurs without any segregation.

Since primary crystal nucleation involves a very small amount of polymers, complete molecular segregation can only be related to the secondary nucleation on the polymer crystal growth. However, interpreting the molecular segregation is a big challenge for the Lauritzen-Hoffman model. As an insightful effort, Hoffman has proposed a separate secondary nucleation event formed by a whole macromolecule with adjacent chain-foldings and two cilia of free chain-ends [33]. According to the Zachman's estimation [7], these two cilia bring about higher surface free energy on the fold-end surface, implying a higher free-energy barrier for such a secondary nucleation. The similar idea has been discussed by Lindenmeyer and Peterson [34]. Therefore, Wunderlich and Mehta suggested a concept of molecular nucleation [35–37]. In this concept, each macromolecule entering the crystal growth front should incur a higher free-energy barrier than the conventional secondary nucleation. The critical chain length for molecular segregation is determined by the critical nucleation condition. This concept represents a quite speculative answer to the question of how molecular segregation occurs.

In a broad sense, the concept of molecular nucleation can be applied to the primary crystal nucleation too, since the sizes of primary nuclei are usually smaller than the coil size of single macromolecules [37]. As a matter of fact, both experiments and simulations have provided evidences for the crystallization within a single homopolymer chain [38–42].

3.3 Intramolecular Nucleation Model

3.3.1 Primary Crystal Nucleation in a Single Chain

The development of molecular simulations of a simple lattice-polymer model has allowed us to survey the topography of free-energy landscapes for single-chain melting and crystallization [43,44]. Thus, a quantitative thermodynamic description to the phase transitions of a single macromolecule can be verified [45].

Assuming the ground state as a single chain embedded in a fully ordered bulk phase, the potential-energy increase of each molten bond is $\Delta e = E_p(q-2)/2$, where q is the coordination number of a regular lattice, and E_p is the energy loss for each bond forming a parallel pair with its neighbour during crystallization. Here, the first factor of two is the number of connected bonds on the chain that has no relation with the lateral packing and should be subtracted from the total amount of parallel neighbours (equal to the coordination number), and the second factor of two is a symmetric factor for pair interactions. On the other hand, the increase of the conformational entropy of each molten bond can be estimated as $\Delta s = ln(q-1)$. Here, each molten bond is assumed as a piece of unperturbed conformation and hence can be estimated in analogy to a non-reversing random walk with $q-1$ random directions in the lattice. Then, according to Eq. (3.1), the free energy of a single-chain crystallite is given by

$$\Delta F = n\Delta f + \sigma(N-n)^{2/3} \qquad (3.6)$$

with $\Delta f = \Delta e - T\Delta s = E_p(q-2)/2 - kTln(q-1)$, where n is the number of molten bonds and N is the total chain length of the single macromolecule.

From Eq. (3.6), one can estimate the critical free-energy barrier for single-chain crystallization, which is the free-energy difference between the maximum transient state and the initial disordered state ($n = N$), as given by

$$\Delta F_c = \frac{4\sigma^3}{27\Delta f^2} \qquad (3.7)$$

The result about the supercooling dependence is consistent with the above estimation for primary crystal nucleation, implying the feasibility of a molecular-nucleation process in primary crystal nucleation of bulk polymers. Furthermore, this free-energy barrier is independent of chain length, in accord with the experimental observations on the primary nucleation rate of bulk polymers.

From Eq. (3.6), one can also estimate the free-energy barrier for single-chain melting, which is the free-energy difference between the maximum transient state and the initial fully ordered state ($n = 0$), as given by

$$\Delta F_m = N\Delta f - \sigma N^{2/3} + \frac{4\sigma^3}{27\Delta f^2} \qquad (3.8)$$

The melting barrier shows a significant chain-length dependence.

Indeed, the molecular simulations of single-chain melting and crystallization on the temperature scanning are in agreement with the predictions from Eqs. (3.7) and (3.8) [45]. As demonstrated in Fig. 3.2, the crystallization temperatures of long single chains appear insensitive to chain lengths on cooling, while the melting temperatures are quite sensitive to chain lengths upon heating back. Further more, the free-energy estimations show a constant barrier for crystallization under a fixed temperature, but variable barriers for melting with different chain lengths, as demonstrated in Fig. 3.3.

The thermodynamic equilibrium between ordered and disordered states is approached by the equivalence of two barriers for melting and crystallization respectively. In this case, the equilibrium temperature and the free-energy barrier can be estimated from Eqs. (3.7) and (3.8), as given by

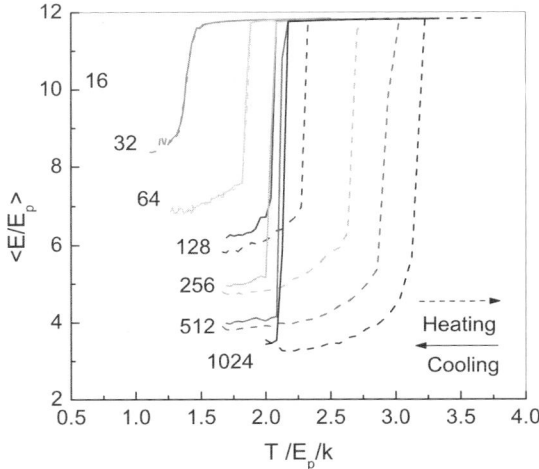

Fig. 3.2. Simulation results of heating and cooling curves (*dashed* and *solid lines* respectively) of the potential energy for single chains with the denoted chain lengths (number of monomers). The overlapping of the short-chain curves implies that their free energy barriers for phase transitions are lower than the thermal energy level. The potential energy is defined as the average amount of non-parallel bonds packing around each bond. The details of heating and cooling programs can be found in [45]

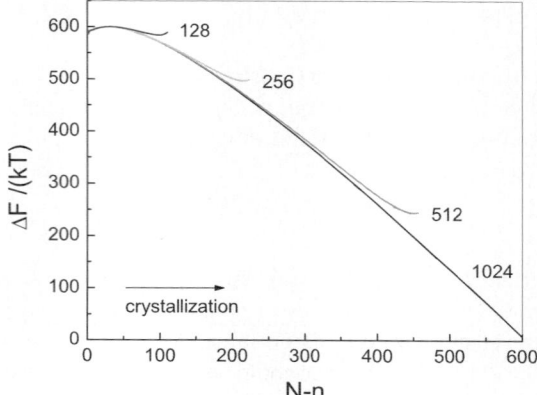

Fig. 3.3. Simulation estimations for the free-energy changes of single-chain systems with the denoted chain length (number of monomers) at the fixed temperature $T = 2.174E_p/k$. The curves are vertically shifted to meet at their tops [45]

$$\Delta f_e = \frac{\sigma}{N^{1/3}}$$

$$\Delta F_e = \frac{4}{27}\sigma N^{2/3} \tag{3.9}$$

The height of equilibrium free-energy barrier could be fitted into the simulation estimation for the single chain with the chain length 1024 by adjusting the fitting parameter $\sigma = 15E_p$, when the partial surface melting of single-chain crystallites has been disregarded, as demonstrated in Fig. 3.4. This fitting parameter is well applicable to the results of other chain lengths, as demonstrated in Fig. 3.5.

With the increase of chain length, the supercooling needed to initiate the spontaneous polymer crystallization increases, as has been discussed by Wunderlich [37]. Spontaneous primary crystal nucleation implies that the thermal-energy level has to match with the free-energy barrier for the intramolecular crystal nucleation. According to Eq. (3.7), the latter should be invariant with chain length at fixed temperature. Therefore, the increase of supercooling with chain length can probably be related to the corresponding increase of the equilibrium melting point for extended-chain crystals of bulk polymers, which is used to be the reference of the supercooling. In principle, concerning the thermodynamic driving force for crystal nucleation, the supercooling should pragmatically be referenced with the melting point of infinite-length crystals. This is because during the nucleation process the nuclei are too small to be able to measure the stability of extended-chain crystals that defines the equilibrium melting point. On the other hand, Eq. (3.7) is only applicable to chain-folding crystallization of long chains. When the chain length becomes small enough, the intermolecular crystal nucleation and the transient mesomorphic phase gradually becomes dominant in the nucleation process. The

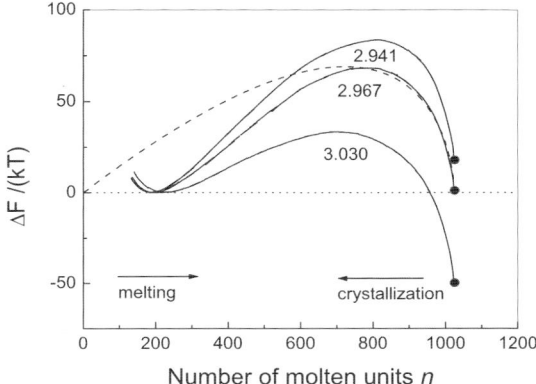

Fig. 3.4. Simulation calculation (*solid lines*) of the free-energy curves vs. the number of molten bonds for a single 1024-mer at the denoted temperatures (E_p/k). The *dashed line* is calculated from Eq. (3.6) with $q = 26$, the fitting parameter $\sigma = 15E_p$ and the fitted equilibrium melting temperature $3.2657E_p/k$. [45]

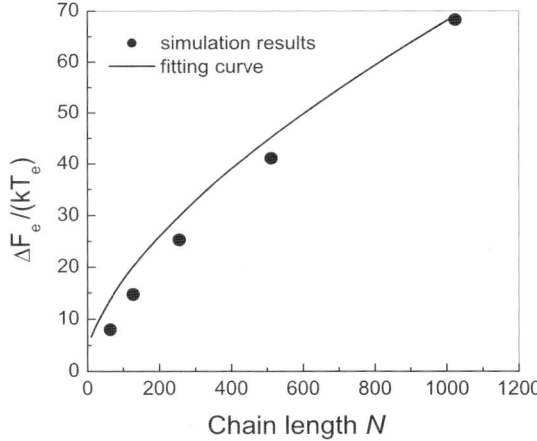

Fig. 3.5. Heights of equilibrium free-energy barriers vs. the chain lengths. The *solid curve* is calculated from Eq. (3.9) with $q = 26$ and $\sigma = 15E_p$ [45]

transient mesophase of alkanes can be regarded as an optimization of intermolecular crystal nucleation to lower the bundle-end surface free energy [46]. These complexities may also change the chain length dependence of the initiating supercooling.

3.3.2 Secondary Crystal Nucleation in a Single Chain

The success of primary molecular nucleation encouraged us to go ahead to consider also secondary molecular nucleation [45]. Assuming that the initiation of crystal growth on either a new layer or a new macromolecule is controlled by secondary nucleation within a single chain, we then focus our attention on two-dimensional crystallization of single chains on a smooth crystal growth front. Similar to Eq. (3.6), the free energy is estimated as

$$\Delta F_{2D} = n\Delta f_{2D} + \sigma_{2D}(N-n)^{1/2} \tag{3.10}$$

Therefore, the free energy barrier for crystallization becomes

$$\Delta F_c = \frac{\sigma_{2D}^2}{4\Delta f_{2D}} \tag{3.11}$$

This result reflects two basic experimental facts of polymer crystal growth from the melt. The first one is that the critical free-energy barrier has a reciprocal dependence on supercooling, and the second one is that it is independent of chain length. On the other hand, the free energy barrier for melting shows a strong chain-length dependence, as given by [45].

$$\Delta F_m = N\Delta f_{2D} - \sigma_{2D}N^{1/2} + \frac{\sigma_{2D}^2}{4\Delta f_{2D}} \tag{3.12}$$

The equilibrium condition is thus

$$\Delta f_{e2D} = \frac{\sigma_{2D}}{N_c^{1/2}}$$

$$\Delta F_{e2D} = \frac{\sigma_{2D}^2}{4\Delta f_{e2D}} = \frac{1}{4}\sigma_{2D}N_c^{1/2} \tag{3.13}$$

Under this equilibrium condition, the chain length can be regarded as a critical parameter N_c for molecular segregation. In the trials of secondary nucleation, shorter chains have their melting barrier lower than their crystallization barrier and hence will be spontaneously excluded from the crystal growth front, while longer chains will survive on the crystal growth front, as schematically demonstrated in Fig. 3.6. Note that the critical chain length is deduced at the equilibrium condition rather than the critical condition proposed in the speculative concept of molecular nucleation.

Since the melting point of a two-dimensional folded-chain crystal of a single chain T_{m2D} is far below the equilibrium melting point of an extended-chain crystal of bulk polymers T_m^0, this molecular segregation occurs at a supercooling of short-chain fractions and thus provides an explanation for the experimental observations. In addition, when the chain length of short-chain fractions goes beyond, but close to, the critical chain length N_c, the short-chain fractions still have a possibility to be excluded from the crystal growth front, making an incomplete molecular segregation. These considerations provide a unified scheme to interpret all three observed scenarios of molecular segregation [47]. The shortest chains have their equilibrium melting point T_m^0 lower than the crystallization temperature T_c, hence their segregation corresponds

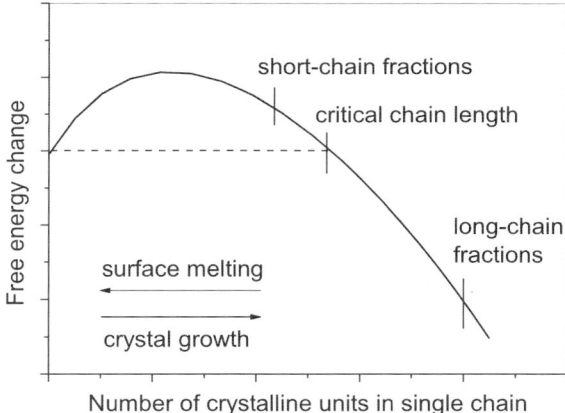

Fig. 3.6. Schematic demonstration of the free-energy barrier for secondary nucleation of single chains on the crystal growth front. The *dashed line* indicates the equilibrium condition for the critical chain length [47]

to the first thermodynamic scenario. The slightly longer chains have their T_m^0 higher than T_c but their T_{m2D} still lower than T_c, so they will be completely excluded from the crystal growth front by their failures of secondary molecular nucleation, corresponding to the second scenario. Much longer chains have their T_{m2D} higher than T_c but close to T_c, so they may partially excluded from the crystal growth front, corresponding to the third scenario. Molecular simulations have provided a general evidence for the second and the third scenarios of bulk polymers [47].

Furthermore, the experimental observations of the critical molar mass for crystallization fractionation in PE melt show a very clear scaling relationship with the crystallization temperature, following the prediction of Eq. (3.13), as demonstrated in Fig. 3.7. The higher the temperature, the larger the critical chain length. Therefore, on cooling from the high temperature, the long-chain fractions of a polydisperse polymer will meet the critical chain length first, and contribute to the dominant crystals, while the short-chain fractions meet the critical chain length at lower temperatures, and contribute to the subsidiary lamellar crystals in the spherulite.

The molecular segregation even exhibits an upper limit of molar mass in crystallization fractionation, above which no segregation occurs anymore. This upper limit was first proposed by Mehta and Wunderlich [36], then verified by the experiments reported by Glaser and Mandelkern [48]. The occurrence of such an upper limit can be explained as a consequence of mismatch between the equilibrium free-energy barrier at the critical chain length and the thermal-energy level kT_{m2D}. According to Eq. (3.13), the former increases with the chain length up to infinity, while the latter approaches a finite value with $\Delta f_e \to 0$. As demonstrated in Fig. 3.8, there exists a crossover providing

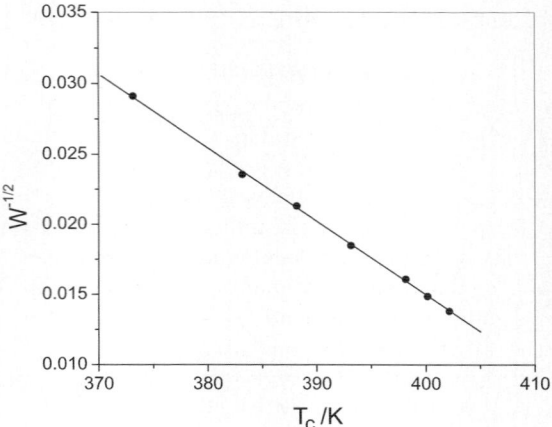

Fig. 3.7. Equilibrium critical molar mass $W^{-1/2}$ of PE melt in the crystallization fractionation, as a function of the crystallization temperature T_c. The data points are drawn from [5], pp 101 in Table V-7, Type M series, excluding the lower tail of molar-mass fractions. The linear regression (*solid line*) gives a correlation coefficient of 0.9997 [45]

the upper limit of critical chain length for molecular segregation. Above this upper limit, the free-energy barrier is too high and the thermal energy will never be able to catch up with it, so the spontaneous exclusion of short-chain fractions will not happen anymore in the trials of secondary molecular nucleation. Therefore, no molecular segregation occurs above this upper limit.

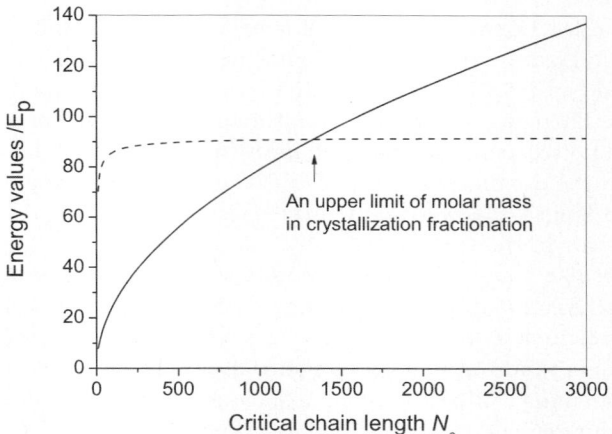

Fig. 3.8. Demonstration of a crossover between the equilibrium free-energy barrier ΔF_{e2D} (*solid line*) and the thermal-energy level $10kT_{m2D}$ (*dashed line*), both vs the critical chain lengths. For the details of calculation, see [47]

3.3.3 From Molecular Nucleation to Intramolecular Nucleation

From the above discussions, one may get the impression that molecular nucleation is quite appropriate to describe the rate-determining step of both primary crystal nucleation and crystal growth. Furthermore, the situation for molecular nucleation can be quite flexible. According to Eq. (3.10), the substrate for secondary molecular nucleation is not necessarily very smooth, because a terrace step crossing over the nuclei will not affect its lateral surface free energy much. Each event of molecular nucleation should be activated by a single macromolecule. Some other macromolecules can be involved in the same event of molecular nucleation and may act in a passive way.

For long chains, after molecular nucleation has been completed, the lateral spreading of the surface nuclei along the chain will not necessarily stop at the

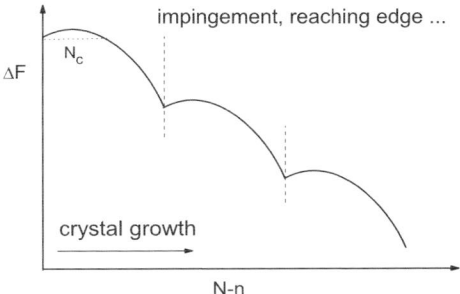

Fig. 3.9. Depicted free-energy curve of a single macromolecule participating the crystal growth through multiple events of intramolecular nucleation [45]

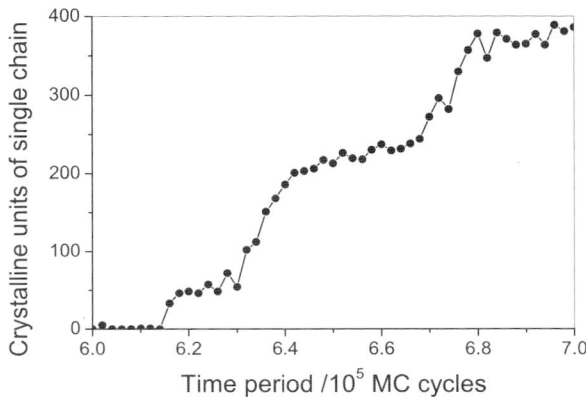

Fig. 3.10. Crystallization process of a single chain on the growth front of a small crystallite traced by the molecular simulations of a semi-dilute polymer solution. The crystalline units are the bonds containing five or more parallel neighbors. In contrast, the crystallization process on a very large growth front exhibits a continuous rather than the stepwise increase of the crystalline units [49]

chain ends but rather earlier, considering the fact that the lateral spreading can be terminated by many possibilities, such as the impingement with other surface crystallites on the same crystal growth front, the growth reaching the edges of a small growth front, or the growth restricted by a slow feeding due to chain entanglement in the melt. The remaining amorphous part of the macromolecule may require more events of molecular nucleation to continue the crystal growth along the chain, as depicted in Fig. 3.9. As a matter of fact, the finite-size effect of the crystal growth front has been observed in our molecular simulations, where the stepwise crystal growth of single macromolecules has been found, as demonstrated in Fig. 3.10 [49].

Therefore, in order to consider the multiple nucleation cases of a single chain, it is necessary to extend the concept of molecular nucleation to the concept of intramolecular nucleation [45]. In this sense, each polymer segment participating in crystal growth should be involved in an event of intramolecular nucleation either actively or passively, while each event of intramolecular nucleation involves only a partial length of macromolecule longer than the critical chain length of molecular segregation at the crystallization temperature. Molecular segregation is only an extreme case of intramolecular nucleation involving chains with a total length shorter than the critical chain length. For longer chains, both the barrier height and the intramolecular size of the critical nucleus are independent of the total chain lengths. For secondary nucleation on a small crystal growth front local thermal fluctuations are not so significant. Thus, it is more reasonable for thermal fluctuations to measure in the scale of the critical chain length, rather than in that the total length of a very long macromolecule.

According to this intramolecular-nucleation model, when the melt of long-chain macromolecules is quenched to low temperatures for fast crystallization, each macromolecule may perform multiple local intramolecular nucleation events and hence will be included in several lamellae or several positions of the same lamellae, with only little changes of their unperturbed coil-size scaling. At each position, intramolecular nucleation yields folded-chain clusters. This picture is quite consistent with Hoffman's proposition of a variable-cluster model for the conformation of macromolecules in the semi-crystalline state [50].

The regime-transition phenomena of polymer crystal growth have been well explained on the basis of the Lauritzen-Hoffman model [14, 15]. Nevertheless, the assumptions about the details of secondary crystal nucleation can be replaced by the intramolecular crystal nucleation, without a substantial loss of semi-quantitative predictions about regime transitions.

3.4 Concluding Remarks

From the classical Lauritzen-Hoffman model to the molecular-nucleation concept, and then the intramolecular-nucleation model, polymer crystal nucleation via chain folding plays an essential role in the mechanism of polymer crystallization. The consideration of the nucleation details evolves from the single stem to the coexistence of the single stem and the single macromolecule, and then to the local segments along the single macromolecule. Intramolecular crystal nucleation is a unique pathway of polymer crystallization which distinguishes itself from that of small molecules. According to the chain-folding principle of polymer crystallization, intramolecular nucleation can be regarded as the origin of adjacent chain-folding in lamellar crystallites.

Intramolecular crystal nucleation does not exclude the fact that intermolecular crystal nucleation can be quite important in crystallization of very short chains, very rigid chains, or polymer crystallization induced by stretching or polymerization, etc. The competition of these two paths and the resulting crystal morphology are worthy of further studies.

Here, the intramolecular crystal nucleation has been discussed mainly in association with the properties of critical free-energy barrier and in comparison with some basic experimental observations. The contributions of other terms in the rate equation, especially the prefactor, can be further considered to explain more experimental facts. Even those subsequent processes following crystal nucleation, such as crystal thickening, perfection, recrystallization, and transient mesophase transitions should be further considered to cover the complete picture of the microscopic mechanism of polymer crystallization. Some discussions on these processes can be found in other contributions of this book.

Acknowledgment

The financial supports from Nanjing University (Talent Introducing Fund No. 0205004108 and 0205004215), from Department of Education of China (NCET-04-0448), and from National Natural Science Foundation of China (NSFC Grant No. 20474027, 20674036), are appreciated.

References

[1] J. W. Gibbs: *The Scientific Work of J. Willard Gibbs*, Vol. 1, (Longmans Green, New York, 1906) p. 219
[2] M. Volmer, A. Weber: Z. Phys. Chem. (Munich) **119**, 227 (1926)
[3] K. F. Kelton: in *Crystal Nucleation in Liquids and Glasses*, vol. 45, ed by H. Ehrenreich, D. Turnbull, (Academic Press: Boston, 1991)
[4] F. P. Price: in *Nucleation*, ed by A. E. Zettlemoyer, (Dekker, New York, 1969) ch 8
[5] B. Wunderlich: *Macromolecular Physics*, Vol. 2 Crystal Nucleation, Growth, Annealing (Academic Press, New York, 1976)
[6] J. D. Hoffman, R. L. Miller: Polymer **38**, 3151 (1997)
[7] H. G. Zachman: Kolloid Z. Z. Polym. **216-217**, 180 (1967)

[8] A. Keller: Philos. Mag. **2**, 1171 (1957)
[9] D. C. Bassett, A. Keller: Philos. Mag. **6**, 1053 (1961)
[10] M. Dosiere, M.-C. Colet, J. J. Point: in *Morphology of Polymers*, ed by B. Sedlacek, (de Gruyter, Berlin, 1986) p. 171
[11] J. I. Lauritzen, J. D. Hoffman: J. Res. Natl. Bur. Stand. Sect. A **64**, 73 (1960)
[12] J. D. Hoffman, J. I. Lauritzen: J. Res. Natl. Bur. Stand. Sect. A **65**, 297 (1961)
[13] J. D. Hoffman, G. T. Davis, J. I. Lauritzen: in *Treatise on Solid State Chemistry*, Vol. 3: Crystalline and Noncrystalline Solids, ed by N. B. Hannay, (Plenum Press, New York, 1976) p. 497
[14] J. D. Hoffman, R. L. Miller: Polymer **38**, 3151 (1997)
[15] J. P. Armistead, J. D. Hoffman: Macromolecules **35**, 3895 (2002)
[16] S. Z. D. Cheng, B. Lotz: Polymer **46**, 8662(2005)
[17] R. Becker: Ann. Physik **32**, 128 (1938)
[18] D. Turnbull, J. C. Fisher: J. Chem. Phys. **17**, 71 (1949)
[19] S. Umemoto, R. Hayashi, R. Kawano, T. Kikutani, N. Okui: J. Macromol. Sci. **B42**, 421 (2003)
[20] S. Umemoto, N. Okui: Polymer **43**, 1423 (2002)
[21] N. Okui, S. Umemoto: in *Lecture Notes in Physics*, Vol. 606, ed by G. Reiter, J.-U. Sommer, (Springer-Verlag, Berlin, Heidelberg, 2003) p. 343
[22] M. Nishi, M. Hikosaka, S. K. Ghosh, A. Toda, K. Yamada: Polym. J. **31**, 749 (1999)
[23] S. K. Ghosh, M. Hikosaka, A. Toda: Colloid Polym. Sci. **279**, 382 (2001)
[24] B. Wunderlich, L. Melillo: Makromol. Chem. **118**, 250 (1968)
[25] G. Ungar: in *Crystallization of Polymers*, ed by M. Dosiere, (Kluwer, Dordrecht, 1993)
[26] H. D. Keith, F. J. Padden, Jr.: J. Appl. Phys. **34**, 2409 (1963)
[27] H. D. Keith, F. J. Padden, Jr.: J. Appl. Phys. **35**, 1270; 1286 (1964)
[28] D. C. Bassett: *Principles of Polymer Morphology*, (Cambridge University Press, Cambridge, 1981) p. 103
[29] U. W. Gedde: *Polymer Physics* (Chapman and Hall, London, 1995) p. 189
[30] S. Z. D. Cheng, B. Wunderlich: J. Polym. Sci., Part B: Phys. Ed. **24**, 577; 595 (1986)
[31] S. Z. D. Cheng, H.-S. Bu, B. Wunderlich: J. Polym. Sci., Part B: Phys. Ed. **26**, 1947 (1988)
[32] U. W. Gedde: Prog. Colloid Polym. Sci. **87**, 8 (1992)
[33] J. D. Hoffman: SPE Trans. **4**, 315 (1964); see also [5]
[34] P. H. Lindenmeyer, J. M. Peterson: J. Appl. Phys. **39**, 4929 (1968)
[35] B. Wunderlich, A. Mehta: J. Polym. Sci. Part B: Polym. Phys. **12**, 255 (1974)
[36] A. Mehta, B. Wunderlich: Colloid Polym. Sci. **253**, 193 (1975)
[37] B. Wunderlich: Faraday Discuss. **68**, 239 (1979)
[38] H. S. Bu, Y. W. Pang, D. D. Song, T. Y. Yu, T. M. Voll, G. Czornnyj, B. Wunderlich: J. Polym. Sci., Part B: Polym. Phys. **29**, 139 (1991)
[39] L. Z. Liu, F. Y. Su, H. S. Zhu, H. Li, E. L. Zhou, R. F. Yan, R. Y. Qian: J. Macromol. Sci., Phys. B **36**, 195 (1997)
[40] T. A. Kavassalis, P. R. Sundararajan: Macromolecules **26**, 4144 (1993)
[41] X. Z. Yang, R. Y. Qian: Macromol. Theory Simul. **5**, 75 (1996)
[42] S. Fujiwara, T. Sato: J. Chem. Phys. **107**, 613 (1997)
[43] W.-B. Hu, D. Frenkel, V. B. F. Mathot: J. Chem. Phys. **118**, 3455 (2003)
[44] W.-B. Hu, D. Frenkel: J. Phys. Chem. B **110**, 3734 (2006)

[45] W.-B. Hu, D. Frenkel, V. B. F. Mathot: Macromolecules **36**, 8178 (2003)
[46] H. Kraack, M. Deutsch, E. B. Sirota: Macromolecules **33**, 6174 (2000)
[47] W.-B. Hu: Macromolecules **38**, 8712 (2005)
[48] R. H. Glaser, L. Mandelkern: J. Polym. Sci., Part B: Polym. Phys. **26**, 221 (1988)
[49] W.-B. Hu, D. Frenkel, V. B. F. Mathot: Macromolecules **36**, 549 (2003)
[50] J. D. Hoffman: Polymer **24**, 3 (1983)

4

Kinetic Theory of Crystal Nucleation Under Transient Molecular Orientation

Leszek Jarecki

Institute of Fundamental Technological Research, Polish Academy of Sciences,
Swietokrzyska 21, 00-049 Warsaw, Poland
ljarecki@ippt.gov.pl

Abstract. Kinetic theory of crystal nucleation is proposed for flexible chain polymers subjected to flow deformation with transient molecular deformation and orientation. Significant transient effects in the kinetics of oriented nucleation are expected in melt processing involving high deformation rates, like in high-speed melt spinning.

Transient effects in the kinetics of oriented nucleation are considered for melt processing in a wide range of deformation rates using a theory of non-linear chain statistics with transient effects. Inverse Langevin elastic free energy of a polymer chain in a Padé approximation, averaged with transient distribution of the chain end-to-end vectors, as well as Peterlin's approximation for the modulus of non-linear elasticity are used. The effects of transient orientation distribution of the chain segments is also considered.

Time evolution of the nucleation potential, critical nuclei size and the nucleation kinetics are discussed as governed by transient free energy of the system subjected to flow deformation and controlled by the relaxation time of the polymer chains. Transient orientation distribution of the chain segments results in time-dependent angular distribution of the nucleation rate and controls ultimate crystalline orientation. Angular distribution of the transient nucleation rate is derived in a quasi steady-state approximation, well-founded by much longer relaxation time of the entire chain macromolecule than an access time of an individual chain segment to the surface of a growing nucleus.

Rotational convection and angular Brownian motion of clusters, as well as athermal nucleation resulting from transient free energy of the system are considered. One concludes that the cluster growth mechanism dominates the other mechanisms of oriented nucleation in the systems with transient chain deformation and orientation. Example computations illustrating transient effects in oriented nucleation are presented for the case of uniaxial elongational flow.

4.1 Introduction

In the modelling of industrial processing of flexible chain polymer melts we often need physically sensible description of the crystallization kinetics. Usually, the melt is subjected to time-dependent deformation rates (fibre spinning,

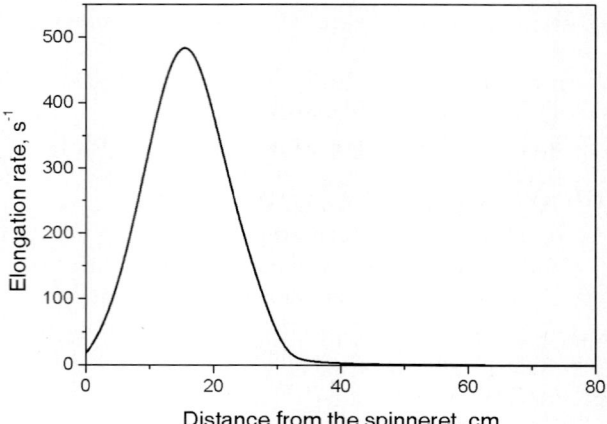

Fig. 4.1. Elongation rate vs. processing distance in high-speed melt spinning of PET fibres at take-up speed 4800 m/min [1]

electro-spinning, film blowing, etc.) producing time-dependent chain deformation and orientation, with transient effects resulting from chain relaxation. In high-speed melt spinning, the polymer is subjected to elongation rates of several hundred times per second (Fig. 4.1), strongly varying along steady-state processing line, and the processing time is on the order of the relaxation time of chains in the system.

Tensile stresses in high-speed melt spinning lead to high chain deformation and orientation increasing crystallization rates by several orders of magnitude and producing highly oriented crystalline structure (Fig. 4.2). Final structure determines ultimate properties of the fibres and evolution of the structure during processing is subject of main interest [2–4]. Analysis of the kinetics of

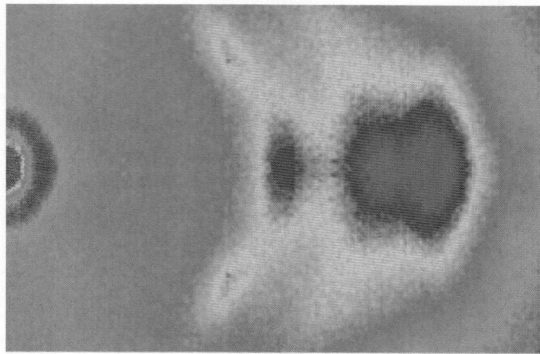

Fig. 4.2. WAXS pattern of as-spun PET fibre obtained in high-speed spinning at take-up speed 4800 m/min [1]

crystal nucleation [5–7] indicate significant effects of molecular deformation and orientation.

In the present paper a kinetic theory of crystal nucleation is considered for the polymers subjected to time-dependent deformation rates, with transient effects of the chain relaxation. The considerations provide a theory useful in modelling fast polymer processing with stress-induced crystallization, like high-speed melt spinning, melt blowing, electro-spinning, etc.

Time evolution of the chain deformation results in evolution of free energy of the system, introducing transient effects in the kinetics of crystal nucleation and crystallization. It has been shown in earlier papers [8, 9] that crystallization influences rheological properties of the polymer fluid and couples structure development with the processing dynamics. Theoretical analysis of chain deformation in the systems subjected to axial elongational flows [10–13] shows considerable transient effects for the processes with high elongation rates ($\dot{e}\tau > 1/2$, $\dot{e}$ – elongation rate, τ – chain relaxation time).

4.2 Time Evolution of the Chain Distribution Function

For fast flow deformations of polymer fluids, a non-linear theory of chain deformation and orientation is considered. To account for non-linear effects and finite chain extensibility, inverse Langevin chain statistics is assumed. Time evolution of chain distribution function in the systems with inverse Langevin chain statistics has been discussed in earlier papers [12, 13] providing physically sensible stress-orientation behaviour in the entire range of the deformation rates and chain deformations.

Time evolution equation for the distribution function of the chain end-to-end vectors, $W(\boldsymbol{h}, t)$, in the systems subjected to time-dependent flow deformation reads

$$\frac{\partial W}{\partial (t/\tau)} - \mathrm{div}\left[\frac{Na^2}{6}\nabla W + W\left(\frac{1}{2}E(h/Na)\boldsymbol{h} - \tau\dot{\boldsymbol{e}}(t)\,\boldsymbol{h}\right)\right] = 0 \qquad (4.1)$$

where the divergence operator is defined in the space of end-to-end vectors, $\boldsymbol{h}$, N is number of statistical (Kuhn) segments in a chain, a – length of the segment, and $\dot{\boldsymbol{e}}(t)$ is time-dependent deformation rate tensor, uniform for polymer chains assigned to a material point of the fluid. For the axial flow deformations we have

$$\dot{\boldsymbol{e}}(t) = \begin{bmatrix} \dot{e}_1(t) & 0 & 0 \\ 0 & \dot{e}_2(t) & 0 \\ 0 & 0 & \dot{e}_3(t) \end{bmatrix} \qquad (4.2)$$

where $\dot{e}_i(t)$ – time-dependent elongation rate applied to the system along the i-th axis.

The polymer chains are considered in (4.1) as non-linear elastic dumbbells embedded in a viscous continuum subjected to the flow deformation. Time dependence of the distribution function is controlled by the transient term which scales with the relaxation time of the chain ends, $\tau = Na^2/(6D)$, where D is diffusion coefficient of the chain ends, and by time-dependent deformation rate tensor, $\dot{e}(t)$. The evolution equation (4.1) accounts for Brownian motion of the chain ends, non-linear elastic forces between the chain ends, and flow convection.

The elastic forces are controlled by the chain modulus $3kTE(h/Na)/Na^2$ where

$$E(h/Na) = \frac{\mathsf{L}^{-1}(h/Na)}{3h/Na} \qquad (4.3)$$

is a non-linear function of the chain extension, h/Na, introduced by Peterlin [14], h – the chain end-to-end distance, $\mathsf{L}^{-1}(x)$ – inverse Langevin function. In the limit of relaxed and unstressed chains, $h/Na \to 0$, we have $\mathsf{L}^{-1}(h/Na) = 3h/Na$ and $E = 1$. In the limit of fully extended chains, $h/Na \to 1$, inverse Langevin function and E tend to infinity. The Peterlin's function E represents a dimensionless, non-linear modulus of elasticity of a chain with limited extensibility, $h/Na \leq 1$, in the units of Gaussian modulus, $3kT/Na^2$. Then, the elastic force between the chain ends reads

$$\boldsymbol{f}(\boldsymbol{h}) = \frac{3kT}{Na^2} E(h/Na) \, \boldsymbol{h} \, . \qquad (4.4)$$

Non-linearity of the elastic force term in (4.1) can be formally eliminated by introducing Peterlin's approximation [14] which represents function $E(h/Na)$ by the value for an average chain extension at any instant of time, t, in the system

$$E(h/Na) \cong \bar{E}(t) = E\left[\langle h^2 \rangle^{1/2}(t)/Na\right] \qquad (4.5)$$

where time-dependent average value

$$\langle h^2 \rangle(t) = \int h^2 \, W(\boldsymbol{h}, t) \, \mathrm{d}^3 \boldsymbol{h} \qquad (4.6)$$

determines time-evolution of the "average" dimensionless modulus $\bar{E}(t)$. Approximation (4.5) introduces linearity to the elastic force term in (4.1) with the average modulus independent of the individual chain extension at any instant of time.

Consequence of the linearity introduced to (4.1) by the Peterlin's approximation is an affine evolution of the end-to-end vectors' distribution function, $W(\boldsymbol{h}, t)$, when assuming initial Gaussian distribution

$$W(\boldsymbol{h}, t=0) = W_0(\boldsymbol{h}) = Const \, \exp\left(-\frac{3h^2}{2Na^2}\right), \qquad (4.7)$$

or any other initial, affinely transformed Gaussian distribution. Solution of (4.1) with the Peterlin's approximation (4.5) and initial Gaussian distribution (4.7) is the following affine evolution of the initial distribution function [12]

$$W(\boldsymbol{h},t) = \det\left[\boldsymbol{\Lambda}^{-1}(t)\right] W_0\left[\boldsymbol{\Lambda}^{-1}(t)\boldsymbol{h}\right] \tag{4.8}$$

where

$$\boldsymbol{\Lambda}(t) = \begin{bmatrix} \lambda_1(t) & 0 & 0 \\ 0 & \lambda_2(t) & 0 \\ 0 & 0 & \lambda_3(t) \end{bmatrix} \tag{4.9}$$

is a time-dependent molecular displacement gradient tensor defining evolution of the chain end-to-end vectors in the system

$$\boldsymbol{h}(t) = \boldsymbol{\Lambda}(t)\,\boldsymbol{h}(t=0) \tag{4.10}$$

Components λ_i of the tensor are chain elongation coefficients and satisfy the following set of differential equations

$$\frac{\mathrm{d}\lambda_i^2}{\mathrm{d}(t/\tau)} + \left[\bar{E}(t) - 2\dot{e}_i(t)\,\tau\right]\lambda_i^2 - 1 = 0 \qquad (i=1,2,3)\,. \tag{4.11}$$

In the case of initial Gaussian distribution of the relaxed chains, the initial condition is $\lambda_i(t=0) = 1$.

Time evolution of the "average" modulus $\bar{E}(t)$ is determined by the evolution of the average chain extension

$$\langle h^2\rangle(t)/N^2 a^2 = \frac{1}{3N}\left[\lambda_1^2(t) + \lambda_2^2(t) + \lambda_3^2(t)\right]. \tag{4.12}$$

With a Padè approximation of the inverse Langevin function [15],

$$\mathsf{L}^{-1}(x) \cong x\,\frac{3-x^2}{1-x^2}\,, \tag{4.13}$$

the evolution of the "average" modulus can be expressed by a closed-form analytical formula

$$\bar{E}(t) \cong \frac{1}{3} + \frac{2}{3 - \left[\lambda_1^2(t) + \lambda_2^2(t) + \lambda_3^2(t)\right]/N}\,. \tag{4.14}$$

The formula provides good approximation of the average modulus of elasticity of the system at any instant of time. It allows formulation of a closed-form non-linear theory of the molecular orientation accounting for limited chain extensibility, valid in the entire range of the chain extensions between the Gaussian limit, $\mathsf{L}^{-1}(h/Na) = 3h/Na$, and full chain extension, $h/Na \to 1$.

Time evolution of the molecular orientation under axial flow deformation can be computed from (11) using the initial condition $\lambda_i(t=0) = 1$. The analytical and numerical solutions calculated at fixed elongation rates are discussed in the earlier papers [12,13]. The chain elongation coefficients $\lambda_i(t)$ satisfy the condition of affine molecular deformation, $\langle h_i^2\rangle(t) = \lambda_i^2(t)\langle h_i^2\rangle_0$ between the average values $\langle h^2\rangle$ at the initial state and at any instant of time. But $\lambda_i(t)$ differ from the macroscopic elongation coefficients, $\exp\left[\int \dot{e}_i(t)\,\mathrm{d}t\right]$

and the affine deformation of chains should be called pseudo-affine, because it is different from macroscopic affine deformation of the system. For relaxed Gaussian chains at the initial state $\langle h_i^2 \rangle_0 = Na^2/3$.

Calculations performed in [12] for the case of fixed elongation rates, $\dot{e}_i(t) = \dot{e}_i = const_i$, indicate that at the beginning of the process chain deformation follows the macroscopic affine deformation mode

$$t/\tau \to 0: \qquad \lambda_i(t/\tau) \to \lambda_i^{\text{aff}}(t) = \exp(\dot{e}_i t) \qquad (4.15)$$

and next it evolves approaching a steady-state limit

$$t/\tau \to \infty: \qquad \lambda_i(t/\tau) \to \frac{1}{\left(\bar{E}^{\text{st}} - 2\dot{e}_i \tau\right)^{1/2}} \qquad (4.16)$$

where $\bar{E}^{\text{st}}$ is the Peterlin's modulus at the steady state limit. The modulus satisfies the following equation

$$\frac{3\left(\bar{E}^{\text{st}} - 1\right)}{3\bar{E}^{\text{st}} - 1} = \frac{1}{3N} \sum_{i=1}^{3} \frac{1}{\bar{E}^{\text{st}} - 2\dot{e}_i \tau}. \qquad (4.17)$$

The example indicates that, due to the transient chain deformation between the initial affine and the steady state mode, we expect transient effects in free energy and, in consequence, in the kinetics of crystal nucleation also at fixed deformation rates applied to the system. With a time-dependent deformation rates, $\dot{e}_i(t)$, the transient effects are more complex. But they are tractable in terms of the present model where a numerical solution of (4.11) is needed for specified time dependent elongation rates.

Figure 4.3 shows the steady-state modulus, $\bar{E}^{\text{st}}$, vs. the reduced elongation rate, $\dot{e}_3 \tau$, computed from (4.17) for uniaxial elongational flow with fixed elongation rate, $\dot{e}_3 = const$. The modulus deviates from the modulus of the Gaussian system at the elongation rates $\dot{e}_3 \tau > 0.5$.

At fixed elongation rates an approximate analytical solution of (4.11) is obtained [12] which shows time evolution of the chain elongation coefficients

$$\lambda_i^2(t/\tau) = \frac{1}{\bar{E} - 2\dot{e}_i \tau} \left\{ 1 - \left(\bar{E} - 1 + 2\dot{e}_i \tau\right) \exp\left[-(\bar{E} - 2\dot{e}_i \tau) \frac{t}{\tau}\right] \right\} \qquad (4.18)$$

where the modulus $\bar{E}$ evolves in time from unity for initial Gaussian system to steady-state value $\bar{E}^{\text{st}}$. At any instant of time, t/τ, the modulus can be determined from the following self-consistent equation

$$\frac{3\left(\bar{E} - 1\right)}{3\bar{E} - 1} = \frac{1}{3N} \sum_{i=1}^{3} \frac{1}{\bar{E} - 2\dot{e}_i \tau} \left\{ 1 - \left(\bar{E} - 1 + 2\dot{e}_i \tau\right) \exp\left[-\left(\bar{E} - 2\dot{e}_i \tau\right) \frac{t}{\tau}\right] \right\}. \qquad (4.19)$$

Figure 4.4 shows time evolution of the chain elongation coefficient between the initial Gaussian state and the final steady-state, computed from (4.18),

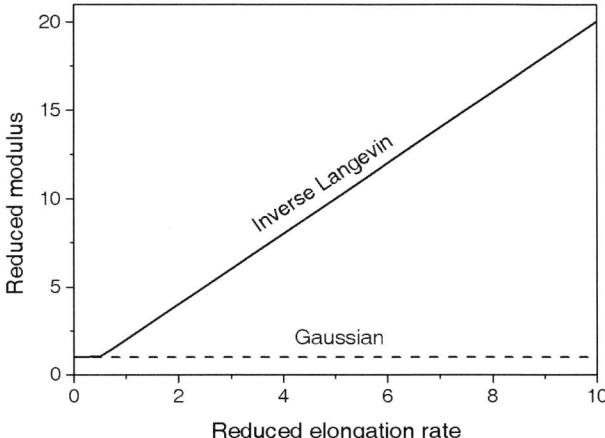

Fig. 4.3. Reduced modulus at the steady-state limit, $\bar{E}^{\mathrm{st}}$, vs. reduced elongation rate, $\dot{e}_3\tau$, computed from (4.17) for the non-linear and Gaussian systems subjected to uniaxial elongational flow and $N = 100$

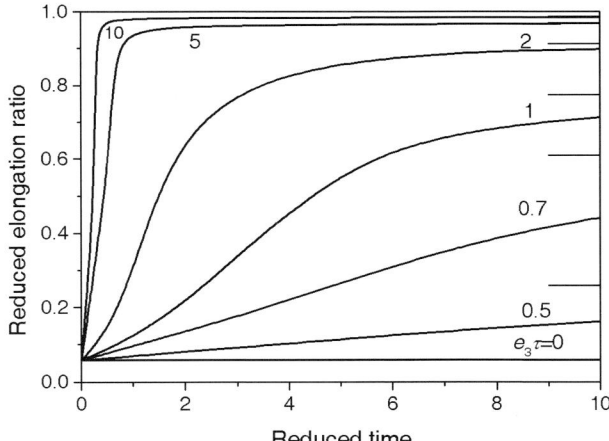

Fig. 4.4. Reduced chain elongation coefficient, $\lambda_3/\lambda_3^{\mathrm{max}}$, vs. reduced time, t/τ computed from (4.18), (4.19) for the uniaxial elongational flow with fixed elongation rates and $N = 100$. Steady-state levels of the elongation coefficients indicated

(4.19) for the uniaxial elongational flow at several fixed elongation rates $\dot{e}_3\tau$ between 0.5 and 10. Values of the elongation coefficient are reduced by the elongation coefficient λ_3^{max} at the full chain extension at which $\langle h_3^2 \rangle = N^2 a^2$ With the Gaussian initial distribution we have $\lambda_3^{\mathrm{max}} = \sqrt{3N}$. High steady-state limits of the chain extensions are predicted for $\dot{e}_3\tau > 1$, and the steady-state is achieved the earlier, the faster is the elongation rate.

The role of the transient effects in the molecular orientation, elastic free energy and, in consequence, in the kinetics of crystal nucleation is illustrated for the uniaxial elongational flow with fixed deformation rate, $\dot{e}_3\tau = const$, which is less complex than the processes with time-dependent deformation rates.

4.3 Free Energy and Orientation Distribution of the Chain Segments

The elastic free energy of an individual polymer chain at the chain extension h/Na reads [16]

$$F_{el}(h/Na) = NkT \int_0^{h/Na} \mathsf{L}^{-1}(x)\,\mathrm{d}x \qquad (4.20)$$

and the orientation distribution of statistical segments of the chain [7, 16],

$$w_0(\vartheta) = \frac{\mathsf{L}^{-1}(h/Na)}{4\pi\,\sinh[\mathsf{L}^{-1}(h/Na)]} \cosh[\mathsf{L}^{-1}(h/Na)\cos\vartheta]\,, \qquad (4.21)$$

ϑ is the angle between a segment and the chain end-to-end vector, $\cos\vartheta = \boldsymbol{a}\cdot\boldsymbol{h}/(ah)$ where $\boldsymbol{a}$ and $\boldsymbol{h}$ are the segment and end-to-end vectors.

Time evolution of the average elastic free energy per unit volume, $f_{el}(t)$, and the orientation distribution of chain segments in the system, $w(\boldsymbol{a},t)$, are controlled by time-dependent distribution of chain end-to-end vectors, $W(\boldsymbol{h},t)$,

$$f_{el}(t) = \nu \int F_{el}(h/Na)\,W(\boldsymbol{h},t)\,\mathrm{d}^3\boldsymbol{h}\,, \qquad (4.22)$$

$$w(\boldsymbol{a},t) = \int w_0(\vartheta)\,W(\boldsymbol{h},t)\,\mathrm{d}^3\boldsymbol{h}\,, \qquad (4.23)$$

where ν – the number of chains per unit volume, and the orientation distribution of the segments is normalised to unity

$$\int w(\boldsymbol{a},t)\,\mathrm{d}^2(\boldsymbol{a}/a) = 1\,, \qquad (4.24)$$

$\boldsymbol{a}/a$ – the unit vector assigned to a segment.

Using Padè approximation of the inverse Langevin function (4.13) the elastic free energy of a non-Gaussian chain with the chain extension h/Na expresses by the following closed formula,

$$F_{el} \cong NkT\left\{\frac{1}{2}\left(\frac{h}{Na}\right)^2 - \ln\left[1-\left(\frac{h}{Na}\right)^2\right]\right\}\,, \qquad (4.25)$$

and the time-dependent average free energy per unit volume of the system reads

$$f_{\text{el}}(t) \cong \nu NkT \left\{ \frac{1}{2} \frac{\langle h \rangle^2(t)}{N^2 a^2} - \left\langle \ln\left[1 - \left(\frac{h}{Na}\right)^2\right] \right\rangle(t) \right\}, \quad (4.26)$$

where $\langle \cdot \rangle$ denotes averages calculated with the time-dependent distribution of $\boldsymbol{h}$ vectors. For pseudo-affine deformation of chains in the flow, $\langle h^2 \rangle(t)$ expresses by (4.12).

The last term in (4.26) which introduces the non-linearity can be approximated by

$$\left\langle \ln\left[1 - \left(\frac{h}{Na}\right)^2\right] \right\rangle \cong \ln\left(1 - \frac{\langle h^2 \rangle}{N^2 a^2}\right). \quad (4.27)$$

The approximation is equivalent to the Peterlin's approximation (4.5) where each chain in the system is represented by a chain with the average extension $\langle h^2 \rangle^{1/2}/Na$. At small chain extensions the approximation (4.27) leads to $\langle h^2 \rangle/N^2 a^2$, while at high chain extensions it diverges to infinity at $\langle h^2 \rangle/N^2 a^2 \to 1$.

With the Peterlin's and Padè approximations, the average elastic free energy reads

$$f_{\text{el}}(t) \cong \frac{\nu NkT}{2} \left\{ \frac{1}{3N} \left[\lambda_1^2(t) + \lambda_2^2(t) + \lambda_3^2(t)\right] \right.$$
$$\left. - 2\ln\left[1 - \frac{1}{3N}\left[\lambda_1^2(t) + \lambda_2^2(t) + \lambda_3^2(t)\right]\right] \right\}. \quad (4.28)$$

In the Gaussian limit we have $\lambda_1^2(t) + \lambda_2^2(t) + \lambda_3^2(t) \ll 3N$ and

$$f_{\text{el}}^{\text{Gauss}}(t) = \frac{\nu kT}{2}\left[\lambda_1^2(t) + \lambda_2^2(t) + \lambda_3^2(t)\right], \quad (4.29)$$

while at full extension $\lambda_1^2(t) + \lambda_2^2(t) + \lambda_3^2(t) \to 3N$ and $f_{\text{el}}(t)$ tends to infinity.

Figure 4.5 illustrates time evolution of the reduced elastic free energy of deformation, $[f_{\text{el}}(t) - f_{\text{el}}(t=0)]/\nu NkT$, computed from (4.18), (4.19), (4.28) for the uniaxial elongational flow with constant elongation rates. The deformation free energy approaches steady-state values the earlier, the faster is the deformation. For very high elongation rates, $\dot{\varepsilon}_3 \tau > 5$, free energy approaches steady-state level, $f_{\text{el}}^{\text{st}}$, within a time of the order of the chain relaxation time, τ.

Figure 4.6 shows reduced steady-state elastic free energy of deformation, $[f_{\text{el}}^{\text{st}} - f_{\text{el}}(t=0)]/\nu NkT$, vs. the processing time, t, computed from (4.28) for melt spinning of PET with the axial velocity gradient shown in Fig. 4.1. Temperature dependence of the chain relaxation time, $\tau(T(t))$, is accounted for assuming Arrhenius temperature dependence of the relaxation time and using the axial temperature profile reported in [1]. Reduced elongation rate,

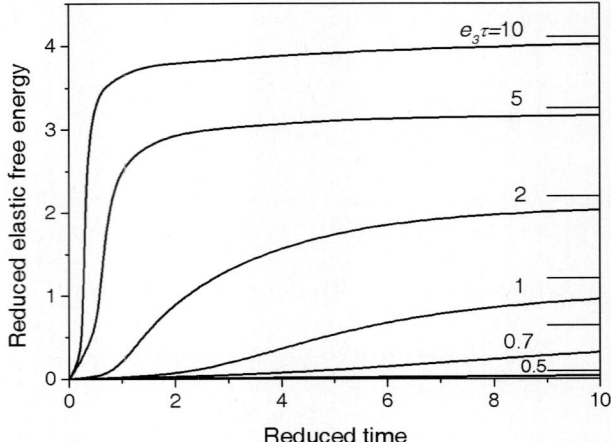

Fig. 4.5. Reduced average elastic free energy of deformation, $[f_{\text{el}}(t) - f_{\text{el}}(t = 0)]/\nu NkT$, vs. reduced time, t/τ, computed from (4.18), (4.19), (4.28) for uniaxial elongational flow with constant elongation rates and $N = 100$. Steady-state limits of free energy indicated

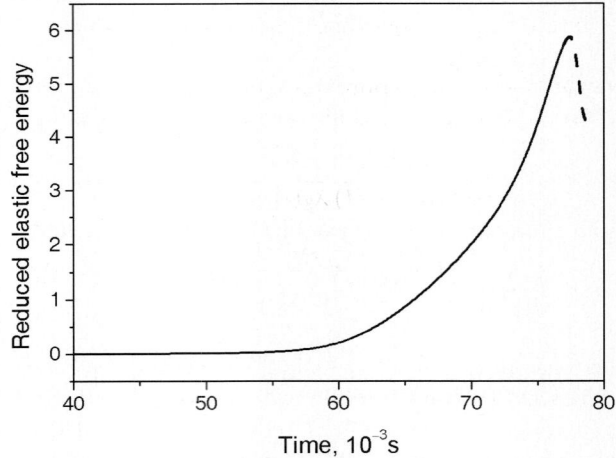

Fig. 4.6. Reduced steady-state elastic free energy of deformation, $[f_{\text{el}}^{\text{st}} - f_{\text{el}}(t = 0)]/\nu NkT$, vs. processing time, t, computed from (4.16), (4.17), (4.28) for melt spinning of PET with axial velocity gradient presented in Fig. 4.1 and $N = 100$

$\dot{e}_3\tau$, vs. the processing time, t, is shown in Fig. 4.7. Relaxation time of PET melt at temperature 300°C at the spinneret output is assumed as 0.003 s.

Solid lines in Figs. 4.6 and 4.7 show parts of the plots before crystallization. When the melt starts to crystallize, the plots achieve maximum at about 1% of crystallinity, and next steeply drop down (dashed lines). Present theory

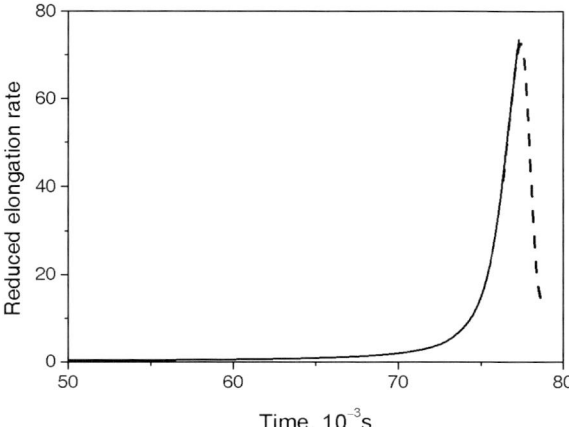

Fig. 4.7. Reduced elongation rate, $\dot{e}_3\tau$, vs. processing time, t, calculated for melt spinning of PET with axial velocity gradient presented in Fig. 4.1 and $N = 100$

concerns elongational flow and chain deformation in pure melt and does not consider effects of crystallization on chain deformation and relaxation time. The present theory fits the solid-line part of the plots up to the maximum point.

Using Peterlin's and Padè approximations and (4.23), time-dependent average orientation distribution of chain segments in the system reads

$$w(\vartheta, \varphi, t) \cong \frac{1}{(4\pi)^2 \, \lambda_1(t) \, \lambda_2(t) \, \lambda_3(t)} \frac{f(\lambda_1, \lambda_2, \lambda_3)}{\sinh f(\lambda_1, \lambda_2, \lambda_3)}$$

$$\int_0^\pi \int_0^{2\pi} \left[\left(\frac{\cos^2 \phi}{\lambda_1^2} + \frac{\sin^2 \phi}{\lambda_2^2} \right) \sin^2 \theta + \frac{1}{\lambda_1^2} \cos^2 \theta \right]^{-\frac{3}{2}}$$

$$\times \cosh\left[f(\lambda_1, \lambda_2, \lambda_3) \, g(\vartheta, \varphi, \theta, \phi)\right] \sin \theta \, d\theta \, d\phi \quad (4.30)$$

where ϑ, φ are the orientation angles in the external coordinates, and the functions

$$f(\lambda_1, \lambda_2, \lambda_3) = \frac{3 - \sum_{i=1}^{3} \lambda_i^2 / 3N}{1 - \sum_{i=1}^{3} \lambda_i^2 / 3N} \left(\sum_{i=1}^{3} \lambda_i^2 / 3N \right)^{\frac{1}{2}}, \quad (4.31)$$

$$g(\vartheta, \varphi, \theta, \phi) = \sin \vartheta \, \cos \varphi \, \sin \theta \, \cos \phi + \sin \vartheta \, \sin \varphi \, \sin \theta \, \sin \phi + \cos \vartheta \, \cos \theta \,. \quad (4.32)$$

The time-dependent orientation distribution of chain segments results from time evolution of the chain elongation coefficients, $\lambda_i(t)$.

Figure 4.8 illustrates time evolution of the average orientation distribution of segments, $4\pi w(\vartheta, \varphi)$, between the initial isotropic state at $t/\tau = 0$ and the

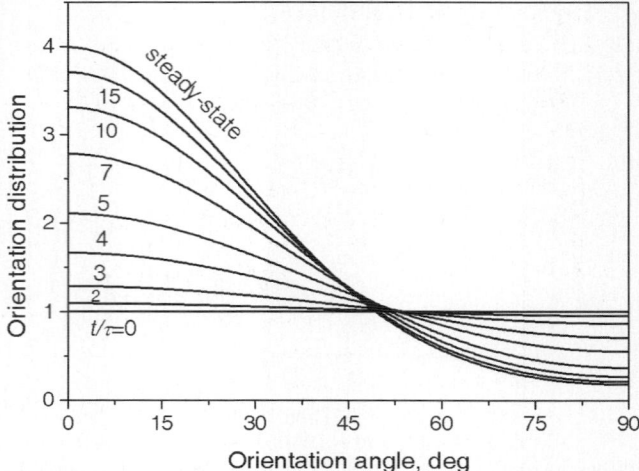

Fig. 4.8. Time evolution of the reduced orientation distributions of chain segments, $4\pi w(\vartheta,\varphi)$, vs. orientation angle, ϑ, between the initial isotropic state and the steady-state limit, computed from (4.30) for uniaxial elongational flow with fixed elongation rate $\dot{e}_3\tau = 1$ and $N = 100$

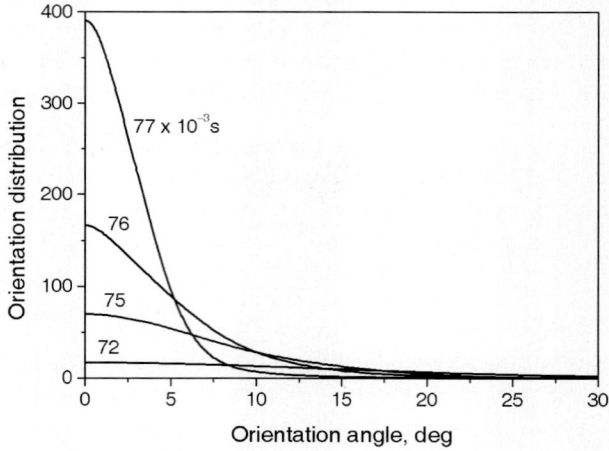

Fig. 4.9. Steady-state limits of the reduced orientation distribution of the chain segments, $4\pi w(\vartheta,\varphi)$, vs. orientation angle, ϑ, computed from (4.30) at several values of the processing time for melt spinning of PET with the axial elongation rate shown in Fig. 4.1 and $N = 100$

steady-state limit at $t/\tau \to \infty$, computed from (4.30) for uniaxial elongational flow with fixed elongation rate $\dot{e}_3\tau = 1$. Figure 4.9 shows steady-state limits of the distribution at any instant of the processing time for melt spinning with the axial velocity gradient presented in Fig. 4.1. The orientation distributions are axially symmetrical vs. the angle φ, and they are plotted vs. ϑ. The plots in Fig. 4.9 show sharp increase of segmental orientation within a short processing time of the order of several microseconds.

4.4 Crystal Nucleation Rate

Molecular deformation and orientation affect free energy of the melt, the degree of undercooling and, in consequence, the kinetics of crystal nucleation and orientation distribution of the resulting crystals [1,2,17–25]. Earlier kinetic theories of nucleation have been formulated for spherical elements and isotropic systems [26–33]. The theory originally proposed by Turnbull and Fisher [30] for nucleation in metals was applied by several authors to nucleation in solutions, condensation of vapours, microcracks' formation in solids, etc. The theory was successful also in applications for unstressed polymers [34–36]. But when considering nucleation in polymers subjected to orienting stresses, the classical approach fails and does not explain high crystalline orientation accompanying significant enhancement in the nucleation rate [20–23].

Discussion of transient effects in the nucleation kinetics in polymers subjected to orienting stresses bases on a kinetic theory of nucleation proposed in [20–22] for the systems of oriented asymmetric elements. The theory converges to the classical approach in the limit of isotropic orientation. This fact explains apparent validity of the classical theory in the applications to unoriented polymers.

Theory of nucleation in systems of orientable elements follows the discrete approach proposed by Frank and Tosi [33] or Hoffman and Lauritzen [35,36], and considers kinetics of addition and dissociation reactions between clusters β_g composed of g elements and single elements α_1. Fundamental assumption of the theory concerns consistency of orientations of the elements taking part in the association reaction. The assumption introduces a disorientation tolerance angle, Δ, which defines a range of single elements' orientations effective for the association at collision with the cluster, as illustrated in Fig. 4.10, and represents angular cross-section of the association reaction [38]. Collisions with the elements from outside of the tolerance range are ineffective, and the disoriented elements play the same role as solvent particles influencing the transition entropy.

In the systems of flexible chain polymers, asymmetric single elements are identified with the uniaxial statistical chain segments. Considering a cylindrical cluster of uniaxial elements oriented at $\boldsymbol{\theta} = (\theta, \Phi)$, the condition of compliant orientation reduces fraction of the chain segments involved in the association process to $\langle w(\boldsymbol{\theta}) \rangle \Delta$ where

$$\langle w(\boldsymbol{\theta}) \rangle = \frac{1}{\Delta} \iint_{\Delta(\boldsymbol{\theta})} w(\vartheta, \varphi) \sin \vartheta \, d\vartheta d\varphi \qquad (4.33)$$

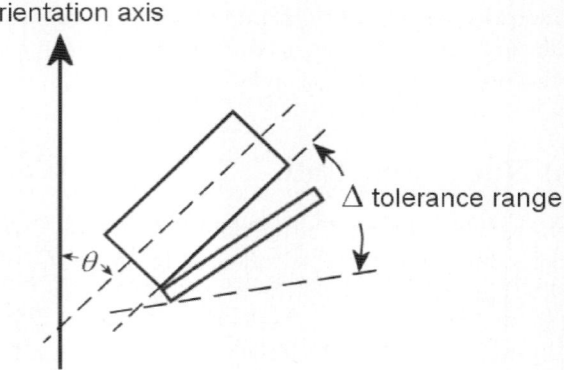

Fig. 4.10. Configuration at a collision of a single element with the cluster oriented at θ within the tolerance range Δ

is an average value of the orientation distribution in the cross-section range $\Delta(\boldsymbol{\theta})$. At various orientations $\boldsymbol{\theta}$ classes of the association-dissociation reaction are considered in which substrates and the products exhibit compliant orientation within the range $\Delta(\boldsymbol{\theta})$.

Under time-dependent chain deformation and orientation of the chain segments, the angular distribution of the rate of production of clusters β_g at any instant of time t by the association-dissociation reaction reads

$$j_g(\boldsymbol{\theta},t) = k_{g-1}^{0+}(t)\, n_{g-1}(\boldsymbol{\theta},t)\, \langle w(\boldsymbol{\theta},t)\rangle\, \Delta - k_g^{0-}(t)\, n_g(\boldsymbol{\theta},t), \qquad g=2,3,4,\ldots, \quad (4.34)$$

where $n_g(\boldsymbol{\theta},t)$ is time-dependent orientation distribution of clusters composed of g single elements in a unit volume, $k_g^{0+}(t)$, $k_g^{0-}(t)$ – association and dissociation rate constants controlled by time-dependent free energy, $\langle w(\boldsymbol{\theta},t)\rangle \Delta$ – time-dependent fraction of asymmetric single elements oriented within the tolerance range Δ. Example computations performed for uniaxial single elements of molecular mass 150–200 and aspect ratio of 10 indicate reduced tolerance angle $\Delta/4\pi$ of the order of 10^{-4} for single elements identified with the statistical chain segments [37]. For such narrow tolerance angle one assumes $\langle w(\boldsymbol{\theta},t)\rangle \approx w(\boldsymbol{\theta},t)$.

In the case of Δ covering the entire orientation space we have $\langle w(\boldsymbol{\theta},t)\rangle \Delta = 1$ and (4.34) reduces to the form used in classical nucleation theory. For unoriented systems we have $\langle w(\boldsymbol{\theta},t)\rangle \Delta = \Delta/4\pi$, independently of Δ, and (4.34) also assumes the classical form, but with the association rate constant reduced by the tolerance range, $k_g^+(t) = k_g^{0+}(t)\Delta/4\pi$, and $k_g^-(t) = k_g^{0-}(t)$. In terms of the rate constants defined for the unoriented system, (4.34) for the oriented systems reads

$$j_g(\boldsymbol{\theta},t) = k_{g-1}^+(t)\, n_{g-1}(\boldsymbol{\theta},t)\, 4\pi\, w(\boldsymbol{\theta},t) - k_g^-(t)\, n_g(\boldsymbol{\theta},t)\,. \quad (4.35)$$

The ratio of the rate constants, k_{g-1}^{+}/k_{g}^{-}, is usually approximated by the ratio at equilibrium

$$\frac{k_{g-1}^{+}(t)}{k_{g}^{-}(t)} \cong \exp\left(-\frac{\Delta\mu_g(t)}{kT}\right) \quad (4.36)$$

where $\Delta\mu_g(t) \equiv \mu_g - \mu_{g-1} - \mu_1(t)$ is the time-dependent potential of association, μ_g – chemical potential of cluster β_g, $\mu_1(t) = \mu_1(t=0) + \delta f(t)$ – time-dependent chemical potential of a single element in the system, $\mu_1(t=0)$ – chemical potential of single element in the initial relaxed system. One assumes chemical potentials of clusters independent of time. For single elements identified with the statistical chain segments, $\delta f(t)$ is the instantaneous average elastic free energy of deformation per statistical segment

$$\delta f(t) = [f_{\mathrm{el}}(t) - f_{\mathrm{el}}(t=0)]/\nu N . \quad (4.37)$$

In the kinetic equation (4.35), transient free energy of the elastic chain deformation controls ratio of the rate constants (4.36) while effects of molecular orientation are accounted for by the concentration factor $4\pi\, w(\boldsymbol{\theta}, t)$. The concentration factor reduces to unity for the case of isotropic systems, assumes values above unity in the range of enhanced orientation, and below unity in the range of reduced orientation. Equation (4.35) introduces effects of molecular deformation and orientation of chain segments.

The distribution of the number of clusters $n(g, \boldsymbol{\theta}, t)$, per unit volume, in the space of clusters' volume, g, and orientations, $\boldsymbol{\theta}$, satisfies the evolution equation

$$\frac{\partial n}{\partial t} = -\mathrm{Div}\,\boldsymbol{j} \quad (4.38)$$

where $\boldsymbol{j} = (n\dot{g}, n\dot{\boldsymbol{\theta}})$ represents components of the flux of clusters growth in the space of cluster volume, $n\dot{g}$, and rotation in the angular space, $n\dot{\boldsymbol{\theta}}$. The divergence operator is defined in a continualized space of size variable g (number of single elements in a cluster) and the orientation angle. Diffusion of the clusters in the space of translations is not considered.

Using continual variable g, the flux of cluster growth assumes a form of the diffusion equation with a potential drift [21]

$$n\dot{g} = j_{\mathrm{gr}}(g, \boldsymbol{\theta}, t) = -D_{\mathrm{gr}}(g)\left[\frac{\partial n}{\partial g} + n\frac{\partial}{\partial g}\frac{\Delta F(g, \boldsymbol{\theta}, t)}{kT}\right] \quad (4.39)$$

where $D_{\mathrm{gr}}(g)$ – "growth diffusion" coefficient proportional to $g^{2/3}$ and responsible for thermally activated transport of single elements to the cluster surface, $\Delta F(g, \boldsymbol{\theta}, t)$ – free energy of formation of a cluster composed of g single elements and oriented at $\boldsymbol{\theta}$ at the moment of time t. With temperature-dependence proposed by the absolute reaction rate theory [38], the "growth diffusion" coefficient reads

$$D_{\mathrm{gr}}(g) = k^{+} g^{2/3} = C\nu_T \exp\left(-\frac{E_D}{kT}\right) g^{2/3} \quad (4.40)$$

where ν_T – thermal frequency, E_D – activation energy of viscous transport, C – a geometric constant. Potential energy of formation of a cluster

$$\Delta F(g, \boldsymbol{\theta}, t) = \Delta F^0(g) - g[\delta f_{\text{el}}(t) + kT \ln 4\pi w(\boldsymbol{\theta}, t)] \qquad (4.41)$$

is controlled by time-dependent elastic free energy of chain deformation, $\delta f_{\text{el}}(t)$, and the reduced orientation distribution of single elements, $4\pi w(\boldsymbol{\theta}, t)$. $\Delta F^0(g)$ is free energy of cluster formation in the unstressed and relaxed system. Positive elastic free energy of chain deformation, $\delta f_{\text{el}}(t) > 0$, and enhanced orientation of chain segments, $4\pi w(\boldsymbol{\theta}, t) > 1$, reduce free energy of cluster formation, ΔF, and should enhance the kinetics of crystal nucleation.

For non-interacting cylindrical Brownian clusters, the flux of cluster rotation reads [25]

$$n\dot{\boldsymbol{\theta}} = j_{\text{rot}}(g, \boldsymbol{\theta}, t) = -D_{\text{rot}}(g)\, \boldsymbol{\nabla}_\theta n(g, \boldsymbol{\theta}, t) + n(g, \boldsymbol{\theta}, t)\, \dot{\boldsymbol{\theta}}_0 \qquad (4.42)$$

where D_{rot} is the rotational diffusion constant around the axis perpendicular to the cluster main axis, proportional to the cluster volume, $D_{\text{rot}} = D_{\text{rot}}^0 g^{-1}$, $\boldsymbol{\nabla}_\theta$ – gradient operator in the space of orientations, $\dot{\boldsymbol{\theta}}_0$ – angular velocity of rotational convection of a cluster by the flow field. Deformation energy of individual clusters is relatively low [39,40] and is neglected in this theory.

When the cylindrical clusters are approximated by ellipsoids of revolution, angular velocity of rotational convection of the clusters in the elongational flow reads [41,42]

$$\dot{\boldsymbol{\theta}}_0 = -\nabla_\theta \Phi(\boldsymbol{\theta}) \qquad (4.43)$$

where Φ – hydrodynamic potential dependent on orientation of main axis $\boldsymbol{\theta}$ of the ellipsoid in the flow. For time-dependent uniaxial elongational flow along the x_3 axis the hydrodynamic potential reads

$$\Phi(\boldsymbol{\theta}, t) = -\frac{\dot{e}_3(t)}{4} \frac{p^2 - 1}{p^2 + 1} (3\cos^2\theta - 1), \qquad (4.44)$$

where θ is the angle between the ellipsoid main axis and the flow direction, p – the axial ratio. When considering cylindrical clusters with the most probable aspect ratio, $p = \sigma_e/\sigma_s$, the hydrodynamic potential expresses by surface energy densities of the cluster, σ_e and σ_s.

In the case of uniaxial elongational flow, the evolution equation for the distribution of cylindrical cluster, $n(g, \theta, t)$, accounting for the flux of cluster growth, rotational diffusion and rotational flow convection reads

$$\frac{\partial n}{\partial (t/\tau_0)} = \frac{\partial}{\partial g}\left[g^{2/3}\left(\frac{\partial n}{\partial g} + n\frac{\partial}{\partial g}\frac{\Delta F(g,\theta,t)}{kT}\right)\right]$$
$$+ \epsilon \frac{1}{\sin\theta}\frac{\partial}{\partial \theta}\left[\sin\theta \frac{\partial n}{\partial \theta} + n\sin\theta \frac{\partial}{\partial \theta}\frac{\Phi(\theta,t)}{D_{\text{rot}}(g)}\right] \qquad (4.45)$$

where

4 Kinetic Theory of Crystal Nucleation

$$\tau_0 = \frac{1}{k^+(T)} = \frac{1}{C\nu_T}\exp\left(\frac{E_a}{kT}\right) \quad (4.46)$$

characterizes the access time of a single kinetic element (statistical chain segment) to the cluster active surface through a viscous motion barrier, and

$$\epsilon = \tau_0 D_{\rm rot}(g) \sim g^{-1} \quad (4.47)$$

is a small parameter of the order of g^{-1} because the rotational diffusion constant is inversely proportional to the cluster volume, $D_{\rm rot}(g) \sim g^{-1}$. For large clusters, in particular in the vicinity of the critical cluster size we have $g* \gg 1$. The reduced thermodynamic and flow potentials, $\Delta F(g,\theta,t)/kT$, $\Phi(\theta,t)/D_{\rm rot}(g)$, are of the same order with respect to the cluster volume g.

The characteristic time of segmental motion in the vicinity of cluster surface, τ_0, is much shorter than the relaxation time of the entire polymer chain composed of N segments, $\tau = Na^2/6D$. In the systems of unentangled chains, the chain relaxation time is proportional to N^2, while in the entangled systems it is proportional to N^3 [43]. The cluster distribution approaches the steady state much earlier than molecular conformation of the chains subjected to flow deformation. With $\tau_0/\tau \sim N^{-2}$, the transient term in (4.45) can be omitted as scaled with much shorter relaxation time, in comparison with the relaxation time of the entire chains controlling the potential of cluster growth. This allows to consider quasi steady-state distribution of clusters at any instant of time scaled with the chain relaxation time t/τ. Quasi steady-state distribution of clusters, controlled by the flux of growth under instantaneous thermodynamic driving force and neglected contribution of the rotational flux, satisfies the following equation,

$$\frac{\partial}{\partial g}\left[g^{2/3}\left(\frac{\partial n_{\rm st}}{\partial g} + n_{\rm st}\frac{\partial}{\partial g}\frac{\Delta F(g,\theta,t)}{kT}\right)\right] = 0, \quad (4.48)$$

where $n_{\rm st}(g,\theta;t)$ represents volume and orientation distribution of the number of clusters per unit volume at time t. Quasi steady-state cluster distribution, parameterised with the time t, determines time-dependent nucleation rate in the transient system and reads

$$n_{\rm st}(g,\theta,t) = n_1(\theta,t)\exp\left[-\frac{\Delta F(g,\theta,t)}{kT}\right]\frac{I(g,\theta,t)}{I(1,\theta,t)} \quad (4.49)$$

where

$$I(g,\theta,t) = \int_g^\infty i^{-2/3}\exp\left[\frac{\Delta F(i,\theta,t)}{kT}\right]di. \quad (4.50)$$

Angular distribution of quasi steady-state nucleation rate by growth is determined by the flux of growth of critical clusters, $g = g^*$, at which $\partial\Delta F(g,\theta,t)/\partial g|_{g=g^*} = 0$. Then, the angular distribution of the nucleation flux by cluster growth is

$$j_{\text{gr}}^{\text{st}}(\theta,t) = -D_{\text{gr}}(g^*)\frac{\partial n_{\text{st}}}{\partial g}\bigg|_{g=g^*} = D_{\text{gr}}(g^*)\frac{n_1(\theta,t)}{I(1,\theta,t)}. \tag{4.51}$$

Using the maximum term approximation for the integral in (4.50) we have

$$j_{\text{gr}}^{\text{st}}(\theta,t) = C n_1 \nu_T g^{*2/3} \exp\left[-\frac{E_D + \Delta F^*(\theta,t)}{kT}\right] \tag{4.52}$$

where n_1 is volume concentration of the chain segments, $\Delta F^*(\theta,t)$ – orientation- and time-dependent critical free energy of cluster formation. For cylindrical clusters, time evolution of the critical free energy and the critical cluster volume read

$$\Delta F^*(\theta,t) = 8\pi v^2 \frac{\sigma_s^2 \sigma_e}{[v\Delta f_0 - \delta f_{\text{el}}(t) - kT \ln 4\pi w(\theta,t)]^2} \tag{4.53}$$

$$g^*(\theta,t) = -16\pi v^2 \frac{\sigma_s^2 \sigma_e}{[v\Delta f_0 - \delta f_{\text{el}}(t) - kT \ln 4\pi w(\theta,t)]^3} \tag{4.54}$$

where Δf_0 is free energy of crystallization per unit volume (negative) in initial isotropic system, v – volume per single element (statistical segment), σ_s, σ_e – side and end surface free energy of the cluster.

Figure 4.11 illustrates time evolution of the angular distribution of the critical cluster volume vs. orientation angle in the elongational flow with fixed elongation rate, $\dot{e}_3 \tau = 1$, reduced by the critical volume at the initial unstressed system. The plots are calculated from (4.54) at temperature $T = 473\,\text{K}$ using the material data for PET [7,44]: volume of statistical chain segment (two monomer units), $v = 5.86 \times 10^{-28}\,\text{m}^3$, side and end surface energies, $\sigma_s = 10.2 \times 10^{-3}\,\text{J/m}^2$, $\sigma_e = 190 \times 10^{-3}\,\text{J/m}^2$, equilibrium melting temperature, $T_m^0 = 553\,\text{K}$, and the heat of crystallization, $\Delta h = 1.8 \times 10^8\,\text{J/m}^3$.

The critical cluster volume reduces significantly with time due to an increase in segmental orientation for clusters oriented along the flow direction and approaches a steady state. For clusters oriented perpendicularly, the critical volume increases in time over the critical volume in the initial system indicating less favourable conditions for nucleation in the range of disoriented clusters. The reduction of critical volume for highly oriented clusters is associated with reduction in free energy barrier of nucleation.

Promotion of a number of subcritical clusters to critical nuclei without growth, caused by reduction of critical cluster volume in time, is athermal nucleation. The concept of athermal nucleation, introduced by Fisher and Turnbull [43], consists in changing thermodynamic criterion of cluster stability. General expression of athermal nucleation in the systems with time-dependent thermodynamic parameters was derived in [21,45]. Angular distribution of athermal nucleation in the transient system is proportional to the distribution of critical clusters and the time derivative of the critical cluster volume

$$j_{\text{ath}}(\theta,t) = -n_{\text{st}}(g=g^*,\theta,t)\frac{\partial g^*(\theta,t)}{\partial t}. \tag{4.55}$$

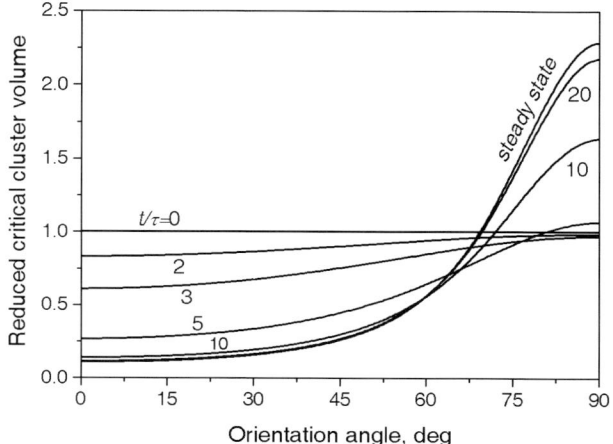

Fig. 4.11. Time evolution of the reduced critical cluster volume, $g^*(\theta, t)/g^*(\theta, t = 0)$, vs. orientation angle, θ, between the initial state and the steady-state calculated for uniaxial elongational flow with fixed elongation rate, $\dot{e}_3\tau = 1$. $N = 100$

The flux of athermal nucleation is positive when associated with a reduction of the critical cluster volume in time due to transient elastic potential of chain deformation and orientation of the chain segments. Fast changes in the critical cluster size may result in athermal nucleation.

Contribution of the athermal nucleation in the transient system can be characterised by the ratio of the athermal flux to the growth flux. With transient effects in the elastic free energy scaled with the chain relaxation time, contribution of the athermal nucleation is of the order of N^{-2} of the growth flux

$$\frac{j_{\text{ath}}(\theta, t)}{j_{\text{gr}}^{\text{st}}(\theta, t)} = -\frac{1}{N^2} \frac{1}{g^{*2/3}} \frac{\partial g^*(\theta, t)}{\partial (t/\tau)}. \qquad (4.56)$$

Low contribution of the athermal nucleation is confirmed by computations performed for the uniaxial elongational flow with fixed elongation rates in the wide range $\dot{e}_3\tau \ll N^2$. For the case of $N = 100$ the range concerns reduced elongation rates below several hundred. Then, athermal nucleation associated with the transient free energy controlled by the chain relaxation time can usually be neglected. The main contribution to the nucleation kinetics in transient systems subjected to flow deformation is controlled by time-dependent orientation distribution of the cluster growth flux.

Figure 4.12 illustrates time-evolution of the angular distribution of quasi steady-state nucleation flux by growth calculated for the uniaxial elongational flow with fixed elongation rate $\dot{e}_3\tau = 1$ between the initial state at $t/\tau = 0$ and the steady-state assuming $N = 100$. Strong angular differentiation of the nucleation rate is predicted which enhances in time by orders of magnitude due to increasing transient free energy of chain deformation and enhancing

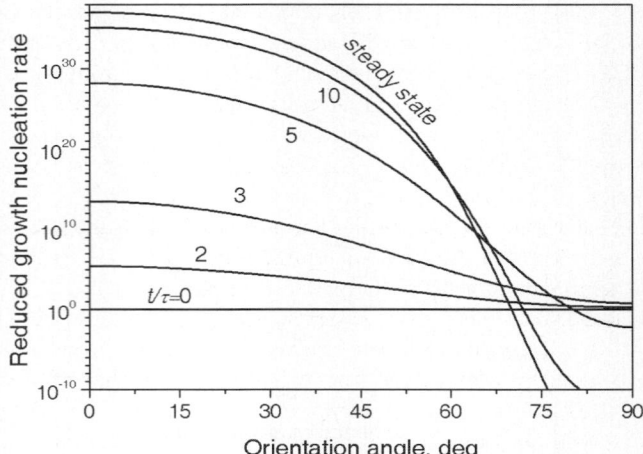

Fig. 4.12. Time evolution of the orientation distribution of the reduced growth nucleation rate, $j_{\mathrm{gr}}(\theta, t)/j_{\mathrm{gr}}(\theta, t = 0)$, vs. orientation angle, θ, between the initial state and the steady-state calculated for elongational flow with fixed elongation rate, $\dot{e}_3\tau = 1$, $N = 100$

segmental orientation. Highest enhancement in the nucleation kinetics is predicted for the orientation along the flow direction, with a reduction in the nucleation rate for perpendicular orientation. Strong transient effects in the nucleation kinetics and in the angular distribution of the nucleation rate, associated with transient chain deformation and orientation distribution of chain segments, are controlled by chain relaxation time and should be considered in the modelling of melt processing with fast deformation rates.

4.5 Conclusions

Transient molecular deformation and orientation in the systems subjected to flow deformation results in transient and orientation dependent crystal nucleation. Quasi steady-state kinetic theory of crystal nucleation is proposed for the polymer systems exhibiting transient molecular deformation controlled by the chain relaxation time. Access time of individual kinetic elements taking part in the nucleation process is much shorter than the chain relaxation time, and a quasi steady-state distribution of clusters is considered. Transient term of the continuity equations for the distribution of the clusters scales with much shorter characteristic time of an individual segment motion, and the distribution approaches quasi steady state at any moment of the time scaled with the chain relaxation time. Quasi steady-state kinetic theory of nucleation in transient polymer systems can be used for elongation rates in a wide range $0 < \dot{e}_3\tau \ll N^2$.

The chain relaxation time controls transient distribution of the chain end-to-end vectors in the system and the transient orientation distribution of chain segments. Time-dependent deformation rate and the chain relaxation introduce time- and orientation-dependent potential of cluster formation and the nucleation rate. The transient orientation-dependent potential of nucleation controls the orientation distribution of the critical cluster energy and the critical cluster size.

Athermal nucleation is induced by time-dependent chain deformation during chain relaxation, tending to an equilibrium steady-state conformations under the flow and depends on the orientation angle.

Time-evolution of orientation distribution of the chain segments results in sharp angular distribution of the nucleation rate. In processes with time-dependent molecular deformation and orientation the flux of cluster growth dominates the rotational flux resulting from Brownian motion and the flow convection, as well as the athermal flux in the nucleation kinetics.

References

[1] A. Blim: Effects of spinning temperature on the structure and dynamics of melt spinning of polyester fibres. PhD Thesis, Institute of Fundamental Technological Research, Warsaw (2004)
[2] A. Ziabicki: *Fundamentals of Fiber Formation* (J. Wiley, London 1976)
[3] A. Ziabicki, H. Kawai: *High-Speed Fiber Spinning* (J. Wiley, New York 1985)
[4] J.A. Cuculo, P.A. Tucker, G.Y. Chen, C.Y. Lin, J. Denton: Intern. Polymer Processing **47**, 85 (1989)
[5] K. Kobayashi, T. Nagasawa: J. Macromol. Sci. Phys. **B4**, 331 (1970)
[6] L. Jarecki: Effects of molecular orientation on thermodynamics of polymer crystallization. PhD Thesis, Institute of Fundamental Technological Research, Warsaw (1974)
[7] A. Ziabicki, L. Jarecki: The theory of molecular orientation and oriented crystallization in high-speed spinning. In: *High-Speed Fiber Spinning*, ed by A. Ziabicki, H. Kawai (J. Wiley, New York 1985) pp 225–269
[8] A. Ziabicki, L. Jarecki, A. Wasiak: Comput. Theoret. Polymer Sci. **8**, 143 (1988)
[9] L. Jarecki, A. Ziabicki, A. Blim: Comput. Theoret. Polymer Sci. **10**, 63 (2000)
[10] L. Jarecki, A. Ziabicki: Polymer **43**, 2549 (2002)
[11] L. Jarecki, A. Ziabicki: Polymer **43**, 4063 (2002)
[12] A. Ziabicki, L. Jarecki, A. Schoene: Polymer **45**, 5735 (2004)
[13] A. Schoene, A. Ziabicki, L. Jarecki: Polymer **46**, 3927 (2005)
[14] A. Peterlin: Polymer Lett. **4**, 287 (1966)
[15] A. Cohen: Rheol. Acta **30**, 270 (1991)
[16] W. Kuhn, F. Grün: Kolloid Z. **95**, 172 (1941)
[17] W.R Krigbaum, R.J. Roe: J. Polymer Sci. **A2**, 4391 (1964)
[18] J.I. Lauritzen, E.A. DiMarzio, E. Passaglia: J. Chem. Phys. **45**, 4444 (1966)
[19] K. Kobayashi, T. Nagasawa: J. Macromol. Sci. Phys. **B4**, 331 (1970)
[20] A. Ziabicki: J. Chem. Phys. **66**, 1638 (1977)
[21] A. Ziabicki: J. Chem. Phys. **85**, 3042 (1986)

[22] A. Ziabicki, L. Jarecki: Inst. Fund. Technol. Res. Report, No 1 (1982)
[23] A. Ziabicki, L. Jarecki: Colloid Polymer Sci. **256**, 332 (1978)
[24] P. Sajkiewicz, A. Wasiak: Colloid Polymer Sci. **277**, 646 (1999)
[25] L. Jarecki: Colloid Polymer Sci. **269**, 11 (1991)
[26] M. Volmer, A. Veber: Z. Phys. Chem. **119**, 277 (1926)
[27] R. Kaischev, I.N. Stransky: Z. Phys. Chem. Abt B, **26**, 317 (1934)
[28] R. Becker, W. Döring: Ann. Phys. **24**, 719 (1935)
[29] R. Becker, W. Döring: Ann. Phys. **32**, 128 (1938)
[30] D. Turnbull, J.C. Fisher: J. Chem. Phys. **17**, 71 (1949)
[31] J. Frenkel: Kinetic theory of liquids (Oxford University, London 1946)
[32] J.B. Zeldovich: Acta Physicochim. USSR **18**, 1 (1943)
[33] F.C. Frank, M. Tosi: Proc. Royal Soc. (London) **263**, 323 (1961)
[34] F.P. Price: J. Polymer Sci. **37**, 71 (1959)
[35] J.D. Hoffman, J.J. Weeks, W.M. Murphy: J. Res. Natl. Bureau Stds. **63A**, 67 (1959)
[36] J.D Hoffman, J.I. Lauritzen: J. Res. Natl. Bureau Stds. **65A**, 297 (1961)
[37] A. Ziabicki, L. Jarecki: J. Chem. Phys. **101**, 2267 (1994)
[38] S. Glasstone, K.J. Laidler, H. Eyring: *The Theory of Rate Processes* (McGraw & Hill, New York 1941)
[39] L. Jarecki, A. Ziabicki: Polymer **18**, 1015 (1977)
[40] L. Jarecki, A, Ziabicki: Thermodynamically controlled crystal orientation in stressed polymers. In: *Flow-Induced Crystallization in Polymers*, ed by R.L. Miller (Gordon & Breach, New York 1977) pp 319–330
[41] G.B Jeffery: Proc. Royal Soc. London, **A102**, 161 (1922)
[42] M. Doi, S.F. Edwards: *The Theory of Polymer Dynamics* (Clarendon Press, Oxford 1986)
[43] J.C. Fisher, J.H. Holomon, D. Turnbull: J. Appl. Phys. **80**, 5751 (1948)
[44] L.H. Palys, P.J. Phillips: J. Polymer Sci. (Phys.) **18**, 829 (1980)
[45] A. Ziabicki: J. Chem. Phys. **48**, 4368 (1968)

5

Precursor of Primary Nucleation in Isotactic Polystyrene Induced by Shear Flow

Toshiji Kanaya, Yoshiyuki Takayama, Yoshiko Ogino, Go Matsuba and Koji Nishida

Institute for Chemical Research, Kyoto University, Uji, Kyoto-fu 611-0011, Japan
kanaya@scl.kyoto-u.ac.jp

Abstract. We performed depolarized light scattering (DPLS) and polarized optical microscope (POM) measurements on the structure formation process or the crystallization process of isotactic polystyrene (iPS) under shear flow below and above the nominal melting temperature T_m. It was found in the DPLS measurements that an anisotropic oriented structure on a μm scale was formed even above the nominal melting temperature. This was also confirmed by POM measurements. This oriented structure must be a precursor of primary nucleation, at least, in the early stage of the formation process. The structure and its formation mechanism are discussed based on the analysis of the DPLS data.

5.1 Introduction

In the last decade many experimental and theoretical works were performed on polymer crystallization under flow to elucidate the formation mechanism of the so-called shish-kebab structure, which consists of the long central fiber core (shish) and the periodically aligned lamella crystals (kebab) on the shish. It is believed that the shish is formed by crystallization of completely stretched polymer chains and the kebabs are folded chain lamella crystals and grow in the direction normal to the shish. Recent development of advanced characterization techniques such as synchrotron radiation (SR) X-ray scattering, neutron scattering and light scattering has shed light on the basic features of polymer crystallization under flow. Some of these works have focused on structure formation in the early stage of the crystallization under flow using "short term shearing" technique [1] because it often governs or at least affects the final structure deeply. In-situ small-angle and wide-angle X-ray scattering (SAXS and WAXS) studies on isotactic polypropylene (iPP) and polyethylene (PE) after pulse shear [2,3] have shown that a scaffold or network of oriented structures is formed prior to full crystallization. In-situ birefringence measurements [4–6] on iPP after short term shearing also suggested formation of a precursor of the shish-kebab structure. These studies as well as other ones [7–10] have demonstrated that a precursor of the shish is formed in the very early stage during the crystallization process after the shear. Furthermore, Hsiao and coworkers have shown in SAXS measurements on iPP [11] that shear induced oriented structures or aggregates of polymer molecules were developed even above the nominal melting temperature. In this work we performed depolarized light scattering and optical microscope measurements on isotactic polystyrene (iPS) under shear flow below and above the nominal melting temperatures T_m. The aim of this work is to see if oriented structures are induced by shear flow even above the nominal melting temperature and discuss the relation to the formation of the shish-kebab structure on the basis of information of the oriented structure.

5.2 Experimental

In this study we used isotactic polystyrene (iPS) with molecular weight $M_w = 400,000$ and polydispersity of $M_w/M_n = 2.0$, where M_w and M_n are the weight-average and number-average molecular weights, respectively. The nominal melting temperature T_m was determined in DSC measurements with a heating rate of $10°C/\text{min}$ is $223°C$. Depolarized light scattering (DPLS) measurements were performed using a home-made apparatus with He-Ne laser (80 meW) as a light source and a screen and CCD camera as a detector system. The range of length of scattering vector Q in this experiment is 4.5×10^{-5} to $3.5\times10^{-5} \text{Å}^{-1}$, where Q is given by $Q = \frac{4\pi n \sin\theta}{\lambda}$ (2θ and n being scattering angle and the refractive index, respectively). Optical microscope (OM) measurements were performed using Olympus BX50 with video attachment. A Linkam CSS-450 high temperature shear cell was used to control the temperature of the sample and the shear conditions. The sample was placed between the two quartz plates for the measurements.

The temperature protocol for the shear experiments is shown in Fig. 5.1: (a) the iPS sample was heated up to $272°C$ from room temperature at a rate of $30°C/\text{min}$, (b) held at $272°C$ for 5min, (c) cooled down to a given

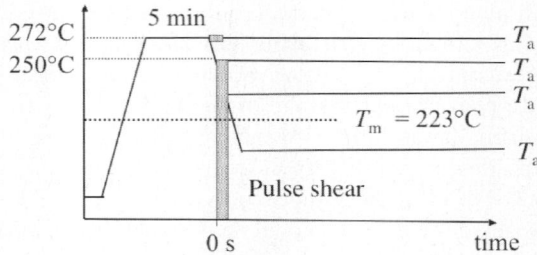

Fig. 5.1. Temperature protocol for the shear experiments on isotactic polystyrene (iPS)

annealing temperature T_a at a rate of 30°C/min, and then (d) held at T_a for the measurements. The iPS melt was subjected to a pulse shear at 250°C for all the measurements except for high temperature measurements above 270°C. In the high temperature measurements above 270°C, the samples were kept at T_a for 5 min and subjected to a pulse shear and observed at T_a. The shear rate was 30 s^{-1} and the shear strain was 12000 % for all the measurements.

5.3 Results and Discussion

Figure 5.2 shows the time evolution of 2D DPLS patterns of iPS after applying a pulse shear at various annealing temperatures below and above the nominal melting temperature (=223°C). At 210°C, a streak-like scattering was observed in a direction normal to the flow in the early stage of crystallization, and then isotropic scattering, which was from spherulites, appeared at about 500 s to cover the anisotropic scattering as seen in the first row in Fig. 5.2.

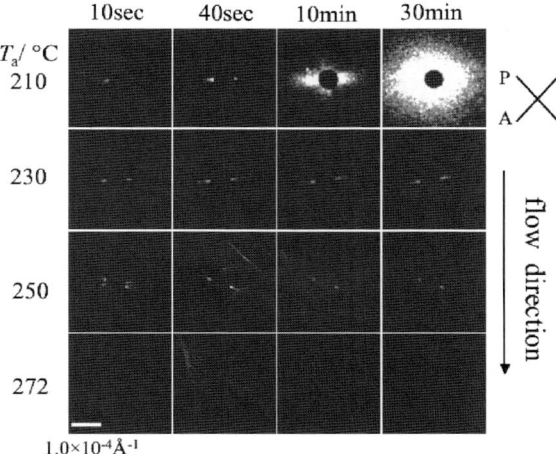

Fig. 5.2. Time evolution of 2D depolarized light scattering (DPLS) pattern from iPS during the annealing process at various temperature after applying a pulse shear with shear rate 30 s^{-1} and shear strain 12000 %. Note that the nominal melting temperature of iPS is 223°C

This streak-like scattering suggests that there are long oriented structures aligned along the flow, which are induced by the shear flow. Even above the nominal melting temperatures such as 230 and 250°C, we observed similar streak-like scattering normal to the flow direction although isotropic scattering was not developed, at least for 1 h. These results directly suggest that long oriented structures can be formed by pulse shear flow even above the nominal melting temperature. On the other hand, above about 270°C, we did not observe any anisotropic scattering, suggesting that there must be a critical temperature for the formation of the oriented structures. In order to evaluate the time evolution of the DPLS intensity quantitatively, we have calculated sector-averaged DPLS intensity in two azimuthal angle ranges between -30 and $+30°$ and between $+60$ and $+120°$, where the azimuthal angle $= 0$ corresponds to the normal direction to the flow. The former and the latter are the scattering intensities normal and parallel to the flow direction, respectively. The parallel intensity is almost zero except for the scattering at 210°C, which arises from the isotropic spherulite scattering. In order to evaluate the anisotropic scattering intensity normal to the flow we subtracted the parallel intensity from the normal one at 210°C. The anisotropic intensity normal to the flow was plotted against the annealing time in Fig. 5.3.

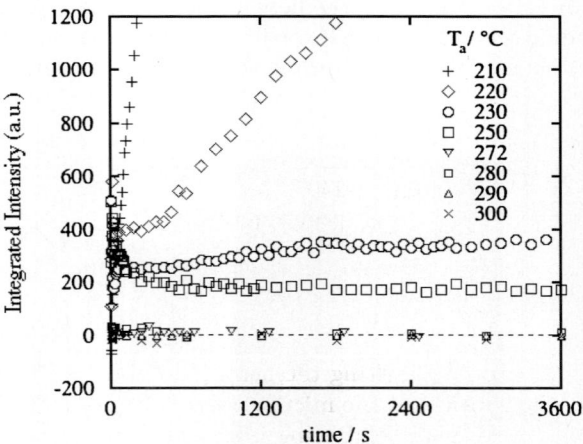

Fig. 5.3. Time evolution of DPLS intensity normal to the flow direction after subtraction of that parallel to the flow direction

At 210 and 220°C below the nominal melting temperature T_m ($=223°$C) the normal intensity corresponding to the streak-like scattering increases with the annealing time. This increase in intensity is due to both the increase in the brightness as well as the increase in the number of the streak-like scattering. On the other hand, at temperatures above the nominal melting temperature T_m such as 230 and 250°C, the intensity is almost independent of the annealing temperature and the annealing time, showing no growth of the streak-like scattering. The scattering intensity is rather stable and survives at least for more than several hours. Above about 270°C, the scattering intensity was almost in the background level, showing that no anisotropic structure formation occurs due to the shear flow. This suggests that the critical temperature for anisotropic oriented structure formation is at around 270°C.

After 1 h annealing at each temperature, we raised the sample temperature at a rate of 30°C/min and monitored the DPLS intensity to see the melting of the oriented structure. In Fig. 5.4, the DPLS intensity along the streak-like scattering (the normal intensity) is shown as a function of temperature.

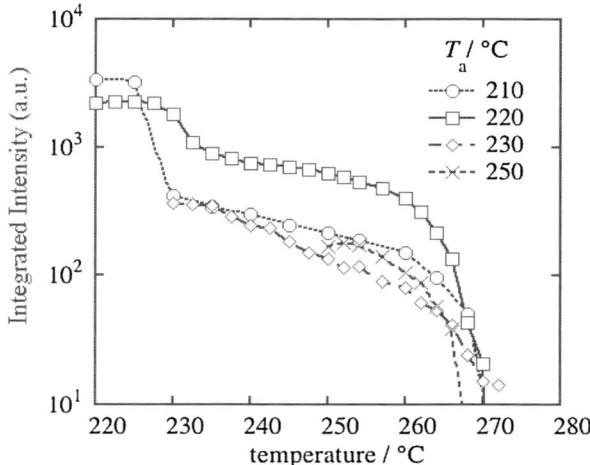

Fig. 5.4. DPLS intensity normal to the flow direction during the heating process after 1 h annealing at various annealing temperatures

At around the nominal melting temperature T_m (=223°C), the intensity decreases abruptly. This is not the melting of isotropic spherulites because we have subtracted the isotropic scattering from the normal intensity, and must correspond to the melting of lamellar structures grown inside the long oriented structures. Above the nominal melting temperature, the intensity due to the streak-like scattering slightly decreased with increasing the temperature and suddenly disappeared or melted at around 270°C, irrespective of the annealing temperature T_a. What we have to consider is if this oriented structure is a crystal or not. Before going to this problem, we would like to discuss the characteristic features of the streak-like scattering.

We found two characteristic features of the streak-like scattering in the DPLS measurements. One is the multi-streak scattering. In the 2D scattering pattern, we often observed two or three streaks in the scattering screen. For example, the 2D DPLS pattern at 250°C at 10 s shows three streaks (see Fig. 5.2). It is interesting to point out that the streak-like scattering often vanishes and often appears during annealing. The vanishing is seen in the 2D DPLS pattern at 250°C. This multi-streak feature suggests that there are only one or a few oriented structures in the incident beam area, and each streak corresponds to an individual oriented structure. The second characteristic feature is a spotted pattern along the streak, which is not always but sometimes

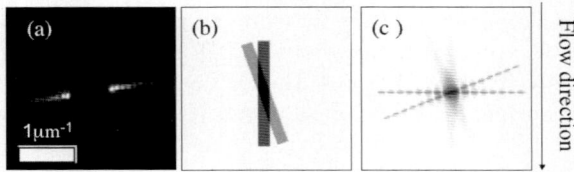

Fig. 5.5. (a): spotted scattering pattern observed at 230°C at 16 min after pulse shear, (b) schematic sketch of two cylinders in an incident beam area, (c) FFT image of two cylinders

observed. Roughly speaking, the observation probability is several % in all the measurements. An example of a spotted scattering pattern is shown in Fig. 5.5(a), which was observed at 230°C at 16 min after pulse shear. What does the spotted scattering pattern mean? As expected from the multi-streak pattern, we must observe an individual anisotropic oriented structure. In this case, the streak-like scattering is directly related to the form factor of an individual oriented structure: the length and the width of the streak-like scattering are related to the width and length of the long anisotropic oriented structure, respectively. Assuming a cylinder form, the observed spotted pattern means that the width (or the radius) of the cylinder does not fluctuate very much along the cylinder. We illustrated a schematic sketch of two cylinders in a incident beam area in Fig. 5.5(b), and the corresponding scattering pattern obtained by a fast Fourier transform (FFT) in Fig. 5.5(c). In fact, a spotted pattern along the streak-like scattering is clearly observed in a direction normal to the orientation, which qualitatively agrees with our observation.

In Fig. 5.6 we plotted the scattering intensity along the streak direction observed at 230°C at 16 min after the pulse shear, which shows about 9 peaks with identical interval ΔQ. According to the theoretical calculation on a cylinder [12], the interval ΔQ is related to the radius of a cylinder R through $R = \pi/\Delta Q$. The observed interval is $\Delta Q = 1\times 10^{-5}$Å, giving a radius of 31μm. Using the value of 31μm, we have calculated the form factor of a cylinder aligned along the vertical direction and plotted the horizontal scattering intensity in Fig. 5.6 by a thin solid line although the theoretical curve was not convoluted with the resolution function of the scattering apparatus. Agreement between the observed data and theoretical curve is good, suggesting that our model for the long anisotropic oriented structure is one plausible solution.

The analysis mentioned above predicts that the long anisotropic oriented structure has a several tens μm in width. Such large anisotropic oriented structures can be observed using polarized optical microscope (POM). Then, we tried to see such structure in POM observations. Some examples of the POM pictures at 230°C are shown in Fig. 5.7 after enhancement of the contrast. At 6s after a pulse shear we observed a very large cylinder-like structure with about 60 μm diameter, but the radius fluctuated along the cylinder

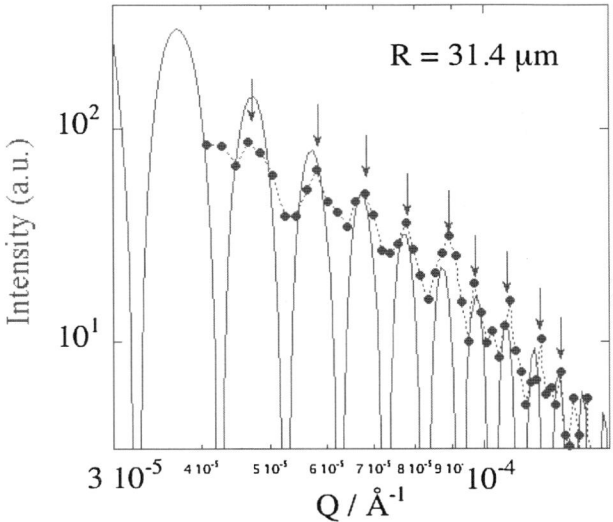

Fig. 5.6. DPLS intensity normal to the flow direction at 230°C at 16 min after pulse shear. Solid thin line is a horizontal component of form factor of a cylinder aligned along the vertical direction

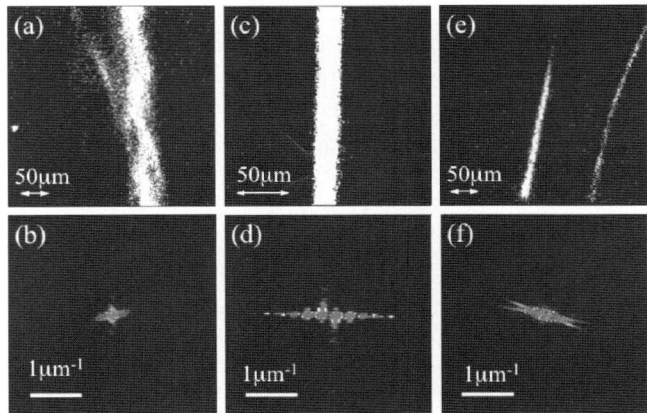

Fig. 5.7. Polarized optical microscope (POM) images from iPS during the annealing process at 230°C after pulse shear (*upper row*) and the corresponding FFT images (*lower row*). (**a**), (**b**): at 6 s, (**c**), (**d**): at 20 min, (**e**), (**f**): at ∼20 min after pulse shear

(Fig. 5.7(a)). The corresponding FFT image in Fig. 5.7(b) shows a short and broad streak normal to the flow direction, corresponding to the large radius and the obscure edge, respectively. At 20 min after the pulse shear, we observed a cylinder-like structure with clear edge (Fig. 5.7(c)). In this case the FFT image clearly shows the spotted scattering pattern in Fig. 5.7(d), agreeing with the observations. At the almost same annealing time (about 20 min), we observed two long but thin cylinder-like structures in a different field of view (Fig. 5.7(e)), resulting in the two-streak scattering pattern in the FFT image in Fig. 5.7(f). This also agrees with our observations in the DPLS measurements. These POM results strongly support our interpretation of the DPLS data for iPS.

Now we would like to discuss if the observed anisotropic oriented structure of iPS is a crystal or not. At temperatures below the nominal melting temperature such as 210°C we observed the growth of spherulites. In this situation we have no reasons to insist that it is not a crystal. On the other hand, above the nominal melting temperature we do not know if it is a crystal or not. There are three reports [13–15] on the equilibrium melting temperature T_m^0 of iPS, which are 243°C [13], 242°C [14] and 289°C [15]. The difference between the first two values and the third one is very large. If the first two values are correct, we expect that the anisotropic oriented structure is a precursor because no crystals could exist above the equilibrium melting temperature T_m^0. However, it is not conclusive at the moment because the difference among these values of T_m^0 is too large. One thing that we have to mention is that the streak-like scattering often disappears in the early stage of the structure formation. If the anisotropic structure is a crystal, it never disappears, implying that it is a precursor at least in the early stage of the structure formation. However, after 1 h annealing, the apparent melting temperature of the anisotropic oriented structure is almost the same (see Fig. 5.4) even if it is annealed below or above the nominal temperature. During annealing below the nominal temperature, the oriented structure must crystallize. Therefore, the melting at the same temperature ($\sim 270°C$) suggests that the final structure after 1 h annealing is a crystal and the melting temperature ($\sim 270°C$) is close to the equilibrium temperature.

Finally the relation between the anisotropic structure and the shish structure is discussed. In the transmitted electron microscope (TEM) measurements, shish structures with diameter of about several nm have been reported in some polymers [16]. It is clear that such a shish structure is different from the anisotropic oriented structure observed here because the spatial scale is very much different. Judging from the size of the anisotropic oriented structure, it must be a bundle of shish-kebab structures at least at the final stage of crystallization. Why is the μm scale structure formed in the initial stage of crystallization under shear flow? One speculative answer is that μm scale phase separation occurs in the initial stage of crystallization between the oriented and unoriented regions, and then crystallization of stretched polymer chains occurs in the oriented domain after aggregation of the stretched chains.

This is a possible scenario but too bold at the moment. We need more data to conclude the formation mechanism of the shish-kebab structure.

5.4 Conclusion

In this work we studied the structure formation of iPS under shear flow below and above the nominal melting temperature using depolarized light scattering (DPLS) and polarized optical microscope (POM) measurements and found that long oriented structures were formed even above the nominal melting temperature. We concluded that it was a precursor of nucleation, at least in the early stage of the structure formation. The characteristic features of the DPLS suggested that we observed an individual oriented cylinder-like structure, and by analyzing the DPLS data we obtained the diameter of the cylinder to be 31μm at 230°C although the POM observations implied that there was a wide distribution of the diameters of the oriented structure. We speculated that the long oriented structure was formed due to phase separation between the oriented and unoriented regions and the shish structures or the extended chain crystals were formed inside the oriented region after aggregation of the extended chains. It is worth confirming this picture in future works.

References

[1] P. Jerschow, H. Janeschitz-Kriegel: Int. Polym. Process **12**, 72 (1997)
[2] R. H. Somani, L. Young, B. H. Hsiao, P. K. Agarwal, H. A. Fruitwala, A. H. Tsuo: Macromolecules **35**, 9096 (2002)
[3] L. Yang, R. H. Somani, I. Sics, B. H. Hsiao, R. Kolb, H. Fruitwala, C. Ong: Macromolecules **37**, 4845 (2004)
[4] G. Kumaraswamy, A. M. Issaian, J. A. Kornfield: Macromolecules **32**, 7537 (1999)
[5] G. Kumaraswamy, R. K. Verma, A. M. Issaian, P. Wang, J. A. Kornfield, F. Yeh, B. Hsiao, R. H. Olley: Polymer **41**, 8934 (2000)
[6] G. Kumaraswamy, J. A. Kornfield, F. Yeh, B. Hsiao: Macromolecules **35**, 1762 (2002)
[7] K. Nogami, S. Murakami, K. Katayama, K. Kobayashi: Bull. Inst. Chem. Res., Kyoto Univ. **55**, 227 (1977)
[8] N. V. Pogodina, S. K. Siddiquee, J. W. V. Egmond, H. H. Winter: Macromolecules **32**, 1167 (1999)
[9] N. V. Pogodina, V. P. Lavrenko, S. Srinivas, H. H. Winter: Polymer **42**, 9031 (2001)
[10] A. Elmoumni, H. H. Winter, A. J. Waddon, H. Fruitwala: Macromolecules **36**, 6453 (2003)
[11] R. H. Somani, L. Yang, B. S. Hsiao: Physica A **304**, 145 (2002)
[12] M. Shibayama, S. Nomura, T. Hashimoto, E. L. Thomas: J. Appl. Phys. **64**, 4188 (1989)
[13] P. J. Lemstra, T. Kooistra, G. J. Challa: Polym. Sci, Part A-2 **10**, 823 (1972)

[14] J. Petermann, H. Gleiter; Polym. Lett. Ed. **15**, 649(1977)
[15] M. Al-Hussein, G. Strobl: Macromolecules **35**, 1672 (2002)
[16] For example, A. Keller, J. W. H. Kolnaar. In: ed. by H. E. H. Meijer: *Processing of Polymers.*(New York: VCH; 1997) pp 189–268

6

Structure Formation and Glass Transition in Oriented Poly(Ethylene Terephthalate)

Koji Fukao[1], Satoshi Fujii[1], Yasuo Saruyama[1] and Naoki Tsurutani[2]

[1] Department of Polymer Science, Kyoto Institute of Technology, Matsugasaki, Sakyo-ku, Kyoto 606-8585, Japan
fukao@kit.ac.jp
[2] Department of Physics, Kyoto University, Kyoto 606-8502, Japan
turutani@scphys.kyoto-u.ac.jp

Abstract. The ordering process and glass transition of poly(ethylene terephthalate) (PET) in oriented glassy states have been investigated using real time X-ray scattering experiment with synchrotron radiation sources and (temperature modulated) differential scanning calorimetry. The time evolution of the X-ray scattering patterns observed during the isothermal annealing process could well be reproduced using the kinetics of structure formation from the nematic-like structure to the crystal-like structure by way of the smectic structure. During the heating process a corresponding change in X-ray scattering patterns was observed and a continuous structural change from the oriented glassy state to the crystalline state through the smectic-like structure was confirmed. This structural change is accompanied with a continuous increase in the glass transition temperature, which is elucidated using differential scanning calorimetry.

6.1 Introduction

Many recent experiments have shown that a dynamical change occurs prior to the formation of usual crystalline order during crystallization process [1–4]. The dynamical change has been discussed in relation to the structure formation in the early stage of polymer crystallization. Such dynamical change is believed to be essential for understanding the mechanism of polymer crystallization. In our previous work, we investigated the dynamics of the α-process, segmental motion of a polymer chain, especially in the early stage of crystallization of poly(ethylene terephthalate) (PET) using the simultaneous time-resolved measurements of small-angle X-ray scattering (SAXS), wide-angle X-ray scattering (WAXS) and dielectric relaxation spectroscopy (DRS) [3,4]. In Fig. 6.1, the results obtained with the three different methods are shown for isothermal crystallization at 97.5°C from quenched glassy states. We have found that the dynamics of the α-process changes drastically before the formation of crystalline structure starts. Figure 6.1(a) shows that at isothermal

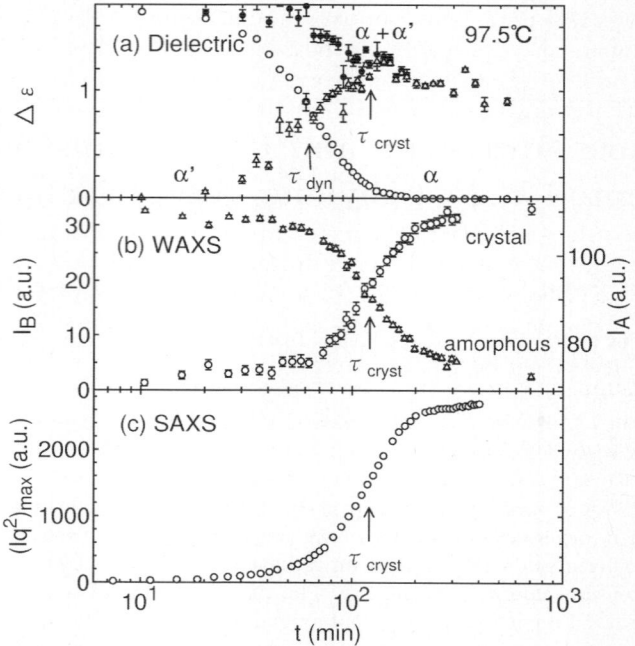

Fig. 6.1. (a) Time evolution of the dielectric strength of the α-process (○) and the α'-process (△). The sum of dielectric strengths of the two processes is also plotted with the symbol ●. (b) Time evolution of peak intensity of Bragg reflections ($0\bar{1}1$ + 010) (I_B, ○) and intensity of amorphous halo (I_A, △) (c) Time evolution of peak intensity of SAXS profile $Iq^2 vs. q$. All data are obtained during the crystallization process at 97.5°C. The *arrows* indicate the two different characteristic times (τ_{dyn} and τ_{cryst})

process, the strength of the α-process decreases with increasing annealing time, while a new relaxation process (α'-process) appears and its strength increases with time. Figures 6.1(b) and (c) show time evolution of scattering intensities associated with the structure formation over the region of WAXS and SAXS, respectively. Comparing the time scale of structure formation, τ_{cryst}, with that of change in dynamics, τ_{dyn}, we have found that there are two different time scales that are associated with polymer crystallization. This result suggests that the ordering process of polymers cannot be described by only one order parameter, i.e., density, but another order parameter must be required. The candidate of the second order parameter is proposed to be orientational order parameter or conformational order parameter of polymer chains [5].

Oriented PET in the glassy states is a useful system when investigating the effect of orientational order on polymer crystallization mechanism. The crystallization rate of PET is slow enough for time-resolved measurements to be performed, and PET films can well be uniaxially drawn by cold drawing at room temperature. Recent X-ray measurements on oriented PET have revealed that a smectic structure appears between totally amorphous states, such as melt and glass, and triclinic crystalline states during the isothermal annealing process below the glass transition temperature or the uniaxially drawing process of PET melts [6–12]. The existence of an intermediate structure such as the smectic structure can be regarded as a result coming from the change in orientational order of polymer chains by cold drawing, because the smectic structure cannot be observed during the crystallization process from unoriented amorphous states of PET. A similar smectic ordering induced by shear deformation is observed also in isotactic polypropylene [13].

Our previous work [3,4] showed that there is a drastic change in dynamics of the segmental motion during the crystallization process. From this result we can expect that the glass transition behavior changes during the ordering process from the oriented glassy states. Such information may be very important for understanding the mechanism of the ordering process of these systems. The glass transition in confined geometry such as in crystalline states has been paid much attention in recent years [14].

In this paper, we investigated the ordering process from oriented glassy states of PET during isothermal annealing process on the basis of the measurements on time evolution of scattering intensities of a reflection charactering the smectic structure and also those of a 4-point pattern in the SAXS region. Using a simple kinetic model, we analyze the observed time evolution of the scattering intensities and discuss the kinetics of the formation and decay of the smectic structure during the isothermal annealing process from the oriented glassy states of PET [15]. We also investigated the ordering process during the heating process. For this purpose, we investigated not only X-ray scattering patterns, but also thermal properties of oriented and unoriented glassy states of PET after annealing treatment using differential scanning calorimetry (DSC) and temperature modulated DSC (TMDSC).

6.2 Experiments

Original PET samples are kindly supplied by Toray, Co. Ltd. The as-received PET is unoriented amorphous films with thickness of 0.2 mm. Oriented glassy states of PET are obtained by drawing the original samples at room temperature. The drawing rate and the final draw ratio are controlled to be 4 mm/min and about 4, respectively.

In order to check the degree of chain orientation and overall diffraction pattern of the drawn PET, wide-angle X-ray diffraction photographs were taken for the drawn PET after annealing at 70°C for one hour in an oil bath. For this purpose, we used monochromatic CuK_α X-ray from a conventional rotating anode type of X-ray generator with graphite monochromater. A vacuum camera was used in order to improve the signal-to-noise ratio. An imaging plate was also used as a recording media.

For time-resolved X-ray measurements, the oriented glassy states of PET were mounted on the sample holder. The sample holder can be moved quickly from the room temperature side to the higher temperature side which is surrounded by a Cu heater block in a vacuum chamber. After a temperature jump from room temperature to an annealing (crystallization) temperature, change in the X-ray scattering pattern with time was measured during the isothermal annealing process. Annealing temperature was controlled to be a temperature between 63.6°C and 76.3°C. For real-time X-ray measurements during the heating process, the sample was mounted onto a hot-stage (LK-600PM, Linkam). Under N_2 flow, the measurements were done during the heating process at the rate of 10 K/min.

X-ray measurements for $q < 7.5$ nm^{-1} (SAXS and intermediate angle X-ray scattering regions) were performed on RIKEN Beamline I (BL45XU) at SPring8, while those for $q < 30$nm^{-1} (WAXS) were done on BL40B2. Here, q is the scattering vector and is defined as $q = 4\pi \sin\theta/\lambda$, where 2θ is the scattering angle. The wavelength of X-rays, λ, is 0.10 nm and 0.8265 nm for BL45XU and BL40B2, respectively. The detector used in our measurements was a combination of X-ray image intensifier and CCD camera. The camera lengths were 682 mm and 142.5 mm for the measurements on BL45XU and BL40B2, respectively. The obtained scattering data were corrected in order to remove the variation in intensities of incident X-rays and the contribution of background scattering.

For DSC and TMDSC measurements we used commercial instruments DSC10A (Rigaku) and MDSC2920 (TA Instrument), respectively. A segment of the film (~1.2mg) was cut from the cold-drawn PET film and put into an aluminum pan with an aluminum lid. For the DSC measurements, the heating rate was 20K/min. As an annealing treatment, the sample was heated up to an annealing temperature T_a ($T_a = 65 - 215$°C) at the rate of 20K/min and then cooled down to room temperature at 1K/min. After this annealing process, the DSC measurements were done. For TMDSC measurements, the heating rate was 1 K/min and the period and amplitude of the temperature modulation were 30 sec and 0.5K, respectively.

6.3 Structural Change at Isothermal Annealing Process

6.3.1 X-ray Diffraction Patterns

Wide-angle X-ray diffraction patterns taken on a flat IP are shown in Fig. 6.2 in order to see the overall diffraction pattern of oriented PET. The sample in Fig. 6.2 was annealed at 70°C for one hour. The polymer chains are highly oriented along the draw axis. There are intense broad scattering areas on the equator and sharper streak-like scattering on the meridian. In particular, on

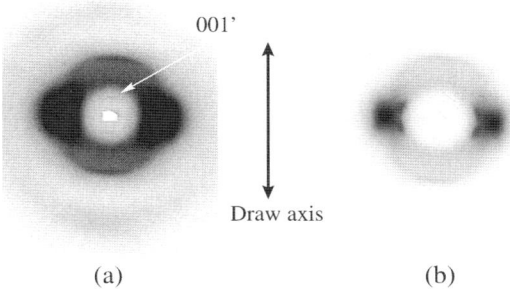

Fig. 6.2. Diffraction patterns in the WAXS region of oriented PET. This sample was annealed at 70°C for one hour. The picture (**a**) is the same as the picture (**b**) except the full scale in intensity. The *arrow* shows the 001' reflection

the meridian there is *a very sharp diffraction spot* near the beam center. We will refer to this reflection as the 001' reflection. The existence of the 001' reflection in the diffraction patterns of the oriented PET was reported by Bonart in 1966 [16]. The lattice spacing corresponding to the 001' reflection was 1.03 nm in the present measurements. In the literature, it is reported that the lattice spacing ranges from 1.02 nm to 1.07 nm depending on the conditions of drawing [6, 8]. The lattice spacing corresponds to the repeating unit of PET monomers along the chain axis. The existence of the 001' reflection on the meridian suggests that PET chains are aligned parallel to each other, and that there is a domain, within which the atomic positions along the chain axis are the same among the neighboring chains. This structure can be regarded as the one similar to the smectic structure of liquid crystals.

As shown in Fig. 6.2(b), we can see four relatively sharp spots just below and just above the equator, in addition to the intense diffuse spots. This suggests that crystalline order is already formed in part at 70°C, below the glass transition temperature of the bulk sample, although the crystalline structure is accompanied by a large degree of disorder.

Figure 6.3 shows time evolution of X-ray scattering patterns of oriented glassy states of PET, observed after the temperature jump from room temperature to 71.2°C using X-ray source from the synchrotron radiation at SPring 8. The draw axis (the z-axis) is shown as an arrow in the figure. The upper row shows the scattering patterns for the q-range of $q < 7.5$ nm^{-1}, and the lower one shows the scattering pattern of the SAXS region of the same scattering patterns as the upper one. The isotropic rings in X-ray scattering patterns are contributions from the scattering from the Kapton films used for the windows of X-ray path. The upper row of figures shows that there is a weak and sharp diffraction spot around $q_z \approx 6.1$ nm^{-1} on the meridian, where q_z is the z-component of the scattering vector. This reflection is the 001' reflection. The

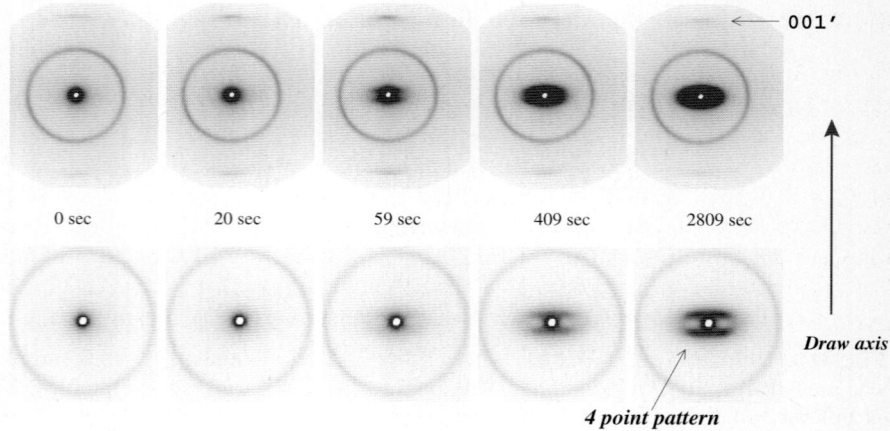

Fig. 6.3. Time evolution of X-ray diffraction patterns of a drawn PET sample annealed at 71.2°C. Diffraction patterns at 0, 20, 59, 409 and 2809 sec are shown. Patterns on the *upper line* are those for $q < 7.5$ nm^{-1}, and those on the *lower line* are for $q < 3.6$ nm^{-1}. The arrow shows the direction of the draw axis (the z-axis).

line width along the direction perpendicular to the meridian can be regarded as a measure of the lateral size of the smectic domain.

The intensity of the reflection 001' increases with time, reaching a maximum at 59sec, and then it begins to decrease gradually with time. The present result shows that the fraction of the smectic structure increases and then decreases with time.

In the lower diffraction pattern in Fig. 6.3, we find that there is almost no SAXS intensity in the beginning of the annealing. However, as time elapses, the intensity increases appreciably in the SAXS region and finally shows a typical 4-point pattern of SAXS. The existence of the 4-point pattern suggests that there is a tilted lamellar structure that is typical of higher order structure of triclinic crystalline PET. In the temperature range around this annealing temperature, 71.2°C, a wide-angle X-ray diffraction measurement revealed that there are very few sharp reflections on the equator, as shown in Fig. 6.2(b). Therefore, the polymer chains in this tilted lamellar structure have a highly disordered (crystalline) structure. Because this highly disordered structure is expected to change into the perfect triclinic structure, we call this structure a *pre-crystalline structure*. According to Asano's model [6], this tilted lamellar structure is formed by tilting PET chains from the draw direction. Here, we can regard the intensity of the 4-point patterns of SAXS as a measure of the fraction of the tilted lamellar structure or the *pre-crystalline* structures.

6.3.2 Integrated Intensity as a Function of Annealing Time

In Fig. 6.4, we show the time evolution of the integrated intensities of the 001' reflection and the 4-point patterns at various annealing temperatures. The integrated intensities of the 001' reflection and the 4-point patterns of SAXS were evaluated from the one-dimensional profile of the scattering intensity at $q_z \approx 6.9$ nm^{-1} and $q_z \approx 0.5$ nm^{-1}, respectively. The contributions from the background and the parasite scattering were subtracted before evaluating the integrated intensities. In Fig. 6.4 it is shown that the intensity of the 001' reflection has a maximum at a time that depends on the annealing temperature. On the other hand, the intensity of the 4-point patterns increases monotonically with time during the isothermal annealing process. From this result, it is found that the fraction of the pre-crystalline structure increases monotonically with time, while the fraction of the smectic structure increases and then decreases with time.

The X-ray diffraction patterns of as-drawn PET samples have a weak 001' reflection on the meridian. This implies that the initial state of oriented PET is a mixture of the nematic structure and the smectic structure, and also that two ordering processes coexist during the isothermal annealing process on PET. (Polymer chains in the nematic structure have a preferred orientation

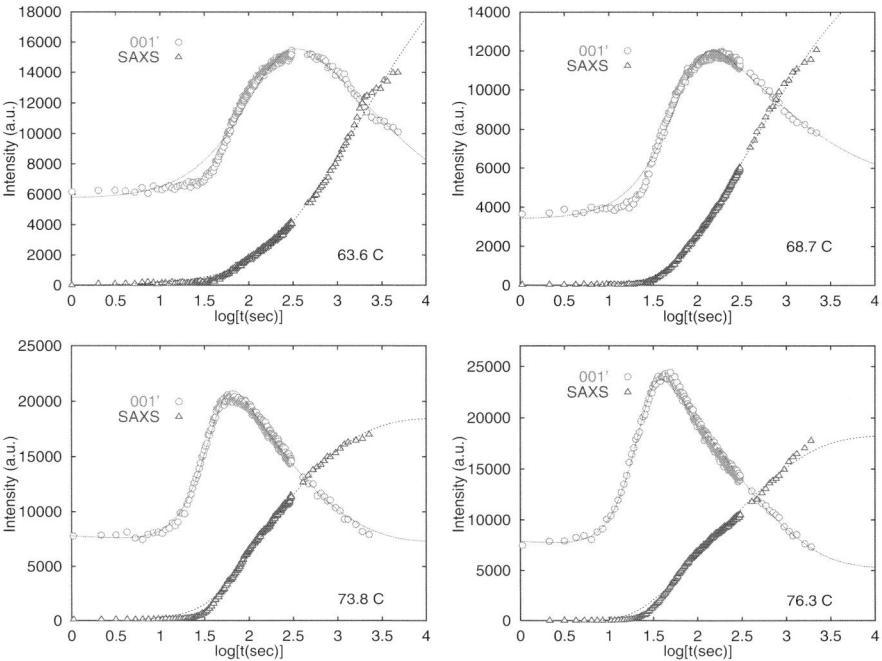

Fig. 6.4. Time evolution of integrated intensity of the 001' reflection (○) and the 4-point patterns of SAXS (△) at various annealing temperatures (63.6°C–76.3°C)

along the draw axis and are randomly aligned along any direction normal to the draw axis.) One is the ordering process from the nematic structure to the smectic structure, and the other is the ordering process from the smectic structure to the pre-crystalline structure. The kinetics of the two ordering processes compete.

It should be noted here that the intensity of the 001' reflection does not seem to decay to zero, but to a finite value, depending on annealing temperature.

6.3.3 Kinetic Model Analysis

Kinetic Model

In order to reproduce the observed time dependence of the intensity of the 001' reflection and the 4-point pattern, we here introduce a kinetic model for the structure formation of PET from the oriented glassy states. In the model, we adopt the following assumptions.

1) The present polymeric system is divided into n_0 equivalent partial regions. There are only three different states for a partial region, that is, *nematic structure*, *smectic structure* and *pre-crystalline structure*. Every partial region takes one of the three states. The number of the nematic regions, smectic regions and pre-crystalline regions at time t are denoted by $n_n(t)$, $n_s(t)$, and $n_t(t)$, respectively.

2) The intensity of the 001' reflection at time t, $I_{001'}(t)$, is proportional to the number of smectic regions at time t, while the integrated intensity of the 4-point pattern in the SAXS regions, $I_{SAXS}(t)$, is proportional to the number of the pre-crystalline regions. Here, the following relations are satisfied:

$$I_{001'}(t) = C \cdot n_s(t) \tag{6.1}$$

$$I_{SAXS}(t) = \tilde{C} \cdot n_t(t), \tag{6.2}$$

where C and $\tilde{C}$ are the constants.

If we assume that the kinetic process from the nematic structure to the smectic structure and from the smectic structure to the pre-crystalline structure are given by the rate constants $\tilde{k}_{ns}$ and $\tilde{k}_{st}$, respectively, the following coupled equations are obtained:

$$\frac{dn_n}{dt} = -\tilde{k}_{ns}(t)n_n \tag{6.3}$$

$$\frac{dn_s}{dt} = \tilde{k}_{ns}(t)n_n - \tilde{k}_{st}(t)n_s \tag{6.4}$$

$$\frac{dn_t}{dt} = \tilde{k}_{st}(t)n_s. \tag{6.5}$$

It should be noted that the kinetic process from the nematic structure directly to the pre-crystalline structure without passing the smectic structure is prohibited for simplicity in this model.

Furthermore, it is assumed that the rate constants $\tilde{k}_{ns}$ and $\tilde{k}_{st}$ depend not only on temperature but also on time and are given by the following equations:

$$\tilde{k}_{ns}(t) = k_{ns}^0 \alpha t^{\alpha-1} \tag{6.6}$$
$$\tilde{k}_{st}(t) = k_{st}^0 \beta t^{\beta-1}, \tag{6.7}$$

where k_{ns}^0 and k_{st}^0 are constants, and α and β are exponents of the power-law with respect to time.

The solution of the above kinetic equations is as follows on condition that $n_n + n_s + n_t \equiv n_0$ (constant):

$$n_n(t) = n_n^0 e^{-k_{ns}^0 t^\alpha} \tag{6.8}$$

$$n_s(t) = -n_n^0 e^{-k_{ns}^0 t^\alpha}$$
$$+ e^{-k_{st}^0 t^\beta} \left[\left(1 + \beta k_{st}^0 \int_0^t s^{\beta-1} e^{-k_{ns}^0 s^\alpha + k_{st}^0 s^\beta} ds \right) n_n^0 + n_s^0 \right] \tag{6.9}$$

$$n_t(t) = n_0 - e^{-k_{st}^0 t^\beta} \left[\left(1 + \beta k_{st}^0 \int_0^t s^{\beta-1} e^{-k_{ns}^0 s^\alpha + k_{st}^0 s^\beta} ds \right) n_n^0 + n_s^0 \right], \tag{6.10}$$

where n_n^0 and n_s^0 are the numbers of the nematic and smectic regions at $t = 0$, respectively.

In case of $k_{ns}^0 \to \infty$, there is no difference between nematic structure and the smectic structure. As a result, the kinetic process can be regarded as the process between the two structures (the nematic, or smectic structure, and the pre-crystalline structures), and approximate expressions for $n_n(t)$, $n_s(t)$ and $n_t(t)$ can be obtained from Eqs. (6.8)–(6.10) as follows:

$$n_n(t) \approx 0 \tag{6.11}$$
$$n_s(t) \approx e^{-k_{st}^0 t^\beta} (n_n^0 + n_s^0) \tag{6.12}$$
$$n_t(t) \approx n_0 - e^{-k_{st}^0 t^\beta} (n_n^0 + n_s^0). \tag{6.13}$$

The time evolution expressed by Eq. (6.13) is the same as that given by the Avrami equation [17]. This means that the present kinetic model naturally includes the time evolution expressed by the Avrami equation, although the existence of the intermediate state between the initial structure and the final one is taken into account.

In the above procedure, we develop the kinetic model on the assumption that all regions should be in the pre-crystalline structure in the final stage of the kinetic process. However, this assumption is clearly oversimplified. It is more likely that some nematic regions and smectic regions do not transform to any other structures, and that there are still unoriented regions which remain unchanged during the annealing process. Because the existence of the unchanged smectic affects the intensity of the 001' reflection, the contribution of the number of the unchanged smectic regions, n_s^∞, is taken into account by replacing $n_s(t)$ with $n_s(t)+n_s^\infty$ in Eq. (1). Thus, we use the following equation instead of Eq. (1),

$$I_{001'}(t) = C \cdot (n_s(t) + n_s^\infty) . \tag{6.14}$$

In order to reproduce the observed intensity of the 001' reflection and the 4-point pattern in the SAXS region, we performed a nonlinear least square fit by using Eqs. (2), (6.8)–(6.10), and (14). The solid and dotted curves in Fig. 6.4 were obtained by the above fitting procedure. In Fig. 6.4, it is found that the time evolution of the intensities of both the 001' reflection and the 4-point pattern in the SAXS region can be well reproduced simultaneously by the present kinetic model. In the next sections, we will discuss the several parameters obtained by the fitting.

Characteristic Time

The times characterizing the kinetic processes from the nematic structure to the smectic structure, τ_{ns}, and from the smectic structure to the pre-crystalline structure, τ_{st}, can be defined in the following way:

$$\tau_{ns} = \left(\frac{1}{k_{ns}^0}\right)^{1/\alpha}$$

$$\tau_{st} = \left(\frac{1}{k_{st}^0}\right)^{1/\beta} .$$

By using the fitting parameters α, β, k_{ns}^0 and k_{st}^0 at various temperatures, we obtained the temperature dependence of τ_{ns} and τ_{st}, as shown in Fig. 6.5. In this figure, it is found that the functional form of τ_{ns} and τ_{st} with respect to temperature can be expressed by the Arrhenius type of the activation process, although the temperature range investigated here was restricted. According to the Arrhenius law, the characteristic times τ_{ns} and τ_{st} can be given by the following equations:

$$\tau_{ns} = \tau_{ns}^0 \exp\left(\frac{\Delta U_{ns}}{T}\right), \quad \tau_{st} = \tau_{st}^0 \exp\left(\frac{\Delta U_{st}}{T}\right),$$

where τ_{ns}^0 and τ_{st}^0 are constants, ΔU_{ns} and ΔU_{st} are the activation energy from the nematic structure to the smectic structure and from the smectic structure to the pre-crystalline structure, respectively. From the slope of the straight line in Fig. 6.5, the values of the activation energies can be evaluated as follows:

$$\Delta U_{ns} = (14.0 \pm 2.5) \text{ kcal/mol} \tag{6.15}$$
$$\Delta U_{st} = (21.9 \pm 4.3) \text{ kcal/mol} \tag{6.16}$$

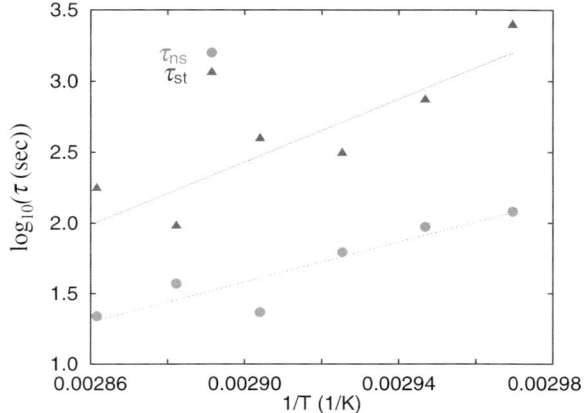

Fig. 6.5. Arrhenius plot of the characteristic times of the transition from nematic structure to smectic structure, τ_{ns}, and the one from smectic structure to pre-crystalline structure, τ_{st}

According to the results by dielectric and mechanical relaxation measurements on PET in [18], we can estimate the averaged activation energy for the α- and β-relaxation processes as follows:

$$\Delta U_\alpha \sim 137 \text{ kcal/mol}, \qquad (6.17)$$

$$\Delta U_\beta \sim 15 \text{ kcal/mol}, \qquad (6.18)$$

where ΔU_α and ΔU_β are the activation energies for the α- and β-relaxation processes, respectively.

Comparing ΔU_{ns} and ΔU_{st} with ΔU_α and ΔU_β, we find that the values of ΔU_{ns} and ΔU_{st} are much smaller than the activation energy of the α-relaxation process, but they are in relatively good agreement with the activation energy of the β-relaxation process. The structural changes discussed in this paper occur below the glass transition temperature of the bulk sample, and hence it is expected that the α-relaxation process, segmental motion, is prohibited and has no appreciable contribution to the structural change below the glass transition temperature. The observed values of ΔU_{ns} and ΔU_{st} supports this expectation. Furthermore, the β-relaxation process, i.e., the local mode relaxation process, may be a microscopic origin for the structural changes from the nematic structure to the smectic structure and from the smectic structure to the pre-crystalline structure, although the detailed mechanism is still to be elucidated. Real time relaxation measurements during the structural changes are highly desired for this purpose.

Exponent of the Kinetic Equation

The exponents α and β which determine the kinetics of the structural change of oriented PET are shown in Fig. 6.6. The values of α and β are obtained by the fitting procedure mentioned above. In Fig. 6.6, it is found that the exponent α for the kinetics from the nematic structure to the smectic structure decreases from 2.3 to 1 with decreasing temperature within the temperature range investigated here. In the case of the Avrami law [17], the crystalline fraction ϕ_c is given by

$$\phi_c(t) \sim 1 - \exp(-It^d) ,$$

where I is the nucleation rate, and d is the exponent. The exponent d is equal to 2 if the mechanism of structural change can be described by a 2-dimensional heterogeneous nucleation and growth. Hence, a possible mechanism for the structural change from the nematic structure to the smectic structure that is consistent with the observed exponent α may be as follows. In a nematic domain, there are smectic regions as heterogeneous nuclei. During the isothermal annealing process, these smectic regions grow two-dimensionally along the direction perpendicular to the chain axis. At a lower temperature, this lateral growth of the smectic domain is hindered, and as a result, the effective dimensionality of the lateral growth is decreased. Analysis of the overall behavior of the diffraction profile along the direction perpendicular to the meridian as a diffuse scattering according to the procedure similar to that shown in Refs. [19, 20] will reveal the nature of the growth of the smectic domain.

As for the structural change from the smectic structure to the pre-crystalline structure, the exponent β is about 0.5 and is almost independent of temperature. The value of the exponent $\beta(\approx 0.5)$ reminds us of the

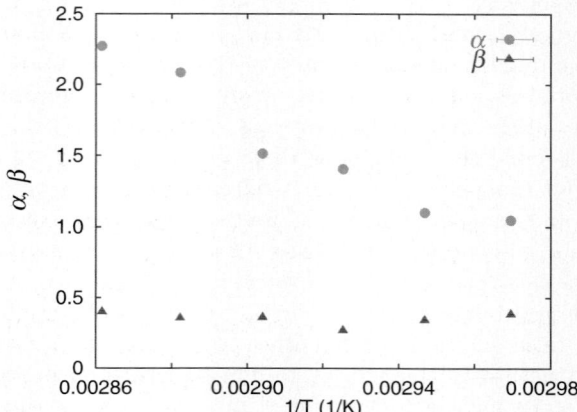

Fig. 6.6. Temperature dependence of the exponent α and β for the transition from the nematic structure to the smectic structure, α, and the one from the smectic structure to the pre-crystalline structure, β

stretched exponential-type of relaxation functions, $\phi(t) = \exp(-(t/\tau)^{\beta_{\text{KWW}}})$ $(0 < \beta_{\text{KWW}} < 1)$ [21]. This function can often be observed in disordered materials and is known as anomalous (slow) relaxation [22]. Hence, this structural formation process is not described by the growth of the pre-crystalline domain, but a slower diffusion process should be taken into account. According to this diffusion picture, it is conjectured that a virtual *particle* diffuses in the real space as a diffusant, and the site through which the particle passes has to change its structure into the pre-crystalline structure if the site belongs to the region of the smectic structure.

6.4 Structural and Thermal Change During the Heating Process

6.4.1 X-ray Scattering Patterns During the Heating Process

In the previous section, the structural change during the isothermal annealing process has been discussed. A similar but larger structural change can be expected during the heating process from room temperature to melting temperature. Figure 6.7 shows the change in the scattering patterns around the position of the 001' reflection on the meridian observed during the heating process from 33°C to 180°C at the rate of 5K/min. At 33°C there is the sharp 001' reflection on the meridian. The intensity of the 001' reflection increases with increasing temperature up to 71°C, and then it begins to decrease. At 100°C the intensities of two off-meridional positions located symmetrically with respect to the meridian are stronger than those around them and two weak spots can be recoginzed. At 180°C the intensities of the spots are enhanced. The two spots correspond to the 001 reflections of the triclinic crystal structure of PET. This result suggests that a continuous structural change from the nematic-like structure to the crystalline structure through the smectic-like structure occurs during the heating process, which is consistent with the results observed during the isothermal annealing process.

We also performed real-time WAXS measurements during the heating process of oriented glassy states of PET, in order to investigate the detailed properties of the continuous structural change. Before WAXS measurements, the samples were annealed at T_a. As an example, in Fig. 6.8, we show the intensity profile along the equator at various temperatures during the heating process at the rate of 10 K/min in the case of T_a=125°C. Figure 6.8(a) shows that the scattering profile does not change below T_a, while Fig. 6.8(b) shows that the intensities of the Bragg reflections such as 010, $\bar{1}$10, and 100 increase with increasing temperature above T_a. From this result, it is found that the ordered structures formed through the annealing procedure before the X-ray scattering measurements are fixed up to the annealing temperature T_a, and a further ordering process occurs only if the temperature proceeds T_a.

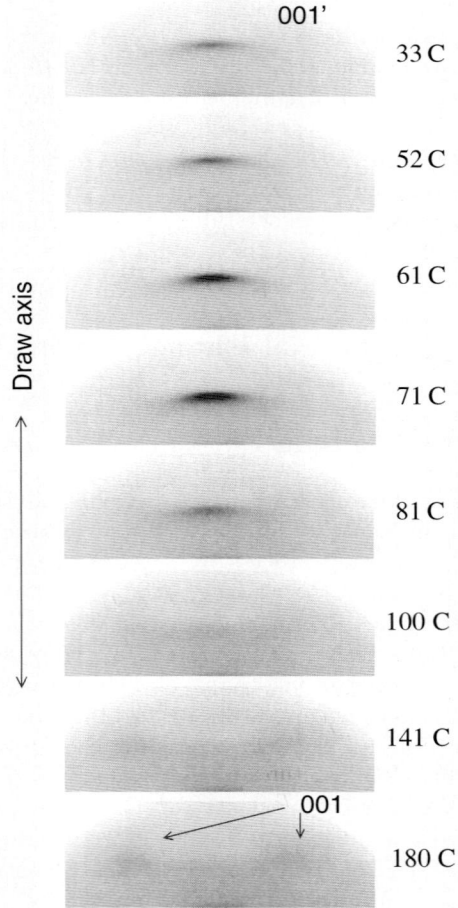

Fig. 6.7. The temperature change in diffraction patterns around the 001' reflection observed during the heating process from 33°C to 180°C at the rate of 5 K/min. The 001' reflection of the smectic structure changes into two 001 reflections of the triclinic crystalline structure of PET with increasing temperature

6.4.2 Thermal Properties of Oriented PET

DSC Measurements

The X-ray scattering measurements revealed that the ordering process in oriented glassy states of PET are generated by a continuous structural change. Here, we will show the corresponding thermal properties observed by DSC. For the DSC measurements, we used the samples of oriented glassy PET that

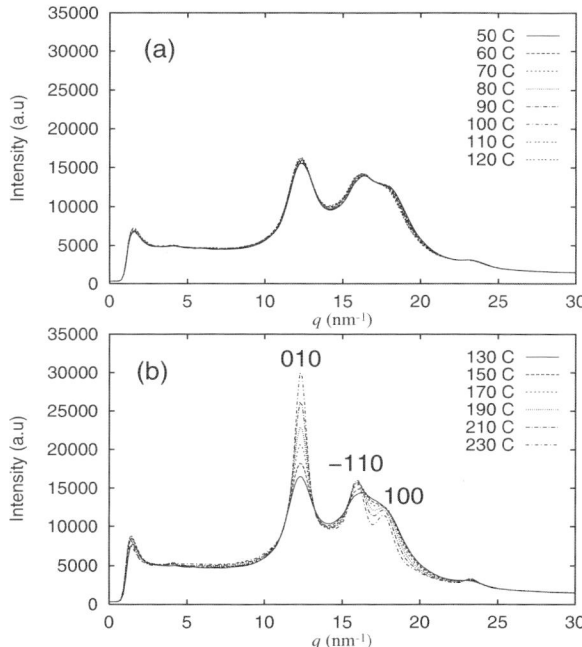

Fig. 6.8. X-ray scattering profile along the equator as a function of the scattering vector at various temperatures during the heating process in oriented glassy states of PET annealed at T_a. Here, $T_a = 125°C$ and the heating rate is 10K/min. The *upper figure* shows the results for temperatures between 50 and 120°C, and the *lower one* shows those for temperatures above $T_a (\equiv 125°C)$

were drawn at room temperature in the same way as in the X-ray measurements. Figure 6.9 shows DSC thermograms (a) for the unoriented glassy states of PET and (b) for the oriented glassy states of PET without any annealing before measurements. In the case of the unoriented sample, we can observe three typical contributions; the first one is due to the glass transition around 76°C, the second one is due to the crystallization at 145°C, and the third one is due to the melting transition. The exothermic peak temperature for crystallization strongly depends on the heating rate. On the other hand, the DSC thermogram of the oriented sample has two essential differences from that of the unoriented samples. 1) There appears to be no appreciable exothermic signal due to crystallization. 2) There is an anomalous change in the total heat flow at a temperature T'_g just below the glass transition temperature T_g of the unoriented samples. ($T_g = 76°C$ for unoriented PET and $T'_g = 63°C$ for non-annealed oriented PET.) A step-like *increase* in the exothermic total heat flow is observed at T'_g. If this change were attributed to the change in heat capacity, the heat capacity would show a step-like *decrease* at T'_g.

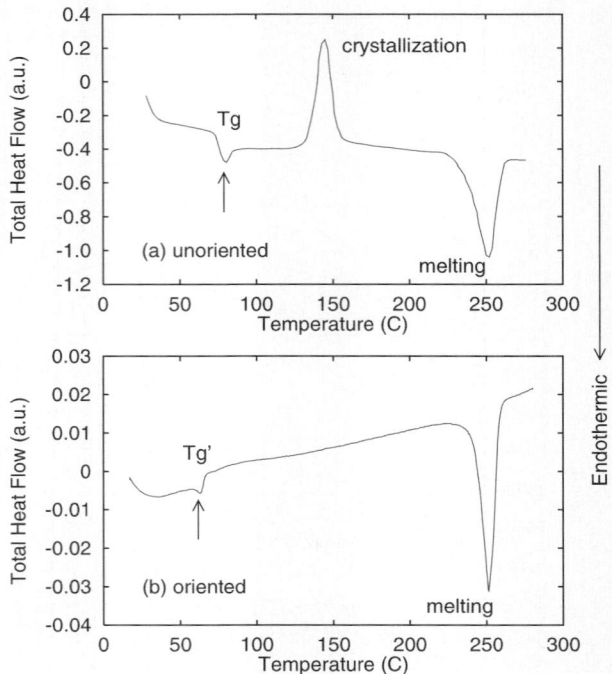

Fig. 6.9. Temperature change in the total heat flow observed during the heating process at 20K/min **(a)** for the unoriented glassy states of PET and **(b)** for the oriented glassy states of PET without any annealing. The negative direction of the total heat flow corresponds to the endothermic heat flow

Figure 6.10 shows DSC thermograms of the oriented PET for various annealing temperatures T_a. The samples were annealed at T_a before the DSC measurements. The measurements were done during the heating process at the rate of 20 K/min. The melting temperature is almost independent of T_a, while the position of the anomaly at T'_g shifts to a higher temperature side with increasing T_a. The value of T'_g is by about 10K larger than the corresponding value of T_a. This result implies that the ordered structure that was formed through the annealing procedure before the DSC measurements is fixed at temperatures below T_a during the heating process. However, if the temperature increases over T_a, the ordering process restarts at $T'_g(> T_a)$ to form a more ordered structure.

As shown in Fig. 6.8, the X-ray scattering patterns show that during the heating process from room temperature, no structural change was observed below T'_g, while an ordering process occurs above T'_g. In this sense, some mobility should be activated at T'_g so that the ordering process starts in a similar way as the segmental motions are activated at T_g in the case of the

6 Structure Formation and Glass Transition in Oriented PET   113

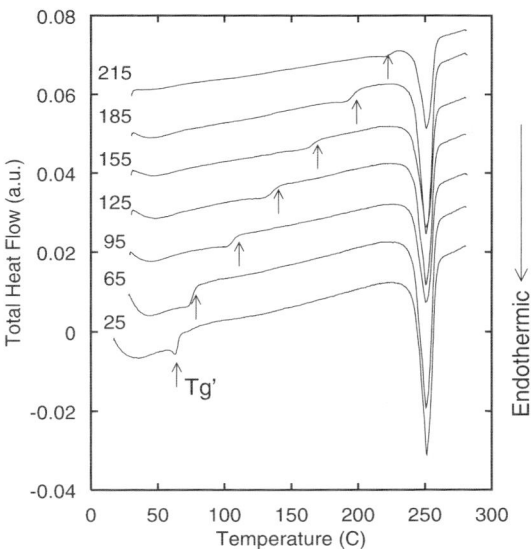

Fig. 6.10. Temperature change in the total heat flow observed during the heating process for the oriented glassy states of PET after annealing at T_a. The *numbers* shown near each curve are annealing temperatures T_a. The *arrows* show the position of the 'glass transition temperature' T_g' for the oriented PET

unoriented PET. Therefore, we can regard the temperature T_g' as the glass transition temperature of the oriented glassy states of PET after the annealing at T_a. However, the observed step-like increase in the exothermic total heat flow at T_g' appears to be totally different from that observed at T_g of the unoriented sample. We try to interpret the observed DSC thermogram around T_g' in the following way: the glass transition occurs at T_g' in the case of the oriented sample and at the same time an ordering process starts. The glass transition induces an endothermic heat flow, while the ordering induces an exothermic heat flow. Hence, both contributions can compete, and as a result the observed DSC thermogram around T_g' can be obtained. As shown in the next section, TMDSC measurements support this interpretation.

Temperature Modulated DSC Measurements

The DSC measurements in the previous section revealed anomalous change in the total heat flow at T_g' where the structural change restarts. We also performed TMDSC measurements in order to investigate the physical origin of the anomalous change at T_g'. In Fig. 6.11, it is found that there is a crossover region where the reversing heat flow changes gradually to a lower value. This region ranges from 80°C to 130°C. The slope of the region above 130°C is

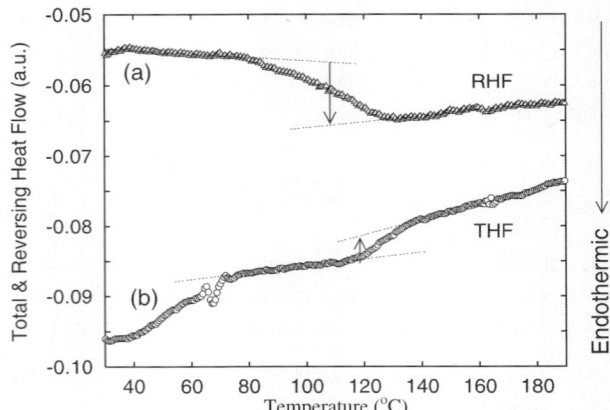

Fig. 6.11. Temperature change in the total and reversing heat flow observed for the oriented PET annealed at $T_a = 125°C$. The curve (a) corresponds to that of the reversing heat flow (RHF) and the curve (b) does to that of the total heat flow (THF).

larger than that below 80°C. Because there should be no contributions from the heat of crystallization, the heat capacity increases in the crossover region and it decreases with increasing temperatures above 130°C. Therefore, we can interpret the temperature dependence of reversing heat flow as follows. The decrease in reversing heat flow between 80°C and 130°C is mainly due to the glass transition because the heat capacity of glassy states is smaller than that of liquid states. On the other hand, the gradual increase in reversing heat flow above 130°C is mainly due to a continuous crystallization that causes the oriented PET to become more ordered structure.

In Fig. 6.11, it is also found that there are two regions where the total heat flow shows anomalous behavior, one is around 70°C and the other is between 110°C and 130°C. The latter one corresponds to the anomalous change in total heat flow observed at T'_g in Fig. 6.10. It can be expected that the observed upper shift of total heat flow at T'_g is mainly due to the heat of crystallization. The position of this anomaly in Fig. 6.11 is different from that in Fig. 6.10. This may be due to the difference in heating rate (20K/min for DSC measurements in Fig. 6.10 and 1K/min for TMDSC measurements in Fig. 6.11.)

Comparing the temperature dependence of the total and reversing heat flow, we can safely regard the temperature T'_g where the total heat flow exhibits an upper shift as the glass transition temperature of the unoriented glassy states of PET. On the other hand, the anomalous change in total heat flow at 70°C in Fig. 6.11, there is no corresponding change in reversing heat flow. This change may be related to some ordering process that has no appreciable contributions to reversing heat flow.

It should be noted that the temperature T'_g where the step-like increase in total heat flow occurs in Fig. 6.10 increases with increasing T_a. This implies that the glass transition temperature of the oriented glassy states of PET increases as the structural order increases during the heating process.

6.5 Concluding Remarks

We investigated the structure formation during the isothermal crystallization (annealing) process from the oriented glassy states of PET using real-time X-ray scattering and thermal measurements. The results obtained are as follows:

1. During the isothermal annealing process at 63.6–76.3°C, the intensity of the 001' reflection from the smectic structure has the maximum at a time, while the intensity of the 4-point pattern in the SAXS region increases monotonically with time.
2. Time dependence of the 001' reflection and the 4-point pattern in the SAXS region can be reproduced by a kinetic model assuming that the fraction of the nematic structure changes into the pre-crystalline structure by way of the smectic structure. A corresponding continuous ordering process could be observed during the heating process.
3. The glass transition temperature of the oriented glassy states increases with increasing degree of ordering.

Although the investigation in this paper mainly concentrated on the time evolution of the 001' reflection and the 4-point pattern in the SAXS region, the measurements of the time dependence of overall diffraction patterns in the WAXS region will be required in order to elucidate the microscopic mechanism of the structure formation in oriented PET. In particular, real-time measurement of the intensity distribution of the meridional reflections of higher order will reveal the detailed mechanism of the structure formation. Real-time relaxation measurements such as dielectric relaxation spectroscopy are also desired for this purpose.

As for the thermal properties, it was revealed that the glass transition temperature of the unoriented glassy states of PET increases extraordinarily with increasing structural order. Recently, it has been reported that there is a rigid amorphous fraction in some polymers [23, 24]. It can be expected that the glass transition temperature of such rigid amorphous fraction is larger than that of usual mobile amorphous fraction, and that it strongly depends on the degree of the ordering.

Acknowledgments

We appreciate A. Koyama, D. Tahara, Y. Miyamoto, Y. Nishikawa, T. Fujisawa, K. Inoue for their helpful collaboration, and Y. Funatsu of Toray

Co., Ltd. for supplying us PET films . The synchrotron radiation experiments were performed at the SPring-8 with the approval of the Japan Synchrotron Radiation Research Institute (JASRI) (Proposal Nos.2001A0306-NDL-np, 2001B0019-NDL-np, 2002A0260-NDL2-np, and 2003B0374-NL2b-np). The work was supported by a Grant-in-Aid for Scientific Research on Priority Areas, Mechanism of Polymer Crystallization (No.12127203) from the Ministry of Education, Culture, Sports, Science and Technology of Japan.

References

[1] M. Imai, K. Mori, T. Mizukami, K. Kaji, T. Kanaya, Polymer **33** (1992) 4451.
[2] C.H. Lee, H. Saito, T. Inoue, Polym. Prep. Jpn. **44** (1995) 735.
[3] K. Fukao and Y. Miyamoto, J. Non-Cryst. Solids, **235-237** (1998) 534.
[4] K. Fukao and Y. Miyamoto, Phys. Rev. Lett., **79** (1997) 4613.
[5] P.D. Olmsted, W.C.K. Poon, T.C.B. McLeish, A.J. Ryan, and N.J. Terrill, Phys. Rev. Lett. **81** (1998) 373.
[6] T. Asano, F.J.B. Calleja, A. Flores, M. Tanigaki, M.F. Mina, C. Sawatari, H. Itagaki, H. Takahashi, I. Hatta, Polymer, **40** (1999) 6475.
[7] D.J. Blundell, A. Mahendrasingam, C. Martin, W. Fuller, D.H. MacKerron, J.L. Harvie, R.J. Oldman, C. Riekel, Polymer, **41** (2000) 7793.
[8] A. Mahendrasingam, D.J. Blundell, C. Martin, W. Fuller, D.H. MacKerron, J.L. Harvie, R.J. Oldman, C. Riekel, Polymer, **41** (2000) 7803.
[9] D.J. Blundell, A. Mahendrasingam, C. Martin, W. Fuller, J. Mat. Sci. **35** (2000) 5057.
[10] G.E. Welsh, D.J. Blundell, A.H. Windle, Macromolecules, **31** (1998) 7562.
[11] G.E. Welsh, D.J. Blundell, A.H. Windle, J. Mat. Sci. **35** (2000) 5225.
[12] D. Kawakami, B.S. Hsiao, S. Ran, C. Burger, B. Fu, I. Sics, B. Chu, T. Kikutani, Polymer **45** (2004) 905.
[13] L. Li and W.H. de Jeu, Faraday Discuss., **128** (2005) 299.
[14] C. Alvarez, I. Sics, A. Nogales, Z. Denchev, S.S. Funari, T.A. Ezquerra, Polymer **45** (2004) 3953.
[15] K. Fukao, A. Koyama, D. Tahara, Y. Kozono, Y. Miyamoto, N. Tsurutani, J. Macromol. Sci. Part.B-Phys. **B42** (2003) 717.
[16] R. Bonart, Kolloid-Z, **213** (1966) 1.
[17] M. Avrami, J. Chem. Phys. **7** (1939) 1103, **8** (1940) 212, **9** (1941) 177.
[18] N.G. McCrum, B.E. Read, and G. Williams, 'Anelastic and Dielectric Effects in Polymeric Solids', (John Wiley and Sons Ltd, London, 1967), p. 506–507, Fig. 13.3.
[19] K. Fukao, J. Chem. Phys. **101** (1994) 7882.
[20] K. Fukao, J. Chem. Phys. **101** (1994) 7893.
[21] G. Williams, D.C. Watts, Trans. Faraday Soc. **66** (1970) 80.
[22] M.F. Shlesinger, E.W. Montroll, Proc. Natl. Acad. Sci. USA **81** (1984) 1280.
[23] H. Xu, B.S. Ince, P. Cebe, J. Polym. Sci. Part B: Polym. Phys. **41** (2003) 3026.
[24] B. Natesan, H. Xu, B.S. Ince, P. Cebe, J. Polym. Sci. Part B: Polym. Phys. **42** (2004) 777.

7

How Do Orientation Fluctuations Evolve to Crystals?

Zhicheng Xiao[1], Jan Ilavsky[2], Gabrielle G. Long[2], and Yvonne A. Akpalu[1,*]

[1] Department of Chemistry and Chemical Biology, Rensselaer Polytechnic Institute, Troy, NY 12180, USA
akpaly@rpi.edu
[2] X-Ray Science Division, Argonne National Laboratory, 9700 S. Cass Avenue, Argonne, IL 60439, USA

Abstract. Light and synchrotron X-ray scattering are used to probe structure formation during isothermal crystallization of an ethylene-1-hexene copolymer (EH064, M_w = 70,000 g/mol, ρ = 0.900 g/cm^3, $M_w/M_n \sim$ 2, 6.4 mole percent hexene) and an ethylene-1-butene copolymer (EB059, M_w = 70,000 g/mol, ρ = 0.905 g/cm^3, $M_w/M_n \sim$ 2, 5.9 mole percent butene). It is shown that clear structural information on size scales ranging from tens of nanometers to several micrometers during early stage crystallization can be obtained by the combined use of small-angle light scattering (SALS) and (USAXS) when crystallizing the polyethylenes at high temperatures (above the peak melting temperature of the polymer and below the theoretical equilibrium melting temperature) required for resolving early stage crystallization without the influence of the crystal growth. Fractal objects with diffuse interfaces are formed initially, where the limiting slope of the scattering profiles increases from around 2 to 4 during early stage crystallization. This indicates that the interfaces of these domains sharpen with time. The interface sharpening process depends on the crystallization temperature and the molecular structure of the polymers. The magnitude of the limiting slope in *log-log* plots of USAXS scattering profiles decreases again as the spherulites are formed and then grow, showing the effect of temperature and molecular structure on the early stage crystallization of polymers.

7.1 Introduction

Polymer crystallization is an industrially and scientifically important phenomenon that has defied detailed molecular description. The chemical structure of the polymeric chain, combined with morphological features acquired during processing governs the properties of semicrystalline polymers. Investigations of the details of liquid-solid transformation have closely paralleled the development of polymer science and interest in understanding the details of crystalline structure had arisen from the well known fact that the crystallinity controls the properties of the material. Polymer crystallization is believed to follow the classical theory of the crystal nuclei into a hierarchy of ordered structures, which involves the growth of lamellar crystals and the aggregation of these lamellae into superstructures such as spherulites and axialites [1–3]. The primary lamellar habit formed is a consequence of the anisotropic growth of crystal nuclei. However, the fundamental mechanisms of polymer crystallization, especially at the early stage, are still poorly understood [4–10]. While considerable theory [2,4,7,8,11–14] of crystallization kinetics has been developed, there is concern as to whether the predicted mechanisms are unique.

For many years, nucleation and growth as a stepwise process has dominated the discussion of polymer crystallization [1–7]. In contrast to this view, Strobl [8]. proposed a multistage process to explain polymer crystallization, while others concluded to the existence of a spinodal-assisted process [13–23] Common to both views is that the crystallization is preceded by an ordered precursor (so-called pre-ordering). Clear structural information about such possible precursors – necessary to verify these hypotheses – is still scarce. As a result, during recent years an important and still open debate has been going on regarding polymer crystallization. Interestingly, pre-ordering was already implied in some rather early studies of polymer crystallization, but did not receive much attention. As early as 1967 Katayama et al. [24] observed a small-angle X-ray scattering (SAXS) peak significantly earlier than the appearance of the corresponding crystalline Bragg peaks in wide-angle X-ray scattering (WAXS). They proposed that density fluctuations occurred before the formation of any crystals. The idea of a multi-stage process dates back to 1967 by Yeh and Geil [25], while Schultz introduced in 1981 a spinodal approach promoting orientation in polymer systems [26]. However, the essential question about the nature of pre-ordering still remained open. The description '' is often not used in a precise way. In spite of this, it is evident that the precursor should possess some ordering intermediate between the liquid and the crystal phase. Moreover, nanofillers can influence these ordering processes as well as subsequent crystal growth.

In spite of the many possible pathways to polymer crystallization, there are two distinct structural characteristics: (I) the formation of a single nucleus and (II) the spatial distribution of nuclei originating processes that lead to an increase in the number of crystal nuclei. X-ray scattering can be used to monitor structural changes during (II) [27,28] (Fig. 7.1).

For I, no changes would be observed in SAXS and WAXS since there is little or no crystal-amorphous density contrast variation during this process [29] On the other hand, polarized light can be used to monitor structural changes for both stages because the clustering of anisotropic crystal nuclei should exhibit depolarized scattering arising from the anisotropy of the constituent polymer chains [30, 31]. The relative arrangement of these nuclei can be on length scales ranging from tens of nanometers (SAXS) to several hundred nanometers (USAXS) or micrometers (SALS). By combining the use of USAXS with traditional SAXS and SALS, the structure changes ranging from nanometer up to micrometer can be studied.

(I) Formation of a single nucleus:

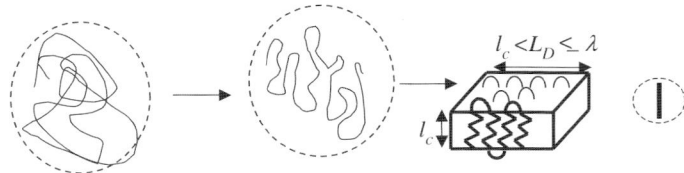

(I) Spatial distribution of nuclei:

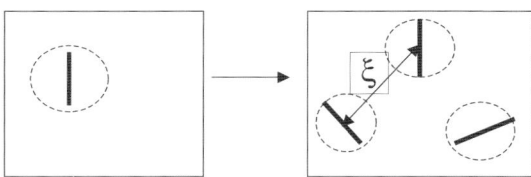

Fig. 7.1. Evolution of orientation fluctuations that occur when an amorphous polymer evolves to a lamellar crystal. This process can be divided into two distinct processes: (I) formation of the anisotropic crystal habit (lamellar crystal) from an amorphous polymer and (II) processes which increase the number of crystal nuclei. l_c and L_D are the crystal thickness and lateral dimension, respectively

In this paper, we use SALS under cross-polarized (H_V) alignment and USAXS to probe structural changes during of polymers. During early stage crystallization of polyethylene/olefin copolymers, fractal objects with diffuse interfaces are formed initially, and the interfaces of these domains sharpen with time. This compaction process is observed for both polyethylene/olefin copolymers, and the rates and pathways depend on both crystallization temperature and the chemical structure of the polymer.

7.2 Materials and Methods

7.2.1 Sample Preparation

The ethylene-1-hexene copolymer (EH064, M_w = 70,000 g/mol, ρ = 0.900 g/cm^3, $M_w/M_n \sim 2$, 6.4 mole percent hexene) and the ethylene-1-butene copolymer (EB059, M_w = 70,000 g/mol, ρ = 0.905 g/cm^3, $M_w/M_n \sim 2$, 5.9 mole percent butene) studied were provided by ExxonMobil. Both copolymers were prepared with a metallocene catalyst. Before any further sample preparation, 3 g of polymer was dissolved in 300 mL of refluxing toluene at 111°C. The solution was poured into an acetone/methanol (50/50) mixture (800 mL) at 0°C. The resulting precipitate was filtered, washed, and dried under vacuum at 40–50°C for 72 h.

7.2.2 Differential Scanning Calorimetery (DSC)

Before the DSC measurements, melting and crystallization procedures were performed on a computer interfaced Instec HCS600V hot stage. Samples were first melted at 180°C for 10 min under nitrogen to remove any thermal history and interlayer melting effect, and then rapidly cooled (at 50°C/min) to a crystallization temperature (T_c). DSC measurements on film samples (∼6 mg) were made using a SEIKO DSC 220C under nitrogen atmosphere. Temperature and enthalpy values were calibrated using indium as a standard. DSC melting curves were obtained by heating samples from −30°C to 160°C at 10°C/min. Data obtained from the first melting-run after different thermal treatments were used for analysis. The mass percent of crystallinity from DSC was estimated by using $w_c^{DSC} = \Delta H_f / \Delta H^0$, where ΔH_f is the heat of fusion obtained after integrating the area under the normalized curve and $\Delta H^0 = 293$ J g^{-1} is the reference heat of fusion for 100% crystalline polyethylene.

7.2.3 Simultaneous WAXS and SAXS

Samples were melt pressed in a vacuum laboratory hot press (Carver Press, Model C) at 160°C for 30 min. The molded films were then allowed to cool to room temperature under vacuum. A dual temperature chamber for the melt crystallization experiments consists of two large thermal chambers maintained at the melt temperature ($T_1 = 160°C$) and the crystallization temperature ($T_2 = 81°C, 83°C, 86°C, 89°C, 92°C$ or $96°C$). After 5–10 min at T_1, the copper sample cell was transferred rapidly (∼2 s) to the other chamber by means of a metal rod connected to a pneumatic device. A detailed description of the arrangement of the sample and of the two detectors used to measure WAXS and SAXS simultaneously has been provided previously [32]. Each polymer sample within the copper cell was 1.5 mm thick and 7 mm in diameter and was contained between two 25 μm thick Kapton films. The actual sample temperature during crystallization (T_2) and melting (T_1) was monitored by means of a thermocouple inserted into the sample cell. The crystallization temperature was usually reached 120 s after transfer without overshooting. Under isothermal conditions the fluctuations in the sample temperature are less than 0.5°C. Unless stated otherwise, all references to time are times elapsed after transferring the sample to the crystallization chamber.

Time-resolved simultaneous SAXS/WAXS data were collected at the Advanced Polymer Beamline at Brookhaven National Laboratory, X27C. The radiation spectrum from the source was monochromated using a double multilayer monochromator and collimated with three 2° tapered tantalum pinholes to give an intense x-ray beam at $\lambda = 1.307$ Å. Two linear position sensitive detectors (European Molecular Biology Laboratory, EMBL) were used to collect the SAXS and WAXS data simultaneously. The usable span of scattering vector magnitudes ($q = (4\pi/\lambda)\sin(\theta/2)$, where λ is the x-ray wavelength and θ is the scattering angle) for SAXS was in the range 0.01 Å^{-1} < q < 0.3 Å^{-1}, while that for WAXS was 0.7 Å^{-1} < q < 2.9 Å^{-1}. Data were collected in 15 s or 30 s time blocks, depending on the crystallization rate. The peak position, peak height and peak width for the crystalline and amorphous reflections in WAXS were extracted by a curve fitting program. A broad Gaussian peak was used to describe the amorphous background. The crystalline peaks (110 and 200) were also fitted with Gaussian functions. For SAXS, the scattering intensity due to thermal fluctuations was subtracted from the SAXS profile $I(q)$ by evaluating the slope of $I(q)q^4$ versus q^4 plots [33] at large wave vectors ($q \gg 0.2$ Å^{-1}).

7.2.4 USAXS Measurements

The USAXS measurements in the q-range 0.0002 A^{-1} < q < 0.3 A^{-1} were carried out at the Advanced Photon Source at Argonne National Laboratory, where the USAXS instrument [34,35] is installed at undulator beamline 33-ID. Monochromatic (12 keV) X-ray flux of $\sim 10^{13}$ photons/second was obtained using a double crystal monochromator equipped with Si (111) crystals, and a pair of mirrors after the monochromator was used to suppress the harmonic content of the beam.

The USAXS instrument uses a 6-reflection symmetric Si (111) channel cut crystal before and another 6-reflection channel cut after the sample to collimate the incoming and analyze the scattered X-ray beam. A windowless ionization chamber using air at ambient pressure and temperature monitors the X-ray beam incident on the sample. Nine decades of scattered beam can be detected by the silicon PIN photodiode detector. The X-ray beam incident on the sample was about 1.5 mm wide by 0.4 mm high.

For each scan, the intensity was measured in a step scan where the analyzer channel cut was rotated to an angle at known offset from the measured peak of the rocking curve, the detector and analyzer vertical positions were adjusted to intercept the scattered beam and the signals from the ionization chamber and photodiode were integrated over the counting time. The range of angles for each scan was from 0.002 degrees above to 2.5 degrees below the peak of the rocking curve. The step size was chosen based on the distance from the angle at the peak of the rocking curve so as to give steps that were roughly equidistant on a logarithmic axis of Q. Data collection for each scan of 150 points required approximately 15 minutes with a 5 second counting time per data point.

The USAXS data were reduced, absolute calibrated, and corrected for instrumental curve using software "Indra"
(http://www.uni.aps.anl.gov/~ilavsky/indra_2.html). Since the vertical angular width of the Bragg reflection of the analyzer optics is much narrower than the horizontal angle subtended by the detector and the sample-detector distance, the collimation of the USAXS instrument is that of a finite horizontal slit. Several methods exist to correct the data for this collimation broadening ("slit smearing"). The iterative method of Lake [36], implemented as part of

provided data evaluation software package "Irena" (http://www.uni.aps.anl.gov/~ilavsky/irena.html), was used to desmear the data.

7.2.5 Small Angle Light Scattering (SALS)

SALS measurements were performed on (90–200 μm thick) film samples sealed between two round glass coverslips. Before the measurements, the sealed samples were heated from room temperature to 180°C, held at this temperature for 5 min and then quenched to the crystallization temperature (T_c). After reaching T_c, the samples were immediately heated to 180°C, held at this temperature for 10 min and then cooled to T_c for measurements. The heating/cooling rate used is about 50°C/min. An Instec HCS600V hot stage was used to control the temperature to within 0.1°C during crystallization measurements. SALS patterns under cross-polarized (H_V) and parallel-polarized (V_V) optical alignments and transmitted light were recorded using a vertical light scattering apparatus described previously [37]. For this work, the usable span of scattering vector magnitudes was in the range 2×1^{-5} Å^{-1} < q < 2.4×10^{-4} Å^{-1}. A mirror attached to the center of the screen was used to reflect the light transmitted by the sample and the intensity of this light was measured by means of a computer controlled optical power meter. All measurements were performed under nitrogen atmosphere.

Experimental scattering intensities from SALS were corrected using procedures previously described [37]. The melt contribution to the corrected H_V scattering intensity was subtracted after accounting for statistical fluctuations. The percent transmission was determined from the ratio of the transmitted light intensity measured with a sample in the beam path to that measured without the sample.

Crystallization mechanisms can be determined by comparing the time evolution of degree of crystallinity and the total integrated scattering intensity or invariant [27, 38]. In this paper, we calculated the degree of crystallinity from WAXS (w_c) and the total integrated scattering intensities or invariant from SAXS (Q_{SAXS}) and H_V SALS (Q_{H_V}). The degree of crystallinity is obtained by using the methods previously described [27]. The uncertainty in w_c by this method is about 2%.

The SAXS invariant (Q_{SAXS}) or H_V SALS invariant (Q_{H_V}) determined from measurements represents fluctuations in the sample and will be referred to as relative invariants.

$$Q_{SAXS} \propto \langle \eta^2 \rangle$$
$$Q_{H_V} \propto \langle \delta^2 \rangle$$

where $\langle \eta^2 \rangle$ represents the nano-scale mean-square density fluctuations of the system, $\langle \delta^2 \rangle$ represents the micro-scale mean-square fluctuations in the averaged anisotropy of the system.

7.3 Results and Discussion

When a crystallizable polymer is cooled below its equilibrium melting temperature, the hierarchical structure formed can be probed by in situ scattering. Crystallization mechanisms can be determined by comparing the time evolution of the degree of crystallinity (w_c) determined from WAXS and the total integrated scattering intensity or invariant during crystallization [27,38] from SAXS.

Figure 7.2 shows WAXS, SAXS and SALS results during isothermal crystallization of EH064. The crystallization kinetics is accurately resolved by X rays at temperatures below the peak melting temperature $T_m^p = 95°C$ (Fig. 7.3). As the crystallization temperature increases close to the peak melting temperature, both the changes in the SAXS invariant and crystallinity determined from WAXS are very small and close to the detection limits. This is because, at these high crystallization temperatures, large thermal fluctuations and the low density contrast between crystal and amorphous phase make it difficult to detect the changes in crystallinity and structure [27,29,39–41]. To resolve early stage crystallization behavior without the influence of crystal

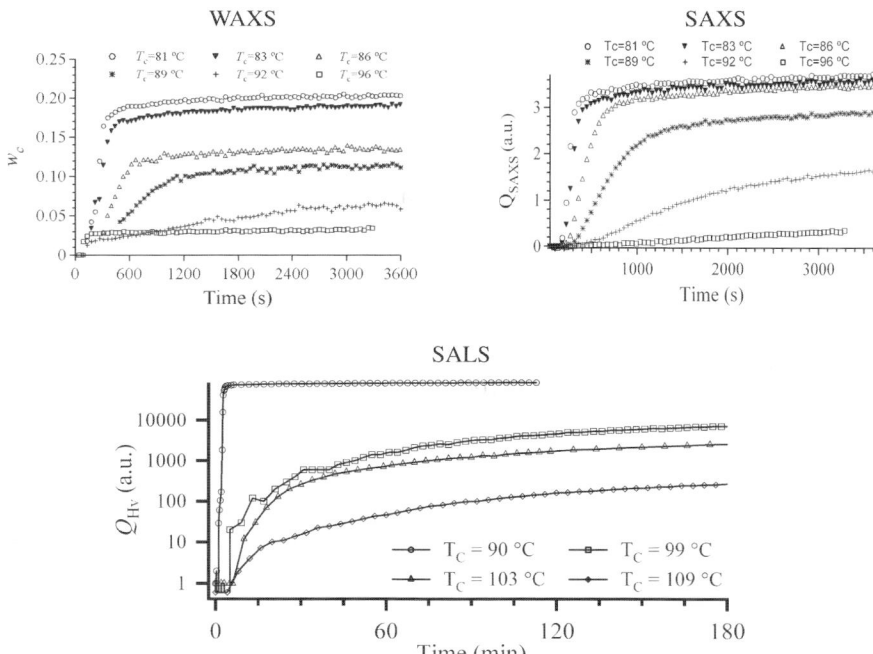

Fig. 7.2. Crystallization behavior of EH064 from WAXS and SAXS. (Time evolutions of WAXS crystallinity (w_c), SAXS invariant (Q_{SAXS}) and H_V SALS invariant (Q_{H_V}) are shown)

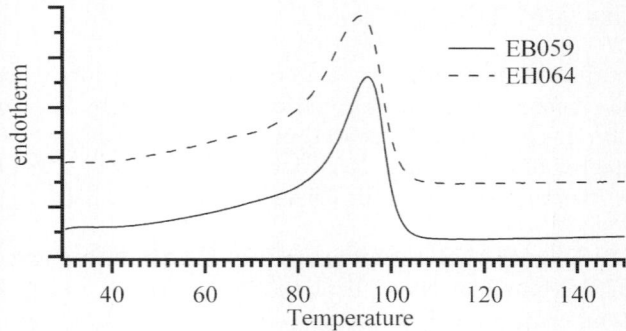

Fig. 7.3. Typical DSC melting curve of polyethylene samples (EH064 and EB059) rapidly crystallized from melt. Heating rate: 10°C/min

growth kinetics, it is necessary to crystallize the polymer at higher temperatures (above the final melting temperature T_m^f). In addition, at these crystallization temperatures, DSC is not a sensitive technique because little or no enthalpy change is usually obtained [42, 43]. Atomic force microscopy [44–46] may be able to resolve early stage crystallization, but to the best of our knowledge, there are no published reports showing this behavior without the influence of crystal growth. On the other hand, if large scale fluctuations are present during early stage crystallization, SALS can be used to investigate the structural changes at these temperatures. In Fig. 7.2, when a polyethylene sample was crystallized at 109°C, we observed a significant increase in scattering intensity. At this temperature, the theoretical crystalline fraction of EH064 is less than 10^{-4} (Fig. 7.4) [47, 48]. Polarized light scattering due to the organization of crystals can be observed whereas the crystallinity can not be directly measured by DSC or X-rays at this temperature. The resulting patterns yield information regarding the nature of the anisotropic domains formed.

Figure 7.5 shows the crystallization behavior of EH064 from SALS at various temperatures. The equilibrium melting temperature of this sample is 136°C [49]. At temperatures below T_m^p, space-filling spherulites are formed as evident from the typical anisotropic H_V and V_V patterns obtained at 90°C. When samples were crystallized at temperatures above T_m^p, the initial H_V SALS patterns exhibit circular symmetry consistent with the formation of anisotropic domains that are randomly oriented with each other. Non-space-filling spherulites are formed at 99°C and 103°C as indicated by anisotropic four-leaf H_V patterns and corresponding isotropic V_V patterns obtained. At 99°C, the maximum attainable crystallinity is about 3 % and it is about 1 % for 103°C (Fig. 7.3). The percentage of transmitted light measured simultaneously with the SALS can also provide information about the change in fluctuations of the system. Figure 7.6 shows the transmission as a function of

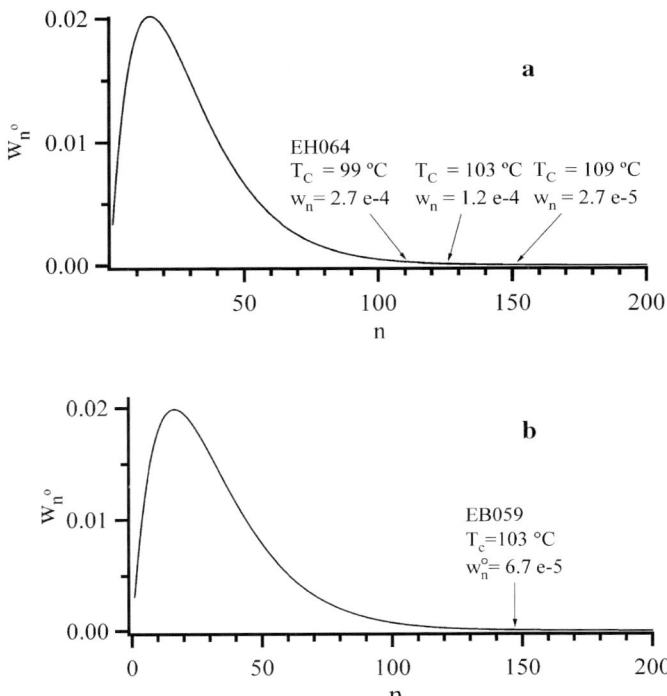

Fig. 7.4. Overall distribution of n-sequences for random polyethylene/olefin copolymers: (**a**) polyethylene/hexane copolymer; (**b**) polyethylene/butane copolymer. The *arrows* indicate the critical length of ethylene sequences which is in equilibrium with undercooled melt at given T_c

time during crystallizations at different temperatures. At temperatures below T_m^p, the transmission reaching a constant after passing through a minimum is consistent with the formation of space-filling spherulites at these temperatures [50, 51]. At temperatures above T_m^p, the slight decrease in transmission with time occurs as the non-space-filling spherulites are formed. As the crystallinity is further decreased at higher temperatures (Fig. 7.4) [47], detailed analysis of the intensity profiles can provide clues about morphology of the anisotropic domains formed.

The circularly averaged intensity of the patterns at 109°C monotonically decreases with scattering vector q (Fig. 7.7). The increase in intensity with time can be associated with the increases in the number and/or the anisotropy of the domains. In Fig. 7.8, the shape of the *log-log* plots of scattering profiles indicates that the domains are fractal objects. Kratky plots in Fig. 7.9 indicate that these domains are diffuse. For , the intensity profiles at high q region (if $q \cdot a \gg 1$, where a the length scale of the scattering objects,) can be fitted to the Porod form [52–54]:

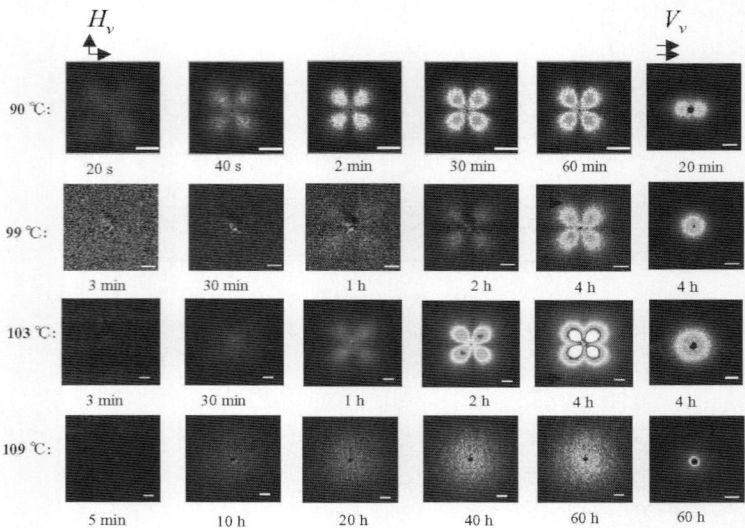

Fig. 7.5. SALS patterns (H_V and V_V) during isothermal crystallization of EH064 at different crystallization temperatures. From DSC, the peak melting temperature (T_m^p) and final melting temperatures (T_m^f) of the sample are $T_m^p = 95°C$, $T_m^f = 103°C$ respectively

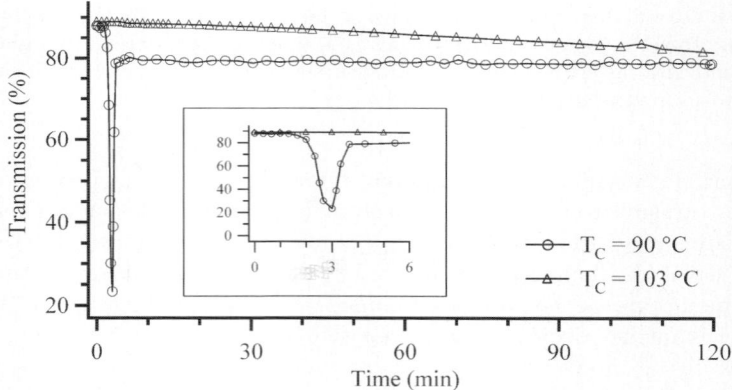

Fig. 7.6. Time evolution of percentage of transmitted light during isothermal crystallization of EH064

$$I = Aq^{-d} \qquad (7.1)$$

where d is the mass fractal dimension ($1 \leq d \leq 3$) of the scattering object, Or

$$I = Aq^{d_s - 6} \qquad (7.2)$$

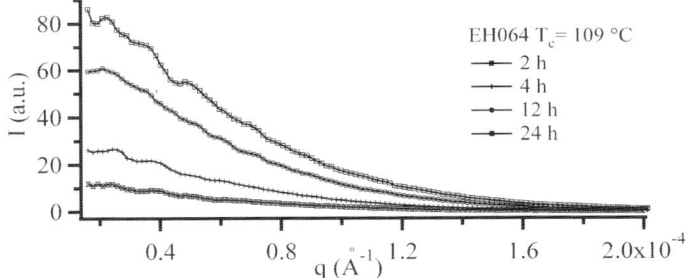

Fig. 7.7. H_V SALS scattering profiles ($I(q)$ vs q) for isothermal crystallization of EH064 at 109°

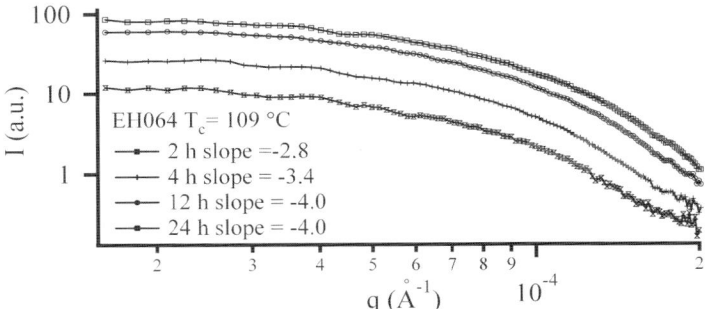

Fig. 7.8. *Log-log* plot of H_V SALS scattering profiles for isothermal crystallization of EH064 at 109°

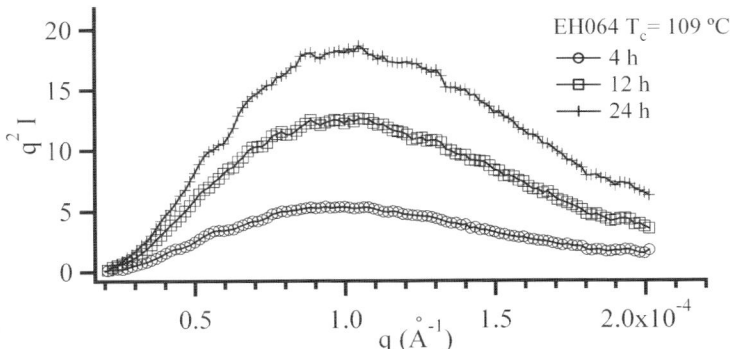

Fig. 7.9. Kratky plots ($I \cdot q^2$ vs q) of H_V SALS scattering for isothermal crystallization of EH064 at 109°

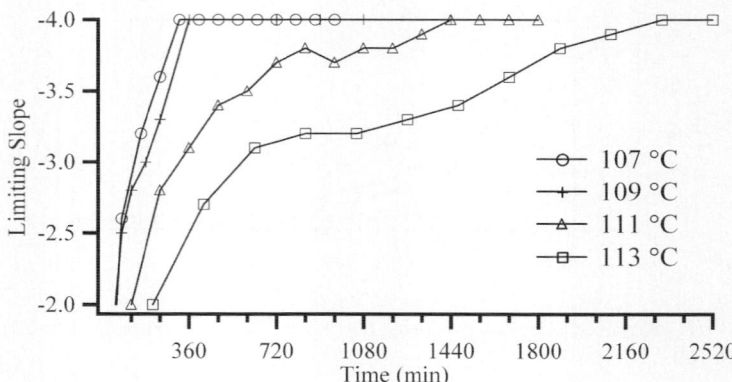

Fig. 7.10. Limiting slope from *log-log* plot of H_V SALS profiles in a function of time when crystallizing EH064 at various crystallization temperatures. Note that the limiting slope for a Gaussian chain (random polymer coil) is 2

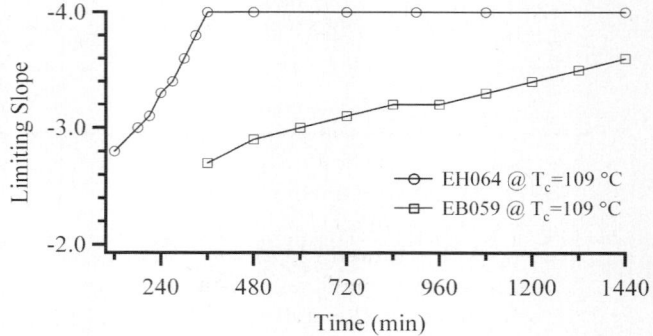

Fig. 7.11. Limiting slope from *log-log* plots of H_V SALS profiles in a function of time during the isothermal crystallization of polyethylenes: EH064 and EB059

where d_s is the surface fractal dimension ($2 \leq d_s \leq 3$). In the case of dense objects with smooth surfaces and sharp boundaries ($d_s = 2$), a slope of -4 in *log-log* plots of scattering profiles is obtained. In Fig. 7.10, the limiting slope from the *log-log* plots decreases from -2 to -4, suggesting that the interfaces of these anisotropic domains become sharper with time. Similar results are obtained at the temperatures above the final melting temperature as shown. It is evident that the detailed interface sharpening process depends on the crystallization temperature. This interface sharpening process is also observed for polyethylenes with other chemical structures (Fig. 7.11). Since the molar mass and density of EH064 and EB059 are essentially the same, the results indicate that early stage crystallization mechanisms for inclusion/exclusion of short chain branches depend on the length of branches.

7 How Do Orientation Fluctuations Evolve to Crystals?

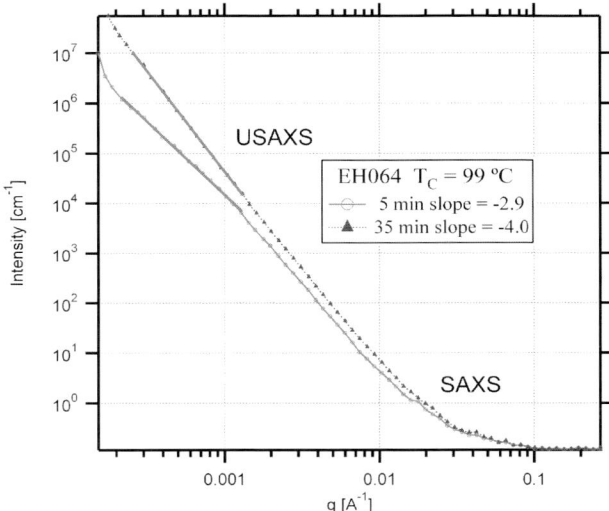

Fig. 7.12. Log-log plot of USAXS & SAXS scattering profiles for isothermal crystallization of EH064 at 99°

Our results show that anisotropic large-scale (size comparable to the wavelength of the light used, 633 nm) fractal domains are formed during early stage crystallization of the polyethylenes studied. However, it is difficult to ascertain whether there is any smaller structure present during early stage crystallization because of the limited sensitivity of simultaneous SAXS and WAXS. Figure 7.12 shows ultra-small-angle X-ray scattering (USAXS) measurements during isothermal crystallization of EH064 at 99°C. Scattering due to fractal objects on size larger than 100 nm is observed in USAXS region. The magnitude of limiting slope of *log-log* plots of USAXS profiles increase from 2.9 (5 min after reaching T_c) to 4.0 (35 min after reaching T_c) initially. This is similar to the observation from SALS measurement (Fig. 7.8). At the same time, no structure change is observed in traditional SAXS region. These results are consistent with the simultaneous SAXS and WAXS measurements for EH064 shown in Fig. 7.2. Owen et al. [55] performed time resolved USAXS measurements during primary crystallization, which showed a of 4 in USAXS region when the spherulites are small. As the spherulites grow larger, the fractal dimension decreases to a limiting value of 2.7. In our USAXS experiments, the magnitude of the limiting slope (fractal dimension) is 4.0 when the spherulites are formed (as indicated by SALS patterns obtain at 99°C in Fig. 7.2) and decreases as the spherulites grow.

Thus we can conclude that early stage crystallization is characterized by an increase in the magnitude of the limiting slope (from 2 to 4) while spherulite growth is characterized by a decrease (from 4 to 2.7). Our results demonstrate that small angle scattering (light and X-rays) can be used to probe the early stage mechanism as a function of crystallization temperature and the molecular structures of the polymers. Early stage crystallization behavior over a wide characteristic length from tens nm to several microns can be determined by combining light with X-rays.

Acknowledgments

The SALS instrument used for this work was developed with National Science Foundation support under NSF DMR-0108976. Simultaneous SAXS and WAXS experiments were carried out at National Synchrotron Light Source, Brookhaven National Laboratory, which is supported by Department of Energy, Division of Materials Sciences and Chemical Sciences. The authors thank NIST for support of the use of beamline X27C. USAXS data was measured at beamline 33-ID at the Advanced Photon Source, Argonne National Laboratory. Use of the Advanced Photon Source is supported by the U. S. Department of Energy, Office of Science, Office of Basic Energy Sciences, under Contract No. W-31-109-Eng-38. We would also thank Dr. D. J. Lohse (ExxonMobil) for providing the copolymers.

References

[1] Lauritzen, J. I.; Hoffman, J. D. *Journal of Research of the National Bureau of Standards Section a-Physics and Chemistry* **1960**, 64, 73-102.
[2] Hoffman, J. D.; Lauritzen, J. I. *Journal of Research of the National Bureau of Standards* **1961**, $A65$, 297.
[3] Mandelkern, L.; Fatou, J. G.; Howard, C. *Journal of Physical Chemistry* **1964**, 68, 3386.
[4] Muthukumar, M. *Philosophical Transactions of the Royal Society of London Series A-Mathematical Physical and Engineering Sciences* **2003**, 361, 539-554.
[5] Armitstead, K.; Goldbeck-Wood, G. *Adv. Polym. Sci.* **1992**, 100, 219-312.
[6] Point, J. J.; Rault, J.; Hoffman, J. D.; Kovacs, A. J.; Mandelkern, L.; Wunderlich, B.; Dimarzio, E. A.; Degennes, P. G.; Klein, J.; Ball, R. C.; Flory, P. J.; Yoon, D. Y.; Guttman, C. M.; Khoury, F.; Voigtmartin, I.; Bassett, D. C.; Frank, W. F. X.; Atkins, E. D. T.; Booth, C.; Uhlmann, D. R.; Grubb, D. T.; Magill, J. H.; Vesely, D.; Keller, A.; Krimm, S.; Samulski, E. T.; Calvert, P. D.; Fischer, E. W.; Stamm, M.; Geil, P. H.; Ullman, R.; Rys, F.; Rigby, D.; Stepto, R. F. T.; Windle, A. H.; Dill, K. A.; Hearle, J. W. S.; Hendra, P. J.; Ward, I. M.; Stejny, J.; Barham, P. J.; Pennings, A. J.; Posthumadeboer, A. *Faraday Discussions* **1979**, 365.
[7] Hoffman, J. D.; Miller, R. L. *Polymer* **1997**, 38, 3151-3212.
[8] Strobl, G. *European Physical Journal E* **2000**, 3, 165-183.
[9] Cheng, S. Z. D.; Li, C. Y.; Zhu, L. *European Physical Journal E* **2000**, 3, 195-197.
[10] Lotz, B. *European Physical Journal E* **2000**, 3, 185-194.

[11] Armistead, K.; Goldbeckwood, G.; Keller, A. *Advances in Polymer Science* **1992**, 100, 221-312.
[12] Keller, A.; Goldbeckwood, G.; Hikosaka, M. *Faraday Discussions* **1993**, 109-128.
[13] Olmsted, P. D.; Poon, W. C. K.; McLeish, T. C. B.; Terrill, N. J.; Ryan, A. J. *Physical Review Letters* **1998**, 81, 373-376.
[14] Ryan, A. J.; Fairclough, J. P. A.; Terrill, N. J.; Olmstead, P. D.; Poon, W. C. K. *Faraday Discussions* **1999**, 13-29.
[15] Imai, M.; Kaji, K.; Kanaya, T. *Physical Review Letters* **1993**, 71, 4162-4165.
[16] Imai, M.; Kaji, K.; Kanaya, T.; Sakai, Y. *Physical Review B* **1995**, 52, 12696-12704.
[17] Matsuba, G.; Kanaya, T.; Saito, M.; Kaji, K.; Nishida, K. *Physical Review E* **2000**, 62, R1497-R1500.
[18] Terrill, N. J.; Fairclough, P. A.; Towns-Andrews, E.; Komanschek, B. U.; Young, R. J.; Ryan, A. J. *Polymer* **1998**, 39, 2381-2385.
[19] Heeley, E. L.; Maidens, A. V.; Olmsted, P. D.; Bras, W.; Dolbnya, I. P.; Fairclough, J. P. A.; Terrill, N. J.; Ryan, A. J. *Macromolecules* **2003**, 36, 3656-3665.
[20] Heeley, E. L.; Poh, C. K.; Li, W.; Maidens, A.; Bras, W.; Dolbnya, I. P.; Gleeson, A. J.; Terrill, N. J.; Fairclough, J. P. A.; Olmsted, P. D.; Ristic, R. I.; Hounslow, M. J.; Ryan, A. J. *Faraday Discussions* **2003**, 122, 343-361.
[21] Ezquerra, T. A.; LopezCabarcos, E.; Hsiao, B. S.; BaltaCalleja, F. J. *Physical Review E* **1996**, 54, 989-992.
[22] Matsuba, G.; Kaji, K.; Nishida, K.; Kanaya, T.; Imai, M. *Macromolecules* **1999**, 32, 8932-8937.
[23] Matsuba, G.; Kaji, K.; Nishida, K.; Kanaya, T.; Imai, M. *Polymer Journal* **1999**, 31, 722-727.
[24] Katayama, K. *Sen'i Gakkaishi* **1967**, 23, S300-S308.
[25] Yeh, G. S. Y.; Geil, P. H. *J. Macromol. Sci., Part B.* **1967**, 1, 235-249.
[26] Schultz, J. M.; Lin, J. S.; Hendricks, R. W.; Petermann, J.; Gohil, R. M. *Journal of Polymer Science, Polymer Physics Edition* **1981**, 19, 609-620.
[27] Akpalu, Y. A.; Amis, E. J. *J Chem. Phys.* **1999**, 111, 8686-8695.
[28] Akpalu, Y. A.; Amis, E. J. *Journal of Chemical Physics* **2000**, 113, 392-403.
[29] Wang, Z. G.; Hsiao, B. S.; Sirota, E. B.; Agarwal, P.; Srinivas, S. *Macromolecules* **2000**, 33, 978-989.
[30] Raman, C. V. *Indian Journal of Physics* **1928**, 2, 387.
[31] Powers, J.; Keedy, D. A.; Stein, R. S. *Journal of Chemical Physics* **1961**, 35, 376.
[32] Hsiao, B. S.; Gardner, K. H.; Wu, D. Q.; Chu, B. *Polymer* **1993**, 34, 3986-3995.
[33] Hsiao, B. S.; Gardner, K. H.; Wu, D. Q.; Chu, B. *Polymer* **1993**, 34, 3996-4003.
[34] Long, G. G.; Allen, A. J.; Ilavsky, J.; Jemianl, P. R.; Zschack, P. *AIP Conference Proceedings* **2000**, 521, 183-187.
[35] Long, G. G.; Jemian, P. R.; Weertman, J. R.; Black, D. R.; Burdette, H. E.; Spal, R. *Journal of Applied Crystallography* **1991**, 24, 30-37.
[36] Lake, J. A. *Acta Crystallographica* **1967**, 23, 191-194.
[37] Akpalu, Y. A.; Lin, Y. Y. *Journal of Polymer Science Part B-Polymer Physics* **2002**, 40, 2714-2727.
[38] Bark, M.; Zachmann, H. G.; Alamo, R.; Mandelkern, L. *Makromolekulare Chemie-Macromolecular Chemistry and Physics* **1992**, 193, 2363-2377.
[39] Gornick, F.; Hoffman, J. D. *Industrial and Engineering Chemistry* **1965**, 57, 16.

[40] Gornick, F.; Hoffman, J. D. *Industrial and Engineering Chemistry* **1966**, 58, 41.
[41] Hoffman, J. D. *Journal of Physics and Chemistry of Solids* **1967**, *S*, 427.
[42] Juhasz, P.; Varga, J.; Belina, K.; Marand, H. *Journal of Thermal Analysis and Calorimetry* **2002**, 69, 561-574.
[43] Akpalu, Y. A.; Li, Y. *PMSE Preprints, American Chemical Society, Division of Polymeric Materials: Science and Engineering* **2004**, 91, 183.
[44] Schultz, J. M.; Miles, M. J. *Journal of Polymer Science Part B-Polymer Physics* **1998**, 36, 2311-2325.
[45] Xu, J.; Guo, B.-H.; Zhang, Z.-M.; Zhou, J.-J.; Jiang, Y.; Yan, S.; Li, L.; Wu, Q.; Chen, G.-Q.; Schultz, J. M. *Macromolecules* **2004**, 37, 4118-4123.
[46] Hobbs, J. K.; Vasilev, C.; Humphris, A. D. L. *Polymer* **2005**, 46, 10226-10236.
[47] Crist, B.; Claudio, E. S. *Macromolecules* **1999**, 32, 8945-8951.
[48] Crist, B.; Howard, P. R. *Macromolecules* **1999**, 32, 3057-3067.
[49] Li, Y.; Akpalu, Y. A. *Macromolecules* **2004**, 37, 7265-7277.
[50] Akpalu, Y.; Kielhorn, L.; Hsiao, B. S.; Stein, R. S.; Russell, T. P.; van Egmond, J.; Muthukumar, M. *Macromolecules* **1999**, 32, 765-770.
[51] Kawai, T.; Strobl, G. *Macromolecules* **2004**, 37, 2249-2255.
[52] Porod, G. *Small-Angle X-ray Scattering*; Academic Press: New York, 1982.
[53] Martin, J. E.; Hurd, A. J. *Journal of Applied Crystallography* **1987**, 20, 61-78.
[54] Beaucage, G.; Rane, S.; Schaefer, D. W.; Long, G.; Fischer, D. *Journal of Polymer Science, Part B: Polymer Physics* **1999**, 37, 1105-1119.
[55] Owen, A.; Bergmann, A. *Polymer International* **2004**, 53, 12-14.

8

Role of Chain Entanglement Network on Formation of Flow-Induced Crystallization Precursor Structure

Benjamin S. Hsiao

Department of Chemistry, Stony Brook University, Stony Brook, NY 11794-3400, USA
bhsiao@notes.cc.sunysb.edu

Abstract. In this article, the role of chain entanglement on the formation of flow-induced crystallization precursor structure in polymer melts was discussed. In particular, recent experimental findings from in-situ rheo-X-ray studies and ex-situ microscopic examinations were described: (1) the shish arise from the stretched chain segments in the entanglement network of the high molecular weight species; (2) the kebabs arise from the crystallization of coiled chain segments following diffusion-control growth; (3) multiple shish was seen in the ultra-high molecular weight polyethylene (UHMWPE) precursor; (4) the shish-kebab reformation is directly related to the relaxation behavior of stretched chain segments confined in a topologically deformed entanglement network. Based on the above results and recent simulation work from other laboratories, a modified molecular mechanism for the shish-kebab formation in entangled melt is presented.

8.1 Introduction

Flow-induced crystallization has long been an important subject in polymer processing because the final properties of any products are directly related to their crystallinity, structure and morphology. These characteristics can be manipulated by variation in processing parameters, such as deformation strain, deformation rate, temperature and pressure [1–8]. The general effects of the application of a flow field on the crystallization behavior, such as crystallization rate and crystallinity, have been reported quite extensively in the literature [3], and they can almost be quantitatively described by recent modeling and simulation capabilities [3,9–11]. In contrast, the topic of structure changes under flow (e.g. polymorphism, morphology and superstructure), which has also received a great deal of attention [2, 12–15], is less clear and cannot be predicted through contemporary theory or modeling [16]. The objective of this work is to provide a framework, based on current opinions and new experimental findings, for future advances in this topic, especially via simulation and modeling routes.

Recently, in-situ rheo-optical (e.g. small-angle X-ray scattering (SAXS), small-angle neutron scattering (SANS), and small-angle light scattering (SALS)) studies carried out in several laboratories, including ours, indicate indicate that the step application of the external flow fields (i.e., shear, elongation and mixed) would cause extension and orientation of only selected chains in entangled polymer melts, whereby the selection process of chain deformation can result in the formation of a crystallization precursor scaffold [17–32]. The topology of a flow-induced scaffold assimilates a polymer network structure, which can lead to gel-like rheological properties [33–36]. It is apparent that the subsequent crystallization behavior and morphological development are directly determined by the initially formed precursor structure. In flexible polymer melts without fillers, the crystallization precursor structure consists of shish-kebab entities, which will be the main focus of this work. In filled polymers, the precursor structure is composed of a network of interacting fillers and overlapping crystalline domains, which will be discussed elsewhere. The content of this article thus includes the current state of opinions on the nature of flow-induced crystallization precursor structures (i.e., shish-kebabs), the unique role of entanglement networks by high molecular weight polymer chains, and the need for to understand the relationship between micro-rheology and local phase transition. It is apparent that the in-depth knowledge of this subject would allow us to control the morphology and end properties of polymer products, and thus is extremely important from both scientific and technological points of view.

8.2 Current Opinions on Flow-Induced Crystallization Precursor Structures

The current opinions on flow-induced crystallization structures from entangled polymer melts can be summarized as follows. Upon the application of a step flow, the external perturbation would discriminate polymer chains with respect to their relaxation times. As deformed long chains take longer times to relax back to the undeformed state, and deformed short chains would recover almost immediately, the resulting chain topology after deformation can resemble a network structure [22], which is illustrated in Fig. 8.1. Diagram A represents the polymer melt before deformation; where all chains are in the 'random coil' state. Diagram B illustrates the melt structure immediately after deformation, where stretched and oriented long chains are singled out with parts of the segments undergoing phase transition (e.g. crystallization) and forming a linear assembly of primary nuclei (or shish). Diagram C represents the formation of folded chain crystalline lamellar structures (kebabs) on the primary nuclei. Such a flow-induced crystallization precursor structure

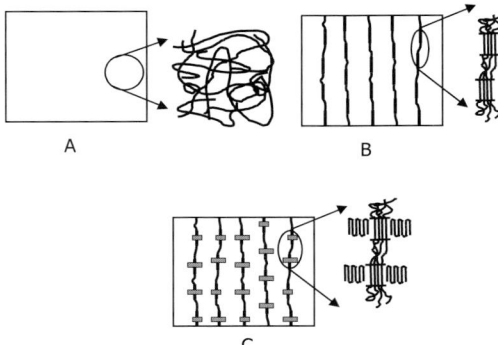

Fig. 8.1. Schematic representation of flow-induced precursor structures at different stages: (**A**) before shear, (**B**) formation of precursor structures containing linear nuclei (shish), (**C**) formation of shish-kebab morphology. (Reprinted with permission from [21])

can in fact be visualized by atomic force microscopy (AFM). Figure 8.2 illustrates a representative AFM image of a sheared bimodal polyethylene (PE) blend, containing 10 wt% UHMWPE and 90 wt% lower molecular weight PE copolymer matrix (with short hexene branches) of low crystallinity, at room temperature. The quenched sheared sample was etched in an acid solution (0.7 wt% permanganate in 2:1 conc. sulfuric acid and phosphoric acid) for 15 min and then microtomed at -140°C. The bright region represents the scaffold based on UHMWPE crystalline shish-kebab structures, which can dictate the subsequent crystallization and morphological development of the matrix if it is more crystallizable.

Based on single chain dynamics in dilute polymer solution, formation of the shish-kebab structure under extensional flow can be best explained by using de Gennes' concept of coil-stretch transition [37]. Keller argued that the coil-stretch transition should also exist in an entangled polymer melt [2,38,39] because the observed flow-induced morphologies in polymer melts were very similar to those observed in polymer dilute solutions. Keller further proposed that the final morphology can serve as a pointer to the pre-existing state of chain extension in flow and vice versa, whereby the shish-kebab morphology is the signature of the coil-stretch transition. Generally, in a polymer with a broad distribution of molecular weights, only the chains longer than the critical molecular weight (M^*) can remain in the stretched state after cessation of flow, while the chains shorter than M^* will relax back to the coiled state in a very short time (Fig. 8.3). The critical molecular weight is related to the elongation rate as $\dot{\varepsilon}_c \propto (M^*)^{-\beta}$. This simple but elegant relationship has recently been verified in polymer melt under shear flow by us using rheo-X-ray techniques [17,18]. However, in the above argument, Keller did not rationalize the obvious consequence of the high molecular weight chains, which would

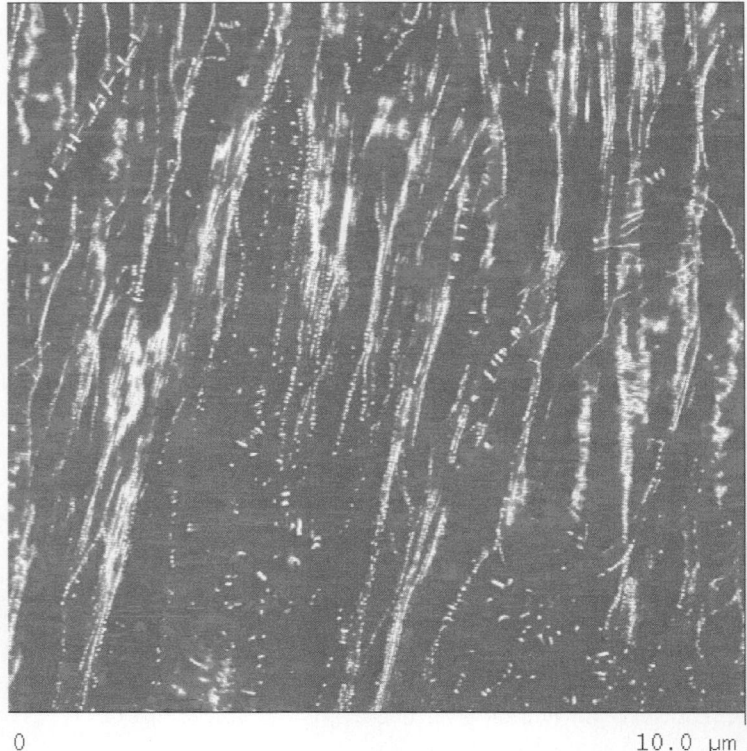

Fig. 8.2. The typical AFM image of a bimodal PE blend containing 10 wt% UHMWPE (bright region) and 90 wt% lower molecular weight PE copolymer matrix with low crystallinity

clearly induce a large number of chain entanglements. It is unlikely that the stretch-coil transition can take place at the level of individual chain because multiple steps of chain disentanglement cannot occur under the typical flow conditions.

Several research groups have tackled the subject of flow-induced shish-kebab formation using simulation approaches [40,41]. For example, Muthukumar et al showed that the shish-kebab structure can be formed by co-existing stretched and coiled chains in a monodispersed system, which simulates the condition in dilute solution under flow [40]. Of course, if one considers the conditions in polymer melts and the chain length of typical high molecular weight species, the use of short monodispersed chains does not reflect the reality. However, if one considers the scenario of chain entanglement, the simulation using monodispersed chains can make perfect sense because the average chain length between the entanglement points is statistically uniform. Furthermore, if the entangled chains in a supercooled state can be considered as

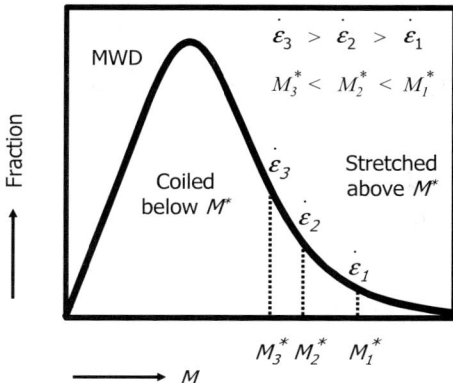

Fig. 8.3. Schematic plot of a molecular weight distribution showing criticality of coil-stretch transition. Increasing $\dot{\varepsilon}$ (elongation rate) increases the amount of stretched chain conformation. (Reprinted with permission from [2])

a network structure with slow dynamics, the flow would induce a deformation network with chain segments oriented and stretched between the entanglement points along the flow direction. Such a deformed network (so-called crystallization precursor scaffold) can obviously affect the subsequent crystallization and morphological development. Hu et al. [47] also demonstrated that even a stretched single aligned chain can act as a "template" for crystallization of kebabs through coiled chains.

In a recent review [6], we reported two interesting findings about the shish-kebab formation during flow-induced crystallization by rheo-X-ray studies. (1) In-situ rheo-X-ray (small-angle X-ray scattering (SAXS) and wide-angle X-ray diffraction (WAXD)) results indicated that, at the early stages of crystallization in sheared polymer melts (e.g. isotactic polypropylene (iPP), polyethylene (PE) and their blends), a scaffold of crystallization precursor structure is formed [17–27]. This scaffold contains shish with extended-chain conformation that can be in either the amorphous [23], mesomorphic [15, 22, 42, 43] or crystalline [2, 3, 44–47] state, and kebabs with folded-chain conformation that are only in the crystalline state. It appears that the main driving force for shish-kebab formation is due to the phase transition of stretched high molecular weight chains. (2) The high resolution scanning electron micrographs (TEM) and SAXS patterns of sheared samples containing 2 wt% of crystallizing ultra-high molecular weight polyethylene (UHMWPE) and 98 wt% of non-crystallizing PE copolymer matrix revealed that the high molecular weight polymer chains in the shish are neither fully stretched nor isolated [48]. Results indicated that the chains were entangled with the neighboring chains. Although the appearance of the stretch-coil transition in the melt seemed to be evidenced by the well-aligned straight shish structure, the region between

Fig. 8.4. SEM micrograph of UHMWPE shish kebab structure with multiple shish induced by shear flow (Reprinted with permission from [45])

kebabs did not necessarily contain one single shish, but multiple short length shish that were clearly separated from each other (Fig. 8.4). The micrographs indicated that the coil-stretch transition occurred only in sections of chains between the kebabs, whereby the individual chains do not disentangle completely in the deformed but still entangled melt.

8.3 Role of High Molecular Weight Species in Flow-Induced Crystallization

The role of high molecular weight chains in flow-induced crystallization has long been of interest in both polymer crystallization and processing communities. For example, Keller et al showed that in polymer melts, the addition of one percent of high molecular weight chains could dramatically increase the number of shish kebab structures under elongation flow [49]. Janeschitz-Kriegl et al [50] reported that in short term shearing at a low degree of supercooling, the crystallization of iPP was highly dependent on the concentration of long chains. In the presence of long chains, shear-induced crystallization was more sensitive to shear treatment and the orientation of surface layers was much higher than the materials without them. Recently, Kornfield et al [30] investigated the blends of two iPP samples with different molecular weight, polydispersity and stereoregularity under a step-shear flow. Their results showed that both crystallization orientation and kinetics increased with increasing long chain content. Several other groups, using similar techniques but under different flow conditions, also reported that the presence of long chains strongly affected the crystallization kinetics and the formation of oriented structures [50–53].

Over the past several years, our research group has also devoted a great deal of effort to investigate the role of high molecular weight chains on the formation of crystallization precursor structure under flow [17–27, 48]. Among our studies, two material systems have been particularly insightful, which are briefly described as follows. The first system involved the bimodal blend of 5 wt% of high molecular weight non-crystalline atactic polypropylene (aPP, M_w = 670,000 g/mol, polydispersity = 2.6) and 95 wt% of low molecular weight crystalline iPP matrix (M_w = 127.000 g/mol and polydispersity = 2.3). The rationale of this blend was as follows. The aPP chain segments cannot be incorporated in the iPP crystal. If the presence of long aPP chains, which do not crystallize, can assist the iPP crystallization under flow, the study would indicate that the iPP crystallization can be initiated from the precursor structure that possesses oriented and extended amorphous chains, which has been predicted by Hu et al. [41] Figures 8.5a shows 2D rheo-SAXS patterns of the pure iPP melt and iPP/aPP blend at 60 min after cessation of shear (shear rate = 60 s^{-1}, duration of shear, t_s = 5 s, T = 145°C). The SAXS patterns showed that the intensity of the meridional maxima arising from the oriented lamellae structure was significantly stronger in the case of blend compared to pure iPP. Note that in SAXS patterns of the iPP and iPP/aPP blends, meridional maxima were superimposed on a diffused scattering ring due to the unoriented crystals. In addition, no equatorial reflections were detected in the SAXS patterns, which

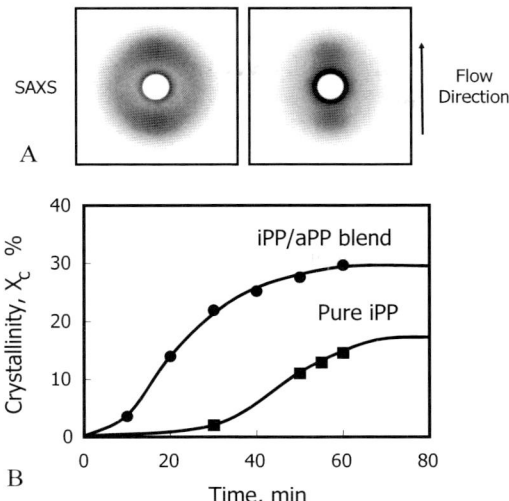

Fig. 8.5. (a) 2D SAXS patterns of the pure iPP and iPP/aPP (5 wt %) blend 60 min after cessation of shear (shear rate = 60 s^{-1}, t_s = 5 s, T = 145°C); (b) corresponding time evolution of the total percent crystallinity in the pure iPP and iPP/aPP blend at 145°C after shear (reprinted with permission from [22])

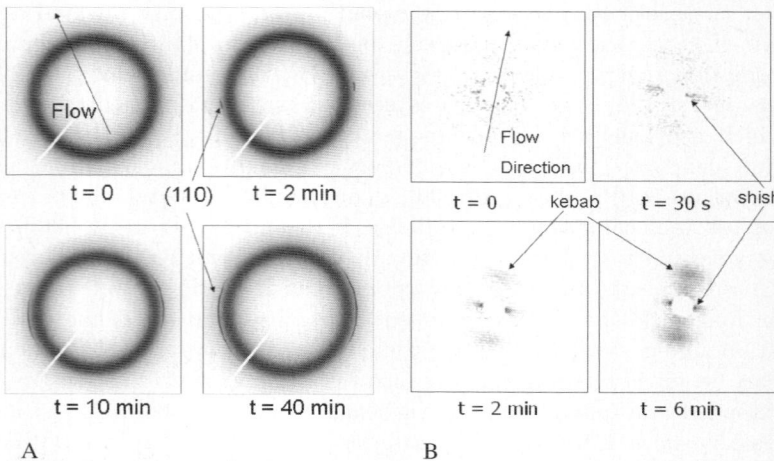

Fig. 8.6. (a) Selected 2D WAXD patterns (the (110) reflection peaks are filled in red to give a better contrast) of LMWPE/HMWPE blend before the application of shear and during isothermal crystallization at 126.5°C after cessation of shear (shear rate = 125 s^{-1}, shear duration = 20 s). (b) Selected 2D SAXS patterns of LMWPE/HMWPE blend before the application of shear and during isothermal crystallization at 126.5°C after the cessation of shear (reprinted with permission from [27])

indicated that the extended-chain shish were probably tiny and farther apart (if crystalline), or that they were non-crystalline with little or no electron density contrast against the surroundings. The latter would make more sense. The time evolution of the total crystalline iPP phase in pure iPP and the iPP/aPP blend at 145°C is shown in Fig. 8.5b. It was seen that for both systems, the volume fraction of the total crystallinity increased with time and subsequently reached a plateau value. The half-time of crystallization was determined as the time at which the relative crystallinity reached a value of 0.5. The estimated value of $t_{1/2}$ for the blend was 21 min compared to 45 min for pure iPP. We concluded that the iPP/aPP blend crystallized much faster than pure iPP. The results of these experiments clearly showed that the addition of higher molecular weight aPP (i.e., with long chain lengths and long relaxation times) affected the extent of both orientation and crystallization kinetics of the lower molecular weight iPP matrix. The degree of iPP crystal orientation was significantly higher (Fig. 8.5a) and its crystallization kinetics substantially improved (Figure 8.5b) in the iPP/aPP blend. These observations indirectly indicated the presence of a shear-induced precursor structure, containing long amorphous aPP chains as non-detectable shish and short iPP crystals as kebabs, which aided in secondary nucleation and thus the crystallization kinetics of the surrounding iPP matrix.

The second system involved the blending of a small amount of crystallizing high molecular weight component in a non-crystallizing lower molecular weight matrix [6, 25, 27]. In one experiment, we have carefully selected a fractionated high molecular polyethylene (HMWPE) (M_w = 1,500,000 g/mol and polydispersity of 1.1) as the scaffolding component and a low molecular polyethylene (LMWPE) copolymer (M_w = 53,000 g/mol and polydispersity of 2.2) as the matrix. As LMWPE is a random copolymer of ethylene (98 mol%) and hexane (2 mol%), it does not crystallize at the experimental temperatures (i.e. $\geq 124°C$). Figure 8.6a shows selective 2D WAXD patterns of LMWPE/HMWPE blend before and during shear-induced crystallization at different times after the cessation of shear (shear rate = 125 s^{-1}, shear duration = 20 s, temperature = 126.5°C). The first appearance of the crystal diffraction (i.e. the equatorial (110) reflection) in WAXD from an orientated structure was seen at 2 min after the cessation of shear. This pattern exhibited a pair of sharp equatorial (110) reflections, which could be attributed to the shish formation with extended-chain crystals. At longer times (e.g., t = 10 min), the azimuthal breadth of the (110) reflection was found to broaden significantly; a closer inspection revealed that it consisted of two discrete peaks with the corresponding azimuthal distribution: (1) the initially formed peak with a narrow azimuthal distribution (point-like), and (2) the subsequently developed peak with a broad azimuthal distribution (arc-like). The evolution of the diffraction pattern suggested the sequential formation of the shish-kebab structure, i.e., shish formed first followed by the growth of kebabs. As no off-axis (110) reflection was observed, we concluded that the subsequently formed kebabs were not twisted because the twisted lamellae (kebabs) should produce four-arc off-axis (110) reflections. At the end of the crystallization (time = 60 min), the total crystallinity estimated from the diffraction profile reached about 1%. This finding supports our hypothesis that LMWPE remains as an amorphous melt under the experimental conditions and the observed crystallinity mainly comes from the HMWPE component. It is reasonable to extrapolate the situation at longer crystallization times, where the crystallinity would be higher but less than 2 % (the composition limit of the blend). Selective 2D SAXS patterns of the LMWPE/HMWPE blend before and after shear (shear rate = 125 s^{-1}, shear duration = 20 s, temperature = 126.5°C) are shown in Fig. 8.6b. The SAXS pattern of the sheared LMWPE/HMWPE blend exhibited a clear equatorial streak arising immediately after the cessation of shear (at t = 30 s). The appearance of the equatorial streak indicated the formation of shish (microfibrils), containing extended crystals formed from the bundles of stretched chain segments parallel to the flow direction. The SAXS results were consistent with the WAXD results. Soon after the shish formation, strong scattering maxima appeared on the meridian. The meridional maximum could be attributed to the kebab growth, resulted from folded-chain crystallization. The oriented scattering features from the shish-kebab structure became stronger with the increase in time, indicating the continuation of the kebab growth.

8.4 New Insights on the Molecular Mechanism of Shish-Kebab Formation

The rheo-SAXS study of the LMWPE/HMWPE blend has revealed several new insights into the molecular mechanism of shish-kebab formation at the initial stage of flow-induced crystallization. These insights can be summarized as follows.

8.4.1 Kebab Growth Follows the Diffusion-Controlled Like Process

The above HMWPE/LMWPE study clearly indicated that shish-kebab structure can be formed from narrowly distributed high molecular weight chains under flow [27], whereby the polydispersity is not a necessity in shish-kebab formation as reported in some other material systems [17–26]. Using an analytical shish-kebab model developed by us recently [54], the average diameter (D) of kebabs was determined from the meridional streak in 2D SAXS (Fig. 8.6b). The kebab growth rate ($G(t) = \mathrm{d}D(t)/\mathrm{d}t$) was found to be varied with time. This is quite different from the spherulite growth study of typical semi-crystalline polymers at the quiescent state, in which a constant lamellar growth rate is often observed if the mass diffusion is not the limiting barrier. The calculated kebab growth rate *versus* time on the double logarithmic scale is shown in Fig. 8.7, which exhibits a slope of –0.76, roughly following the relationship of $\log G \propto (-0.5) \log t$ for the diffusion-controlled growth process. This has also been reported by us in a different rheo-SAXS study of the LMWPE/UHMWPE blend [48] and is consistent with Hobbs' recent AFM observations of the kebab growth under shear in real time [55]. In the simulation study of the shish-kebab formation by Muthukumar et al [40], they argued that coiled chains can attach onto the shish (stretched chains) and

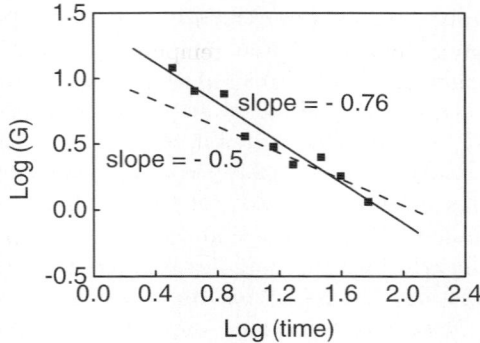

Fig. 8.7. The average growth rate G of the kebab diameter in LMWPE/HMWPE blend after shear. Results indicate the kebab growth follows the diffusion-controlled like process (reprinted with permission from [27])

form kebabs. Although they did not explicitly state that the kebab growth should follow the diffusion pathway, the implication that the kebab growth was significantly influenced by the rate of addition or diffusion of chains is clear.

8.4.2 Thermal Stability of Flow-Induced Shish-Kebab Scaffold

The shish-kebab precursor structure in a once-sheared LMWPE/UHMWPE blend sample (shear rate = 125 s^{-1}, shear duration = 20 s, temperature = 126.5°C) was treated with repeated melting and reformation processes under the confined quiescent state. The goal of this study was to understand the molecular mechanism responsible for the thermal stability of the shish-kebab structure in an entangled melt. Upon cooling, the shish-kebab structure was found to reform very quickly from un-relaxed stretched chain segments, but the corresponding fraction decreased with the increase in temperature. Results indicated that the shish-kebab reformation is directly related to the relaxation behavior of stretched chain segments confined in a topologically deformed entanglement network, which is discussed below.

In the first thermal cycle, the shear melt was heated from 126.5°C at 1°C/min in several steps to 133.5°C. At each intermediate step, the sample was held for 5 min to complete the melting process. After the final step of melting at 133.5°C, the sample was cooled down to 126.5°C at 30°C/min and held there for 60 min. Selected 2D SAXS patterns during this cycle are shown in Fig. 8.8. It was found that with the increase of temperature, the scattered intensity from the shish-kebab structure decreased accordingly due to melting. The cooling step revealed a very interesting observation, i.e., the melted shish-kebab structure reformed substantially, i.e., the overall scattering feature of the reformed SAXS pattern at 126.5°C was similar to that of the initial SAXS pattern (Fig. 8.8). After the first cycle, a sequential cycling thermal treatment was applied on the once sheared sample. In this treatment, the temperature was elevated at a 3°C/min rate to a temperature that was about 2–2.5°C higher than the highest temperature of the previous cycle. At the final temperature of each cycle (2nd and higher), the melt was always equilibrated for 3 min, and then subsequently cooled down at a 10°C/min rate to 124°C, under which the melt was allowed to crystallize for 10 min. The sample was subjected to sequential thermal cycles until the final temperature of 154°C. Figure 8.9 illustrates the final SAXS patterns collected at the end of re-crystallization in all thermal cycles (note that the crystallization temperature was 126.5°C for the first cycle and 124°C for all other cycles). In this figure, the equatorial streak was seen in the first two patterns (i.e., 133.5 and 136°C, respectively), but the later patterns only exhibited a meridional scattering feature. The scattered intensity decreased notably after each cycle. Since crystallization took place at the same temperature (124°C) for the same duration of time (10 min), the decrease in scattered intensity could be attributed to the decreasing number of primary nuclei (from the stretched

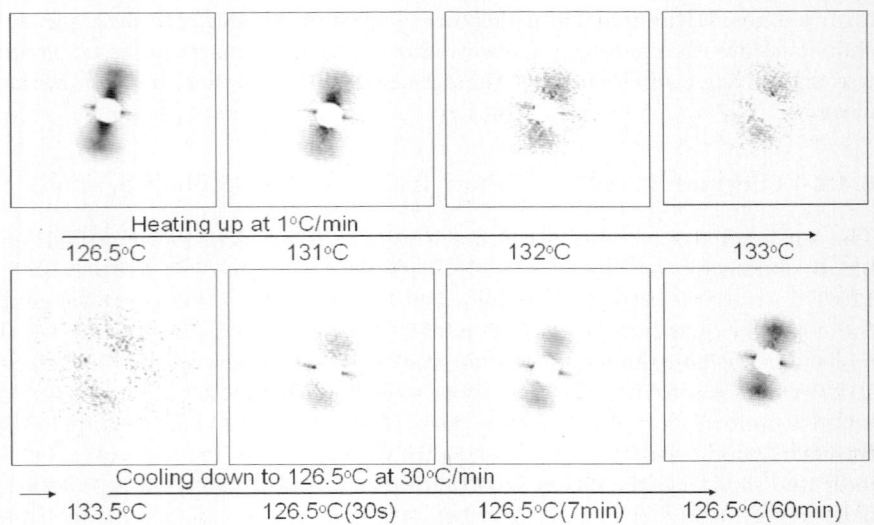

Fig. 8.8. Selected 2D SAXS patterns of the LMWPE/HMWPE blend during the first thermal cycle; the patterns shown below were taken after 5 min hold at the corresponding temperature (reprinted with permission from [27])

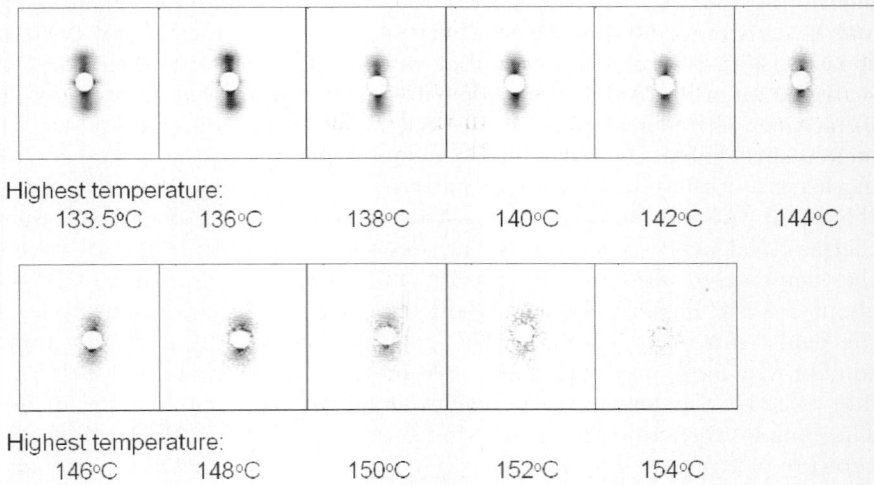

Fig. 8.9. Selected 2D SAXS patterns collected at the end of re-crystallization in each thermal cycle. The temperature shown below the pattern is the highest temperature during the heating stage in corresponding thermal cycle; the crystallization temperature is 126.5°C for the first cycle, and 124°C for others (reprinted with permission from [27])

chain segments) for creation of kebabs. The final SAXS pattern at the end of the eleventh thermal cycle did not show any sign of the shish-kebab structure. This cycle had the highest temperature of 154°C; thus, it appears that here the complete relaxation of the stretched chain segments (or the deformed entanglement network) took place.

If one considers the entangled melt as a network structure, containing entanglement points at dynamic equilibrium as physical crosslinks, then the flow field should generate two populations of chain segments, as illustrated in Fig. 8.10: (1) stretched segments oriented along the flow direction and confined by the entanglement points, and (2) un-oriented and -stretched segments (or coiled segments) perpendicular to the flow direction. The extent of the stretched segments in the entangled melt under flow is a function of both strain and strain rate, unlike the deformation of chemically crosslinked network material that is only a function of strain. Under the supercooled state, the stretched segments can rapidly crystallize into shish with extended-chain conformation and the coiled segments can crystallize into the kebabs with folded-chain conformation. In the molten state, the relaxation time scale for the stretched chain segments (shish) can be orders of magnitude higher than that of coiled segments. Upon cooling, the stretched chain segments would quickly re-crystallize into shish, which can subsequently nucleate microkebabs. Thus, the residual shish-kebab structure at the end of each cycle is a direct reflection of the state of the stretched and coiled chain segments upon heating under the confined planar constraints to the highest temperature of the cycle. In each thermal cycle with a higher temperature, the relaxation of the deformed entanglement network would increase, leading to reduction of the extent of stretched segments. In the last thermal cycle with the holding temperature of 154°C for 3 min, the shish and kebabs reformation did not

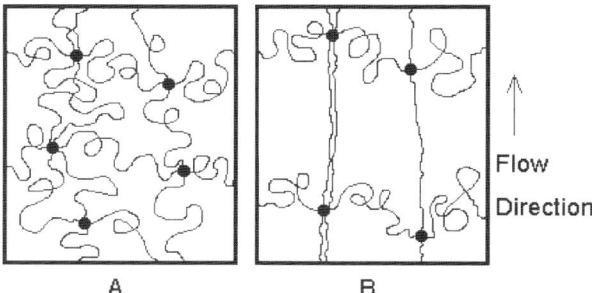

Fig. 8.10. Schematic representation of the entanglement network of HMWPE chains (**A**) under uniaxial deformation (LMWPE chains are not shown here). Upon shearing, some chain segments between entanglements (shown as *round dots*) are stretched along the flow direction, but most segments remained in the coiled state (**B**). The stretched segments form the precursors for the shish formation, and the coiled segments can grow into kebabs (reprinted with permission from [27])

take place, indicating the complete relaxation of the deformed entanglement network.

8.5 Relationship between Micro-rheology and Precursor Morphology

Generally, the morphological development via crystallization is a localized event, depending on the local conformation and nearby topology of crystallizing chains, whereas the rheological behavior is often dealt with as the bulk event using the mean field approach. As most polymer systems are polydispersed, containing a distribution of chain length, the conventional rheological approach of treating the whole assembly of chains will not be suitable for correlation with the morphological development.

Based on the above results, it is clear that the flow-induced crystallization precursor scaffold is mainly composed of long chains with high entanglement density, in which some segments between the entanglement points are stretched, oriented and can undergo a first-order phase transition. The general scaling rule for a single polymer chain that can partake in the formation of such a scaffold under flow (e.g. shear) can be considered as follows. There are several characteristic time scales that determine the state of orientation and extension for chain segments between the entanglement points in the polymer during and after cessation of shear. These time scales include: (1) the imposed shear time scale ($\tau_{\exp}$), which is inversely proportional to the shear rate ($\tau_{\exp} \propto \dot{\gamma}^{-1}$), (2) the Rouse time ($\tau_R$), which is related to the relaxation of the chain segments between the entanglement points, and (3) the reptation time (τ_d), which is related to the relaxation of monodispersed chain [56,57]. In addition, there are other "breathing" modes; such as contour length fluctuation (CLF) and convective constraint release (CCR), that can relax chain or segment orientation. At very low shear rates or $\tau_{\exp} \gg \tau_d$, reptation along with CLF dominate the relaxation process and almost no chain segments should remain oriented; here the stretched state is, obviously, not at all possible. At intermediate shear rates or $\tau_{\exp} \sim \tau_d$, the chain segments between the entanglements can be oriented but not stretched. At very high shear rates or when $\tau_R > \tau_{\exp}$, the chain segments between the entanglements can be both oriented and stretched. In the chosen shear conditions, the scenario of $\tau_R > \tau_{\exp}$ must be met to induce the shish entity, containing oriented and stretched chain sections between the entanglement points. It is apparent that the crystallization process can alter the state of the entanglement topology. That is, in the crystalline region the entanglement is largely eliminated; however, the entanglement density should increase in the amorphous region surrounding the lamellae.

Without question, to correlate the morphological development with rheological behavior, one must consider the micro-rheology of polymer chains, especially those with long relaxation times, that can first participate in the initial stage of flow-induced crystallization (i.e., the formation of crystallization precursor scaffold). The conventional mean-field rheological approach may not reveal the critical role of the local stretching and orientation of long chains dispersed in the matrix of short chains. Thus, the relationship between the micro-rheology and the corresponding micro-morphology of the precursor development under non-equilibrium conditions should be crucial to our understanding of how one can manipulate the initial scaffold in the early stages of flow-induced crystallization. This knowledge shall enable us to develop the capability to predict and to control the final morphology and, thus, final properties via varying processing conditions.

8.6 Concluding Remarks

Although many new physical insights into the subject of flow-induced crystallization, especially at the early stages of crystallization, have been recently obtained using model polymer blends and advanced rheo-optical techniques (e.g. synchrotron SAXS, SANS, SALS and Raman), the major advances in this area will probably take place by modeling and simulation, guided by theories with experimental support, in the future. However, before creditable applications of modeling and simulation to predict the morphology and properties, more experimental results and further theoretical developments are needed. The experimental contributions will mostly be made by using advanced in-situ or one-line techniques, such as synchrotron X-ray and neutron scattering and spectroscopic methods, as well as better model compounds that can simulate different materials systems. The theoretical efforts will probably involve the correlation of chain dynamics and phase transition in a local environment involving chains of long relaxation times and its application to the "heterogeneous" bulk state.

Acknowledgment

The authors gratefully thank the National Science Foundation (DMR-0405432) for the support of this work. The author also acknowledges the assistance of Drs. R. H. Somani, L. Yang, L. Zhu, I. Sics, C. Avila-Orta, C. Burger, S. Toki, X. H. Zong, Mr. J. K. Keum, and Mr. F. Zuo for invaluable experimental contributions.

References

[1] Ward, I. M. Structure and Properties of Oriented Polymers, Wiley, New York, 1975.

[2] Keller, H. W. H. Kolnaar in Processing of Polymers, edited by H. E. H. Meijer, VCH, New York, 1997, 18, 189.
[3] Eder, G., and Janeschitz-Kriegl, H. in Processing of Polymers, edited by H. E. H. Meijer VCH, New York, 1997, 18, 269.
[4] Wilkinson, A. N. and Ryan, A. J. (Eds.), Polymer Processing and Structure Development, Kluwer, Dodrecht, 1998.
[5] Kornfield, J. A., Kumaraswamy, G., Issaian, A. M. Industrial & Engineering Chemistry Research, 2002, 41(25), 6383.
[6] Somani, R. H., Yang, L., Zhu, L., Hsiao, B. S. Polymer, 2005, 46(20), 8587.
[7] Keller, A., Hikosaka, M., Rastogi, S., Toda, A., Barham, P. J., Goldbeck-Wood, G. Journal of Materials Science, 1994, 29(10), 2579.
[8] Bassett, D. C. Advances in Polymer Science (2005), 180 (Interphases and Mesophases in Polymer Crystallization I), 1.
[9] Ziabicki, A. Mathematics in Industry 2002, 2 (Mathematical Modelling for Polymer Processing), 59.
[10] Janeschitz-Kriegl, H., Ratajski, E., Wippel, H. Colloid and Polymer Science 1999, 277, 217.
[11] McHugh, A. J., Kohler, W. H., Shrikhande, P. Plastics, Rubber and Composites 2004, 33, 377.
[12] Rees, D. V., Bassett, D. C. Journal of Materials Science, 1971, 6(7), 1021-35.
[13] Wunderlich, Macromolecular Physics, Academic, New York 1973,Vol. 2.
[14] Haudin, J. M. and Monasse, B. NATO Science Series, Series E: Applied Sciences, 2000, 370 (Structure Development during Polymer Processing), 47.
[15] Li, L., de Jeu, W. H. Advances in Polymer Science, 2005, 181 (Interphases and Mesophases in Polymer Crystallization II), 75.
[16] McLeish, T., Olmsted, P., Hamley, I. Special Publication - Royal Society of Chemistry 2001, 263 (Emerging Themes in Polymer Science), 223.
[17] Somani, R. H., Hsiao, B. S., Nogales, A., Srinivas, S., Tsou, A. H., Sics, I., Balta-Calleja, F., Ezquerra, T. A. Macromolecules 2000, 33, 9385.
[18] Nogales, A., Hsiao, B. S., Somani, R. H., Srinivas, S., Tsou, A. H., Balta-Calleja, F., Ezquerra, T. A. Polymer 2000, 42, 5247.
[19] Somani, R. H., Hsiao, B. S., Nogales, A., Srinivas, S., Tsou, A. H., Balta-Calleja, F., Ezquerra, T. A. Macromolecules 2001, 34, 5902.
[20] Somani, R. H., Yang, L., Hsiao, B. S. Physica A 2002, 304, 145.
[21] Somani, R. H., Yang, L., Sics, I., Hsiao, B. S., Pogodina, N. V., Winter, H. H., Agarwal, P., Fruitwala, H., Tsou, A. Macromol. Symp. 2002, 185, 105.
[22] Somani, R. H., Yang, L., Hsiao, B. S., Agarwal, P., Fruitwala, H., Tsou, A. H. Macromolecules 2002, 35, 9096.
[23] Somani, R. H., Yang, L., Hsiao, B. S., Fruitwala, H. J. Macromol. Sci. Part B-Phys., 2003, B42, 515.
[24] Agarwal, P. K., Somani, R. H., Weng, W., Mehta, A., Yang, L., Ran, S., Liu, L., Hsiao, B. S. Macromolecules 2003, 36, 5226.
[25] Yang, L., Somani, R. H., Sics, I., Hsiao, B. S., Kolb, R., Fruitwala, H., Ong, C. Macromolecules 2004, 37(13), 4845.
[26] Somani, R. H., Yang L., Hsiao, B. S., Sun, T., Pogodina, N. V., Lustiger A. Macromolecules, 2005, 38(4), 1244.
[27] Zuo, F., Keum, J. K., Yang, L., Somani, R. H., Hsiao, B. S. Macromolecules, 2006, 39(6), 2209.
[28] Kumaraswamy, R. K. Varma, A. M. Issaian; J. A. Kornfield, F. Yeh, B. S. Hsiao, Polymer 2000, 41(25), 8931.

[29] Kumaraswamy, J. A. Kornfield, F. Yeh, B. S. Hsiao, Macromolecules 2002, 35 1762.
[30] Seki, D. W. Thurman, J. P. Oberhauser, J. A. Kornfield, Macromolecules 2002, 35, 2583.
[31] Fukushima, H., Ogino, Y., Matsuba, G., Nishida, K., Kanaya, T. Polymer 2005, 46(6), 1878.
[32] Sharma, Lakshmi; Ogino, Y., Kanaya, T. Macromolecular Materials and Engineering, 2004, 289(12), 1059.
[33] Pogodina, V., Siddiquee, S. K., VanEgmond, S. J. W., Winter, H. H. Macromolecules 1999, 32, 1167.
[34] Pogodina; N. V. and Winter, H. H. Macromolecules 1998, 31, 8164.
[35] Pogodina, N. V., Lavrenko, V. P., Winter, H. H., Srinivas, S. Polymer 2001, 42, 9031.
[36] Pogodina, N. V., Winter, H. H., Srinivas S. J. Polym. Sci., Polym. Phys. 1999, 37(24), 3512.
[37] de Gennes, P. G., J. Chem. Phys., 1974, 60, 5030.
[38] Pope, D. P. and Keller, A. Colloid Polym. Sci., 1978, 256, 751.
[39] Miles, M. J. and Keller, A. Polymer, 1980, 21, 1295.
[40] Dukovski, I. and Muthukumar, M. J. Chem. Phys., 2003, 118, 6648.
[41] Hu, W., Frenkel, D., Mathot, V. B. F. Macromolecules 2002, 35, 7172.
[42] Li, L. and de Jeu, W. H. Macromolecules 2003;36:4862.
[43] Li, L. and de Jeu, W. H. Phys. Rev. Lett. 2004;92:075506.
[44] Pennings, J., van der Mark, J. M. A. A., Kiel, A. M. Kolloid Z. Z. Polym. 1970, 237, 336.
[45] Mackley, M. R. and Keller, A. Polymer, 1973, 14, 16.
[46] Pope, P. and Keller, A. Colloid Polym. Sci., 1978, 256, 751.
[47] Miles, M. J. and Keller, A. Polymer, 1980, 21, 1295.
[48] Hsiao, B. S., Yang, L., Somani, R. H., Avila-Orta, C. A., Zhu, L. Phys. Rev. Lett, 2005, 94(11) 117802,1.
[49] Bashir, Z., Odell, J. A., Keller, A. J. Mater. Sci., 1972, B6, 493
[50] Jerschow, P., Janeschitz-Kriegl, H. International Polymer Processing, 1997, 12, 72.
[51] Lagasse, R. R. and Maxwell, B. Polym. Eng. Sci., 1976, 16, 189.
[52] Duplay, C., Price, F., Stein, R. J. Polym. Sci. (Polym Symp.), 1978, 63, 77.
[53] Vleeshouwers, S. and Meijer, H. Rheol. Acta, 1996, 35, 391.
[54] Keum, J. K., Burger, C., Hsiao, B. S., Somani, R., Yang, L., Kolb, R., Chen, H., Lue, C. T. Prog. Colloid Polym. Sci., 2005, 130, 114.
[55] Hobbs, J. K., Humphris, A. D. L., Miles, M. J. Macromolecules, 2001, 34, 5508.
[56] Milner, S. T. and McLeish, T. C. B. Phys. Rev. Lett., 1998, 31, 725.
[57] Bent, J., Hutchings, L. R., Richards, R. W., Gough, T., Spares, R., Coates, P. D., Grillo, I., Harlen, O. G., Read, J. D., Garham, R. S., Likhtman, A. E., Groves, D. J., Nicholson, T. M., McLeish, T. C. B. Science, 2003, 301, 1692.

9

Full Dissolution and Crystallization of Polyamide 6 and Polyamide 4.6 in Water and Ethanol

Marjoleine G.M. Wevers[1,2,3], Vincent B.F. Mathot[2,3*], Thijs F.J. Pijpers[2], Bart Goderis[2], and Gabriel Groeninckx[2]

[1] DSM Research, P.O. Box 18, 6160 MD Geleen, The Netherlands
[2] Laboratory of Macromolecular Structural Chemistry, Division of Molecular and Nanomaterials, Department of Chemistry, Katholieke Universiteit Leuven, Celestijnenlaan 200F, 3001 Heverlee, Belgium
[3] SciTe, Ridder Vosstraat 6, 6162 AX Geleen, The Netherlands, www.scite.nl
vincent.mathot@scite.nl
* corresponding author

Abstract. The full dissolution and crystallization of PA6 in water and PA4.6 in water and ethanol under pressure are described. Dissolution of PA6 in water is very fast and effective: it is completed during heating at 5°C/min in a DSC without stirring. It drastically lowers subsequent crystallization and melting temperatures. The maximum depression of the crystallization and melting temperatures is approximately 60°C. This temperature depression of the transitions is independent of concentration over a large range (10–70 m% PA6 in water). Dissolving PA6 in water during a DSC cycle causes a moderate shift of the molar mass distribution to lower values. The DSC based crystallinities at 110°C for PA6-water are fairly independent of concentration but higher values are obtained compared to pure PA6. From WAXD-measurements and crystal structure calculations in the case of PA6 it is concluded that α-type crystallites grow from the melt as well as from water based solutions. Furthermore, water does not enter the crystallites. PA4.6 in water and ethanol shows a similar behavior as PA6 but the transition temperature depressions are larger and the plateau in the temperature – m% plot is narrower for both solvents.

9.1 Introduction

It is well known that polyamides can contain a considerable amount of water because of the amide functionality [1, 2]. The absorption of water influences several properties. It lowers the glass transition temperature (T_g) [3–10]: the higher the level of water absorption, the larger the drop of T_g [11]. A water absorption of 12.4 m% (mass percentage) in Polyamide 4.6 (PA4.6) results in a drop of T_g from 80 down to -40°C [8, 12] and a water absorption of 11 m% in Polyamide 6 (PA6) results in a drop of T_g from 50 down to –30°C [8]. The uptake of water also affects the stiffness, yield strength and ductility [8, 9, 13]. It results in a dimensional change [14] and influences electrical properties [13]. Such changes are usually attributed to the plasticizing effect of water [10, 15, 16]. High moisture levels also can lead to degradation, foaming and blistering [16, 17].

It has now been firmly established that the absorbed water molecules penetrate only into the amorphous regions of the polyamide thereby loosening the hydrogen bonds [3–7, 9, 10, 18–26] between the chains. Also, there have been suggestions that water can affect the crystallites as well [27–29].

A chemical equilibrium between condensation and chain scission is established when a polymer melt receives a given amount of water. The actual water content after equilibration depends on the temperature, pressure, time and the types of end groups [8, 16, 30]. If more H_2O is added to the equilibrated mixture, a new equilibrium sets in with a more pronounced degradation through hydrolytic chain scission with reduction of the molar mass and consequent deterioration of the final properties [30]). The reverse direction is taken (i.e. polycondensation) if H_2O is removed from the equilibrium melt [16].

Puffr and Šebenda developed the basic ideas about the mechanisms of water sorption in polyamides [31]. They deduced a two-step model for water sorption. In PA6 three molecules of water are bound on two neighboring amide groups in an accessible region. The first water molecule forms a double hydrogen bond between the carbonyl groups. Therefore, this water may be assessed as firmly (or tightly) bound water whose activity is low. Practically, it is nearly impossible to remove it completely. The second and third water molecules join the already existing H-bonds from the NH-groups to other CO-groups with a negligible thermal effect. Consequently, this water is denoted as loosely (or weakly) bond. Starkweather [32] later on extended this two-step model by introducing clustering: additional water molecules attach to firmly or loosely bound water molecules and form structures called "clusters". A possible alternative to clustering is the existence of freezable unbound water in cavities of several tenths of angströms. However, it was pointed out by Chatzi et al. [33] that usually bulk polymers do not have the required cavities for the full development of the cage-like structures of liquid water [1].

With respect to crystallization, an increasing crystallinity is found by annealing using boiling water [5, 10] or using steam [23, 24, 34–36]. The reason could be that water leads through loosening of the hydrogen bonds to a higher mobility of the molecules and segments thereof influencing crystallization. Because obviously crystallization and melting in the presence of *real* water has not been studied in the past – only Murthy et al. found a melting point depression of 20°C after annealing in water at 120 or 140°C [5] – the influence of water on possible transitions and morphologies of polyamides is still quite unclear.

Because of all these mentioned undesired phenomena, polyamides are typically dried by producers at e.g. 105°C at vacuum to remove the water completely before delivery to customers.

Instead of studying the uptake of water to the (limited) amounts mentioned, we performed research with respect *to fully dissolving polyamides in water*. To this end, we used a concept developed at DSM Research in the early nineties [37, 38], (and followed recently [39]): to dissolve polyamides in water and other solvents under pressure. This dissolution decreases the melting- and crystallization temperatures drastically, as will be shown further on. In this way it is attempted to turn the negative aspect of water uptake into a positive one by using water as an environment-friendly solvent under pressure to realize ways to operate at much lower temperatures than normally is used. Operating at much lower temperatures could be an advantage for processing of polyamides: the processing temperature range is shortened by which energy could be saved. Also foaming with water in an environmentally friendly way could become feasible. Another interesting application could be the use of additives with a lower thermal stability.

The scientific objective of the present research endeavor is the understanding of dissolution and crystallization of polyamides in water and other solvents. In this paper we report on results obtained by SEC, DSC and WAXD measurements.

9.2 Experimental Section

9.2.1 Materials

The polymers used in this study are PA6 granulate and PA4.6 granulate, as produced by DSM, see Table 9.1. The PA6 and PA4.6 granules were chopped into pieces before measurement. Demineralized water and ethanol were used in the experiments.

Table 9.1. Properties of the polyamides used in the experiments

Property	PA6	PA46
Density at 23°C (kg/m^3)	1130	1180
M_w (kg/mol)	31	40
M_w/M_n	3.8	2
Melting peak temperature (°C) at 5°C/min	222	290
Crystallization peak temperature (°C) at 5°C/min	173	265

9.2.2 Preparation and Characterization of the Samples

Measuring Under Vapor Pressure

It is well known that the vapor pressure of water and ethanol increases on heating. At 100°C at 1 atm, water boils and evaporates and at 78.4°C at 1 atm, the same happens for ethanol [40]. To be able to perform measurements at temperatures much higher than the boiling points, cells are needed withstanding vapor pressures up to 27.9 bar at 230°C for water [41], and up to 63.9 bar at 244°C for ethanol.

DSC Measurements

A PerkinElmer DSC-7 was used to measure crystallization and melting temperatures and heats of crystallization of the samples, and also to prepare samples for measurements with other techniques. The measurements were performed at 5°C/min in high-pressure pans, allowing having liquid water above 100°C by preventing water evaporation during the DSC measurements. Thus, high-pressure pans enabled the study of the dissolution of the polyamides in water under their own vapor pressure. Very pure nitrogen was used as purge gas and the pans were closed under gaseous nitrogen to avoid any oxidation of the samples during the measurements. For the polyamides used, the chopped granules were mixed with water at room temperature in the DSC-pans and closed at room temperature. For PA6 up to 50 m% in water, the sample has been heated first from 50°C to T_{melt}= 200°C, followed by isothermal stabilization for 15 min; cooling to 50°C; isothermal stabilization for 5 min and finally by heating to 200°C, all scans at a rate of 5°C/min. Above 50 m% PA6 in water, to dissolve the polymer and also to avoid self-nucleation by the polymer, a T_{melt} of 30°C above the dissolving temperature was taken. The isothermal stabilization step of 15 min was chosen throughout on the basis of experiments varying the stabilization times. After 15 min stabilization time – in contrast to shorter times – a single crystallization peak was observed instead of a double crystallization peak. Longer times for stabilization gave no further changes. For PA4.6, the procedure was similar, except for the temperature, which was raised to 230°C for water and to 290°C for ethanol. All DSC curves have been normalized to the mass of the polymer. The small upward curvature of the DSC curve at increasing temperatures, most probably caused by the evaporation of water at constant volume at its own vapor pressure, has not been corrected for.

The crystallinity was determined following the method described by Mathot [42], which takes into account the temperature dependence of the enthalpy of crystallization [43]:

$$W^c(T) = \frac{[A_1 - A_2]_T}{\Delta h(T)} \cdot 100 \tag{9.1}$$

where $W^c(T)$ is the percentage crystallinity, $\Delta h(T)$ the enthalpy difference between fully amorphous and fully crystalline polymer as function of temperature, and $[A_1 - A_2]_T$ the transition enthalpy. As an introductory approach, the $\Delta h(T)$ values used in the calculations are taken from the ATHAS databank [44]. Accordingly, the heat involved in the dissolution of the crystals is neglected and – as the reference melting-dissolution enthalpy is for sure too large – too small crystallinities are obtained. The values for PA6 polymer were used. The crystallinity was calculated at 110°C, at which temperature crystallization is virtually complete.

X-ray Measurements

Wide-Angle X-ray Diffraction (WAXD) measurements were performed at the European Synchrotron Radiation Facility station BM26B,(Grenoble, France). WAXD data were collected on a curved microstrip gas chamber detector, covering a scattering angle range from 8° to 34°. The angular calibration of the detector was performed with a silicon/CHOL mixture. The WAXD intensities were normalized to that of the primary beam as measured by an ionization chamber placed downstream from the sample.

The samples were encapsulated in specially made high-pressure aluminum cells. The cells were mounted in an in-house designed, temperature-controlled holder. The holder provides temperature control by which heating and cooling ramps (0°C to 250°C) as well as isothermal steps can be programmed. For the X-ray measurements, the same temperature program as for DSC has been taken. The measurements have been performed at 5°C/min scanning rate, during which consecutive diffraction patterns were collected over time spans of 12 seconds, yielding a temperature resolution of 1°C in the cooling and heating runs. Caffeine (T_m = 220°C), Benzoic acid (T_m = 120°C) and Benzophenone (T_m = 48.1°C) were used as temperature calibrants. The temperature was controlled within ±1°C. The X-ray beam energy was 13 keV (wavelength, λ = 0.954 Å). Such a high-energy beam was chosen as an optimal compromise between absorption due to the water in the sample and the aluminum of the cell on the one hand and scattering by the sample on the other. The intensities recorded at the WAXD detector are reported as if they were taken using λ = 1.54 Å, the wavelength of a Cu-source for comparison with literature data. In the present setup, intensities could only be measured at scattering angles, $2\theta_{\lambda=1.54Å}°$, above 13–14°.

The structure of PA6 has been investigated in detail and at least two crystal polymorphs, referred to as α and γ have been identified [45, 46]. The crystal packing of the α-polymorph consists of polymer sheets parallel to the (a,b)-plane. Within these sheets the PA6 chains align anti-parallel, and are held together by hydrogen bonds. In the γ-form the polymer chains form sheets parallel to the (b,c)-plane but the individual chains align parallel. A third modification of PA6, the β-form, has also been proposed [45, 47–49]. Other authors state that the various structures found are just α or γ-structures with various degrees of perfection and can be viewed as intermediate structures between the crystalline forms with respect to the H-bond setting and chain conformation [50–52].

The two crystalline phases can be identified by their distinct X-ray diffraction patterns. The unit cells of the α and γ phases are different, the unit cell for α is characterized by: a = 9.56; b = 17.24; c = 8.01; $\alpha = \gamma = 90°$; $\beta = 67.5°$ [45] and the unit cell for γ by: a = 9.33; b = 16.88; c = 4.78; $\alpha = \gamma = 90°$; $\beta = 121°$ [46]. Therefore, the two most intense reflections of the α and γ phases appear at slightly different angles and are at, respectively, 20° (200,α_1)/24°(002 + 202,α_2), and 22°(100,γ_1)/23° (201 + 200,γ_2). γ_1 and γ_2 show up as a single peak with a small shoulder [53]. The notation of Holmes et al. [45] is used. It has to be remarked that the data in the Cambridge Structural Database in Conquest version 1.7 (software for search and information retrieval) does not correspond totally with the original article. The notation of Malta et al. [54] – who re-examined the crystal structure of the α-phase – is not used, because they switched the b and c-axis, compared to Holmes et al. who used the 2nd convention (b-unique setting) [45].

Hydrolysis of PA6 by Water as Studied by SEC

For the study of hydrolysis of PA6 by water using Size Exclusion Chromatography (SEC) measurements, 50 m% PA6 and water were mixed at room temperature and subjected to a DSC cycle.

The temperature profile used was heating at 5°C/min from 50 to 210°C; isothermal stabilization for 15 minutes; cooling at 5°C/min from 210 to 50°C, followed by isothermal stabilization for 5 minutes. The samples were then dried in a vacuum oven at 90°C for 16 hours, and after that cooled and stored at room temperature in an excicator with drying agent (Engelhard KC Trockenperlen Orange). SEC has been performed with a HP 1090 Liquid Chromatograph (Hewlett Packard) coupled with an HP 1047A differential refractometer (measuring temperature 35°C); a Viscotek H502B viscometer (measuring temperature 38°C); Viscotek data manager DM400 and a multi angle laser light scattering detector (Wyatt Technology). Columns used: three PSS PFG linear XL, 8*300 mm (measuring temperature 35°C). The eluent is hexafluoroisopropanol with 0.1% kaliumtrifluoroacetate, flowrate: 0.4 ml/min.

9.3 Results and Discussion

9.3.1 Dissolution and Crystallization of PA6 in Water by DSC

In the first cooling curve at 5°C/min, see Fig. 9.1, a single peak ($T_{cr,peak}$ = 172°C) with a small high-temperature shoulder (at the start of crystallization) is visible for pure PA6. The onset of crystallization is at 184°C and the end of major crystallization is at approx. 166°C. When subsequently heated with

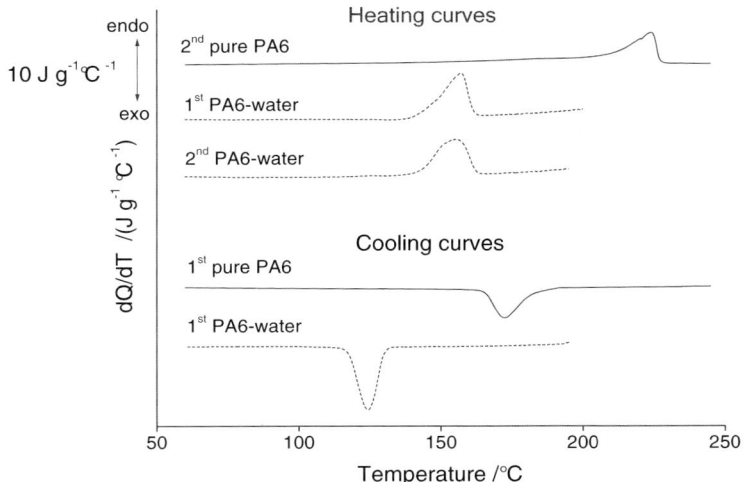

Fig. 9.1. DSC curves of 58 m% PA6 in water and pure PA6

5°C/min, pure PA6 starts to melt appreciably from approx. 211°C on and ends melting at 227°C. The DSC heating curve is single-peaked with $T_{m,peak} = 222°C$.

As a characteristic example for PA6-water systems the dissolution of 58 m% PA6 in water is described first. Under its own vapor pressure, this PA6 is dissolved in water at temperatures above 164°C, see Fig. 9.1. The fact that full dissolution is already seen in the first heating curve at much lower temperatures compared with pure PA6 means that the dissolution is very quick and effective, even without mixing.

Dissolution of the polyamide in water also drastically lowers subsequent crystallization and melting with respect to temperature, as is seen in Fig. 9.1. Such depressions of the transition temperatures for the PA6 in water at 58 m% turn out to occur in the whole concentration range, see Fig. 9.2. The maximum depression of the crystallization and melting temperatures is approximately 60°C (130°C and 165°C for PA6 in water compared to 184°C and 227°C for pure PA6 for onset crystallization and end melting respectively). This temperature depression of the transitions is independent of the concentration over a large concentration range (10–70 m% PA6 in water), as shown in Fig. 9.2, what could be an indication of demixing.

Furthermore, the DSC-curve of 23 m% PA6 in water, see Fig. 9.3, shows a double-peaked dissolution curve during second heating. This turns out to be characteristic for the low concentration range, in between 5 and 50 m% PA6 in water. The double-peaked shape most probably indicates recrystallization. As shown in Fig. 9.3, at 96 m% PA6 in water, there is just one crystallization and dissolution peak. This peak is broadened compared to pure PA6, possibly

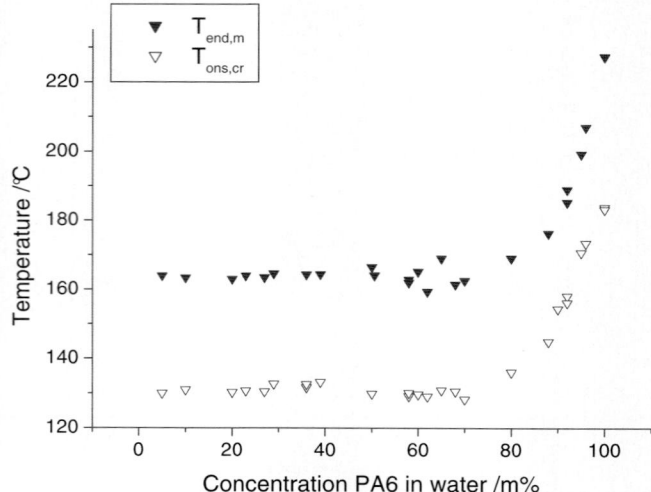

Fig. 9.2. Onset crystallization and end melting temperature as function of the concentration of PA6 in water/m%

indicating reorganization phenomena. This behavior is characteristic for the high concentration range, from 80 m% PA6 in water onwards.

9.3.2 Influence of Dissolution of PA6 on Molar Mass Distribution by SEC

As water hydrolyses the amide bond, it can be expected that dissolving PA6 in water could cause a decrease of molar mass. Because in case of a drastic lowering of the molar mass, this change would influence crystallization, morphology and dissolution / melting, the changes with respect to molar mass have to be checked. SEC measurements were performed to see the effect of dissolution in water on the molar mass distribution of PA6. Usually, when calculating molar mass distributions of PA6 for industrial purposes, the oligomer fraction present in PA6 is not taken into account. As the oligomer fraction would increase in case of decreasing molar mass by dissolving in water, in calculating molar mass parameters in the present study also the oligomer fraction has been taken into account.

Dissolving PA6 in water during a DSC cycle indeed causes a shift of the molar mass distribution to lower values, as shown in Fig. 9.4. So, shortening of the polyamide chains by scission takes place. However, after a cycle still a polyamide results with a fairly high molar mass. On the other hand, as is well known from industrial practice, also processing from the molten state shifts the molar mass distribution towards lower values. This processing-induced equilibrium molar mass may well be situated in between the molar mass associated with dissolution-induced scission and that of the native material. It

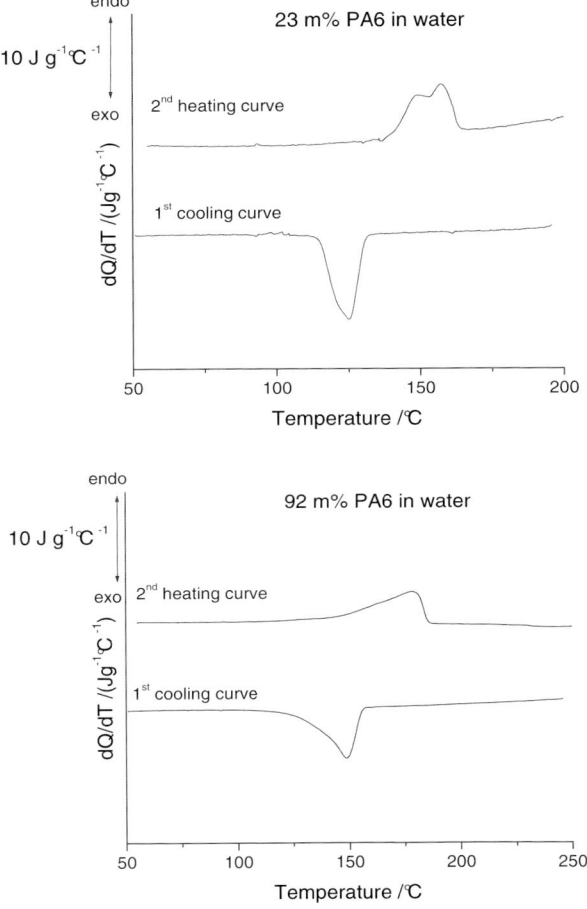

Fig. 9.3. DSC curves of 23 m% PA6 in water (**a**) and 96 m% PA6 in water (**b**)

can be noticed from Fig. 9.4 that after dissolution the molar mass distribution becomes narrower, compared to the native material, pointing at a preferential cutting of the longer chains (Table 9.2).

Branching information can be obtained via the Mark-Houwink relation:

$$[\eta]_{HFP}^{38°C} = K.M^a \qquad (9.2)$$

In a good solvent the slope 'a' reaches 0.7 for linear polymers. Lower values indicate worse solvent quality or possible long chain branching. The resulting Mark-Houwink plot gives no indication for long chain branching. Because of the hydrolysis during a DSC cycle in the presence of water, as revealed by SEC, all DSC and X-ray measurements where performed on fresh samples.

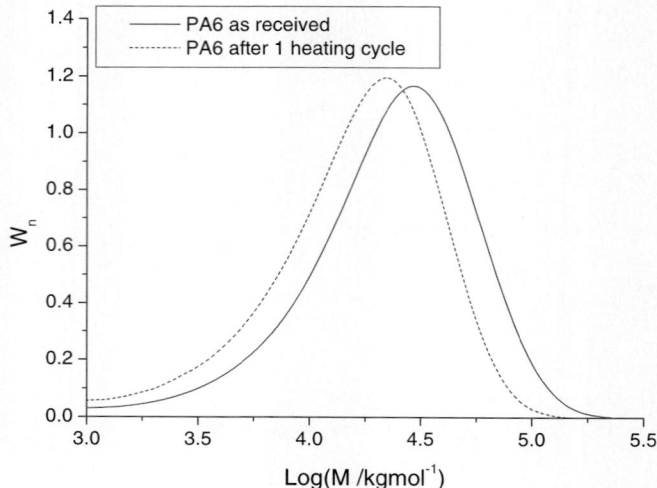

Fig. 9.4. Molar mass distribution oligomers included, of pure PA6 and 50 m% PA6 in water after 1 heating/cooling cycle with DSC after drying

Table 9.2. Characteristic values of the molar mass distribution for native, pure PA6 and 50 m% PA6 in water after 1 heating/cooling cycle by DSC and subsequent drying

Polyamide	M_n(kg/mol)	M_w(kg/mol)	M_w/M_n	Mark-Houwink slope 'a'
Pure PA6	8.0	31	3.8	0.666
PA6 after 1 heating / cooling cycle	7.6	22	2.8	0.676

9.3.3 Influence of Dissolution of PA6 on the Crystallinity Found by DSC

Temperature-dependent crystallinities by DSC have been calculated for the whole concentration range. The values are shown in Fig. 9.5 for crystallinities at 110°C, obtained from DSC cooling curves. At this temperature both PA6 and PA4.6 in water are found to be crystallized to their maximum extent, see Figs. 9.1 and 9.11 respectively. When the crystallinities for pure PA6 and PA6-water are compared, the crystallinities of the PA6-water systems are fairly independent of concentration. Furthermore clearly higher values result after crystallization in water (an average value of approx. 39 % for PA6-water systems compared to approx. 27 % for pure PA6). It has to be emphasized that the determination of the crystallinities from heating curves is quite difficult because of curvature of the DSC curves, see e.g. Fig. 9.3. This curvature is

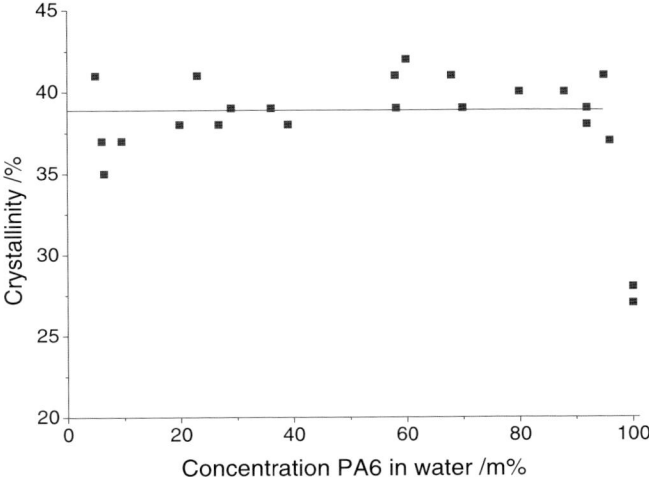

Fig. 9.5. Crystallinity from DSC-cooling at 110°C as function of the concentration PA6 in water

most probably due to the influence of evaporation of water at constant volume at its own vapor pressure at increasing temperatures [55] and is almost absent in cooling; which is why the crystallinities calculated from DSC cooling curves are judged to be the most trustable ones.

Obviously, crystallization in water leads to remarkable higher crystallinities (that are probably even higher than calculated here for reasons mentioned in the experimental section). Most probably this effect is caused by the weakening and breaking of hydrogen bonds by water uptake, leading to higher mobility of the polyamide chains and so making crystallization easier.

9.3.4 Dissolution and Crystallization of PA6 by WAXD

In Fig. 9.6, the WAXD-pattern of the 2nd heating curve of 49 m% PA6 in water is shown. At room temperature, the pattern displays two sharp peaks at $2\theta = 19.6°[\alpha_1(200)]$ and at $2\theta = 24.0°[\alpha_2(002)+(202)]$ on top of a halo due to water and non-crystalline PA6. These sharp peaks point to the α-structure [45, 56]. When the mixture is heated, the position of the (200) reflection remains identical while the position of the (002) + (202) peak shifts to lower angles, as illustrated in Fig. 9.7. This behavior is also seen in the pure PA6, and is typical for polyamides. At 165°C, all PA6 dissolves in water and accordingly the halo amplifies. When the polyamide-water mixture is cooled at 5°C/min, the two α-peaks start to appear at 130°C (Fig. 9.8). These temperatures are consistent with the melting and crystallization temperatures found from the DSC-measurements. This is the case for the whole concentration range.

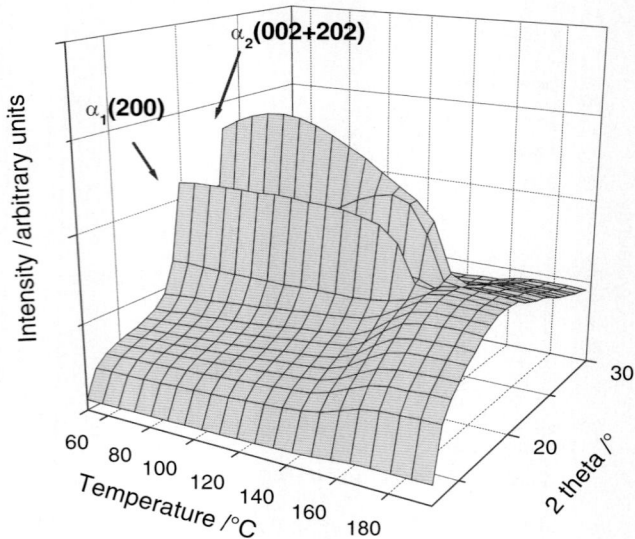

Fig. 9.6. WAXD patterns of the monoclinic α-crystal structure: intensity as function of 2θ and temperature for 49 m% PA6 in water, 2nd heating curve

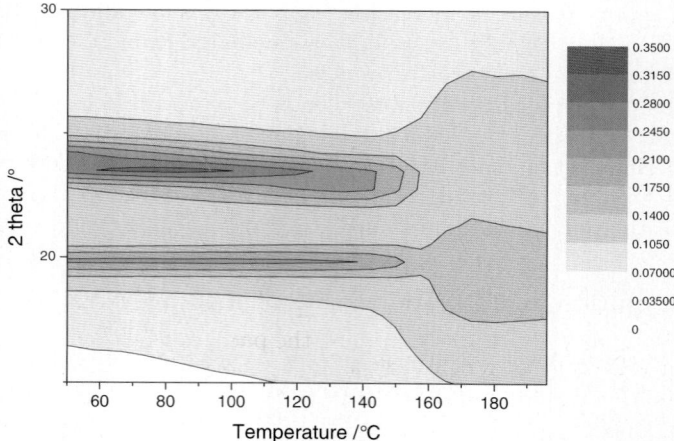

Fig. 9.7. WAXD contour plot of the monoclinic α-crystal structure: 2θ as function of temperature for 49 m% PA6 in water, 2nd heating curve

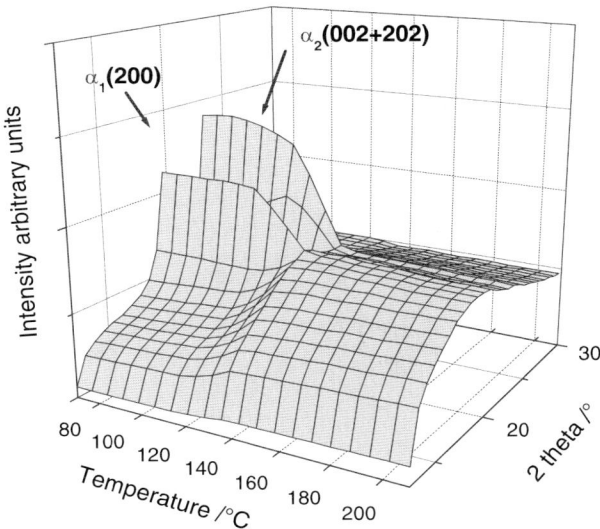

Fig. 9.8. WAXD patterns of the monoclinic crystal structure: intensity as function of 2θ and temperature for 49 m% PA6 in water, 1st cooling curve

Figure 9.9 illustrates that no other structure besides the α-polymorph is found – as for pure PA6 with peaks at $20°(\alpha_1)$ and $23.8°(\alpha_2)$ – when 48% PA6 in water is crystallized at the same cooling rate of $5°C/\min$. The $\alpha_1(200)$ peak, to a first approximation, represents the distance between hydrogen-bonded chains (constituting hydrogen-bonded sheets); and obviously does not depend on the water concentration, see Fig. 9.10. Though within the experimental error, the position of the peak has not changed, the α_1-peak of pure PA6 is clearly broader, see Fig. 9.9, compared to the same peak of PA6 crystals in the presence of water. This suggests that the size of the crystallite grains in a direction along the sheets is larger when water is present, or in other words that the sheets are larger. Alternatively, the packing of the chain segments in the crystallites could be better in the case of water.

The second peak $\alpha_2[(002) + (202)]$, corresponds in general terms to the distance between hydrogen-bonded sheets, and increases slightly from $23.7°$ to $24.1°$ when water is present, thereby excluding the possibility that water is present between the sheets. On the contrary, the conclusion must be that the presence of water outside the crystals results in a slightly *tighter* packing of the sheets. The interaction between the sheets is mainly of the van der Waals type. It seems that the hydrophobic aliphatic segments tend to come nearer when water is present, similar to the collapse of a polymeric chain in a bad solvent.

That the water is not included in the crystallites is confirmed by a calculation on perfect crystals using Platon [57]. In α-PA6 the chains are ordered in

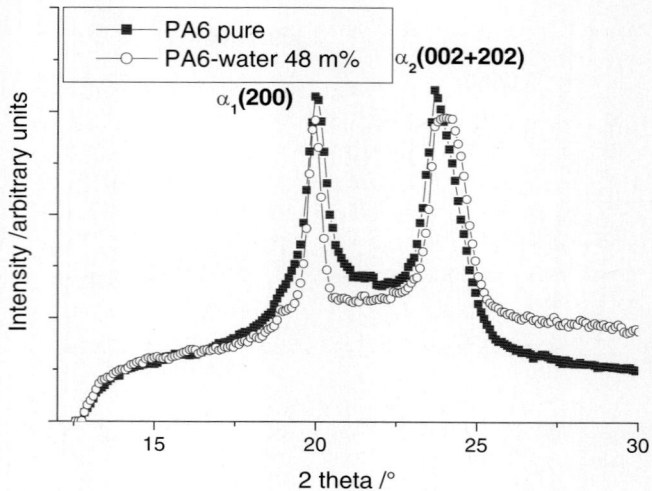

Fig. 9.9. WAXD patterns of intensity as function of 2θ at 50°C for 48 m% PA6 in water and pure PA6

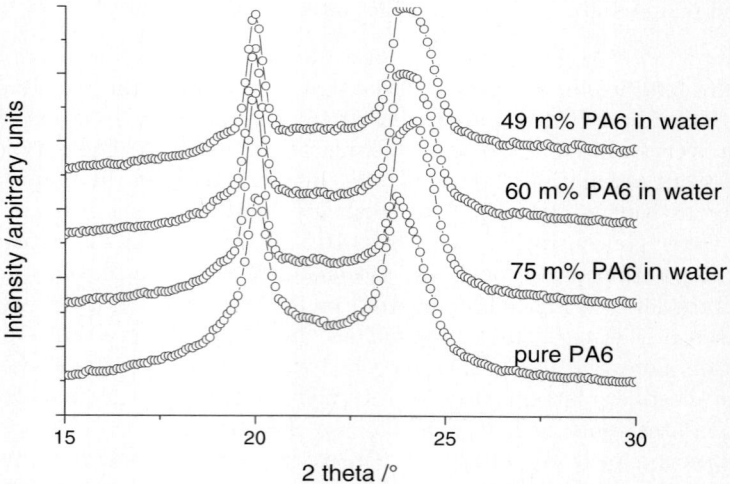

Fig. 9.10. WAXD patterns of intensity as function of 2θ at 50°C for 49 m%, 60 m%, 75 m% PA6 in water and pure PA6 for the 2nd heating curve

sheets in the (a,b)-plane and in these sheets the PA6 chains are anti-parallel, coupled by hydrogen bonding. The hydrogen bond between the C=O and N-H groups of two anti parallel polymer chains in the alpha form is an intra-sheet distance and gives rise to the WAXD peak (200) in both pure PA6 and the PA6-water systems. If a H_2O is placed between the hydrogen bond donor and acceptor, a shift of the WAXD peak (200) to lower 2θ angles is expected, due to the expansion of the unit cell in the a-direction, which shift however is not found.

In addition, calculations of the void volumes of pure α-crystals indicate no solvent accessible areas. These calculations are carried out with the program Platon [58], using a 1.2 Å probe to scan the van der Waals surface of the polymer. The used van der Waals radii for C, N and O are 1.70Å, 1.55 Å and 1.52Å respectively. Since no solvent accessible areas were found and no shift (within the experimental error) of the WAXD peak (200) was observed, we conclude that no water is located inside the crystallites of the α-form.

9.3.5 Dissolution of Other Polyamides and in Various Solvents

The remarkable lowering of the transition temperatures of PA6 in water cannot only be obtained with PA6, but also with other polyamides like PA6.6 and PA4.6 – see Fig. 9.11 and 9.12 for PA4.6 – and with other hydrogen bonded solvents, like ethanol, see Fig. 9.12.

For the DSC-curves of PA4.6-water, in general the same behavior as in PA6-water systems is observed. These also show double-peaked dissolution curves in the low concentration range (in between 5 and 50 m% PA4.6 in water), what could be indicative of recrystallization. In the high concentration range (from 80 m% onwards) the dissolution peaks are broadened too, possibly indicating reorganization phenomena.

Thus, it is possible to dissolve PA4.6 in ethanol and water, as shown in Fig. 9.12. Compared to PA6, the maximum crystallization and melting point depression is larger: $\triangle T_{\max} = 100°C$ for PA4.6 in water and $\triangle T_{\max} = 80°C$

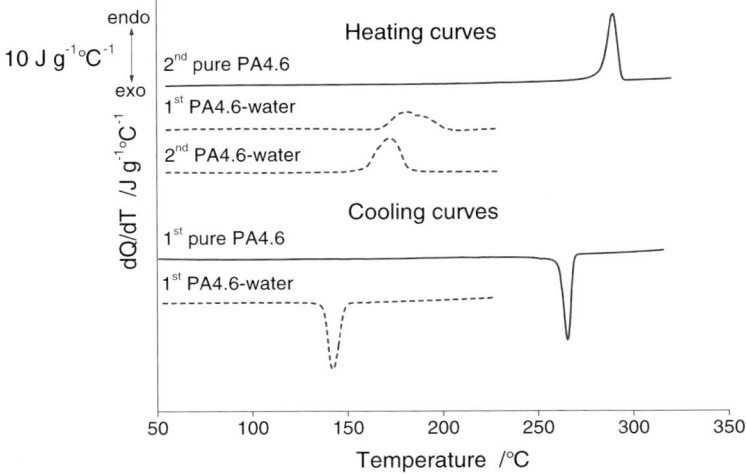

Fig. 9.11. DSC curves of 49 m% PA4.6 in water and pure PA4.6

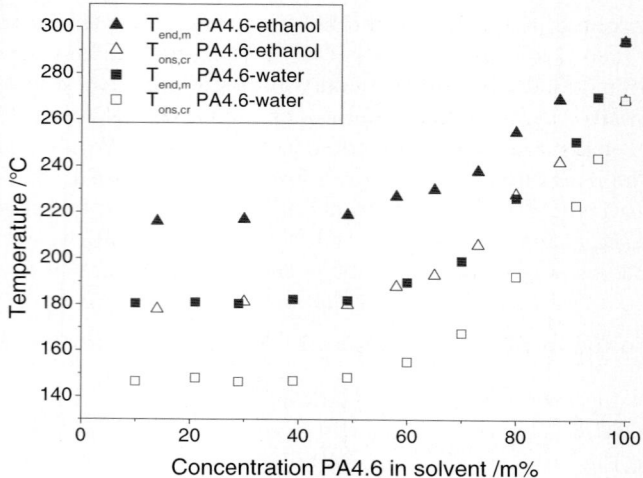

Fig. 9.12. Onset crystallization and end melting temperature as function of the concentration PA4.6 in solvent /m%

for PA4.6 in ethanol. The plateau in the Temperature – m% plot on the other hand, is smaller with respect to concentration: from 10 to 50 m% (Fig. 9.12).

Interpretation of the DSC-curves of the PA4.6-ethanol mixture turns out to be complicated because of the critical temperature of ethanol at 243.1°C.

9.4 Conclusions

The results show clearly that PA6 is soluble in water over the whole concentration range and that water acts as a crystallization- and melting point suppressor. Chain scission takes place, but after a cooling/heating cycle still a polyamide of high molar mass results. The morphology of PA6 α-crystals is preserved and water does not enter the crystallites. PA4.6 has a similar behavior as PA6 and is soluble in several solvents (ethanol and water), though the transition temperature depressions are larger. When the PA4.6-water and the PA4.6-ethanol sytems are compared, water is the better solvent.

Acknowledgements

MW thanks the European Community for a Marie Curie Industry Host Fellowship grant, contract nr HPMI-CT- 2001-00112 through DSM Research. The use of the Dubble beamline at the ESRF, Grenoble, France is appreciated much. BG is a post-doctoral fellow of the Fund for Scientific Research Flanders (FWO-Vlaanderen). The authors thank Jan Servaes of the Katholieke Universiteit Leuven, University Technical Department, Belgium, for the design and construction of the pressure cells and holders. Koen Robeyns (Katholieke Universiteit Leuven, Molecular and Structural Biology) is acknowledged for the calculations and the helpful discussions on possible inclusion of water in the crystallites. Tiny Frijns-Bruls and Erik Geladé (DSM Resolve) are thanked for performing the SEC-measurements.

References

[1] B. Frank, P. Frübing, P. Pissis, *J Polym Sci B Polym Phys* **34**, 1853 (1996).
[2] K. Kawasaki, Y. Sekita, *J Polym Sci A* **3**, 2437 (1964).
[3] J. L. Hutchison, N. S. Murthy, E. T. Samulski, *Macromolecules* **29**, 5551 (1996).
[4] B. Knopp, U. W. Suter, *Macromolecules* **30**, 6114 (1997).
[5] N. S. Murthy, M. K. Akkapeddi, W. J. Orts, *Macromolecules* **31**, 142 (1998).
[6] C. A. Fyfe, L. H. Randall, N. E. Burlinson, *J Polym Sci A Polym Chem* **31**, 159 (1993).
[7] L. P. Razumovskii, V. S. Markin, G. Y. Zaikov, *Polym Sci USSR* **27**, 751 (1985).
[8] J. Aerts, et al., *Thechnische Thermoplaste, Polyamide part 3* (Hanser Verlag: München, 1998), chap. 3, pp. 549–638.
[9] H. K. Reimschuessel, *J Polym Sci Polym Chem* **16**, 1229 (1978).
[10] N. S. Murthy, M. Stamm, J. P. Sibilia, S. Krimm, *Macromolecules* **22**, 1261 (1989).
[11] G. J. Kettle, *Polymer* **18**, 742 (1977).
[12] P. Adriaensens, et al., *Polymer* **42**, 7943 (2001).
[13] N. Jia, H. A. Fraenkel, *J Reinf Plast Compos* **23**, 729 (2004).
[14] K. Inoue, S. Hoshino, *J Polym Sci Polym Phys* **14**, 1513 (1976).
[15] R. J. Hernandez, R. Gavara, *J Polym Sci B Polym Phys* **32**, 2367 (1994).
[16] Y. P. Khanna, P. Han, E. D. Day, *Polym Eng Sci* **36**, 1745 (1996).
[17] F. P. La Mantia, R. Scaffaro, *Polym Degrad Stab* **75**, 473 (2002).
[18] Y. S. Papir, S. Kapur, C. E. Rogers, E. Baer, *J Polym Sci A2* **10**, 1305 (1972).
[19] D. C. Prevorsek, R. H. Butler, H. K. Reimschuessel, *J Polym Sci A2* **9**, 867 (1971).
[20] H. W. Starkweather, *J Appl Polym Sci* **2**, 129 (1959).
[21] J. Hirschinger, H. Miura, K. H. Gardner, A. D. English, *Macromolecules* **23**, 2153 (1990).
[22] V. M. Litvinov, J. P. Penning, *Macromol Chem Phys* **205**, 1721 (2004).
[23] A. Koshimo, *J Appl Polym Sci* **9**, 55 (1965).
[24] A. Koshimo, T. Tagawa, *J Appl Polym Sci* **9**, 117 (1965).
[25] N. S. Murthy, W. J. Orts, *J Polym Sci B Polym Phys* **32**, 2695 (1994).
[26] J. Pleštil, J. Baldrian, Y. M. Ostanevich, V. Y. Bezzabotnov, *J Polym Sci B Polym phys* **29**, 509 (1991).
[27] N. S. Murthy, S. M. Aharoni, A. B. Szollosi, *J Polym Sci Polym Phys* **23**, 2549 (1985).
[28] G. A. Campbell, *J Polym Sci Polym Lett* **7**, 629 (1969).
[29] H. M. Heuvel, R. Huisman, *J Appl Polym Sci* **23**, 713 (1981).
[30] K. Nagasubramanian, H. K. Reimschuessel, *J Appl Polym Sci* **17**, 1663 (1973).
[31] R. Puffr, J. Šebenda, *J Polym Sci C* **16**, 79 (1967).
[32] H. W. Starkweather, J. R. Barkley, *J Polym Sci Polym Phys* **19**, 1211 (1981).

[33] E. G. Chatzi, H. Ishida, J. L. Koenig, *Appl Spectrosc* **40**, 847 (1986).
[34] M. Tsuruta, A.Koshimo, T. Tagawa, *J Appl Polym Sci* **9**, 31 (1965).
[35] M. Tsuruta, A. Koshimo, *J Appl Polym Sci* **9**, 39 (1965).
[36] A. Koshimo, *J Appl Polym Sci* **9**, 81 (1965).
[37] M. F. J. Pijpers, V. B. F. Mathot, R. L. Scherrenberg, Kristallisatie- en smelttemperatuur depressie van polyamiden door water, methanol and andere oplosmiddelen onder verhoogde (damp)druk, *Tech. Rep. RC95-12799*, DSM Research (1995).
[38] R. Korbee, A. V. Geenen, Patent No. WO 99/297467 pp. 1–16 (1999).
[39] S. Rastogi, A. E. Terry, E. Vinken, *Macromolecules* **37**, 8825 (2004).
[40] N. Kosaric, et al., *Ulmann's encyclopedia of industrial chemistry* (Wiley-VCH Verlag GmbH: Weinheim, 2000 electronic release), chap. Ethanol-physical properties, 6th edn.
[41] E. U. Frank, G. Wiegand, N. Dahmen, *Ullmann's encyclopedia of industrial chemistry* (Wiley-VCH Verlag GmbH: Weinheim, 2000 electronic release), chap. Water at high pressure and temperature, 6th edn.
[42] V. B. F. Mathot, *Calorimetry and thermal analysis of polymers* (Hanser Publishers: Münich, 1994), Chap. 5, pp. 105–167.
[43] V. Mathot, T. Pijpers, M. Steinmetz, G. van der Plaats, *Proceedings of the 25th North American Thermal Analysis Society Conference* (McClean: Virginia, 1997), p. 64. The temperature dependent crystallinity software program which runs under Windows and is not instrument specific, has been developed by DSM Research BV and Anatech BV jointly, and is availabe trough PerkinElmer.
[44] http://web.utk.edu/~athas/databank/intro.html.
[45] D. R. Holmes, C. W. Bunn, D. J. Smith, *J Polym Sci* **17**, 159 (1955).
[46] H. Arimoto, M. Ishibashi, M. Hirai, Y. Chatani, *J Polym Sci A* **3**, 317 (1965).
[47] A. Ziabicki, *Koll-Z* **167**, 132 (1959).
[48] L. G. Roldan, H. S. Kaufman, *J Polym Sci B Polym Lett* **1**, 603 (1963).
[49] F. Auriemma, V. Petraccone, L. Parravicini, P. Corradini, *Macromolecules* **30**, 7554 (1997).
[50] J. Gianchandani, J. E. Spruiell, E. S. Clark, *J Appl Polym Sci* **27**, 3527 (1982).
[51] J. P. Parker, P. H. Lindenmeyer, *J Appl Polym Sci* **21**, 821 (1977).
[52] R. Tol, V. B. F. Mathot, H. Reynaers, B. Goderis, G. Groeninckx, *Polymer* **46**, 2966 (2005).
[53] G. Gurato, A. Fichera, F. Z. Grandi, R. Zannetti, P. Canal, *Makromol Chem* **175**, 953 (1974).
[54] V. Malta, G. Cojazzi, A. Fichera, D. Ajo, R. Zanetti, *Eur Polym J* **15**, 765 (1979).
[55] J. Ibarretxe Uriguen, L. Bremer, V. Mathot, G. Groeninckx, *Polymer* **45**, 5961 (2004).
[56] N. S. Murthy, S. A. Curran, S. M. Aharoni, H. Minor, *Macromolecules* **24**, 3215 (1991).
[57] http://www.cryst.chem.uu.nl/platon/pl000000.html.
[58] http://www.cryst.chem.uu.nl/platon/pl000302.html.

10

Small Angle Scattering Study of Polyethylene Crystallization from Solutions

Howard Wang

Department of Materials Science and Engineering, Michigan Technological University, Houghton, MI 49931
wangh@mtu.edu

Abstract. Crystallization of low molecular weight polyethylene from concentrated solutions has been studied using small angle neutron scattering (SANS). The detection sensitivity of the volume fraction degree of crystallinity is estimated to be 10^{-5}, allowing for measuring the structure and kinetics during the very early stages of crystal growth. SANS spectra for both the early and late stages of crystallization can be satisfactorily interpreted with a lamellar crystal model; there is no evidence of diverging or spinodal-decomposition-like density fluctuations during the early stage of crystallization in polyethylene solutions. A possible explanation of the dominant wavevector in small angle x-ray scattering that led to the proposal of "spinodal decomposition" mechanism for early stage crystallization is suggested.

10.1 Introduction

The early stage crystallization in polymers has been a topic of interest since the beginning of the polymer science. After decades of intensive studies and spirited debates, a kinetic picture of nucleation and growth (NG) has generally been accepted [1–5]. In the classical view of the homogeneous nucleation, density fluctuations in supercooled liquids result in small clusters of ordered segments; those sub-critical-sized nuclei dissolve back into the liquid phase, whereas nuclei larger than the critical nucleus grow indefinitely. In recent years, there have been renewed interests in the mechanism of the structure development during the induction period of crystallization due to the proposal of a new mechanism with a rather different picture [6–9], in which supercooled melt undergoes "spinodal decomposition" (SD) resulting in spatial separation of polymer chains into domains where chains have better conformational order and domains with chains like those in ordinary melts. This preordering process assists the initial nucleation of crystals from the melt. The SD proposal triggered more experimental and theoretical studies [10–17] as well as rethinking and discussions on the route toward polymer crystallization [18–21].

After more than two decades since the first notion of SD in polymer crystallization literatures [22] and one decade after the proposal of the SD mechanism for early stages of crystallization [7,8], new evidences are continually reported in recent years to support either SD [23–26] or NG [27–29] in polymer crystallization. Heeley et al. [23] gave detailed analysis of small angle x-ray scattering (SAXS) spectra on the early stage of isotactic polypropylene using Cahn-Hilliard theory and obtained a spinodal temperature, which is below the melting temperature. Ania el al. [24] revealed a characteristic wavelength of 14.7 nm of long-range density fluctuations growing with time during the induction period of polyamide 6, 6 crystallization using SAXS. Using optical microscopy, Nishida et al. [25] observed bicontinuous patterns, which were regarded as characteristic for SD, as PET melt was rapidly quenched below a stability limit. Zhang et al. [26] compared the difference between cold and melt crystallization in isotactic polystyrene and suggested that the melt crystallization process of polymer be explained by the SD theory [12].

On the other hand, there are reports backing the classical NG picture. Chen et al. [27] studied the nucleation process of poly(ethylene oxide) (PEO) crystallization from both the isotropic and structured melt using simultaneous SAXS and wide angle x-ray diffraction (WAXD). They suggested the existence of primary nuclei in the melt due to localized large amplitude density fluctuation and calculated the size of nuclei as a function of time. Panine et al. [28] investigated the early stage of melt crystallization in low-density polyethylene using improved x-ray detection and found the behavior of SAXS data in the very early stages was not consistent with spinodal decomposition mechanism. Owen et al. [29] studied the room temperature crystallization of poly[(R)-3-hydroxybutyrate] using both ultra-small angle x-ray scattering (USAXS) and SAXS and found early development of structures was essentially lamellae with different forms of fractal aggregates.

It is clear that SAXS using synchrotron x-ray radiations has been playing an important role in the debate of the early stage crystallization mechanism. Much of the current controversies stemmed from different interpretations of SAXS signals prior to the appearance of the WAXD peaks. It is desirable that the same problem be viewed from a different angle. Recent experiments show that small angle neutron scattering (SANS) using composition contrast is a useful tool complementing SAXS for studying polymer crystallization [30,31]. In this chapter, we report additional evidences of structural development during the early stages of crystallization in polymer solutions. While studies on oligomeric solutions do not directly address the controversies in mechanisms of melt crystallization, we hope to cast insights in this matter from the measurement point of view. Furthermore, the well-established optical microscopy observation of the spherulitic growth in polymer melts over length scales of many microns could not be satisfactorily accommodated in the SD mechanism, which involves density fluctuations of ten nanometers. In this chapter, we point out an alternative picture to reconcile the scattering and morphology measurements.

10.2 Experiment

The polyethylene (PE) with $M_w = 2.1$ kg/mol and PDI ~ 1.15 was obtained from the Pressure Chemical Co. PE solutions with volume fractions of $\phi_0 = 0.10$, 0.13 and 0.24 (in coil solutions) in deuterated o-xylene (>99 at% deuterium, C/D/N Isotopes Inc.) were measured using the NG3 30m SANS instrument at the NIST Center for Neutron Research of the National Institute of Standards and Technology. Incident neutrons of wavelength $\lambda = 6$ Å and two sample-to-detector distances of 1.33 and 6.5 m yielded a range of scattering wavevectors, 0.006 Å$^{-1} < Q < 0.4$ Å^{-1}. In the cooling study, the samples were first equilibrated at 120.0°C then sequentially cooled to lower temperatures; SANS spectra were collected over a period of 27 min after the solution being isothermally stored at each temperature for ca. 170 min. The structure in solution could be considered arrested during the measurement. For isothermal crystallization study, the $\phi_0 = 0.10$ solution was first homogenized at 100.0°C for 20 min, and then cooled to 90.0°C for SANS measurement. Time intervals for the data acquisition varied from 400 s at the initial stage to 20 min at the later time of the isothermal crystallization. The time label for each SANS spectrum in an isothermal series was set to the ending time of the data acquisition. The temperature stability of the stage was within ±0.2°C, and the temperature accuracy at the sample was ±0.5°C. After the correction for background and detector efficiency, and the conversion to an absolute scale using the direct beam intensity, the 2D SANS intensity was circularly averaged to yield the total scattering cross section of the sample.

10.3 Results and Discussion

Typical SANS spectra of the $\phi_0 = 0.24$ solution after cooling from 120.0°C to various temperatures were shown in Fig. 10.1. The symbols are experimental data and the curves through symbols are the best model fitting. In the model as developed previously [30], lamellar crystals are assumed to coexist with coil chains; they contribute to scattering neutron independently. The low temperature (85.0, 90.0 and 95.0°C) spectra show both the characteristic form and structure factors of stacked lamellae. The structure parameters obtained from the model fitting are listed in Table 10.1, indicating extended-chain crystals. The degree of crystallinity, which is defined as the volume fraction (ϕ_{cry}) of the crystalline phase in this manuscript, is estimated following a procedure described later.

A simplistic view of the scattering features can be described as follows. The asymptotic Q^{-2} power law at the low-Q is characteristic for sheet-like 2D structure, and the rapid fall of the intensity for about 2 orders of magnitude at ca. 0.03 Å^{-1} implies the loss of self-correlation beyond the lamellar thickness. Those are considered the "form factor" of individual sheets with a uniform thickness. On the other hand, the peaks at ca. 0.025, 0.05 and 0.08 Å^{-1} are the

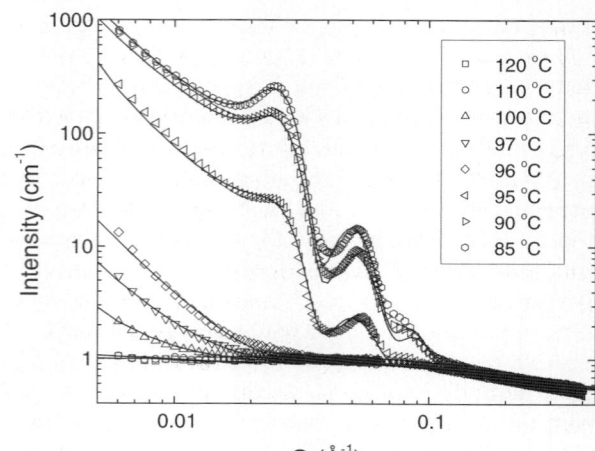

Fig. 10.1. SANS spectra of the $\phi_0 = 0.24$ solution after cooling from 120.0°C to various temperatures. The symbols are experimental data and the curves through symbols are the best model fitting. The low temperature (85.0°C, 90.0°C and 95.0°C) spectra show both the characteristic form and structure factors of stacked lamellae

Table 10.1. The structure parameters obtained from model fitting

T_c (°C)	Long period, L_{lam} (Å)	Crystal thickness, L_{cry} (Å)	Crystallinity ϕ_{cry}
85.0	230	180	0.21 (assumed)
90.0	223	175	0.12
95.0	213	174	0.05

first three diffraction orders due to the density correlation among the stacking lamellae, which depict the "structure factor" of lamellar stacks.

Because of the very low noise of the SANS spectra, the spectra of the coil solution (110.0 and 120.0°C) show clean coil behavior, whereas those of SAXS typically show large excess scattering in the same low-Q region from both the beam-stop spill-over and density fluctuations due to random thermal motions. The cleanness of the SANS spectra of homogenous coil solution allows for detecting very slight structural change during early stages of crystallization.

At temperatures 96.0, 97.0 and 100.0°C, the scattering intensity arising from the solution background at the low-Q indicates the development of new structures from the solution. Those spectra could be emulated by multiplying a constant, 0.05, 0.02, and 0.008 for 96.0, 97.0 and 100.0°C, respectively, to the lamellar scattering intensities at 95.0°C, and summing with a relatively invariable coil solution spectrum as that at 120.0°C. That implies a simple rescaling of the intensity by ϕ_{cry} without invoking a different kind of structure for the early stage. Furthermore, the high-Q cut-off of the scattering signals from nascent crystals at ca. 0.03 Å^{-1} implies that the minimum dimension of the crystal in this study is around the extended chain length, which could also be the size of nuclei. For a period during the early stage crystallization, because of the rapid drop of the intensity beyond the Q-value corresponding to the lamellar thickness, the scattering from nascent crystals at higher-Q is overwhelmed by the coil scattering, resulting in an apparent cut-off.

The detection sensitivity in this study can be estimated. Assuming PE being fully crystallized at 85.0°C, the ϕ_{cry} of 0.21 (assuming crystal density of 1.0 g/cm^3) is equivalent to the intensity of ca. 10^3 cm^{-1} at 0.006 Å^{-1}. For the coil solution at 120.0°C, $\phi_{cry} = 0$, the measured intensity is ca. 1 cm^{-1} with 4% statistical errors. So the detection sensitivity of SANS for the volume fraction crystallinity, ϕ_{det}, is estimated to be ca. 10^{-5} (4% × 1 ÷ 10^3 × 0.21 = 8 × 10^{-6}), comparable to that of light scattering while more than one decade better than that of SAXS. This is however a conservative estimate since the crystallinity at 85.0°C would be less than ideal, and all crystals are not in perfect lamellar stacks, but the order of magnitude should not be affected.

The low detection limit allows for the measurement of crystallization kinetics at the early stage. Figure 10.2 shows time sequences of selected SANS spectra for the $\phi_0 = 0.10$ solution at 90.0°C. The solid curves are best model fitting. The inset shows the resulting ϕ_{cry} from fitting, which increases linearly with time. Note that the range of the very low crystallinity is not readily accessible by SAXS, and is rare in polymer crystallization literatures. It re-

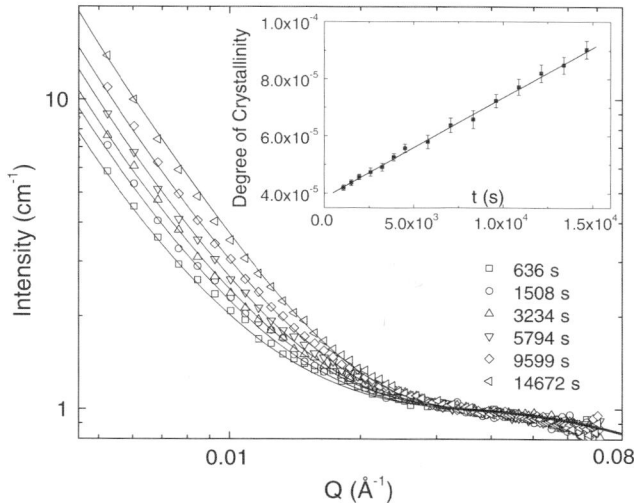

Fig. 10.2. Time sequences of SANS spectra showing crystallization kinetics in $\phi_0 = 0.10$ solution during the isothermal storage at 90°C. The inset shows the evolution of the volume fraction degree of crystallinity obtained from model fitting. It increases linearly with time. Note that the measurement of the rather small crystallinity quantity is not readily achievable by SAXS

quires further investigation to reveal the implication of the linear kinetics at such low ϕ_{cry}. Those measurements make it possible to probe the structural development during the "induction period" of polymer crystallization.

The scattering invariant, which can be experimentally obtained as the integrated total intensity, describes the mean square fluctuations of the scattering length density (SLD), $Q_{inv} = \int I(Q)Q^2 dQ = \phi(\Delta\rho)^2$, which is independent of the forms of fluctuations. So a small volume fraction of condensed crystal phases with a high contrast to the background can yield the same scattering invariant as small amplitude fluctuations throughout the entire sample volume. That concept is schematically illustrated in Fig. 10.3. For an initially homogeneous system with a constant SLD of ρ_0 at arbitrary positions (dashed lines), there are two distinct routes for developing heterogeneity as shown by the solid-line profiles in Figs. 10.3a and 10.3b. In the former, sporadically-grown crystals of 10 nm in thickness and microns in separation have high SLD contrast with the surrounding solution; whereas in the latter, small undulations of SLD around ρ_0 fill the entire system. Those two routes correspond to the NG and SD mechanisms, respectively, and can be compared by their scattering invariants.

The detection limit for compositional fluctuations can be estimated from ϕ_{det} using the equality $\phi_{det}(\rho_{cry}-\rho_{sol})^2 = \phi_{fluc}\Delta\rho_{fluc}^2$, where ρ_{cry} and ρ_{sol} are the SLDs of the crystal and the solution phase, respectively, and ϕ_{fluc}

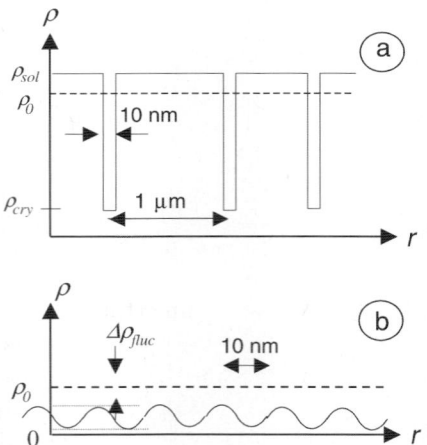

Fig. 10.3. Schematic illustration of the invariant argument. Initially the system is homogeneous with a constant SLD of ρ_0 at arbitrary positions (*dashed lines*). *Solid-line* profiles indicate two distinct routes of heterogeneity development: (**a**) sporadically-grown crystals of 10 nm in thickness and microns in separation having high SLD contrast with surrounding solution; (**b**) small undulations of SLD around ρ_0 filling the entire system. The length labels indicate the order of magnitude of corresponding features

and $\Delta\rho_{fluc}$ are the volume fraction and scattering contrast of fluctuating heterogeneities, respectively. Using $\phi_{det}= 10^{-5}$, $|\rho_{cry}-\rho_{sol}| = 1$, $\phi_{fluc} = 0.5$ the detection limit of the amplitude of wave-like fluctuations is estimated to be 0.2 %, which is taken as half of the $\Delta\rho_{fluc}$. There is no evidence of ***diverging*** (or spontaneous, or spinodal-decomposition like) compositional fluctuations during the early stage of crystallization, which would at a point exceed the 0.2 % detection limit and be measurable. On the other hand, the possibility of ***non-diverging*** (standing-wave like) fluctuations with amplitude less than 0.2 % still exists.

Based on the scattering features, we can also cast some insights in the SAXS data that led to Kaji et al.'s proposal of the "spinodal decomposition" mechanism during the early stage of polymer crystallization. As aforementioned, one difference between SAXS and SANS spectra is that the former usually carries large background excess intensities at the low-Q. A standard practice to extract the structural signal is by measuring the excess scattering from the melt, and subsequently subtracting that from crystallization spectra. This is illustrated in Fig. 10.4 using SANS spectra of $\phi_0 = 0.13$ solution as an example. The filled squares and open circles are experimental data of the solution at 120.0°C and 90.0°C, representing the homogeneous coil solution and the early stage crystallization states, respectively. The Q-range in this plot, 0.005 to 0.05 Å^{-1}, is also typical of those reported in SAXS studies. The coil solution spectrum shows very clean flat intensity within the range of the plot, indicating neither crystal structures in the sample nor Q-dependent background noise in the measurement. The spectrum at 90.0°C shows intensities in the low-Q region due to the crystal formation in the solution. The characteristics of the spectrum are identical to the form factor of nascent lamellar crystals, which shows an intensity cut-off around the Q-value corresponding to the lamellar crystal thickness and the asymptotic power-law at the lower-Q. The power exponent of –2.2 is slightly higher than that of ideal 2D objects possibly due to both the finite size and fractal surface roughness, whose effect diminishes as crystals grow to large dimension and crystal facets perfect at later times.

To simulate SAXS spectra, we assume the low-Q excess intensity (due mainly to the beam-stop spill-over, stray photons, thermal fluctuations etc), I_{ex}, as the dotted curve in Fig. 10.4, which drops rapidly from 100 times the white background at $Q = 0.006$ Å^{-1} to below the background at ca. 0.02 Å^{-1}. That is typical in reported SAXS studies [6–9,22,23]. The SAXS spectra of the sample are generated by adding the I_{ex} to the corresponding SANS spectra, shown in the same plot as the dash-dotted curve for the coil solution and the dashed curve for the early stage crystals, respectively. It is clear that at the early stage, the Q^{-2} power law is overwhelmed by the much steeper SAXS excess intensity. In principle, true spectrum can be recovered if both spectra with and without crystals are measured accurately and with good statistics. However, in practice obtaining a small difference from two large numbers has always been tricky. For SAXS, it is difficult to measure beam

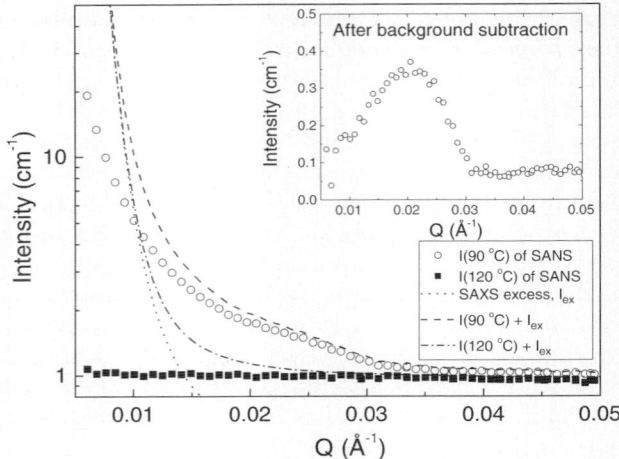

Fig. 10.4. Illustration of excess signals and their subtraction in typical small angle x-ray scattering experiment. The *filled squares* and *open circles* are SANS spectra of $\phi_0 = 0.13$ solution in the coil and the early stage crystallization states, respectively. The *dotted curve* is assumed SAXS excess intensity at the low-Q, and the *dash-dotted* and *dashed curves* are simulated SAXS spectra for the melt and the early stage crystals by adding the dotted curve to the corresponding SANS spectra. The inset shows that over-subtracting the dashed curve could result in a prominent intensity peak

intensity accurately because of the decaying and fluctuating synchrotron light intensity at the ring-exit and the limited accuracy of ionization chambers. The subtraction could therefore be arbitrary. Slight over-subtraction of the spectra in correcting the background is possible. An over-subtracted spectrum from the dashed curve is shown in the inset of Fig. 10.4, which has a prominent peak at 0.022 Å^{-1}. The comparison between the features in SANS and SAXS spectra leads to a speculation that the SAXS scattering peak regarded as the dominant "spinodal decomposition" wavelength is possibly a result of the background over-subtraction. As a matter of fact, clean and stand-alone peaks were never reported in SAXS literatures, they all appeared in certain forms of "shoulder" intensity, similar to that of the dashed curve in Fig. 10.4. The validity of this speculation, however, needs to be examined with further experiments.

As shown in Fig. 10.3 and discussed above, features of similar length scales but with higher contrast with the matrix could result in equivalent scattering intensity at a much lower volume fraction. That offers an alternative picture for resolving the difficulties in bridging the length scale gap between the scattering and the optical microscopy: the scattering during the early stage is attributed to isolated individual lamellae, which could be microns apart while giving scattering features similar to features regarded as SD. While this study is carried out on solutions, the speculations on melt crystallization phenomena based on the analogy in scattering features need to be tested experimentally.

10.4 Conclusion

We have learned several things from the current SANS study of polyethylene crystallization from concentrated solutions. (1) The detection sensitivity of the volume fraction degree of crystallinity is estimated to be 10^{-5}. That allows for measuring the kinetics during the induction stage of polymer crystallization. (2) There is no evidence of diverging or spinodal-decomposition-like density fluctuations in polymer crystallization from solution. (3) By comparing the SANS spectra in this study and the literature SAXS data, it is speculated that the scattering intensity peak in the latter, which is regarded as the signature of the "spinodal decomposition" mechanism, is a result of the background over-subtraction. The data in this study support the nucleation and growth mechanism for polymer crystallization from solution.

Acknowledgment

We acknowledge the NSF Career Award DMR-0348895 and the support of the National Institute of Standards and Technology in providing the neutron research facilities used in this work. We also acknowledge insightful discussions with Drs. Boualem Hammouda (NIST), Stephen Z. D. Cheng (Akron), and Benjamin S. Hsiao (Stony Brook).

References

[1] Wunderlich, B. "Macromolecular Physics", Acdemic Press, New York, 1973.
[2] Hoffman, J.D., Davis, G.T., Laurizten JI. in "Treatise on Solid State Chemistry", V3, Crystalline and Non-crystalline Solids, Hannay NB. Ed., Plenum, New York, 1976.
[3] Khoury, F.A., Passaglia, E. in "Treatise on Solid State Chemistry", V3, Crystalline and Noncrystalline Solids, Hannay, N.B. Ed., Plenum, New York, 1976.
[4] Hoffman, J.D., Miller, R.L. Polymer, 38, 3151, 1997.
[5] Cheng, S.Z.D., Lotz, B. Phil. Trans. R. Soc. Lond. A, 361, 517, 2003.
[6] Imai, M., Mori, K., Mizukami, T., Kaji, K., Kanaya, T. Polymer, 33, 4451, 1992.
[7] Imai, M., Mori, K., Kizukami, T., Kaji, K., Kanaya, T. Polymer, 33, 4457, 1992.
[8] Imai, M., Kaji, K., Kanaya, T. Phys Rev Lett, 71, 4162, 1993.
[9] Imai, M., Kaji, K., Kanaya, T., Sakai, Y., Phys. Rev. B., 52, 12696, 1995.
[10] Terrill, N.J., Fairclough, P.A., Towns-Andrews, E., Komanschek, B.U., Young, R.J., Ryan, A.J., Polymer, 39, 2381, 1998.

[11] Olmsted, P.D., Poon, W.C.K., McLeish, T.C.B., Terrill, N.J., Ryan, A.J., Phys. Rev. Lett., 81, 373, 1998.
[12] Matsuba, G., Kaji, K., Nishida, K., Kanaya, T., Imai, M., Polym. J, 31, 722, 1999.
[13] Matsuba, G., Kaji, K., Nishida, K., Kanaya, T., Imai, M., Macromolecules, 32, 8932, 1999.
[14] Akpalu, Y.A., Amis, E.J., J. Chem. Phys., 111, 8686, 1999.
[15] Wang, Z.G., Hsiao, B.S., Sirota, E.B., Agarwal, P., Srinivas, S., Macromolecules, 33, 978, 2000.
[16] Wang, W., Schultz, J.M., Hsiao, B.S., Macromolecules, 30, 4544, 1997.
[17] Muthukumar, M., Welch, P., Polymer 2000;41:8833, ibid,; Phys. Rev. Lett., 87, 218302, 2001.
[18] Strobl, G., European Physical Journal E, 3, 165, 2000.
[19] Lotz, B., European Physical Journal E, 3, 185, 2000.
[20] Cheng, S.Z.D., Li, C.Y., Zhu, L., European Physical Journal E, 3, 195, 2000.
[21] Muthukumar, M., European Physical Journal E, 3, 199, 2000.
[22] Petermann, J., Gohil, R.M., Schultz, J.M., Hendricks, R.W., Lin, J.S., J. Polym. Sci. B. Polym. Phys.20, 523, 1982.
[23] Heeley, E.L., Maidens, A.V., Olmsted, P.D., Bras, W. Dolbnya, I.P., Fairclough, J.P.A., Terrill, N.J., Ryan, A. J., Macromolecules, 36, 3656, 2003.
[24] Ania, F., Flores, A., Calleja, F.J.B., J. Macromol Sci.-Pys., B42, 653, 2003.
[25] Nishida, K., Kaji, K., Kanaya, T., Matsuba, G., Konish, T., J. Polym. Sci. B. Polym. Phys., 32, 1817, 2004.
[26] Zhang, J.M., Duan, Y.X., Sato, H., Men, D.Y., Yan, S., Noda, I., Ozaki, Y., J. Phys. Chem. B, 109, 5586, 2005.
[27] Chen, E.Q., Wang, X., Zhang, A., Mann, I., Harris, F.W., Cheng, S.Z.D., Hsiao, B.S., Yeh, F., Macromol. Rappid Commun., 22, 611, 2001.
[28] Panine, P., Urban, V., Boesecke, P., Narayanan, T.J., Appl. Cryst. 2003;36:991.
[29] Owen, A., Bergmann, A., Polymer International, 53, 12, 2004.
[30] Wang, H., J. Polym. Sci. Part B: Polym. Phys.42, 3133, 2004.
[31] Zeng, X., Ungar, G., Spells, S.J., King, S.M., Macromolecules, 38, 7201, 2005.

11
Morphologies of Polymer Crystals in Thin Films

Günter Reiter[1], Ioan Botiz[1], Laetitia Graveleau[1], Nikolay Grozev[1], Krystyna Albrecht[2], Ahmed Mourran[2], and Martin Möller[2]

[1] Institut de Chimie des Surfaces et Interfaces, ICSI-CNRS, 15, rue Jean Starcky, B.P. 2488, 68057 Mulhouse Cedex, France
[2] DWI an der RWTH Aachen, Pauwelsstr. 8, 52056 Aachen, Germany
g.reiter@uha.fr

Abstract. We present microscopy investigations on the morphology of crystals of poly-2-vinylpyridine-polyethyleneoxid diblock copolymers ($P_2VP - PEO$) in thin films with thicknesses below and above the thickness of a single lamella. For crystallisation temperatures above 45°C, nucleation is highly unlikely. Thus, the resulting morphologies represent essentially single crystals, allowing us to relate morphology to the kinetics of crystal growth. In several examples we demonstrate the influence of thermal history and film thickness on molecular orientation and pattern formation during crystal growth. We discuss the analogies and differences between crystallisation of small molecules and polymers.

11.1 Introduction

The extraordinary beauty and the vast diversity of the possible morphologies represents one of the main reasons why crystallization attracts our interest. Of course, many properties (mechanical, optical, electrical, ...) depend on the detailed structure of crystalline materials. Thus, it is highly important to understand how crystals grow and how the various morphologies result from growth processes which can be controlled via parameters like temperature or concentration.

At first glance, forming a crystal starting from individual molecules seems to be an easy task. One simply has to assemble these molecules according to the "construction manual" defined by the characteristic parameters of the unit cell of the crystal. However, if the crystallization process advances at a finite growth rate, complications arise due to the necessary transport of the molecules towards the "construction site" and the time needed for the integration into the crystal. As a consequence, the growth front becomes unstable and the resulting crystals are less perfect in the sense that their morphology is not any longer exhibiting the rather simple form of the unit cell. Fascinating patterns are created as can be seen frequently in Nature, for example in snowflakes [1]. Such instabilities of the growth front can be reduced in frequency, or eliminated completely, if the growth process proceeds at low rates. Then, each molecule arriving at the front will have enough time to "examine" the crystal surface and to find the "ideal" location for its integration in the crystal. "Ideal" means that the growth front stays smooth as in such case each molecule will profit from having the maximum possible number of neighbouring molecules, thereby reducing the free energy of the system. This reflects the fact that the most stable state of a crystal is exhibiting well-defined facets which one can observe even on macroscopic scales. The two key processes which determine if growth is sufficiently slow to allow for the formation of ideal crystals are:

- The transport of the molecules to the crystal. In order to build a crystal, enough molecules have to be available at the growth front. Transport can be controlled by the concentration of molecules in solution or by their rate of diffusion.
- The probability to attach a molecule at a crystal surface. This probability should not be equal to unity as this would mean that the molecules are fixed at the place of the crystal where they arrive first. This would not allow for the elimination of defects like cavities, as almost no molecules will be able to diffuse into "fjord-like" channels.

The attachment probability can be influenced by several parameters. For example, one can control the attachment probability and consequently the growth rate via temperature. Crystallization is only possible below the melting point of the system. Exactly at the melting temperature, the probability for attaching a molecule to the crystal is so low that the molecule will not stay attached. Thus, the crystal will not grow. Attachment may also be in competition with the displacement of other molecules like solvent molecules, additives or impurities. For example, if an additive is strongly adsorbed onto the crystal surface the crystal will not grow as fast as in cases when these additional molecules are not present.

If the probability of attachment is high (low chance of desorbing molecules from a crystal), crystal growth is dominated by diffusion of the molecules to the crystal front. In this so-called "diffusion limited aggregation" (DLA) regime [2–14] the resulting complex patterns (e.g. "tree-like", dendritic,...) can be characterized according to their morphology and their average density upon increasing size [4]. This density, while being less than the density of an "ideal" faceted crystal, may either be constant (= compact) or decreasing as one moves away from the nucleation site (= fractal).

In the absence of anisotropy of the attachment probability for the various crystal faces, in the absence of anisotropic diffusion or without any selective influence of additives, random growth will be the result, exhibiting no preferred overall directions. Locally, however, the characteristic parameters of the unit cell (like the characteristic angles between the crystal axes) will be respected, which may also be visible on macroscopic scale, for example via a six-fold symmetry of the resulting fractal pattern (compare snowflakes [1] or for thin films of isotactic polystyrene [10, 11, 15]).

It should be noted that all non-ideal, kinetically controlled structures like snowflakes represent out-of-equilibrium objects. That means that they depend on the conditions under which they were formed and also on how they evolved in time after growth (history dependence). Qualitatively, one may state that the faster the patterns were formed, the more they consist of small subunits (thinner branches or smaller objects constituting the pattern [6–8]), the more fragile they are and the more they tend to change with time ("ageing effect") [16, 17].

All the aspects and processes discussed above apply also to the crystallisation of polymers [18–31]. However, in addition, the connectivity of the monomers causes several restrictions. One of the most obvious ones is represented by the quasi-two-dimensional lamellar shape of polymer crystals. A lamella is formed by crystalline segments of the chain (the stems) arranged vertically and limited (on top and below) by amorphous (fold) surfaces. Therefore, polymer crystals grow essentially only in two dimensions. Growth in the third dimension is rather difficult, in particular when the polymer contains non-crystallisable units which will segregate to the surfaces of the crystals. Growth in the third dimension necessitates deviations from the perfect lamellar structure, e.g. screw dislocations. The topic of polymer crystallisation has been the subject of a tremendous amount of studies over the last 60 years [5–8, 10–65].

In the here presented experimental study of crystal growth and morphology development we will control the number of available molecules by using thin films of varying thickness [12]. Thickness also affects the transport process as a thickness-dependent depletion zone forms between crystal front and molten film. The attachment probability will be mainly controlled by the crystallization temperature. Finally, the growth rate depends on both, film thickness and temperature. We mainly focus on the formation of mono-lamellar single crystals. By using block copolymers with one amorphous block, we take advantage of the additional confinement effect imposed by the rather thick glassy layers on the surfaces of the lamellae formed by the amorphous block. Moreover, we have chosen a polymer where the amorphous block also is responsible, in comparison to the corresponding crystallisable homopolymer, for lowering the probability of nucleation. Thus, by effectively suppressing nucleation, we can clearly distinguish between nucleation events and growth processes. We aim at understanding how the various morphologies observed in such thin polymer films can be related to growth processes. In addition, we also try to identify patterns where the polymer aspects become evident or even dominant.

11.2 Experimental Section

Thin Films of P$_2$VP-PEO Diblock Copolymers

For our studies we used a poly-2-vinylpyridine-polyethyleneoxid block copolymer ($P_2VP-PEO$, $M_w = 16400 + 46600$ g/mol, $M_w/M_n < 1.14$) which was synthesized in the group of Martin Möller.

The crystallizable block is polyethyleneoxide, a well investigated polymer [6, 7, 12, 23, 59]. Various studies have also been performed with block copolymers [32, 40–42]. The maximum length of the crystalline PEO block in the fully extended state is: $L = l_u N = 295$ nm with $l_u = 0.2783$ nm [22] and $N = 1060$ is the number of monomers.

Samples were prepared from dilute toluene solutions by spincoating thin films onto the wafers. The polymer adsorbed onto the oxide surface of silicon wafers which were cleaned in a water-saturated UV-ozone atmosphere. By this cleaning procedure, we created a surface with a high density of hydroxyl groups. The initial film thickness has been determined by ellipsometry. Crystallization and subsequent thermal treatment (annealing) were performed directly under an optical microscope in an inert atmosphere (nitrogen flow).

Online inspection of the crystallization process by optical microscopy was used to determine the growth rates. Atomic force microscopy (AFM), a technique often employed for the investigation of polymer crystallization (see for example: [12, 16, 37, 38, 48, 51–58, 60–65]), was used to visualize the resulting crystal morphology in more detail.

Optical Microscopy Measurements

The samples were placed onto an enclosed hotplate, purged with nitrogen, under a Leitz-Metallux 3 or an Olympus BX51 optical microscope. The crystallisation temperature at the hot stage was controlled to within 0.1 degrees. No polarization or phase contrast was used. Contrast is due to the interference of the reflected white light at the substrate/film and film/air interface, resulting in well-defined interference colours which could be calibrated with a resolution of about 10 nanometers. This is sufficient to allow to visualize the growth of crystalline structures even in monolayer regions. For the thicker films around 100 nm (light blue interference colour) studied in this work, the interference patterns became the lighter the thicker the films (or crystals) were. For films around 20-40 nm (brown interference colour), the interference patterns became the darker the thicker the films (or crystals) were. We have followed the progression of the crystal growth front in real time by capturing images by a CCD camera. It should be noted that in all of the thin films studied here, only very few nucleation sites have been observed on the whole sample. At temperatures above about 45°C, no nucleations events were observed for many hours (after cooling down from higher temperatures). In order to allow nonetheless for crystallisation at such high temperatures, the sample was either shortly cooled to lower temperatures or a sort of self-seeding was employed by melting a sample, which was first crystallized at rather low temperatures like room temperature, at temperatures around 65°C for about 1 to 2 minutes and then lowering the temperature to the chosen crystallisation temperature. Experiments at temperatures above about 45°C allowed us to follow the progression of the crystallisation front of single crystals over distances of many 100 micrometers without the interference with any nucleation events.

Atomic Force Microscopy (AFM)

Measurements were performed with a Nanoscope IIIa/ Dimension 3000 (Digital Instruments) in the tapping mode at ambient conditions, using the electronic extender module allowing simultaneously the phase detection and height imaging. We used Si-tips (model TESP) with a resonant frequency of about 160–190 kHz. Scan-rates were between 0.2 and 4 Hz. The free oscillation amplitude of the oscillating cantilever was typically around 50 nm, the setpoint amplitude (damped amplitude, when the tip was in intermittent contact) was slightly lower. Topographic (height mode) and viscoelastic (phase-mode) data were recorded simultaneously.

11.3 Results and Discussion

11.3.1 Changes in Morphology with Crystallization Temperature

It is of course well-known that the morphology of a crystal depends in an clearly visible way on the crystallisation temperature, or more precisely the degree of undercooling. Primarily, this is caused by the temperature dependant rate at which molecules get attached and integrated into a crystalline structure. The slower this rate, the "simpler" are the crystalline structures in terms of morphology. However, we want to emphasize that even crystals of rather complex morphology like dendrites may still be single crystals where all molecules are perfectly ordered with respect to each other.

We investigated crystallization of polymers by following the process in real time under a microscope. In a first approach, we have chosen films of thickness significantly larger than the thickness of one lamella. In Fig. 11.1, a typical result, reminiscent of spherulitic growth, is shown.

At a crystallization temperature (T_c) of 45°C, growth is isotropic in all directions, expressed by a constant radial growth rate of 8.85 μm/min. We note that the orientations of the characteristic features ("fingers") are fluctuating randomly around the radial direction. Although the pattern (see Fig. 11.1C) clearly indicates the competition between the individual fingers, none of these fingers is able to "run away" from all others. At a scale larger than the size of the fingers, the growth front is comparatively smooth. Ahead of this front, a

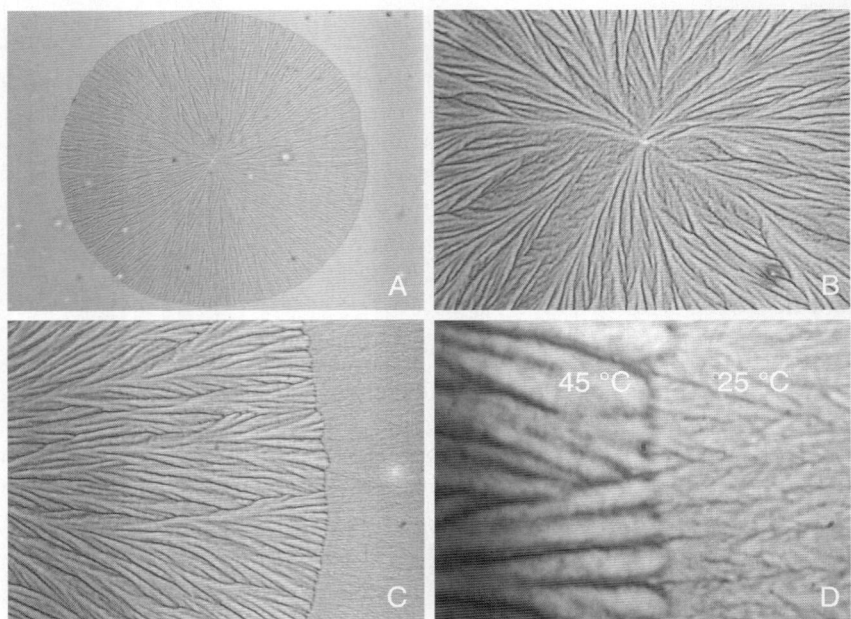

Fig. 11.1. Typical micrographs obtained for crystallization in a 108 nm thick film, taken after growing the crystal for 130 min at 45°C at a radial growth rate of 8.85 μm/min and then quenching the sample to room temperature. The images have the following sizes: (**A**) 865 × 650 μm², (**B**) +(**C**) 174 × 131 μm², and (**D**) 35 × 26 μm²

depletion zone can be observed, particularly visible in the higher resolution of Fig. 11.1D. The finer fingers in the right part of Fig. 11.1D result from faster growth at room temperature. At this thickness the lighter parts (light blue, resulting from interference of white light) are thicker than the darker (dark blue) ones.

At the higher temperature of Fig. 11.2 ($T_c = 50°C$), and in contrast to Fig. 11.1, growth is not isotropic but is clearly "guided" by four orthogonal directions. Perpendicular to these main axis, side branches are formed, from which several generations of always orthogonal branches depart. It is interesting that the hierarchy of side-branches leads to rather smooth lateral faces building the frontiers of square-shaped single-crystal. It should be noted that the thickness of the diagonals is slightly higher than that of all side-branches. The tips of the diagonals can be considered as the driving points of crystallization in this system. As in Fig. 11.1, we can clearly observe a depletion zone ahead of the crystal front. Interestingly, at lengthscales significantly larger than the width of the individual branches, these side-branches seem to create

Fig. 11.2. Typical micrographs obtained for crystallization in a 108 nm thick film. Image (**A**) was taken at the beginning of crystallization while images (**B**)–(**D**) were taken after growing the crystal for 160 min at 50°C at a rate of 3.88 μm/min (along the diagonal) and then quenching the sample to room temperature. The size of all images is: 174×131 μm^2. At this thickness the lighter parts (*light blue*) are thicker than the darker (*dark blue*) ones

a super-structure via the alternating dominance of every fifth or so orthogonal side-branch of the same generation.

We note that such a transition from a circular to a quadratic envelope of the crystals has also been reproduced by computer simulations [13,14]. There, this transition is due to the reduction of the growth front nucleation probability. While the disk-like pattern consists of multiple crystals, the square-shaped pattern represents a single crystal. We thus assume that, for the given film thickness, we observed a transition from a polycrystalline structure to a single crystal within the temperature interval from 45 to 50°C.

When increasing the temperature to $T_c = 55$°C (Fig. 11.3), growth proceeded similar to Fig. 11.2. This can be seen from the dominance of the diagonals. However, the hierarchy of side-branches is less visible because there are fewer but wider side-branches and the gaps between them have been almost completely filled. This was possible as, at such slow growth, transport was not the limiting step. In addition, as more perfect crystals could be formed, the probability for the nucleation of screw dislocations was reduced. It should be

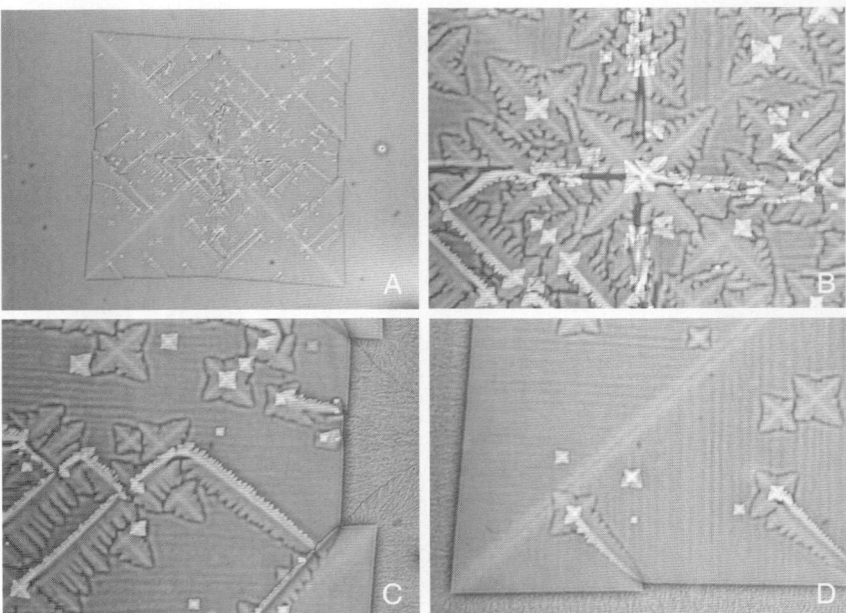

Fig. 11.3. Typical micrographs obtained for crystallization in a 108 nm thick film, taken after growing the crystal for 855 min at 55°C at a rate of about 1 μm/min (along the diagonal) and then quenching the sample to room temperature. The images have the following sizes: (**A**) 865 × 650 μm^2, (**B**)–(**D**) 174 × 131 μm^2. At this thickness the lighter parts (*light blue*) are thicker than the darker (*dark blue*) ones

realized that a 108 nm thick film is thicker than one lamella and thus much more molecules than needed for the formation of a completely filled lamella are available. We also want to draw the attention to the undulations (ripples) clearly seen in Fig. 11.3D, which form parallel lines at an angle of 45 degrees to the leading diagonal. The spacing between these ripples is highly constant. In addition, these ripples are not perturbed by spirals which most likely resulted from dislocations originating at the junction line between side-branches. Thus, we may conclude that these ripples can cross several side-branches. This may indicate that the ripples are not necessarily originating from the crystal and need not to be crystalline either.

11.3.2 Dependence of Morphology on Initial Film Thickness

Looking at Fig. 11.4, it is certainly intriguing to note that the same kind of square shaped envelope as in Figs. 11.2 and 11.3 is also observed for much thinner films, even thinner than the height of the lamella. (Depending on crystallisation temperature, the lamella thickness varied between about 20 nm

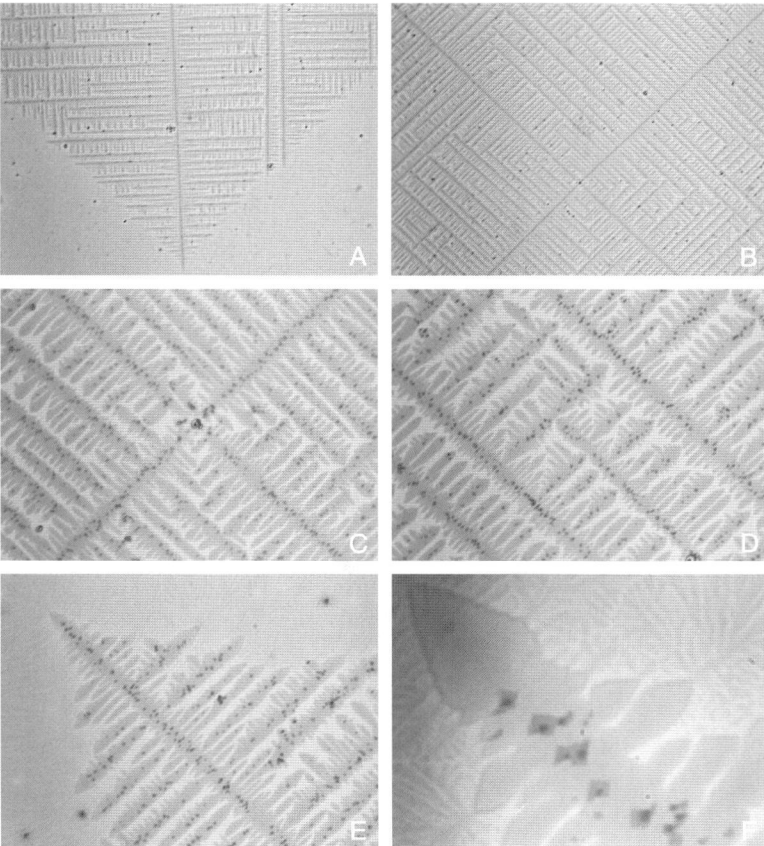

Fig. 11.4. Typical micrographs obtained for crystallization in a 37 nm thick film, taken after growing the crystal for 16 hours at 50°C at a rate of 0.7 µm/min (along the diagonal) and then quenching the sample to room temperature. The images have the following sizes: (**A**) and (**B**) 865 × 650 µm², (**C**)–(**E**) 174 × 131 µm², and (**F**) 35 × 26 µm². At this thickness the darker parts (*brown*) are thicker than the lighter ones

for $T_c = 25°C$ and about 45 nm for $T_c = 55°C$, as measured by AFM. Moreover, the dendritic structure does not differ qualitatively for thick and thin films. However, for the thin film a clear depletion zone, exposing even the substrate, can be seen in between the individual branches. Therefore, we can easily distinguish the multiple generations of side branches, all mutually orthogonal to each other.

When using thicker films (see Fig. 11.5), we qualitatively can find many similarities to the previous Figs. 11.2 to 11.4. The conditions (film thickness and crystallisation temperature) are similar to Fig. 11.3. However, this sample

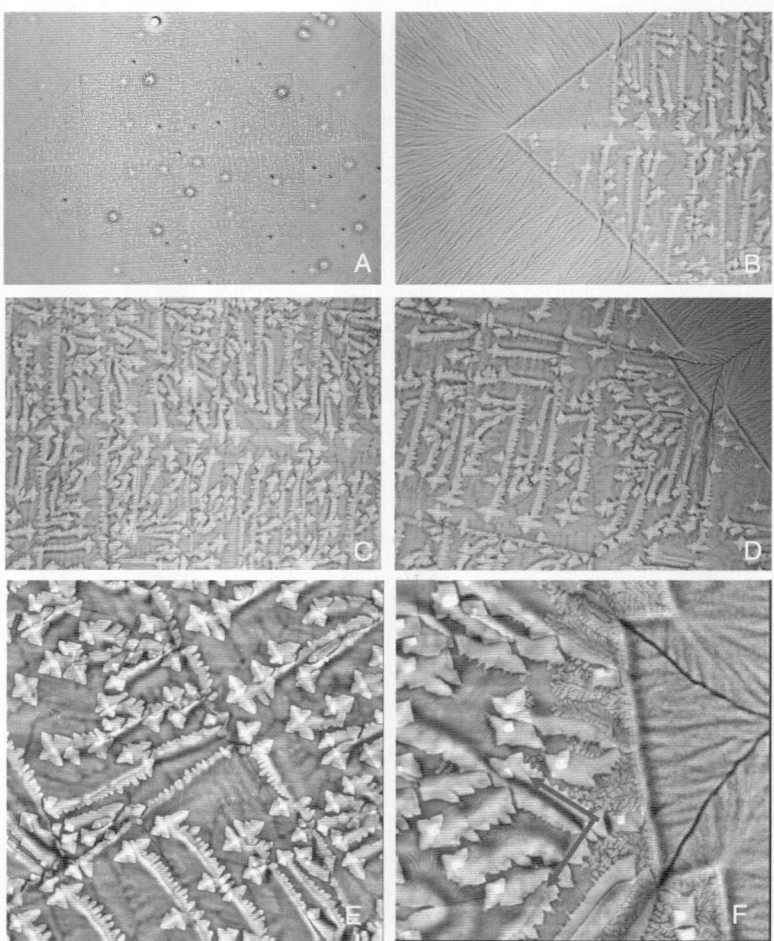

Fig. 11.5. Optical micrographs (**A–D**) and AFM images (**E–F**) obtained for crystallization in a 97 nm thick film, taken after growing the crystal for 893 min at 55°C at a rate of about 0.7 μm/min (along the diagonal) and then quenching the sample to room temperature. The images have the following sizes: A) 865×650 μm^2, (**B**)–(**D**) 174×131 μm^2, (**E**) 70×70 μm^2, and (**F**) 40×40 μm^2. At this thickness the lighter parts in A–D (*light blue*) are thicker than the darker (*dark blue*) ones

showed for an unknown reason a stronger tendency for spiral formation. Thus, the underlying dendritic pattern got nicely decorated with spirals. The interesting point is that due to the square-like geometry of the spirals we are able to conclude that all these spirals are uniformly oriented, in clear registry with the underlying morphology of the orthogonal branches. The diagonal of the square-like spirals are parallel to the diagonals of the underlying large square-shaped crystal. We can also deduce that the spirals are mainly (if not exclusively) formed at junction lines between branches. The key features (diagonals, distance between branches, ...) can be easily anticipated from the arrangement of the spirals. We also can identify the importance of the tips of the branches and the crystal-melt interfaces (CMI) with respect to the growth process. This becomes obvious when analysing these regions after a temperature jump to room temperature (see Figs. 11.5D and 11.5F). There, the finer branched patterns originating from crystallisation at room temperature depart in both directions. On top of the crystalline structure grown at 55°C, these branches grew away from the CMI in the direction opposite to the growth direction at 55°C. In addition, crystalline structures depart also in the other direction away from the CMI.

Growing polymer crystals in films significantly thinner than the lamella and at temperatures close to the melting point leads to rather compact crystals as shown in Fig. 11.6. We emphasize that under such conditions crystals only grow very slowly, allowing the perfecting of the crystals. Therefore, the width of the side branches is becoming significantly larger and only few side-branches are formed. It may be anticipated that the same pattern as in Fig. 11.4 could be observed if the crystals grew up to macroscopic sizes of centimetres.

We also want to draw the attention to the undulations (or ripples) visible at the surface, comparable to the ones seen in Fig. 11.3. In agreement with the observations of [15], the distance between these ripples depends on crystallisation temperature. We point out again that these ripples are at an angle of 45 degrees to the main growth direction represented by the diagonals. At the same time, the ripples are orthogonal to the faces of the square-like envelope of the crystal. We note that such ripples are not only seen in the square-shaped crystals of PEO but similar observations have been made also for the hexagonal crystals of isotactic polystyrene [10, 11, 15].

Although the sample shown in Fig. 11.7 was crystallized under almost the same conditions as the one in Fig. 11.6, the fact that it could grow at a higher rate (due to the higher number of available molecules – the film was about 11 nm thicker – the crystal front moved faster) the crystal front is more prone to become unstable. Therefore, the square-shape envelope of the crystal is not established. Nonetheless, some features like the four-fold symmetry and the dominance of the diagonals are reproduced also in this situation. The ripples are also clearly visible.

After cooling down the sample to room temperature the sample crystallized at a much faster rate and thus we observe much finer patterns. At the crystal front flat-on oriented lamellae were formed, reflecting also the orientation of the crystal formed at 55°C. Interestingly, while the flat-on oriented lamellae originating from the tip survived, the ones further away from the tip were stopped by the arrival of apparently faster growing edge-on lamellae. This competition (simultaneous appearance) of flat-on and edge-on lamellae exists also at the boundaries of the crystal shown in Figs. 11.6B and 11.6C.

Fig. 11.6. Typical optical micrograph (**A**) and AFM height images (**B** and **C**) obtained for crystallization in a 26 nm thick film, taken after growing the crystal for 2300 min at 55°C at a rate of about 0.045 μm/min (along the diagonal) and then quenching the sample to room temperature. The images have the following sizes: (A) 90×86 μm^2, (B) 75×75 μm^2, with a height range of 120 nm, and (C) 38×38 μm^2, with a height range of 70 nm. At this thickness the darker parts (*brown*) in (A) are thicker than the lighter ones

11.3.3 The Kinetics of Crystal Growth and the Effect of Changing Temperature

In order to visualize how dendritic crystals grow and how a square-shape envelope is formed from such dendritic structures, we have superposed two images from the same crystal taken at an interval of 95 sec (see Fig. 11.8). As we want to focus on single crystals only, we have chosen a rather thin film of about 40 nm in order to avoid growth front nucleation (GFN) [13, 14]. GFN depends on the number of molecules present. Thus, in such thin films the transition from poly-crystals to single crystals can occur at lower temperatures than in the 108 nm thick films discussed above.

In the centre of Fig. 11.8, we can clearly identify the square consisting of a dentrictic structure of comparatively small side-branches, resulting from

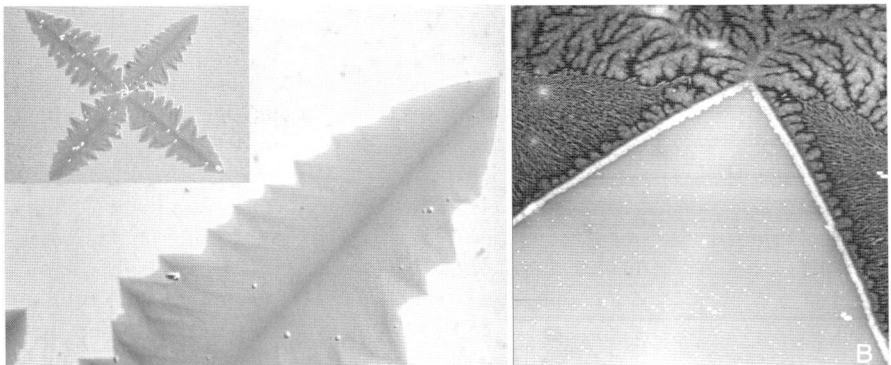

Fig. 11.7. Typical optical micrograph (**A**) and AFM image (**B**) obtained for crystallization in a 37 nm thick film, taken after growing the crystal for 2300 min at 55°C at a rate of about 0.2 µm/min and then quenching the sample to room temperature. The images have the following sizes: (A) 174×131 µm^2 (the size of the inset is: 500×377 µm^2) and (B) 25×25 µm^2, with a height range of 100 nm. At this thickness the darker parts (brown) in (A) are thicker than the lighter ones

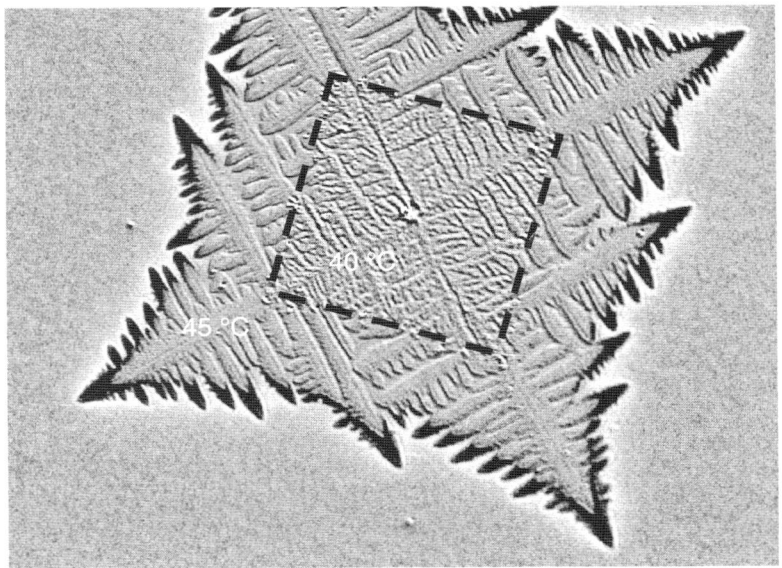

Fig. 11.8. Superposition of two optical micrographs taken at an interval of 95 sec for a 40 nm thick film. For the first image, the sample was first kept at 40°C for 300 seconds and then for 480 seconds at 45°C. The difference between the two images, i.e. the distance the crystal grew, is represented in black. The size of the image is 174×131 µm^2

crystallisation at 40°C at a rate of about 7.3μm/min. The outer part was grown at 45°C at a rate of about 5.5μm/min.

The superposition of two images at an interval of 95 sec allows to clearly identify the points where the crystal grew. Growth proceeded mostly via the primary tips along the diagonals and some of the major side branches. Secondary branches were much less contributing. We also note that the structures are not fully symmetric, neither around one individual diagonal nor when comparing the detailed structures of "trees" along different diagonals. We can conclude from this figure that the envelope of the crystals can only be approximated by a straight line if the number of dominant side-branches is large.

We have already mentioned several times before that faster growing crystals exhibit a higher frequency of side-branching. In Fig. 11.9, we present such a transition from small to wide side-branches in a single experiment by applying a temperature jump from 50°C to 55°C. It can be clearly seen that at the lower temperature four main "dendritic trees" are growing in orthogonal directions. At such conditions, the envelope is rather quadratic. We want to draw the attention also to the junction lines where the trees meet, which are at an angle of 45 degrees to the diagonals of the square. However, after raising the temperature, the number of side-branches is significantly reduced (on the average about 4 branches are fused to one). It may be anticipated that also at this higher crystallization temperature, after appropriate rescaling of the image size, the same qualitative quadratic shape of the crystal would be obtained but after significantly longer crystallization time.

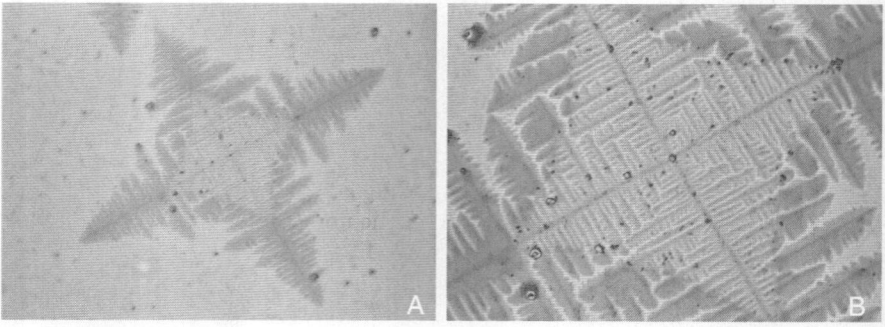

Fig. 11.9. Optical micrographs of a crystal in an about 32 nm thick film which grew first for 24 hours at 50°C and then for 72 hours at 55°C before the sample was cooled to room temperature. The size of the images is (**A**) 430 × 235 μm^2, (**B**) 174 × 131 μm^2

11.3.4 "Decoration" of Flat-On Lamellar Crystals by Ripples and Spirals

Optical microscopy allows to follow crystal growth in real time. Due to the interference effect, which is the advantage of reflecting substrates such as silicon wafers, we can demonstrate that the ripples were formed *during* growth and are not, for example, the consequence of a mechanical instability induced by a temperature jump. In Fig. 11.10, it can been seen that these ripples existed already during growth at 57°C, and were continuously formed as the crystals grew. While the ripples are perpendicular to the (lateral) sides of the squares, the spirals, once nucleated at the growth front, form tails at an angle of 45 degrees. It is highly probable that the spirals, and the subsequent tails, formed at the junction line between two underlying side-branches. This is consistent with our conclusions drawn in the context of Fig. 11.3.

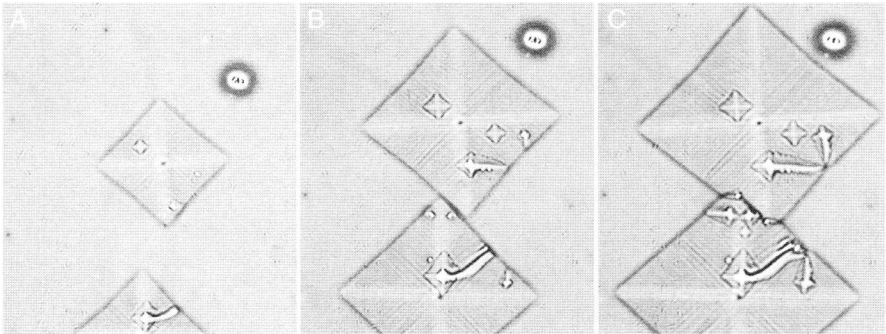

Fig. 11.10. Series of three optical micrographs taken at an interval of (**B**) 69 min and (**C**) 114 min for a 102 nm thick film during crystallization at 57°C. The size of the images is 35×41 μm^2

From Fig. 11.11, we can conclude, taking advantage of the square-shaped spirals, that *all* spirals are oriented along the underlying individual branches. *All* spirals have exactly the same orientation of their diagonals as the dendrites seen in the optical micrograph. Thus, it is not surprising that all spirals are correlated among themselves, as can be nicely seen in Figs. 11.11C and 11.11D.

The process of spiral formation can be deduced from details observed by AFM, as shown in Fig. 11.12. A film of 102 nm is more than twice as thick as one lamellar thickness (about 45 nm at 57°C). Thus, the first lamella, growing close to the substrate is still covered with a large reservoir of non-crystalline – but crystallizable – polymers. From our experiments, it turned out that the junction lines between side branches are regions of high probability for spiral nucleation. These spirals may either form a new lamella, similar in shape as the underlying square-shaped lamella (see Figs. 11.10 and 11.12A). Or, as shown in Figs. 11.10 and 11.12C, such spirals may be accompanied by a

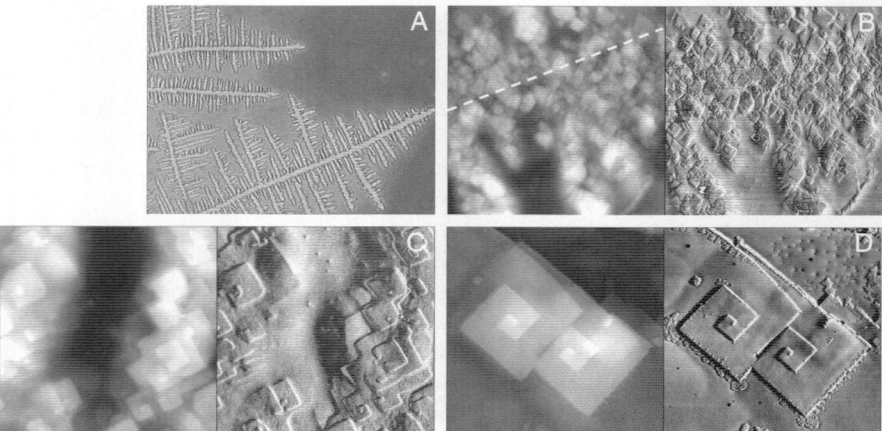

Fig. 11.11. (**A**): Optical micrograph of an about 70 nm thick film taken after growing the crystal for 66 hours at 40°C and then quenching the sample to room temperature. (**B**) and (**C**): AFM images (*left*: topography and *right*: phase) of the same crystal. (**D**): AFM image (left:topography and right: phase) of spirals formed in a 50 nm thick film, crystallized for 72 hours at 55°C. The size of the images is (A) 290 × 218 µm², (B) 15 × 15 µm², (C) 5.5 × 5.5 µm², (D) 15 × 15 µm²

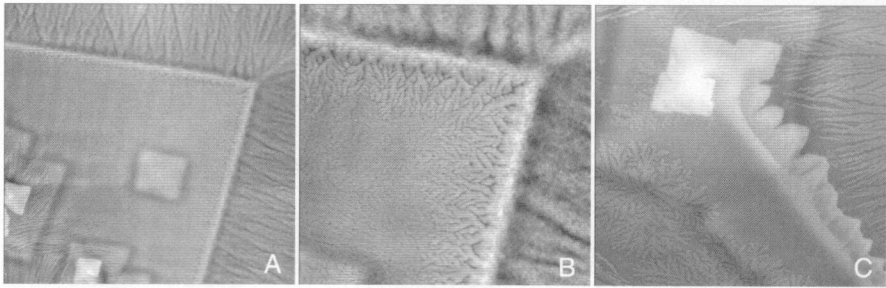

Fig. 11.12. AFM images showing some details on spirals and "secondary" crystalline structures formed on top of a crystal for a 102 nm thick film crystallized at 57°C and then quenched to room temperature. Besides the spirals which grew at 57°C, we also can identify crystals grown at room temperature. The size of the images is (**A**) 50 × 50 µm² (height range: 150 nm), (**B**) 20 × 20 µm² (height range: 80 nm), (**C**) 20 × 20 µm² (height range: 180 nm)

tail which most likely "decorates" the junction line between side-branches. The basic mechanism for why such junction lines represent locations of high probability for spiral nucleation is not yet clear. Interestingly, the growth fronts also present locations of high nucleation probability as can be seen in Fig. 11.12B (see also Fig. 11.5). There, after cooling the sample to room temperature, many flat-on lamellae depart in both directions, backwards and forwards.

11.3.5 Orientation of the Crystalline Lamellae with Respect to the Substrate

As a final remark, we would like to mention that the orientation of the lamellae is not necessarily parallel to the substrate. Actually, as already indicated in Figs. 11.6 and 11.7 for thin films, both orientations of the lamellae can occur simultaneously. Interestingly, also for thicker films, Figs. 11.13 and 11.14 indicate that while at low temperatures most likely edge-on lamellae are formed, flat-on structures are more favourable at higher temperatures. The wedge-like structure seen in optical microscopy can be identified as flat-on lamellae. The fact that such wedges are widening demonstrates that the growth rate of flat-on structures at the higher crystallisation temperatures is slightly higher than the one of the edge-on lamellae. Thus, after sufficiently long time, and assuming that nucleation is not the dominating process, only flat-on lamellae continue to grow. In Fig. 11.13C, we can clearly see the coexistence of both

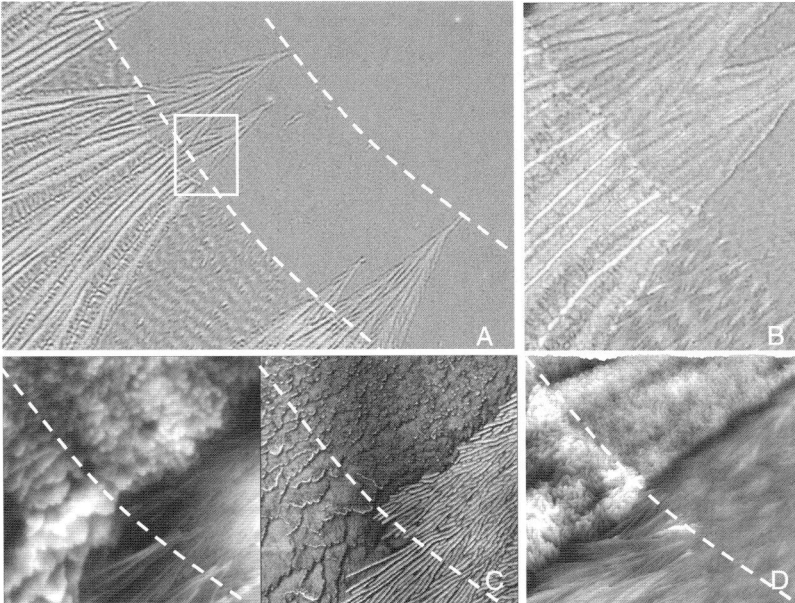

Fig. 11.13. Optical micrographs (**A**+**B**) and corresponding AFM images (**C**+**D**) showing that edge-on and flat-on lamellar structures can be formed simultaneously in a thin film. The temperature was raised in two steps from 40°C to 45°C and finally to 50°C. The film thickness is about 90 nm. The temperature transitions are indicated by broken lines. The lowest temperature is in the upper right corner of (**A**). Parts (**B**) and (**C**) represent the region selected by the white rectangle in A). In part C, both topography and phase image are shown. (**D**) is a 3D-representation of the transition zone between 40°C and 45°C. The size of the images is (A) 200×140 μm^2, (B) 47×63 μm^2, (C) 2×2 μm^2, and (D) 10×10 μm^2

196   G. Reiter et al.

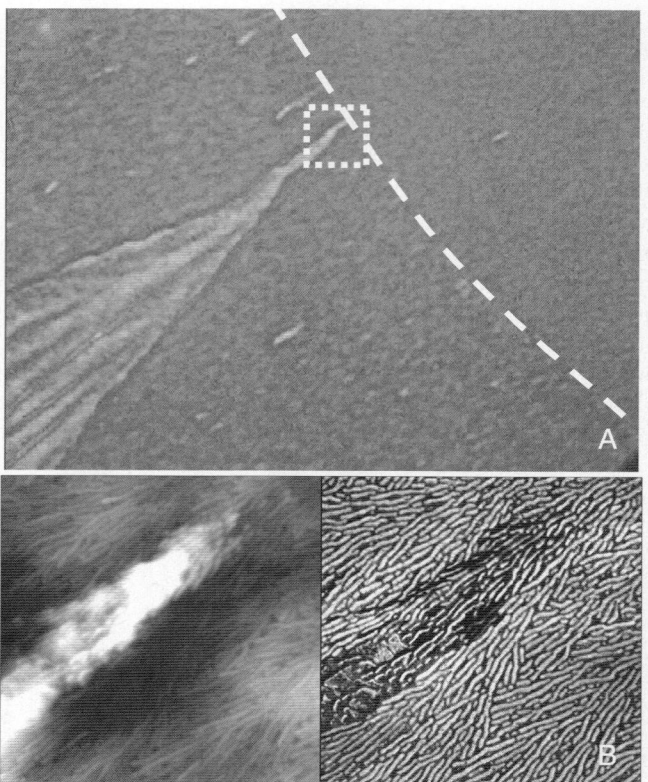

Fig. 11.14. (**A**) Optical micrograph (A) and (**B**) corresponding AFM image for the section indicated by the white square in (A) (including topography and phase information) showing the "birth" of a flat-on lamellar structure embedded within edge-on lamellae after the temperature was raised in a step (indicated by the *broken line*) from 40°C to 45°C. The film thickness is about 90 nm. The size of the images is (A) 63 × 47 µm², (B) 1.5 × 1.5 µm², height range: 40 nm

types of lamellae. It can also be seen that after increasing the temperature, the edge-on lamellae are more straight and the lateral size of the piled-up flat-on lamellae, probably formed by spirals resulting from screw dislocations, is larger. The more pronounced height variations seen in Fig. 11.13D indicate that the flat-on lamellae have the tendency to grow more in the direction normal to the substrate than the edge-on lamellae. The surface of flat-on lamellae is rougher than the edge-on counterparts. Thus, in interpreting optical microscopy images one has to take into account that more light is scattered on flat-on structures and that the edge-on lamellae appear smoother.

In Fig. 11.14 we can identify several attempts for the onset (nucleation) of flat-on lamellae. However, at the temperature of 45°C, even if nucleation of flat-on structures does not seem to be the limiting step, only a few of these attempts finally grew further than a few micrometers. In our opinion, this indicates that the selection mechanism for the orientation is based on kinetics arguments. Only the fastest growing lamellae will "survive". Probably, in films thicker than one lamella, the orientation of polymer stems in the nuclei, which frequently form at the growth front, is not predetermined. In films thinner than one lamella, however, nucleation of edge-on lamellae may be more difficult than flat-on structures. However, if edge-on lamellae are somehow nucleated they may grow significantly faster.

11.4 Conclusions

The here presented examples of morphologies of lamellar polymer crystals strongly suggest that the concept of DLA is the most appropriate way of describing how these crystals have been formed. As this concept can be applied to most crystallising materials, we conclude that polymer crystallisation can be treated in such a unified way. It is not necessary to invoke special approaches to explain the various morphologies of polymer crystals.

Although we did not discuss the problem of nucleation in any depth, it is clear that, for the here studied system, it is extremely difficult to initiate crystallisation. Taking advantage of this difficulty allowed us to separate the nucleation step from the subsequent growth process and thus to analyse exclusively the growth kinetics of independent single crystals.

It is quite intriguing to observe that even the simple square-shaped crystals without any empty sites within their envelope seem to result from dendritic growth. There, the positions and the orientations of the square-shaped spirals decorating such "filled" crystals are indicating the growth pattern of the underlying first lamella. The orientation of the spirals is fully parallel to the orientation of the initial lamella. Moreover, these spirals are not distributed fully randomly on the surface. They are located along the supposed junction lines of the side-branches of the underlying dentritic structure. Looking only at the square-shape envelope of such crystals, one may erroneously assume that they grew by attaching polymers to the sides. However, our results indicate that growth is dictated by the tips along the diagonals. One may distinguish several stages during the growth of such crystals. The crystals grow first along these diagonals. As a consequence of the diffusion limited transport process, side branches form in the direction orthogonal to the diagonals. Successive formation of more and more side branches, also of higher generations leads to the formation of the square shape envelope. At a length-scale much larger than the width of a single side-branch this envelope can be approximated quite well by straight lines.

Experiments with films of different thickness show that the quasi-two-dimensional growth morphologies are qualitatively the same, independent if more or less molecules than necessary for the formation of a single lamella are available. The major difference of thicker films arises from the possibility to add polymers in between side branches by taking them from the "reservoir" on top of the lamella. This allows to close the gaps between side-branches. Our experiments indicate that this growth mode also favours the formation of spirals at the junctions between side-branches. One may speculate that the formation of such spirals, leading to stacked lamellae, is the major mode for growing the polymer crystals in three dimensions, even for single crystals.

The slow growth rates together with the possibility to form simple planar morphologies make polymer crystals ideal model systems for fundamental studies of crystal growth. Thus, combining these model experiments with theoretical concepts, including computer simulations, provides a highly promising approach for improving our understanding of polymer crystallisation and may also shed some light on central questions of crystal growth in general.

Acknowledgements

We are indebted to Dr. Cvetelin Vasilev for some preliminary experiments and to Bernard Lotz, Dimitri Ivanov, Gert Strobl, Wenbing Hu and Jens-Uwe Sommer for fruitful discussions. We acknowledge financial support provided through the the European Community's "Marie-Curie Actions" under contracts MRTN-CT-2003-505027 [POLYAMPHI] and MRTN-CT-2004-005516 [BioPolySurf].

References

[1] K. G. Libbrecht: Rep. Prog. Phys. **68**, 855 (2005).
[2] J. Langer: Rev. Mod. Phys. **52**, 1 (1980).
[3] P. Meakin, *Fractals, scaling and growth far from equilibrium*, Vol. 5 of Cambridge Nonlinear Science Series, Cambridge University Press, Cambridge (1998).
[4] E. Brener, H. Müller-Krumbhaar, D. Temkin: Phys. Rev. E, **54**, 2714 (1996).
[5] Y. Sakai, M. Imai, K. Kaji, M. Tsuji: J. Crystal Growth **203**, 244 (1999).
[6] G. Reiter, J.-U. Sommer: Phys. Rev. Lett. **80**, 3771 (1998).
[7] G. Reiter, J.-U. Sommer: J. Chem. Phys. **112**, 4376 (2000).
[8] J.-U. Sommer, G. Reiter: J. Chem. Phys. **112**, 4384 (2000).
[9] A. Holzwarth, S. Leporatti, H. Riegler: Europhys. Lett. **52**, 653 (2000).
[10] K. Taguchi, H. Miyaji, K. Izumi et al: Polymer **42**, 7443 (2001).
[11] K. Taguchi, H. Miyaji, K. Izumi et al: J. Macromol. Sci. B: Phys **41**, 1033 (2002).
[12] M.V. Massa, K. Dalnoki-Veress, J.A. Forrest: Eur. Phys. J. E **11**, 191 (2003).
[13] L. Granasy, T. Pusztai, T. Börzsönyi et al: Nat. Mat. **3**, 645 (2004).
[14] L. Granasy, T. Pusztai, G. Tegze et al: Phys. Rev. E **72**, 011605 (2005).
[15] K. Taguchi, Y. Miyamoto, H. Miyaji et al: Macromolecules **36**, 5208 (2003).
[16] G. Reiter, G. Castelein, J.-U. Sommer: Phys. Rev. Lett. **86**, 5918 (2001).

[17] J.-U. Sommer, G. Reiter: Europhys. Lett. **56**, 755 (2001).
[18] P. H. Geil, D. H. Reneker: J. Polym. Sci., **51**, 569 (1961).
[19] E.W. Fischer: Kolloid Z. u. Z. Polym., **231**, 458 (1968).
[20] A. Keller: Rep. Prog. Phys. **31**, 623 (1968).
[21] A. Kovacs, A. Gonthier, C. Straupe: J. Polym. Sci.: Polym. Symp. **50**, 283 (1975).
[22] A. Kovacs, C. Straupe, A. Gonthier: J. Polym. Sci.: Polym. Symp. **59**, 31 (1977).
[23] A.Kovacs, C. Straupe, J. Cryst. Growth: **48**, 210 (1980).
[24] G. Strobl: The Physics of Polymers, Springer, Berlin, Heidelberg, N.Y., 1997.
[25] A. Keller, S.Z.D. Cheng: Polymer **39**, 4461 (1998).
[26] K. Armistead and G. Goldbeck-Wood: Adv. Polym. Sci., **100**, 219 (1992).
[27] D. M. Sadler, G. H. Gilmer: Polymer **25**, 1446 (1984).
[28] H. D. Keith, F. J. Padden, Jr., B. Lotz et al: Macromolecules **22**, 2230 (1989).
[29] A. Toda, A. Keller: Colloid Polym. Sci. **271**, 328 (1993).
[30] S. Z. D. Cheng, B. Lotz: Polymer **46**, 8662 (2005)
[31] G. Strobl: Eur. Phys. J. E **18**, 295 (2005).
[32] B. Lotz, A.J. Kovacs: ACS Polym. Prepr. **10**, No. 2, 820 (1969).
[33] C. W. Frank, V. Rao, M. M. Despotopoulou et al: Science **273**, 912 (1996).
[34] G. Strobl, Eur. Phys. J. E: **3**, 165 (2000).
[35] G. Reiter, et al.: Phys. Rev. Lett. **83**, 3844 (1999).
[36] A. Winkel, J. Hobbs, M. Miles: Polymer **41**, 8791 (2000).
[37] C. Basire, D.A. Ivanov: Phys. Rev. Lett. **85**, 5587 (2000).
[38] Y.K. Godovsky, S. Magonov: Langmuir **16**, 3549 (2000).
[39] L. Li, C.-M. Chan, K.L. Yeung: J.-X. Li, K.-M. Ng, Y. Lei, Macromolecules **34**, 316 (2001).
[40] Y.-L. Loo, R.A. Register, A.J. Ryan: Phys. Rev. Lett. **84**, 4120 (2000).
[41] C. De Rosa, C. Park, E.L. Thomas, B. Lotz: Nature **405**, 433 (2000).
[42] G. Reiter, G. Castelein, J.-U. Sommer, A. Röttele, T. Thurn-Albrecht: Phys. Rev. Lett. **87**, 226101 (2001).
[43] P. N. Chaturvedi: J. Mater. Sci. **29**, 3749 (1994).
[44] C. Liu, M. Muthukumar: J. Chem. Phys. **109**, 2536 (1998).
[45] P.D. Olmsted et al.: Phys. Rev. Lett. **81**, 373 (1998).
[46] W. Zhou et al.: Macromolecules **33**, 6861 (2000).
[47] M. Psarski, E. Piorkowska, A. Galeski: Macromolecules **33**, 916 (2000).
[48] F. Zhang, J. Liu, H. Huang et al: Eur. Phys. J. E **8**, 289 (2002).
[49] Y. Kikkawa, H. Abe, T. Iwata et al : Biomacromolecules **3**, 350 (2002).
[50] Y. Kikkawa, H. Abe, M. Fujita et al: Macromol. Chem. Phys. **204**, 1822 (2003).
[51] M. Wang, H.-G. Braun, E. Meyer: Polymer **44**, 5015 (2003).
[52] E.-Q. Chen, A. J. Jing, X. Wenig et al: Polymer **44**, 6051 (2003).
[53] R. Kurimoto, A. Kawaguchi: J. Macromol. Sci. B: Phys **42**, 441 (2003).
[54] Y.-G. Lei, C.-M. Chan, Y. Wang et al: Polymer **44**, 4673 (2003).
[55] R. Mehta, W. Keawwattana, A. L. Guenthner et al: Phys. Rev. E **69**, 061802 (2004).
[56] M. Tian, M. Dosiere, S. Hocquet et al: Macromolecules **37**, 1333 (2004).
[57] A. Tracz, I. Kucinska, J. K. Jeszka: Macromolecules **36**, 10130 (2003).
[58] N. Dubreuil, S. Hocquet, M. Dosière et al: Macromolecules **37** (1), 1 (2004).
[59] M. V. Massa, K. Dalnoki-Veress: Phys. Rev. Lett. **92**, 255509 (2004).
[60] Y. Wang, S. Ge, M. Rafailovich et al: Macromolecules **37**, 3319 (2004).
[61] A. Toda, M. Okamura, M. Hikosaka et al: Polymer **46**, 8708 (2005).

[62] C. Qiao, J. Zhao, S. Jiang et al: J. Polym. Sci., Part B: Polym. Phys. **43**, 1303 (2005).
[63] X.-M. Zhai, W. Wang, Z.-P. Ma et al : Macromolecules **38**, 1717 (2005).
[64] H. Xu, R. Matkar, T. Kyu: Phys. Rev. E **72**, 011804 (2005).
[65] V. H. Mareau, R. E. Prud'homme: Macromolecules **38**, 398 (2005).

12

Crystallization of Frustrated Alkyl Groups in Polymeric Systems Containing Octadecylmethacrylate

Elke Hempel[1], Hendrik Budde[2,3], Siegfried Höring[2], and Mario Beiner[1]

[1] FB Physik, Martin-Luther-Universität Halle-Wittenberg, D-06099 Halle, Germany
beiner@physik.uni-halle.de
[2] FB Chemie, Martin-Luther-Universität Halle-Wittenberg, D-06099 Halle, Germany
[3] Fraunhofer Pilotanlagenzentrum für Polymersynthese und -verarbeitung, Value Park, Bau A74, D-06258 Schkopau, Germany

Abstract. This chapter deals with the crystallization behavior of long frustrated alkyl groups as part of side chain polymers. Results from crystallization experiments on poly(n-octadecylmethacrylate) [PODMA] homopolymers with different molecular weight and on microphase-separated poly(styrene−*block*−octadecylmethacrylate) block copolymers [P(S−*b*−ODMA] by calorimetry and scattering techniques are presented. A phenomenological picture describing the different stages of the side chain crystallization in PODMA is given. The influence of additional constraints in P(S−*b*−ODMA) block copolymers containing PODMA lamellae or cylinders in a glassy environment is studied. It is shown that these block copolymers are systems with a hierarchy of length scales in the nanometer range. The crystallization behavior in PODMA lamellae is basically bulk-like while strong confinement effects are indicated in case of PODMA cylinders with a diameter <20 nm by increasing crystallization times and broader transformation intervals in isothermal crystallization curves. These effects can be explained by changes concerning nucleation mechanism and crystal growth. The reduced degree of crystallinity in small cylinders is discussed based on hypothetical pictures for the internal structure of the PODMA domains.

12.1 Introduction

Most of the work in the field of polymer crystallization in the last decades has been done on long polymer chains like polyethylene, polypropylene, poly(ethylene oxide) or poly(ethylene terephthalate) which form folded-chain crystals with a typical thickness of about 10 nm [1–5]. In all these cases chain folding plays an important role and crystallization starts from a chemically homogeneous melt. More recently, the confined crystallization of polymers in small domains of microphase-separated block copolymers has attracted a lot of attention [6, 7]. The influence of constraints on the crystallization behavior of main chain polymers has been studied in small self-assembled domains with dimensions in the range 10-50 nm. There are different aspects of the crystallization process which can be affected under these conditions. As long as crystallization does not destroy the morphology of the initially microphase-separated block copolymer the influence of confinements on crystal growth and nucleation behavior can be studied [8]. Depending on the morphology the individual crystals can grow only in one (cylinders) or two dimensions (lamellae) [9]. Moreover, a transition from heterogeneous to homogeneous nucleation occurs if the individual compartments are isolated from each other [10, 11]. Nucleation experiments on microphase-separated block copolymers can provide not only average nucleation rates like classical experiments on homogeneous nucleation [12] but also spatially resolved information about the nucleation of individual domains since the crystallizable material is embedded in a solid-like (usually glassy) matrix. Note, that in practically all these cases block copolymers have been used where the crystallizable component is a main chain polymer. Thus, long chains must reorganize in small domains before crystallization can take place. Although there is an incredible number of experimental results and various theoretical approaches basic aspects of the polymer crystallization are still not finally understood. Details of nucleation mechanism and early stages of crystallization [13, 14] are not yet clear and even the question whether the crystal thickness is determined by equilibrium thermodynamics or a consequence of non-equilibrium effects is controversially debated [15, 16].

Compared to the variety of studies on crystallizable main chain polymers and the accumulated phenomenological knowledge about their crystallization behavior relatively less is known about the crystallization of side chain polymers. Only some studies reporting details on the side chain crystallization process can be found in the literature [17–21] although polymers with long alkyl groups in the side chain are studied since the early days of polymer research [22]. Recent experiments on amorphous side chain polymers like methacrylates [23, 24], acrylates [24], itaconates [17, 25] or hairy rod polyimides [26] have shown that there is a nanophase separation of main and side chain parts already in the melt. Long alkyl groups belonging to different monomeric units aggregate in alkyl nanodomains with a typical size in the range 0.5-2 nm [24]. If the alkyl groups are long and flexible enough side chain crystallization can occur [27]. In contrast to the situation in main chain polymers the crystallization process in these side chain polymers starts from a nanostructured melt. The confined side chain crystallization process in microphase-separated block copolymers has not been investigated in great detail so far. Some studies which are more focused on the structure of block copolymers containing liquid-crystalline side chain polymers have been presented recently [28–31]. The results indicate that block copolymers with one component being a side chain polymer are interesting materials which show potentially a hierarchy of length scales in the nanometer range. It seems to be interesting to see to which extent side chain crystallization and nanophase separation are affected by the concurrence of two length scales in the nanometer range in microphase-separated block copolymers with a crystallizable side chain polymer as one component.

The following chapter deals with the side chain crystallization process in poly(n-octadecylmethacrylate) homopolymers [PODMA] with $C = 18$ alkyl carbon per side chain as well as microphase-separated poly(styrene$-b-$octadecylmethacrylate) block copolymers [P(S$-b-$ODMA)] containing lamellar and cylindrical PODMA domains. Results from calorimetric measurements and X-ray as well as synchrotron scattering data will be presented. The chapter is divided in two sections: The first part is related to the crystallization behavior PODMA homopolymers with different molecular weight. The influence of immobile main chains on the crystallization of long frustrated alkyl groups aggregated in alkyl nanodomains with a typical dimension of about 2 nm will be discussed. A four stage picture for the crystallization of frustrated alkyl groups is proposed. The second part deals with the confined side chain crystallization in small (10-25 nm) PODMA domains of microphase-separated P(S$-b-$ODMA) block copolymers. The influence of constraints introduced by the glassy polystyrene phase on the crystallization kinetics is shown. Changes concerning crystal growth, nucleation behavior and degree of crystallinity are discussed and speculative pictures for the internal structure of lamellar and cylindrical PODMA domains are considered.

12.2 Side-chain Crystallization in Poly(*n*-octadecylmethacrylate)

Higher poly(*n*-alkyl methacrylates) have been studied since the early 1950s by various experimental techniques like dilatometry, x-ray scattering, calorimetry, dielectrics and shear spectroscopy [32–36]. Most of the studies deal with relaxation behavior and properties of amorphous members although it was already known at that time that the alkyl groups in the side chains can crystallize if they are sufficiently long. Atactic poly(n-alkyl methacrylates) with $C \geq 12$ alkyl carbons per side chain are able to crystallize. About further details of the side chain crystallization in poly(n-alkyl methacrylates), however, is still not too much known. There are only a few papers in the literature which are focused on this topic [19, 37]. Otherwise, higher poly(n-alkyl methacrylates) belong to the interesting class of materials where long CH_2 sequences can crystallize with the specialty that the microstructure restricts crystallization to sequences with a finite length.

Crystalline systems containing long sequences of CH_2 units are of extraordinary importance in many fields. *Polyethylene* is the mostly used polymer with a variety of technical applications but has a lot of unusual features which are not completely understood. Peculiarities are the large mobility of chain segments in the crystalline state, significantly thickening of the crystals during the crystallization process [38], a transition from orthorhombic to hexagonal packing in "defected" polyethylenes [39] or the occurrence of shish-kebab structures in polyethylene blends crystallized under shear [40,41]. *Linear alkanes* with 10 to 25 carbons are basic ingredients of petroleum and have been studied in any detail [12, 42–45]. Also these molecules show peculiarities namely the occurrence of hexagonally packed mesophases – so called rotator phases – in a narrow temperature interval between liquid and orthorhombic phase [43, 46, 47] and surface induced crystallization at the liquid-to-rotator transition [48, 49]. *Phospholipids* consisting of a hydrophilic head group and long hydrophobic alkyl groups are a third class of materials where alkyl sequences can crystallize in several modifications. Various membranes and vesicles in biological systems are designed by nature based on phospholipids. Somehow the crystallization in side chain polymers with long alkyl groups can be seen as a crystallization of alkanes in the presence of external constraints. Similarities to the situation of short polyethylene sequences stretched between two entanglements are obvious. Interestingly some of the typical features of polyethylene and alkanes seem to survive in crystallizable side chain polymers with long alkyl groups. This may indicate that the crystallization behavior of alkyl groups is modified but not completely changed if they are frustrated since they are bonded to a main chain with lower mobility. Thus, we believe that studies on side chain polymers with long alkyl groups are not only important to understand the crystallization behavior of these systems but may also contribute to a better understanding of peculiarities in other system containing long CH_2 sequences.

In the recent years it has been shown based on x-ray scattering and relaxation spectroscopy data for a series of atactic poly(n-alkyl methacrylates) series that long alkyl groups belonging to different monomeric units and chains aggregate in the melt [23, 24, 50]. Small alkyl nanodomains with a typical dimension in the range 0.5–2 nm are formed in systems with $C = 4$ to $C = 18$ alkyl carbons per side chain. Depending on the length of the alkyl groups or equivalently the size of the alkyl nanodomains the CH_2 units either remain disordered for $C < 12$ and show an independent dynamics in form of a "polyethylene-like glass transition" reflecting the independence of the dynamics of the alkyl groups [24] or undergo for $C > 12$ a crystallization process [32, 51]. The transition between both situation is interesting because early stages of crystallization might be stabilized due to frustration of the alkyl groups by the immobile main chains. In the following part results for the side chain crystallization in poly($n-$octadecylmethacrylate) with $C = 18$ alkyl carbons per side chain will be presented. Atactic PODMA samples with different molecular weights synthesized by anionic polymerization are investigated (Table 12.1).

Scattering data for high molecular weight PODMAs in the molten state show a relatively broad prepeak with a maximum at $q \approx 2.1$ nm^{-1} (Fig. 12.1). According to a Bragg approximation this peak corresponds to a distance

Table 12.1. Characterization of the PODMA homopolymers and parameters describing the crystallization process

Sample*	M_n	M_w/M_n	D_c	$T_m^{max\dagger}$	$T_c^{max\dagger}$	$u^\#$	$n^\ddagger$
	kg/mol		mol%	°C	°C	10^{-3}	
ODMA8	2.8	2.3	29	33	23	34 (23)	4.5
ODMA16	5.3	1.8	30	34	24	34	4.0
ODMA27	9.0	2.1	31	32	25	34 (26)	4.1
ODMA53	18.1	1.5	28	33	23	34 (24)	3.7
ODMA88	29.8	1.8	27	34	24	34	3.4
ODMA100	33.7	1.8	29	35	25	34	3.5

* The numbers give the degree of polymerization P. † Peak maxima from scans with rates of ± 10 K/min. # Taken from scans with rates of ± 10 K/min (± 1 K/min).
‡ Taken from Avrami plots (cf. Fig. 12.12). The uncertainty is about ± 0.2. The width of the transformation interval can be estimated according to $\Delta \log t_c = 1.253/n$ (cf. [37]).

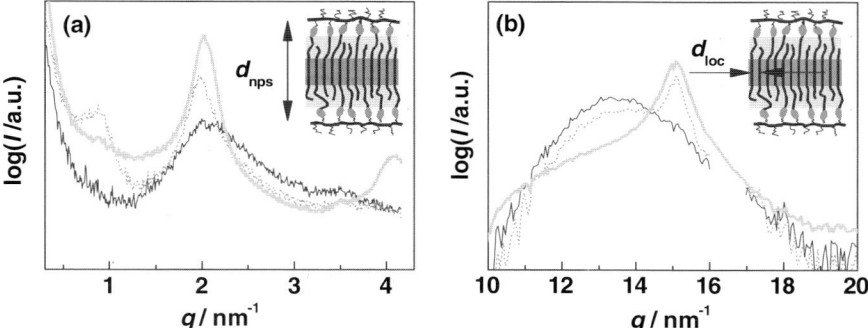

Fig. 12.1. Small (**a**) and wide (**b**) angle scattering data for PODMA27 in the amorphous state (*thin lines*) and two semi-crystalline (*dotted and thick lines*) states measured simultaneously on beam line BM26 at the ESRF in Grenoble. The amorphous state corresponds to a sample which is measured 1min after a quench from 50 to 25°C. The semi-crystalline samples are crystallized at room temperature for 2min (*dotted*) or more than one week (*thick*). Schematic pictures of the structure of semi-crystalline PODMA are shown in the insets (*small ellipses*: carboxyl groups; *light gray*: alkyl nanodomains; *dark gray*: crystalline layer). The length scales d_{nps} and d_{loc} corresponding to the scattering peaks in the small and wide angle ranges are indicated (further details see text)

of about $d_{nps} = 2\pi/q_{max} \approx 2.9$ nm. During the side chain crystallization process the prepeak sharpens and shifts slightly to smaller q values. The peak maximum for the semi-crystalline PODMA samples occurs at $q \approx 2$ nm^{-1} corresponding to $d_{nps} \approx 3.1$ nm. The reduced peak width indicates an increase of the correlation length. The occurrence of a prepeak shows that the

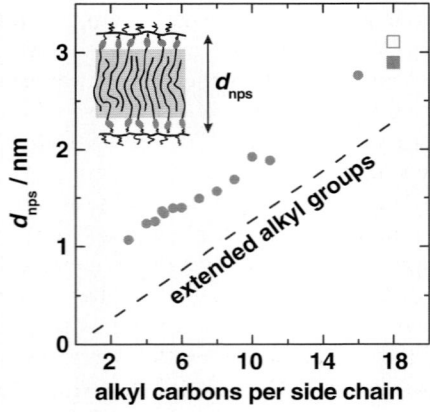

Fig. 12.2. Equivalent Bragg spacing d_{nps} as calculated from the maxima of the pre-peaks for lower amorphous poly(n-alkyl methacrylates) (●) as well as amorphous (■) and semi-crystalline (□) PODMA (cf. Fig. 12.1). The inset shows a schematic picture of the structure of amorphous poly(n-alkyl methacrylates)

crystallization of long alkyl groups in PODMA starts from a pre-structured melt. Long alkyl groups belonging to different monomeric units and chains aggregate already in the melt like in lower poly(n-alkyl methacrylates) where the side chains are unable to crystallize. In PODMA a higher order peak at $q \approx 4.1$ nm^{-1} appears after sufficiently long crystallization under ambient conditions. This indicates lamellar packing, i.e. small stacks formed by alternating layers with high main chain or high side chain concentration. Note, that the scattering contrast in PODMA is basically due to differences in the electron density between carboxyl groups and alkyl nanodomains. Thus, nanophase separation can be already seen in the melt. The prepeak reflects basically the main chain to main chain distance. This situation is only modified during the crystallization process due to an increase of the electron density in the crystalline parts of the alkyl nanodomains. Thus, further details like the thickness of the crystalline layers can not be extracted from x-ray scattering data easily.

A schematic picture of the structure of semi-crystalline PODMA is shown in the insets of Fig. 12.1. Basic idea is that the alkyl groups belonging to different monomeric units form layer-like alkyl nanodomains which are separated by main chains. The periodicity d_{nps} of this lamellar structure reflected by the prepeak in x-ray scattering data is indicated. We assume that the alkyl groups within the alkyl nanodomains are basically interdigitated and not to far away from an extended chain conformation. This can be concluded from the dependence of the d_{nps} value on the number of alkyl carbons per side chain C for atactic poly(n-alkyl methacrylates) (Fig. 12.2). The dependence of d_{nps} on C is nearly linear and the slope is about 0.13 nm [50] per additional CH$_2$ unit in the side chain corresponding to the value reported for extended chain alkanes (0.127 nm/carbon).

The halo in the scattering data for amorphous PODMA around $q = 13.5$ nm^{-1} reflects the average distance between non-bonded alkyl carbons in the melt ($d_{loc} = 0.465$ nm). During crystallization a sharp reflex at $q \approx 15.1$ nm^{-1} develops on top of the amorphous halo. This corresponds to a d_{loc} value of about 0.41 nm which is expectedly significantly smaller than for the amorphous sample. The absence of additional peaks for the semi-crystalline sample shows that the crystalline alkyl segments are hexagonally packed. This corresponds to the situation in the rotator phase of alkanes [43, 46, 47] or in strongly defected polyethylenes [39]. The reduced average distance in the crystalline state is related to the increase in the average density during crystallization ($\Delta\rho \approx 4.5\%$ [32]). Note, that the d_{nps} value increases while the d_{loc} decreases during the crystallization. This indicates that the densification process is highly anisotropic on a molecular scale. The small increase in the main chain to main chain distance d_{nps} during crystallization can be explained based on the increasing trans content in the alkyl groups. The most natural assumption seems to be that the crystalline alkyl sequences are located in the middle of the layer-like alkyl nanodomains (cf. insets Fig. 12.1). The nature of an additional peak in the scattering curves at $q \approx 0.95$ nm^{-1} which occurs temporary in the course of the crystallization process is not finally understood. Possibly, this peak indicates a super structure appearing as a transitient state during crystallization. As already mentioned above, the information about structural details in x-ray scattering data is rather limited. Interestingly, more can be learned from DSC experiments.

Isothermal crystallization experiments in the temperature range between 24 and 32°C have been performed using a Perkin Elmer DSC 7 and a Pyris Diamond DSC. The sample with a mass of about 10mg was annealed at $T = 50°C$ significantly above the melting temperature for ten minutes, quenched with a rate of –40 K/min to the crystallization temperature T_c, isothermally crystallized for a given time t_c and finally reheated with a rate of +10 K/min. The melting peak in the heating scans contains the information about the crystalline material formed during isothermal crystallization. A representative set of heating curves measured after isothermal crystallization at $T_c = 31.5°C$ is shown in Fig. 12.3a. Based on the area of the melting peaks the heat of melting q_m can be determined. In Fig. 12.3b the q_m values are plotted versus crystallization time t_c for different crystallization temperatures T_c. Assuming that the heat of melting for the alkyl groups in PODMA corresponds to that of octadecene ($q_{m,OD} = 61.4$kJ/mol [37]) one can estimate the degree of crystallinity of the alkyl groups according to $D_c = q_m/q_{m,OD}$. The shape of the different isotherms is quite similar. In all cases a dramatic increase of q_m and D_c is observed in a narrow time interval. This sigmoidal increase corresponds to the primary crystallization process and comes to an end at $q_m \approx 45$ J per gram PODMA or analogously at $D_c \approx 25\%$. For larger crystallization times t_c a strong secondary crystallization process is indicated by a linear increase of q_m and D_c on logarithmic time scales. Note, that the secondary crystallization process of PODMA is very pronounced compared to the findings for many other polymers. One should also remember that significant crystal thickening is a special feature of polyethylene.

A clear trend in the isotherms in Fig. 12.3b is that the crystallization kinetics slows down if the crystallization temperature T_c increases. This

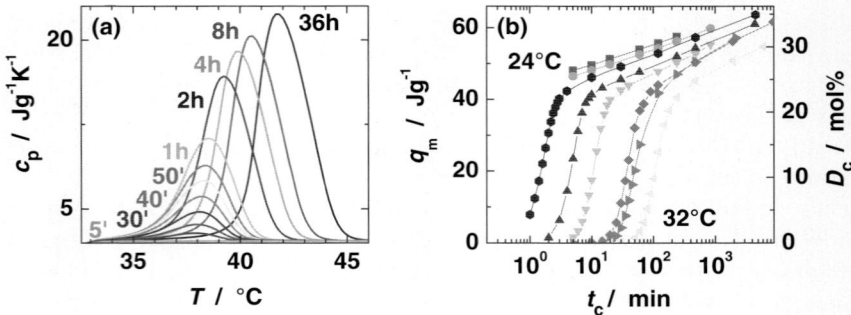

Fig. 12.3. (a) DSC heating curves (+10 K/min) for PODMA measured after isothermal crystallization at $T_c = 31.5°C$ for different crystallization times t_c. (b) Heat of melting q_m and degree of crystallinity D_c vs. logarithm isothermal crystallization time $\log t_c$ measured at different temperatures (from the *top* to the *bottom*: $T_c = 24, 26, 27.7, 29, 30, 31, 31.5, 32°C$). The q_m values are calculated from melting peaks as shown in part (a) for $T = 31.5°C$

temperature dependence indicates that the crystallization kinetics is dominated by the nucleation process. With decreasing crystallization temperature homogeneous nucleation becomes more and more effective. Obviously, this effect overcompensates the strong decrease in mobility with decreasing temperature. Note, that the strong dependence of the characteristic crystallization time on T_c is accompanied by small undercooling values $u = (T_m - T_c)/T_m$ for PODMA homopolymers obtained from DSC experiments with a rate of ±10 K/min and ±1 K/min (cf. Table 12.1). These u values are similar to the values for homogeneously nucleated alkanes in small droplets [42]. Further details of the nucleation behavior will be discussed in the second part of this paper where the crystallization behavior of PODMA homopolymers is compared with findings for microphase-separated P(S−b−ODMA) block copolymers.

An interesting detail in Fig. 12.3a is the fact that the peak position is practically time independent for short crystallization times t_c while a significant shift of the peak maximum is observed for longer times. In Fig. 12.4 the melting temperature T_m corresponding to the peak maximum is plotted versus the degree of crystallinity D_c for different crystallization temperatures T_c. This plot shows clearly that the melting temperature T_m is practically constant during the primary crystallization process ($D_c < 25\%$) and that T_m increases systematically in the course of the secondary crystallization ($D_c > 25\%$). This effect occurs for all investigated crystallization temperatures T_c. In the framework in of the Gibbs-Thompson relation (($T_m^0 - T_m)/T_m^0 \propto 1/L$), where L is the crystal thickness and T_m^0 the melting temperature for a crystal with infinite thickness ($L \to \infty$), this finding indicates that the crystal thickness increases significantly during the secondary crystallization process. According to this picture L is practically constant during the primary crystallization step

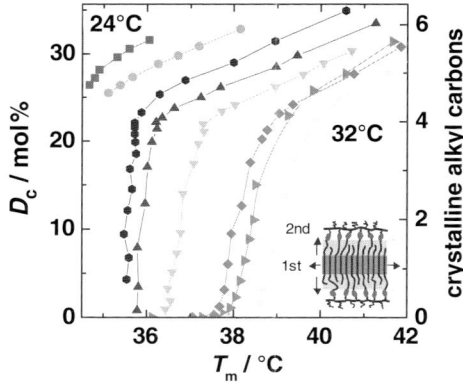

Fig. 12.4. Degree of crystallinity D_c and average number of crystalline alkyl carbons per side chain vs. melting temperature T_m from isothermal crystallization measurements. The symbols correspond to that one used for the different isotherms in Fig. 12.3b. The inset shows schematically the directions of crystal growth during primary (1st) and secondary (2nd) crystallization

although the degree of crystallinity D_c is significantly increasing. Considering a lamellar picture as shown in the inset of Fig.12.4 one can estimate based on the calorimetrically determined D_c values the thickness of crystalline lamellae. From $D_c \approx 25\%$ at the end of the primary crystallization. On gets a stem length of about five alkyl carbons ($L \approx 0.6$ nm).

Calorimetric observations and scattering results presented above allow a relatively detailed description of the side chain crystallization process of long alkyl groups in PODMA homopolymers. We propose a hypothetical picture with four stages: (i) nanophase-separated melt as initial situation for the crystallization of alkyl groups in side chain polymers like PODMA, (ii) nucleation and early stages of crystallization characterized by highly extended alkyl groups, (iii) primary crystallization related to lateral growth of thin crystalline lamellae and (iv) crystal thickening during secondary crystallization. Further details of the different stages are discussed below:

(i) Structural data for amorphous PODMA and other side chain polymers show that long alkyl groups aggregate in the melt. Alkyl nanodomains with a typical size in the range 0.5-2 nm (for PODMA ≈ 2 nm) are formed. Short alkyl groups are frustrated by immobile main chains and unable to crystallize. The dynamics of these amorphous side chain polymers is characterized by an additional polyethylene-like glass transition (α_{PE}) due to cooperative motions within the alkyl nanodomains which are basically decoupled from the main chain dynamics (α). With increasing side chain length frustration effects become less important. If the alkyl groups are long and flexible enough side chain crystallization occurs. Parts of the alkyl groups far away from the main chains are able to reach crystalline order. The length of the alkyl group which is required before crystallization occurs depends on different structural aspects like tacticity but also on the main chain mobility. This can be concluded from a comparison of the crystallization behavior of atactic poly(n-alkyl methacrylates) and poly(n-alkyl acrylates). In case of the poly(n-alkyl acrylates) having more flexible main chains and a significantly lower $T_g(\alpha)$ the octyl member with $C = 8$ alkyl carbons can crystallize while the first methacrylate which can crystallize is the dodecyl member with $C = 12$ alkyl carbons. All results indicate that the alkyl groups in side chain polymers are already aggregated in the melt before crystallization starts to occur. In order to crystallize the alkyl groups have to overcome frustration introduced by the immobile main chain. This special situation may stabilize states which are unstable or transient and hardly detectable in related systems like alkanes or polyethylene.

(ii) At the transition from the disordered to the crystalline state the alkyl groups in side chain polymers like PODMA are probably in a nearly extended conformation with a large trans content. This is indicated by a nearly linear increase of the main chain to main chain distance d_{nps} for the methacrylate series with a slope of about 0.13 nm per additional CH_2 unit similar to the values reported for crystalline alkanes (Fig. 12.2). Another argument supporting this idea is the fact that d_{nps} is only slightly changing during crystallization. The occurrence of highly extended alkyl groups in the amorphous state might be related to some of the unusual features observed for different systems containing long CH_2 sequences. It has been discussed recently that highly stretched chain segments act as nucleation sites in sheared polyethylene blends [13, 40, 52, 53]. Low molecular weight PEs mixed with a small fraction of ultra-long PE chains show this phenomenon quite clearly [41, 54]. Based on these similarities one can speculate that the nucleation mechanism for side chain polymers with alkyl groups is similar. Whether or not this is a more general mechanism for polymeric systems containing long alkyl sequences remains open. The fact that most of these systems show small undercooling and the occurrence of hexagonally packed (meso)phases is at least remarkable.

(iii) Once the primary crystallization has started the degree of crystallinity D_c increases rapidly. Calorimetric results indicate that thin crystalline lamellae with a constant thickness of about 0.6 nm corresponding to a stem length of about five carbons grow laterally within existing alkyl nanodomains (inset Fig. 12.4). This can be concluded from the constant melting temperature during the primary crystallization based on the Gibbs-Thompson relation. The general arrangement of the alkyl groups is obviously unaffected by the crystallization process. The alkyl groups are interdigitated before and after crystallization has taken place. The small increase in the main chain to main chain distance d_{nps} from 2.9 nm to 3.1 nm can be understood as a direct consequence of the increasing trans content in the alkyl groups during crystallization. The most natural assumption is that the crystalline lamellae are located in the middle of the alkyl nanodomains where the constraints introduced by the immobile main chains are minimal. It is an open question so far whether or not the side chain crystallization in atactic PODMAs requires rearrangements on larger scales. This point needs further investigation. Note, that there is a additional peak in the scattering curves at $q \approx 0.95$ nm^{-1}. This peak disappears (or can not be resolved) if crystallization has reached some perfection. The nature of this peak is not finally understood but it might be an indication for the existence of a (temporary) super structure.

(iv) The secondary crystallization in PODMA seems to be related to a thickening of the thin crystalline lamellae in the alkyl nanodomains, i.e. the crystal grows in direction which is basically perpendicular to the direction of growth during primary crystallization (inset Fig. 12.4). This is concluded from the significant increase of melting temperature T_m with the degree of crystallization D_c on logarithmic time scales under isothermal conditions. This effect is observed for all investigated temperatures. Crystal thickening is related to a transition of CH_2 units from the non-crystalline part of the alkyl groups to the hexagonally packed crystal. According to the simple lamellar picture shown in the inset of Fig. 12.4 secondary crystallization should come to an end if the thickness of the crystals approaches the length of the alkyl groups. However, the degree of crystallization D_c in our isothermal crystallization experiments on PODMA was always much smaller than 100%. Secondary crystallization starts at $D_c \approx 25\%$ and the largest D_c value reached in our calorimetric studies after long isothermal crystallization times was $D_c \approx 35\%$. This corresponds to a change in the stem length from nearly five to a bit more than six alkyl carbons. In the framework of our picture this means that there are still six alkyl carbons in the amorphous spacers between the crystalline layer in the center of the alkyl nanodomain and the immobile methacrylate main chains. Studies of the saturation process might be interesting in order to understand details of the interrelation between crystallization tendency of the frustrated alkyl groups and mobility of the main chains.

An additional result of this study is that the crystallization behavior of PODMA is practically independent on the molecular weight of the sample. The overall crystallization behavior with a strongly temperature dependent primary crystallization and a pronounced secondary crystallization process is nearly unaffected in PODMA samples with degrees of polymerization in the range $6 < P < 100$. All parameters describing the primary crystallization process like temperature-dependent crystallization time, degree of crystallinity and width of the transformation interval are very similar for all investigated samples (Table 12.1, [27]). This observation is consistent with a relatively robust structure of the nanophase-separated melt which is also observed for amorphous poly(n-butyl methacrylate) samples with different degree of polymerization ($6 < P < 1000$) [55] and the observation that the dynamics within the disordered alkyl nanodomains is basically decoupled from that of the main chain in several series of side chain polymers. Thus, we expect that the proposed picture for the crystallization of frustrated alkyl groups might be applicable to various side chain polymers containing long alkyl groups.

12.3 Confined Crystallization in Microphase-separated Poly(styrene−*block*−octadecylmethacrylate) Copolymers

An interesting approach to learn more about polymer crystallization in general and about the influence of constraints on the crystallization in nanostructured systems in detail are experiments on microphase-separated block copolymers with crystallizable component. This idea has been applied to various block copolymers with long main chain polymers like polyethylene, poly(ethylene oxide) or poly(ϵ−caprolactone) as crystallizable component. In most of the studies the confining component was amorphous but the mobility has been varied in a wide range. The amorphous component can be either glassy or rubbery during the crystallization of the second component. The situation is called "strong confinement" case if the block copolymer morphology is not affected by the crystallization process [8]. This is the typical situation for systems where the glass temperature T_g of the amorphous block is significantly higher than the crystallization temperature T_c of the crystallizable block. Such systems have been used in the last decades to study details of nucleation behavior and crystal growth. Strong confinement situations have been also observed in strongly segregated block copolymers having a amorphous component which is rubbery during the crystallization ($T_g \leq T_c$). The other extreme – called "breakout behavior" [8] – is characterized by the dominance of the crystalline structure and the disappearance of the original block copolymer morphology in the semi-crystalline state. Breakout behavior is typically observed in weakly segregated block copolymers with an amorphous component which is rubbery during the crystallization [56, 57].

Polymer crystallization in strong confinement has been studied frequently but the confined crystallization in block copolymers containing crystallizable side chain polymers has not been studied in much detail so far. An interesting question is whether or not confinement effects on the crystallization behavior are similar to those in previously studied systems where long main chains do crystallize. As discussed in the previous section (12.2) semi-crystalline side chain polymers are characterized by a lamellar packing of alkyl nanodomains and main chains. The periodicity of this basically lamellar structure depends on the length of the alkyl groups and is reflected by the prepeak in x-ray data. Thus, block copolymers containing side chain polymers as crystallizable component are potentially systems with a hierarchy of length scales in the nanometer range. Two coexisting structural elements are expected: One is due to the microphase-separation of the two blocks (10-50 nm) and a second scale is due to the nanophase-separation of the side chain polymer (2-3 nm). The interrelation between these two length scales in microphase-separated block copolymers with one component being a crystallizable side chain polymer is another interesting phenomenon in these systems. Whether or not both structures survive in these complex systems is not a priori clear because two mechanisms of structure formation concur. According to the picture for the

Table 12.2. Characterization of the P(S-b–ODMA) block copolymers and parameters of the crystallization process

Label*	Φ_{ODMA}	N_{ODMA}	N_S	M_w/M_n	D_c	$T_c^{max\dagger}$	$T_m^{max\dagger}$	T_ω^+	$u^\#$	$n^\ddagger$
					mol%	°C	°C	°C	10^{-3}	
Lam-9 nm	0.39	25	154	1.1	25	20.4	32.4	94	39 (31)	2.8
Lam-17 nm	0.44	52	257	1.1	27	22.8	33.8	102	36 (32)	3.4
Cyl-11 nm	0.18	12	224	1.1	15	8.7	17.7	94	31 (24)	0.7
Cyl-16 nm	0.16	21	448	1.1	30	18.3	31.1	103	42 (-)	1.5
Cyl-24 nm	0.26	38	426	1.1	30	20.6	33.3	103	41 (-)	1.8
PODMA27	1.00	27	-	2.1	31	23.9	34.3	–	34 (26)	4.1

* The numbers indicate thickness of PODMA lamellae or diameter of PODMA cylinders estimated based on Φ_{ODMA} values from NMR. † Peak maxima from scans with rates of $dT/dt = \pm 10$ K/min. $^+$ Dynamic glass temperatures from c_p''-maxima in TMDSC scans (time period $t_p = 60$s, temperature amplitude $T_a = 0.4$K, underlying heating rate +2 K/min). # Taken from scans with rates of ±10 K/min (±1 K/min). ‡ Taken from Avrami plots (Fig. 12.12). The uncertainty is about ±0.2. The width of the transformation interval can be estimated from $\Delta \log t_c = 1.253/n$ (cf. [37]).

side chain crystallization presented in Sect. 12.2 nanophase separation and side chain crystallization are strongly related phenomena.

In the following part we present detailed experiments on microphase-separated poly(styrene−b−octadecylmethacrylate) block copolymers [P(S−b−ODMA] containing PODMA domains with a typical size of 10-25 nm (Table 12.2). Structural aspects of these systems are studied by scattering techniques in different q ranges. It will be shown that nanophase separation still occurs and that long alkyl groups can also crystallize in small PODMA domains. The influence of confinement on different aspects of the crystallization behavior like crystallization kinetics, nucleation mechanism, and degree of crystallinity is studied by calorimetry and scattering techniques.

Scattering data for two representative P(S−b−ODMA) block copolymers from synchrotron scattering experiments on beam line BM26 at the ESRF in Grenoble are shown in Fig. 12.5. All block copolymers in this study are annealed in a first step at 150°C for 24h under vacuum in order to prepare microphase-separated samples. The curve in Fig. 12.5a for a P(S−b−ODMA) block copolymer containing 39vol% ODMA (Lam-9 nm) shows typical features of a block copolymer with lamellar morphology while the data in Fig. 12.5b for a sample containing 18vol% ODMA (Cyl-11 nm) are consistent with a scattering pattern of a copolymer with cylindrical morphology. Repeating units can be calculated based on maximum position of the first scattering peak q_{max} and Bragg equation $d = 2\pi/q_{max}$. By using the ODMA content as obtained from ^{1}H-NMR measurements the size of the PODMA domains was estimated. One gets a thickness of ≈9.5 nm for the PODMA lamellae in the Lam-9 nm sample and a diameter of ≈ 11.5 nm of the PODMA cylinders in the Cyl-11 nm

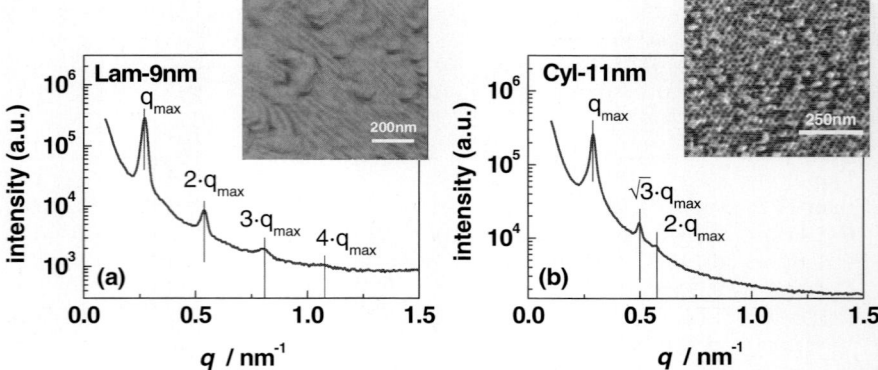

Fig. 12.5. Small angle x-ray scattering data and AFM pictures for microphase-separated P(S-block-ODMA) copolymers with (**a**) lamellar and (**b**) cylindrical morphology. The peak maxima at $q_{\max} = 0.27$ nm^{-1} for Lam-9 nm and $q_{\max} = 0.29$ nm^{-1} for Cyl-11 nm correspond to Bragg spacings of $d_{mps} = 23.3$ nm and $d_{mps} = 21.7$ nm, respectively

sample. Atomic force microscopy (AFM) pictures shown as insets in Fig. 12.5 support the morphologies and domain sizes reported above. Results for several other P(S–b–ODMA) copolymers with lamellar and cylindrical morphology will be presented below. The parameters characterizing these samples are summarized in Table 12.2. Further details about the synthesis of the samples by anionic polymerization are discussed elsewhere [37].

DSC heating scans for PS and PODMA homopolymers as well as two P(S–b–ODMA) block copolymers (Lam-9 nm, Cyl-11 nm) measured at a rate of +10 K/min after cooling with the same rate (-10 K/min) are presented in Fig. 12.6a. The curves show clearly that the crystallization of PODMA side chains occurs at temperatures far below the glass transition of PS. In the heating scans for the P(S–b–ODMA) copolymers the melting peak of PODMA ($T_m \approx 16 - 30°$C) and the glass transition of PS ($T_g \approx 90°$C) coexist. Note, that the melting temperature – defined here based on the maximum of the melting peak – for the PODMA cylinders is smaller than T_m for the homopolymers (Table 12.2). An analysis of the strength of the glass transition in the polystyrene domains shows that Δc_p depends linearly on the weight fraction w_{PS} (Fig. 12.6b). The obtained values are only slightly smaller than the values predicted based on a linear dependence of Δc_p on w_{PS} as determined from ^{1}H-NMR spectra. This confirms that the P(S–b–ODMA) block copolymers are microphase-separated and indicates that side chain crystallization occurs in a strong confinement. There should be no significant change of the block copolymer morphology during crystallization since $T_g - T_c$ is large. The dependence of the dynamic glass temperature T_ω from temperature modulated DSC (TMDSC, $t_p = 60$s) on the number of units in the polystyrene block N_S is a bit stronger than predicted by the Fox-Flory equation $T_g \propto 1/N_S$ for the

12 Crystallization of Frustrated Alkyl Groups   215

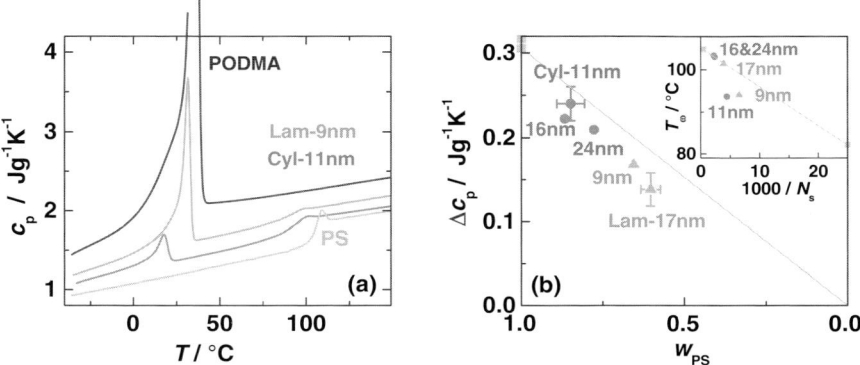

Fig. 12.6. (a) DSC heating curves for two P(S−b−ODMA) block copolymers with lamellar (Lam-9 nm) and cylindrical (Cyl-11 nm) morphology as well as for PODMA and PS homopolymers. Heating rate was +20 K/min. The curves are measured after cooling with −20 K/min. (b) Relaxation strength of the polystyrene phase Δc_p as function of weight fraction PS w_{PS} for five P(S-block-ODMA) copolymers with different morphology. The *dashed line* corresponds to the linear increase of Δc_p with w_{PS}. The inset shows the dynamic glass temperatures T_ω from TMDSC ($t_p = 60$s) depending on the reciprocal number of styrene units is the PS block. The *dotted line* is an interpolation between the T_ω values for two PS homopolymers with different molecular weights based on the Fox-Flory-equation

homopolymer. Similar behavior is reported for several block copolymers and might be due to small differences in chain conformation as well as average density compared to the bulk situation.

Information about internal structure of the PODMA domains and influence of the confinement on the nanophase separation in side chain polymers can be obtained from scattering data for microphase-separated P(S−b−ODMA) block copolymers in a wide q range. Experiments on beam line BM26 at the ESRF are an extraordinary way to perform simultaneous online measurements in the small and wide angle range on one sample. Results for our block copolymers indicate that these samples are indeed systems with a hierarchy of length scale in the nanometer range. Representative data for a sample with lamellar morphology (Lam-17 nm) are shown in Fig. 12.7. In the q range below 1 nm^{-1} (Fig. 12.7a) the typical scattering pattern of a lamellar block copolymer is obtained. Results for the molten and the semi-crystalline state are compared. The scattering curve for the molten sample shows higher order peaks at all integer multiples of q_{max} while only the odd orders have been detected for the semi-crystalline sample. The fact that position and shape of the first order peak at q_{max} is unaffected shows nicely that the crystallization occurs in a strong confinement. The disappearance of even orders might be a reflection of density changes during crystallization and is a typical feature of symmetric block copolymers. In the q range between 1 nm^{-1} and 4 nm^{-1} the prepeak

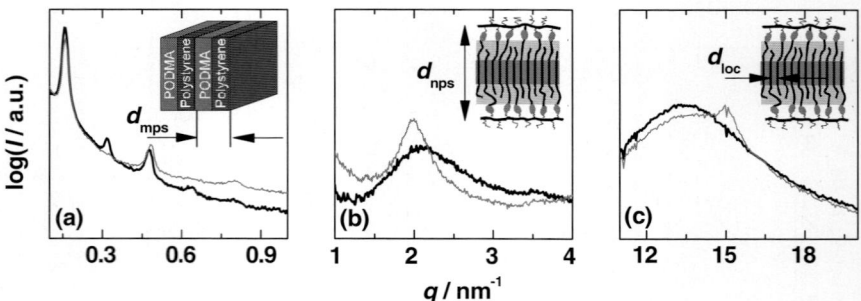

Fig. 12.7. Scattering curves for a lamellar P(S–b–ODMA) block copolymer (Lam-17 nm) measured on beam line BM26 at the ESRF in Grenoble (*thick line*: amorphous – $t_c = 1$min, $T_c = 27°$C; thin line: semi-crystalline – $t_c = 20$min, $T_c = 27°$C). Different structural elements are indicated by different scattering peaks: (**a**) microphase separation of PS and PODMA blocks ($d_{mps} \approx 40$ nm), (**b**) nanophase separation of main and side chains in the PODMA domains ($d_{nps} \approx 3.1$ nm), (**c**) local packing of the alkyl groups ($d_{loc} \approx 0.41$ nm)

of the PODMA is observed. The prepeak reflects basically the main chain to main chain distance in PODMA domains and sharpens during crystallization as discussed above in case of the PODMA homopolymers (cf. Fig. 12.1). Note, that the prepeak positions in lamellar block copolymer and homopolymers are practically identical. This shows that the internal structure of PODMA in lamellae with a thickness of 10-25 nm is comparable to that in bulk PODMA. The scattering data measured in the wide angle range around $q \approx 15$ nm^{-1} (Fig. 12.7c) show that a sharp reflex develops during the side chain crystallization process. The position of this reflex – indicating hexagonally packed alkyl groups – is also comparable to that in PODMA homopolymers. The intensity of the reflex, however, seems to be smaller. This is mainly due to the fact that not only the PODMA component but also the amorphous PS component is contributing to the amorphous halo. Moreover, the isothermal crystallization time in this experiment was relatively short (20min at 27°C) resulting in a degree of crystallinity which is significantly smaller compared to the situation for the homopolymer shown in Fig. 12.1 ($t_c >1$ week at $T_c \approx 25°$C).

A more detailed interpretation of the internal structure of the PODMA domains is possible based on WAXS data for a semi-crystalline block copolymer (Lam-20 nm) measured at room temperature using a Bruker D500 system (Fig. 12.8). Data for polystyrene and semi-crystalline PODMA homopolymers are shown for comparison. It is obvious that the scattering curve for the microphase-separated block copolymer can be interpreted as a superposition of contributions originating from both individual components. The first two peaks (1&2) at $q \approx 2.1$ nm^{-1} and $q \approx 4.3$ nm^{-1} indicate the nanophase separation and the lamellar packing of main and side chains in semi-crystalline

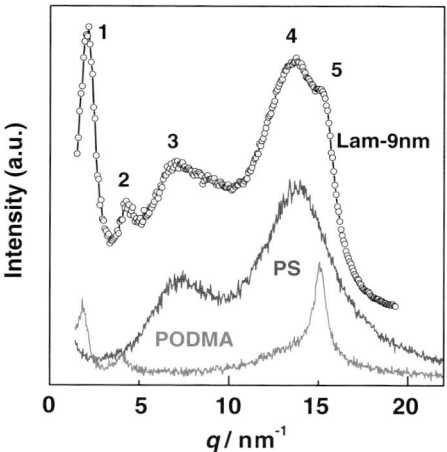

Fig. 12.8. Wide angle scattering curves for a P(S–b–ODMA) block copolymer with lamellar morphology (Lam-9 nm) as well as PS and PODMA homopolymers [58]. The curves are shifted vertically. The peak maxima correspond to equivalent Bragg spacings of $d_1 = d_{nps} = 3.05$ nm (second order peak at q_2), $d_3 = 0.85$ nm, $d_4 = 0.46$ nm, and $d_5 = d_{loc} = 0.41$ nm. Further details are discussed in the text

PODMA. The sharp reflex (5) at $q \approx 15.1$ nm^{-1} belongs to the crystalline parts of the alkyl groups in the PODMA domains. The two other peaks (3&4) at $q \approx 7.1$ nm^{-1} and $q \approx 13.7$ nm^{-1} originate mainly from the polystyrene component although a certain part of the amorphous halo (4) is due to non-crystalline CH$_2$ sequences in the alkyl groups of the PODMA block. The WAXS data support strongly the idea that internal structure and crystallization behavior of PODMA chains in thin lamellae is not influenced by the glassy polystyrene phase in the environment. Note, that this is true for PODMA lamellae with a thickness of about 17 nm as well as those with a thickness of about 9 nm. The structural findings for both samples are quite similar. Interestingly, WAXS data for the cylindrical samples indicate that the nanophase separation in the PODMA domains is surviving also under these more extreme conditions. Although constrained in small cylinders with a diameter of about 10 nm PODMA main and side chains seem to be able to demix on length scales of about 3 nm. All peaks discussed above for the lamellar P(S-b-ODMA) block copolymer have been also found for the samples with cylindrical morphology [59]. The PODMA peaks are less intense due to the volume fraction $0.16 < \phi_{PODMA} < 0.26$ which is significantly smaller compared to the situation in the lamellar samples ($0.39 < \phi_{PODMA} < 0.44$). However, the alkyl groups can crystallize to some extent (Fig. 12.6) and the lamellar order of the individual planes is indicated. Somehow this seems to be surprising since it indicates that both processes of structure formation do not disturb each other too much. Microphase separation in these block copolymers can possibly only occur if the side chain polymers are arranged in an appropriate way within the PODMA cylinders. That alkyl groups belonging to different monomeric units have a strong tendency to aggregate can also be concluded from the fact that random poly(styrene–$stat$–octadecylmethacrylate) copolymers with substantial ODMA content show side chain crystallization [59]. This

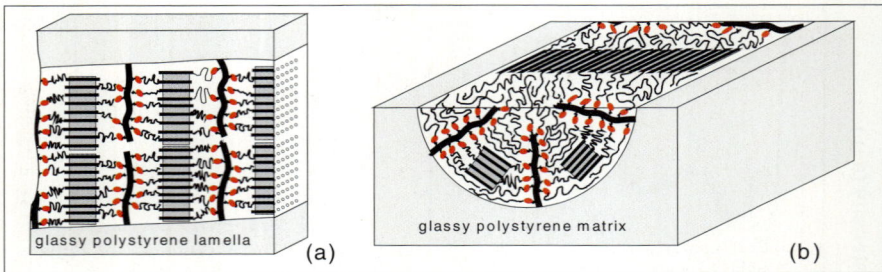

Fig. 12.9. Schematic pictures of the internal structure of the PODMA domains in P(S−b−ODMA) copolymers with (**a**) lamellar and (**b**) cylindrical morphology. The small ellipses represent the carboxyl groups of the metharcylate main chains. The dark gray regions are crystalline

can only happen if the alkyl groups are aggregated and tiny alkyl nanodomains are formed.

Hypothetical pictures which may describe the internal structure of PODMA domains in microphase-separated P(S−b−ODMA) block copolymers with lamellar and cylindrical morphology are shown in Fig. 12.9. It can be seen that the lamellar morphology fits nicely to the lamellar structure of the nanophases within the semi-crystalline PODMA domains. Thus, confinement effects on the side chain crystallization should not be very pronounced in this case. The main effect might be that the lateral growth of the individual crystalline layers is stopped by glassy polystyrene lamellae. However, the lateral size of the individual crystals in POMDA homopolymers is also limited due to constraints produced by immobile main chains in the environment and it is not clear whether or not the constraints in lamellar block copolymers are really stronger. More pronounced confinement effects are expected for PODMA cylinders since there is a conflict between the curved PS-PODMA interfaces and the basically lamellar structure of semi-crystalline PODMA. Otherwise, nanophase separation seems to be preserved and the side chains can crystallize. A speculative structure which fulfills these requirements is shown in Fig. 12.9b. The main chains are aggregated in certain planes and most of the side chain are oriented parallel to the cylinder axis. Within the main chain planes should be only a limited fraction of alkyl groups which can hardly crystallize. Although the PODMA chains are confined nanophase separation and side chain crystallization can still occur since PODMA forms stacks of main and side chain layers within the cylindrical domains. To which extent side chain crystallization is affected by the environment can not be easily seen in the complex scattering data for microphase-separated P(S−b−ODMA) block copolymers. This aspect was studied in more detail by calorimetry.

Isothermal crystallization experiments using DSC support the idea that the crystallization behavior of thin PODMA lamellae in P(S−b−ODMA) block copolymers is comparable to that of PODMA homopolymers. The

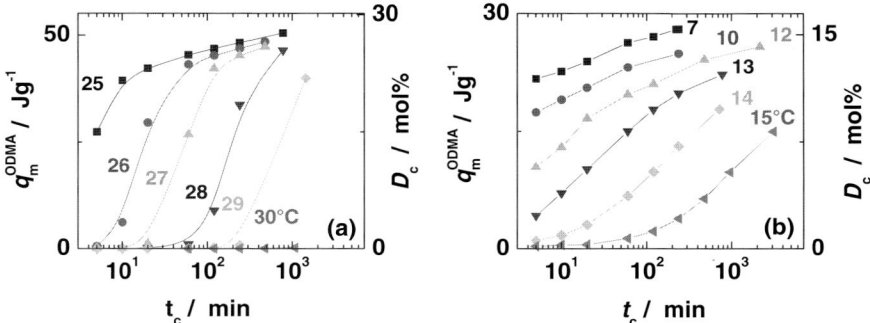

Fig. 12.10. Heat of melting q_m and degree of crystallinity D_c vs. isothermal crystallization time t_c measured at different temperatures for two semi-crystalline block copolymers with (**a**) lamellar (Lam-9 nm) and (**b**) cylindrial (Cyl-11 nm) morphology

applied program is the same which is described in detail in the last Sect. 12.2. The sample is quenched from the melt in the DSC instrument to the isothermal crystallization temperature T_c and hold there for a certain crystallization time t_c. The information about heat of melting per gram ODMA q_m^{ODMA} and degree of crystallinity D_c of the alkyl groups is taken from the melting peak in a subsequent heating scan. The dependence of q_m^{ODMA} and D_c on the isothermal crystallization time t_c is shown for different temperatures in Fig. 12.10a. Obviously, the isotherms for the Lam-9 nm sample are similar to those for the PODMA homopolymer (Fig. 12.3b). Melting temperature, heat of melting per ODMA unit, width of the isothermal transformation interval and overall shape of the isotherms are comparable (Tables 12.1 & 12.2). This can be also seen based on a direct comparison of master curves (Fig. 12.11a) which are constructed from the individual isotherms shown in Fig. 12.10 by a horizontal shift to the isotherm belonging to a certain reference temperature T_{ref}.

A significantly different crystallization behavior is found for small PODMA cylinders with a diameter of about 11 nm in a rigid polystyrene matrix (Cyl-11 nm). The q_m values from isothermal crystallization experiments at different temperatures (Fig. 12.10b) indicate that heat of melting per gram ODMA q_m^{ODMA} and degree of crystallinity are 50% reduced and that the transformation interval is much broader. Compared to homopolymers and lamellar P(S−b−ODMA) block copolymers crystallization in small PODMA cylinders occurs at significantly smaller temperatures. A direct comparison of the master curves in Fig. 12.11a shows these differences clearly. Obviously, strong confinement effects occur in case of small PODMA cylinders (Cyl-11 nm) embedded in a glassy polystryrene matrix. The differences in the width of the transformation interval can be also expressed in terms of Avrami coefficients n as obtained from an Avrami plot (Fig. 12.12a) based on the Avrami equation

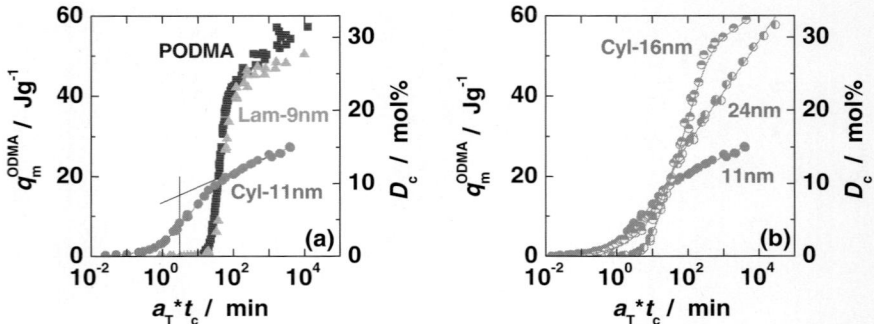

Fig. 12.11. (a) Master curves for PODMA27 ($T_{ref} = 31°C$, *squares*) and two semi-crystalline P(S–b–ODMA) block copolymers with lamellar (Lam-9 nm, $T_{ref} = 27°C$, *triangles*) and cylindrical (Cyl-11 nm, $T_{ref} = 12°C$, *circles*) morphology. The master curves are obtained from isotherms as shown in Figs. 12.3b & 12.10 by a horizontal shift to the given reference temperatures T_{ref}. The tangent construction used to determine the half time $\tau_c(T_{ref})$ is indicated. (b) Master curves for three different block copolymers with cylindrical morphology. The diameter of the PODMA cylinders is indicated (Cyl-16 nm, $T_{ref} = 24.2°C$; Cyl-24 nm, $T_{ref} = 26°C$)

$X = 1 - \exp(-kt_c^n)$ with X being the normalized degree of crystalinity and k a rate constant. One gets values $n < 1$ for cylindrical PODMA domains with a diameter of about 11 nm (cf. Table 12.2) while the values for PODMA homopolymers and lamellar block copolymers are in the range $n = 3 - 4$. Note that values $n \approx 0.7$ have been reported for small, homogenously nucleated alkane droplets dispersed in a solvent [12, 42, 45]. Avrami exponents $n = 3 - 4$ have been interpreted as indication for heterogeneous nucleation and three- or two-dimensional crystal growth while $n \approx 1$ indicates homogenous nucleation and/or one-dimensional crystal growth [6, 60].

In a next step the dependence of the confinement effects in cylindrical block copolymers on the diameter of the PODMA cylinders has been checked. Master curves for two systems containing PODMA cylinders with a diameter of 16 nm and 24 nm are compared with those for small 11 nm cylinders in Fig. 12.11b. The data show that there is a strong dependence on the cylinder diameter. The degree of crystallinity D_c which is reached after long crystallization times is higher for the larger diameters. Moreover, there seem to be ranges in the transformation interval with different slopes indicating time- or temperature-dependent changes in the crystallization behavior. A possible explanation for this effect might be the concurrence of heterogeneous and homogeneous nucleation as reported for other crystallizable block copolymer systems [9]. These changes in the curve shape are also reflected in an Avrami plot as shown in Fig. 12.12b. For the largest PODMA cylinders (Cyl-24 nm) the estimated Avrami coefficient is $n \approx 1.8$. This value is already smaller than n for PODMA lamellae and indicates an influence of the confinement. If

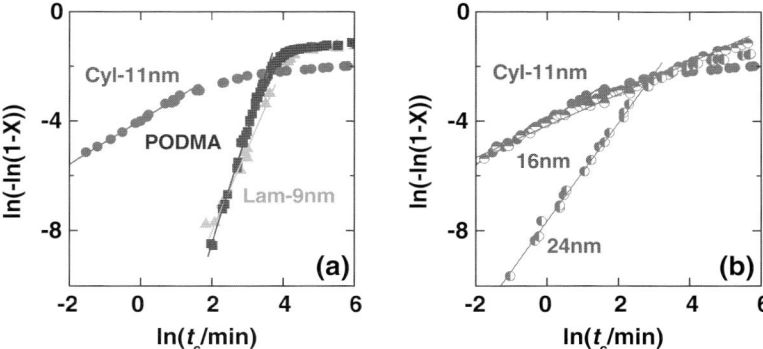

Fig. 12.12. Avrami plots for (**a**) two P(S−b−ODMA) block copolymers (Lam-9 nm,Cyl-11 nm) and a PODMA homopolymer and (**b**) block copolymers containing PODMA cylinders with different diameters (Cyl-11 nm,Cyl-16 nm,Cyl-24 nm) constructed based on master curves as shown in Fig. 12.11. Fits to the Avrami equation are indicated by *solid lines*. The fit parameters are given in Table 12.2

the diameter decreases the confinement effects are more pronounced and the n values approach 0.7. This indicates a transition to one-dimensional growth and dominantly homogeneous nucleation. This seems to be in accordance with the intuitive picture that crystals in small cylinders can basically grow only in one direction and that most of the PODMA cylinders are not connected. However, final conclusions about the nature of the changes in the crystallization kinetics under confinement can not be drawn based on Avrami coefficients. Further parameters have to be included in the discussion in order to understand these systems better and to deconvolute changes in nucleation mechanism and crystal growth.

Figure 12.13 shows temperature-dependent half times τ_c as obtained from individual isotherms as shown in Figs. 12.3 & 12.10. The τ_c values are determined using a tangent construction which is shown in Fig. 12.11a and correspond to those crystallization times at which 50% of the primary crystallization at T_c has appeared. A comparison of the $\tau_c(T_c)$ values for different PODMA containing systems shows that the main trends are similar. For all investigated P(S−b−ODMA) block copolymers a rapid increase of τ_c with increasing temperature T_c is observed like in the homopolymers. The half times for PODMA lamellae are slightly larger but comparable to those for PODMA homopolymers in the temperature range from 22 to 32°C. In case of block copolymers with cylindrical morphology a strong dependence of τ_c on the diameter of the PODMA cylinders is observed. For the largest diameter (Cyl-24 nm) the crystallization times are similar to those for the smallest lamellae (Lam-9 nm) and only ≈ 30 times larger than the those for the homopolymer. With decreasing diameter of the PODMA cylinders the half time increases dramatically. For the Cyl-16 nm sample τ_c is at least 300 times larger in the

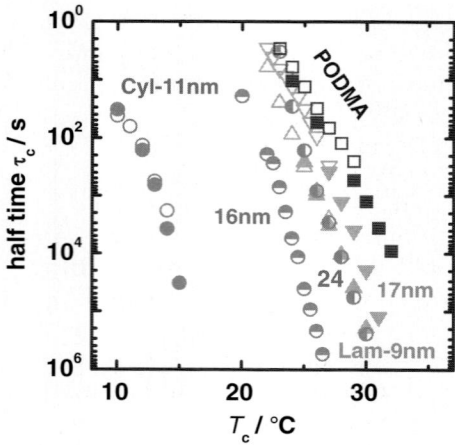

Fig. 12.13. Characteristic crystallization times τ_c vs. crystallization temperature T_c for the homopolymer PODMA27 (*squares*) and block copolymers with lamellar (*triangles*) and cylindrical morphology (*circles*). The curves for different samples are labeled. The *filled and partly-filled symbols* are half times τ_c. The *open symbols* correspond to τ_c values taken from heat flux curves measured during isothermal crystallization (for details see [37])

investigated temperature range while the estimated factor is $> 10^8$ in case of the smallest cylinders (Cyl-11 nm). This trend corresponds to a decrease of the crystallization temperatures in DSC scans measured at constant cooling rate (Table 12.2) and may indicate a transition to homogenous nucleation. Note, that the change in the calorimetrically observed T_c values is significant (T_c(PODMA)−T_c(Cyl-11 nm) > 15 K) but relatively small compared to other microphase-separated block copolymers with crystallizable component where the reduction in T_c due to a change from heterogenous to homogenous nucleation is often about 50K [10, 11]. However, the maximal difference in T_c observed for Lam-11 nm copolymer is comparable to the T_c change in case of homogeneously nucleated alkanes. This indicates common aspects in the crystallization behavior although the temperature dependence of the half time τ_c for octadecane is even stronger than that for PODMA [37].

Summarizing the observations above one can conclude that the crystallization behavior of PODMA homopolymers and thin PODMA lamellae surrounded by glassy polystyrene in microphase-separated P(S−b−ODMA) block copolymers is basically identical. The temperature-dependent half times $\tau_c(T_c)$, width of the transformation interval (n) as well as degree of crystallinity D_c are similar. There is only a slight increase in the τ_c values accompanied by a small increase in the width of the transformation interval in isothermal crystallization curves reflected by a change from $n \approx 4$ for the homopolymer to $n \approx 3$ for PODMA lamellae. This could be interpreted as a transition from three-dimensional to two-dimensional crystal growth due to constraints in lamellar block copolymers. Dramatic changes in the nucleation mechanism, however, seem to be unlikely. The PODMA lamellae are interconnected and not isolated. The strong decrease of τ_c with decreasing T_c may indicate that homogeneous nucleation is always important in PODMA containing systems. This is somehow consistent with the fact that clear growth fronts have never been observed in experiments by optical polarization microscopy on PODMA. Seemingly the number of nuclei is always large. The nearly unaffected degree of crystallinity D_c is consistent with the speculative picture for the internal structure of the PODMA lamellae which is shown in Fig. 12.9a. The lamellar structure of semi-crystalline PODMA fits to the lamellar morphology of lamellar P(S−b−PODMA) block copolymers. Thus, no significant constraints occur in the interfacial regions which would reduce the degree of crystallinity. According to this picture it is understandable that the D_c values for PODMA lamellae and homopolymers are comparable.

More pronounced confinement effects have been observed in case of P(S−b−ODMA) samples containing PODMA cylinders which can crystallize in a glassy polystyrene matrix. The strength of the confinement effects increases if the diameter of the PODMA cylinders decreases: The crystallization kinetics slows down, the width of the transformation interval increases corresponding to a decrease of the Avrami coefficient n and a significant decrease of the D_c values is observed. For the largest cylinders in our series (Cyl-24 nm) the only indication for confinement effects is a significant broadening of the transformation interval corresponding to an Avrami coefficient $n \approx 1.8$ which is much smaller compared to $n \approx 4$ obtained for PODMA homopolymers. Note, that the half times for the Cyl-24 nm and the Lam-9 nm samples are nearly identical. The Avrami coefficient for the Lam-9 nm sample ($n \approx 2.8$), however, is larger than that for the Cyl-24 nm sample. This may indicate that changes in the crystal growth are relevant for the broadening of the transformation interval in case of large cylindrical PODMA domains. With decreasing diameter of the PODMA cylinders the half times increase dramatically and the Avrami coefficient approaches $n \approx 0.7$. This can be explained based on a dominantly homogeneous nucleation process in small cylinders which are basically isolated from each other although a certain fraction of the domains might by interconnected. Thus, heterogeneous and homogeneous nucleation may coexist as discussed for other block copolymers [9]. The reduced degree of crystallinity D_c can be explained based on the speculative picture for the internal structure of cylindrical PODMA domains as shown in Fig. 12.9b. Although the nanophase separation of main and side chains survives inside the cylindrical PODMA domains the D_c values are significantly reduced. The idea is that the alkyl groups close to the main chain planes can not easily crystallize because the curved PS-PODMA interfaces do not fit to the lamellar morphology of semi-crystalline PODMA. Moreover, each imperfection in the position of the main chain in the PS-PODMA interface will lead to a reduced degree of crystallinity D_c since the alkyl groups feel more constraints.

In the light of these results it would be interesting to study the side chain crystallization in well isolated compartments namely in small spherical PODMA domains. Studies of this type have been performed for other block copolymers and give interesting information about the process of homogeneous nucleation [10, 11]. However, we did not succeed so far to prepare microphase-separated P(S−b−ODMA) block copolymers containing small

PODMA spheres. One reason might be the complex internal structure of PODMA domains due to the nanophase-separation tendency in this side chain polymer. Possibly there is a strong (possibly irresolvable) conflict between nanophase-separation tendency and formation of PODMA spheres with small diameter. Whether or not there are really peculiarities in the phase behavior of P(S−b−ODMA) block copolymers is a topic of further investigations. Work on large molecular weight P(S−b−ODMA) samples is in progress. We expect that topological constraints should be less pronounced in case of PODMA spheres with larger diameter since the required curvature of the PS-PODMA interfaces is smaller. Even when PODMA spheres can be formed under these conditions the crystallization tendency of the alkyl groups in spherical PODMA domains should be always severely reduced. However, it is interesting to note in this context that random poly(styrene−stat−octadecylmethacrylate) copolymers are able to crystallize. This shows that side chain crystallization of long alkyl groups can occur under extreme conditions and offers interesting opportunities to influence the mechanical properties of polymeric systems systematically [59].

12.4 Conclusions

We have shown in this chapter that the side chain crystallization in poly(n-octadecylcylmethacrylate) homopolymers can be understood as a crystallization of frustrated alkanes. Due to the fact that long alkyl groups in side chain polymers like higher methacrylates are chemically bonded to a immobile main chain only a certain part of the alkyl groups far away from the main chain is able to reach crystalline order. In PODMA the alkyl groups do crystallize close to room temperature in a hexagonal lattice corresponding to the situation in the rotator phase of alkanes or in highly defected polyethylenes. Scattering data and calorimetric results suggest that the side chain crystallization process in PODMA proceeds in four steps: (i) nanophase separation in the melt = aggregation of alkyl groups belonging to different monomeric units to alkyl nanodomains with a size of about 2 nm; (ii) early stages of crystallization = formation of nuclei in amorphous alkyl nanodomains containing nearly extended alkyl groups; (iii) primary crystallization = lateral growth of a thin layer in the middle of each alkyl nanodomain and (iv) secondary crystallization = thickening of the crystalline layers on logarithmic time scales. Although the situation is surely modified by the existence of immobile main chains several features which are observed in other systems containing long CH_2 sequences seem to be preserved. Otherwise, the constraints may stabilize states which are instable or transient in cases where long CH_2 sequences are less hindered by the environment. This shows that a good understanding of the crystallization process in side chain polymers with long alkyl groups is not only interesting for an understanding of these materials but also for a description of peculiarities in other crystallizable systems.

The results presented in Sect. 12.3 of this chapter demonstrate that microphase-separated poly(styrene−$block$−octadecylmethacrylate) copolymers with lamellar and cylindrical morphology are systems with a hierarchy of length scales in the nanometer range. Self-assembled pattern − formed by classical microphase-separation of incompatible PS and PODMA blocks on a scale of 10–50 nm and due to nanophase separation of main and side chains in the PODMA domains on a scale of 2–3 nm − coexist. Scattering data show in combination with calorimetric results that the crystallization process of PODMA in these systems is a strong confinement case. The polystyrene phase is glassy at those temperatures where side chain crystallization occurs within the PODMA domains. Thus, the crystallization process is unable to change the block copolymer morphology. The influence of constraints introduced by the glassy PS phase on the side chain crystallization in small PODMA lamellae or cylinders was studied in detail by DSC. It is shown that there are only weak confinement effects in case of PODMA lamellae with a thickness of 9–17 nm. The crystallization behavior of the PODMA lamellae is similar to that of PODMA homopolymers. In case of cylindrical POMDA domains the strength of the confinement effects is increasing with decreasing domain size. For large ($d = 24$ nm) cylinders degree of crystallinity D_c and crystallization times τ_c are only slightly affected while for cylinders with a diameter of about 11 nm D_c is 50% reduced and τ_c is $\approx 10^8$ times larger compared to the situation in PODMA homopolymers. The width of the transformation interval during isothermal crystallization is significantly increasing with decreasing diameter, corresponding to a decrease of the Avrami coefficient from $n \approx 1.8$ for the largest ($d = 24$ nm) to $n \approx 0.7$ for the smallest ($d = 11$ nm) cylinders in this study. These results are consistent with a transition from dominantely heterogenous nucleation and three-dimensional crystal growth in bulk PODMA to dominantely homogeneous nucleuation and one-dimensional growth in small cylindrical PODMA domains. The finding that the degree of crystallinity in PODMA lamellae is unaffected while the D_c values in small PODMA cylinders are significantly reduced might be related to the fact that the internally lamellar structure of semi-crystalline PODMA is matching to the lamellar block copolymer morphology while stronger conflicts occur in case of samples with cylindrical morphology. Speculative pictures for the internal structure of the PODMA domains explaining these finding qualitatively have been presented.

In order to quantify the influence of changes in crystal growth and nucleation behavior more seriously experiments on P(S−b−ODMA) block copolymers with spherical morphology and definitively isolated PODMA domains would be helpful. The preparation of such materials seems to be a challenge since small PODMA spheres do not form easily. Possibly, there is a conflict between the internal structure of the PODMA domains and the block copolymer morphology if the diameter is too small. However, work along this line is in progress and we are still hopeful that information about nucleation rates can be derived from such systems. Another topic which has to be settled by additional experiments is the internal structure of the PODMA domains in block copolymers with different morphology. The speculative pictures shown in Fig. 12.9 should be compared with results from scattering experiments on oriented block copolymer samples. In combination with these topics we investigate currently the structure formation in random poly(styrene−$stat$−octadecylmethacrylate) copolymers which show interestingly also side chain crystallization. All these studies on small alkyl nanodomains embedded in a glassy environment should contribute to a better understanding of the nature of early stages of crystallization in case of materials containing long CH_2 sequences.

Acknowledgement

The authors thank Ch. Darko and I. Lieberwirth (Mainz) for assistance with the WAXS measurements, F. Menau and W. Bras (Grenoble) for support at beamline BM26, M. Buschnakowski and S. Henning (Halle) for AFM measurements as well as K. Schröter, Th. Thurn-Albrecht (Halle) and G. Wegner (Mainz) for helpful discussions. This research was supported by German Science Foundation (SFB418) and ESFR Grenoble (SC1779).

References

[1] A. Keller, Phil. Mag. **2**, 1171 (1957).
[2] E.W. Fischer, Z. Naturforschg. **12a**, 753 (1957).
[3] B. Wunderlich, *Macromolecular Physics*. (Academic Press, New York, 1973).
[4] G. Strobl, *The Physics of Polymers*. (Springer, Berlin, 1997).
[5] J.-U. Sommer, G. Reiter, *Polymer Crystallization – Observations, Concepts and Interpretations*. (Springer, Berlin, 2003).
[6] I.W. Hamley, *The Physics of Block Copolymers*. (Oxford University Press, Oxford, 1998).
[7] Y.-L. Loo, and A.J. Ryan, in: I.W. Hamley (ed.) *Developments in Block Copolymer Science and Technology*. (Wiley, New York, 2004).
[8] Y.-L. Loo, R.A. Register, and A.J. Ryan, Macromolecules **35**, 2365 (2002).
[9] Y.-L. Loo, R.A. Register, A.J. Ryan, and G.T. Dee, Macromolecules **34**, 8968 (2001).
[10] G. Reiter, G. Castelein, J.U. Sommer, A. Röttele, and T. Thurn-Albrecht, Phys. Rev. Lett. **87**, 226101 (2001).
[11] Y.-L. Loo, R.A. Register, and A.J. Ryan, Phys. Rev. Lett. **84**, 4120 (2000).
[12] D. Turnbull and R.L. Cormia, J. Chem. Phys. **34**, 820 (1961).
[13] G. Strobl, Eur. Polym. J. E **3**, 165 (2000).
[14] G. Strobl, Eur. Polym. J. E **18**, 295 (2005).
[15] M. Muthukumar, Phil. Trans. R. Soc. Lond. A **361**, 539 (2003).
[16] J.-U. Sommer, *Theoretical Aspects of the Equilibrium State of Chain Crystals*, Lect. Notes in Phys., **714**, 21–45 (Springer, Berlin, 2006).
[17] J.M.G. Cowie, Z. Haq, I.J. McEwen, and J. Velickovic, Polymer **22**, 327 (1981).

[18] J. Clauss, K. Schmidt-Rohr, A. Adam, C. Boeffel, and H.W. Spiess, Macromolecules **20**, 5208 (1992).
[19] M. Mierzwa, G. Floudas, P. Stepanek, and G. Wegner, Phys. Rev. B **62**, 14012 (2000).
[20] G.J.J. Out, A. Turetzkii, and M. Möller, Macromol. Rapid Commun. **16**, 107 (1995).
[21] F. Lopez-Carrasquero et al., Polymer **44**, 4969 (2003).
[22] A. Turner Jones, Macromol. Chem. **71**, 1 (1964).
[23] M. Beiner, K. Schröter, E. Hempel, S. Reissig, and E. Donth, Macromolecules **32**, 6278 (1999).
[24] M. Beiner and H. Huth, Nature Materials **2**, 595 (2003).
[25] V. Arrighi, A. Triolo, I.J. McEwen, P. Holmes, R. Triolo, and H. Amenitsch, Macromolecules **33**, 4989 (2000).
[26] K.W. McCreight, et al., J. Polym. Sci. B: Polym. Phys. **37**, 1633 (1999).
[27] E. Hempel, H. Budde, S. Höring, and M. Beiner, J. Non-cryst. Solids **352**, 5013 (2006).
[28] M. Mao, J. Wang, S.R. Clingman, C.K. Ober, J.T. Chen, and E.L. Thomas, Macromolecules **30**, 2556 (1997).
[29] I.W. Hamley, V. Castelletto, Z.B. Lu, C.T. Imrie, T. Itoh, and A. Al-Hussein, Macromolecules **37**, 4798 (2004).
[30] I.A. Ansari, V. Castelletto, T. Mykhaylyk, I.W. Hamley, Z.B. Lu, T. Itoh, and C.T. Imrie, Macromolecules **36**, 8898 (2003).
[31] O. Ikkala, and G. ten Brinke, Science **295**, 2407 (2002).
[32] S.S. Rogers and L. Mandelkern, J. Phys. Chem. **61**, 985 (1957).
[33] J.D. Ferry *Viscoelastic Properties of Polymers* (Wiley, New York, 1980).
[34] N.G. McCrum, B.E. Read, G. Williams *Anelastic and Dielectric Effects in Polymeric Solids* (Dover Press, New York, 1991).
[35] R.L. Miller, R.F. Boyer, and J. Heijboer, J. Polym. Sci.: Polym. Phys. Ed. **22**, 2021 (1984).
[36] J.L. Gomez Ribelles, M. Monleon Pradas, A. Vidaurre Garayo, F. Romero Colomer, J. Mas Estelles, and J.M. Meseguer Duenas, Macromolecules **28**, 5878 (1995).
[37] E. Hempel, H. Budde, S. Höring, and M. Beiner, Thermochim. Acta **432**, 254 (2005).
[38] M. Hikosaka, K. Amano, S. Rastogi, and A. Keller, J. Mater. Sci. **35**, 5157 (2000).
[39] W. Hu, S. Srinivas, and E.B. Sirota, Macromolecules **35**, 5013 (2002).
[40] P.D. Olmsted, W.C.K. Poon, T.C.B. McLeish, N.J. Terrill, and A.J. Ryan, Phys. Rev. Lett. **81**, 373 (1998).
[41] B.S. Hsiao, L. Yang, R.H. Somani, C.A. Avila-Orta, and L. Zhu, Phys. Rev. Lett. **94**, 117802 (2005)
[42] H. Kraack, E.B. Sirota, and M. Deutsch, J. Chem. Phys. **112**, 6873 (2000).
[43] A.B. Herhold, H.E. King, and E.B. Sirota, J. Chem. Phys. **116**, 9036 (2002).
[44] P. Huber, D. Wallacher, J. Albers, and K. Knorr, Europhys. Lett. **65**, 351 (2004).
[45] R. Montenegro, M. Antonietti, Y. Mastai, and K. Landfester, J. Phys. Chem. B **107**, 5088 (2003).
[46] E.B. Sirota, H.E. King, G.J. Hughes, and W.K. Wan, Phys. Rev. Lett. **68**, 492 (1992)

[47] A.B. Herhold, D. Ertas, A.J. Levine, and H.E. King, Phys. Rev. E **59**, 6946 (1999)
[48] N. Maeda, M.M. Kohonen, and H.K. Christenson, J. Chem. Phys. B **105**, 5906 (2001)
[49] B.M. Ocko, X.Z. Wu, E.B. Sirota, S.K. Sinha, O. Gang, and M. Deutsch, Phys. Rev. E **55**, 3164 (1997)
[50] M. Beiner, Macromol. Rapid Commun. **22**, 869 (2001).
[51] E. Hempel, H. Huth, and M. Beiner, Thermochim. Acta **403**, 105 (2003).
[52] M. Muthukumar, and P. Welch, Polymer **41**, 8833 (2000).
[53] M. Muthukumar, Eur. Polym. J. E **3**, 199 (2000).
[54] L. Yang, R.H. Somani, I. Sics, B.S. Hsiao, R. Kolb, H. Fruitwala, and C. Ong, Macromolecules **37**, 4845 (2004).
[55] S. Hiller, O. Pascui, H. Budde, O. Kabisch, D. Reichert, and M. Beiner, New Journal of Physics **6**, 10 (2004).
[56] R.M. Ho, T.M. Chung, J.C. Tsai, J.C. Kuo, B.S. Hsiao, and I. Sics, Macromol. Rap. Commun. **26**, 107 (2005).
[57] S. Nojima, K. Kato, S. Yamamoto, and T. Ashida, Macromolecules **25**, 2237 (1992).
[58] C. Darko Master Thesis, Universität Halle / MPI-P Mainz (2005).
[59] E. Hempel, H. Budde, S. Höring, and M. Beiner, to be published.
[60] L. Mandelkern, in: J.E. Mark (ed.) *Physical Properties of Polymers.* (Cambridge, Cambridge, 2003).

13

Crystallization in Block Copolymers with More than One Crystallizable Block

Alejandro J. Müller, María Luisa Arnal, Vittoria Balsamo

Materials Science Department, Universidad Simón Bolívar, Aptdo. 89000, Caracas 1080-A, Venezuela
amuller@usb.ve

Abstract. Recent results on the crystallization of block copolymers with more than one crystallizable block are reviewed. The effect that each block has on the nucleation, crystallization kinetics and location of thermal transitions of the other blocks has been considered in detail. Depending on the thermodynamic repulsion between the blocks, the initial melt morphology in weakly segregated double crystalline diblock copolymers can be sequentially transformed by the crystallization of the different blocks. The crystallization kinetics of each block can be dramatically affected by the presence of the other, and by the crystallization temperature; the magnitude of the effect is a function of thermodynamic repulsion. Also the morphology has been investigated and peculiar double crystalline spherulites with intercalated semi-crystalline lamellae of each component have been observed in weakly segregated diblock copolymers. In the case of ABC triblock copolymers with more than one crystallizable block, many interesting effects have been found; among them, self-nucleation, sequential or coincident crystallization, and fractionated crystallization can be mentioned. Additionally, the effect of the topological constrains due to the number of free ends has been studied. Factors like chemical structure, molecular weight, molecular architecture and number of crystallizable blocks provide a very large number of possibilities to tailor the morphology and properties of these interesting novel materials.

13.1 Introduction

Crystallization in block copolymers has attracted much attention, and several reviews about it have been recently published [1–3]. The ability of block copolymers to self-assemble in the melt according to the relative thermodynamic repulsion between its components has been extensively explored both theoretically and experimentally [4, 5]. When one or more block copolymer components can crystallize, a competition between phase segregation and crystallization can lead to major changes in microstructure. These have been described in detail, in particular for the case of diblock copolymers with only one crystallizable block. Copolymers within this category that have been reported in the literature are mainly hydrogenated polybutadiene (HPB or PE), poly(ethylene oxide) (PEO) or poly(ε-caprolactone) (PCL) based materials [3]. Depending on the location of three important transition temperatures, namely, the glass transition temperature, T_g, of the non-crystallizable block, the crystallization temperature, T_c, of the block under consideration, and the order disorder transition, T_{ODT}, of the diblock copolymer, the crystallization can drive structure formation, or it can be confined within the phase segregated microdomains (MDs), although in the latter, distortion of the MDs may occur. All possible cases that depend on the relative location of the transition temperatures have been recently reviewed [2, 3].

In this paper we review recent results (mostly from the year 2000 onwards, with a few exceptions) on less commonly known situations in the crystallization of block copolymers, where both blocks within a diblock copolymer, or more than one block (typically two) within triblock terpolymers can crystallize. As it is expected, the crystallization behaviour of crystalline-crystalline block copolymers is more complicated; for instance, when the copolymers are quenched from a microphase-separated melt into various temperatures below the melting temperatures of the corresponding blocks, different situations can be observed. When the melting temperatures of both blocks are close enough such as in poly(ethylene oxide)-b-poly(ε-caprolactone), a coincident crystallization of both blocks can be obtained by quenching. On the other hand, when the melting temperature of one block is far from the other, a completely different behaviour can be seen; one block crystallizes in advance and produces a specific morphology, which can or cannot be modified upon crystallization of the other block. Such modification depends, among other controlling parameters, on segregation strength, crystallization temperature and molecular weight of the block components. In this work, we concentrate on aspects that range from the melt structure, and how it can be obliterated by crystallization, to the crystallization kinetics of each individual block. The effect that each block has on nucleation and crystallization kinetics, as well as on the corresponding thermal transitions of the other is particularly relevant, and will be considered in detail in those systems for which data is available.

13.2 Double Crystalline AB and ABA Copolymers

In 1972, Perret and Skoulios [6, 7] published the first reports on double crystalline diblock copolymers with PEO and PCL semicrystalline components, i.e., poly(ethylene oxide)-b-poly(ε-caprolactone) (PEO-b-PCL or EOC). Later on, several authors have prepared and characterized this kind of materials or their ABA analogs, i.e., PCL-b-PEO-b-PCL [8–19]. Recently, PEO as well as PCL have also been incorporated in linear [20–23] and star shaped ABC triblock terpolymers [24–27].

Based on X-ray measurements, Nojima et al. [9], Piao et al. [17] and He et al. [18, 19] found that poly(ethylene oxide)-b-poly(ε-caprolactone) diblock copolymers exhibit miscibility in the melt. A similar conclusion was given by Petrova et al. [13] due to the observation of a single T_g by means of Differential Scanning Calorimetry (DSC). Upon crystallization, a microphase separation process takes place, in which alternating lamellae of both components are formed. It is precisely this ability to separate in microphases, together with properties like biocompatibility, amphiphilic character, permeability and controlled biodegradability, that make the investigation of these materials very interesting for a wide range of applications. Among these applications, controlled drug release and tissue engineering can be mentioned. The crystallization process of both blocks is greatly affected by composition, block molecular weight, and architecture of the particular copolymer. The miscibility of the blocks in the melt and the covalent link between them causes a depression of the crystallization and melting temperatures of both components, a phenomenon that is more pronounced for PEO, which is, according to most of the reports, the one that crystallizes at the highest supercooling. Nevertheless, the small difference between the melting and crystallization transitions of the corresponding homopolymers, PCL and PEO, and the significant influence of the molecular weight on these, makes possible the finding of variations in the melting and crystallization order of the blocks, as well as overlapping of the corresponding thermal processes. The latter is observed, especially, when dynamic cooling and heating scans are performed in a DSC.

In the literature many differences can be found in the temperature range used for the study of the crystallization and melting processes of PEO and PCL based AB diblock and ABA triblock copolymers. When the studies are performed above room temperature, an important fraction of the blocks may remain amorphous [8–11, 14, 16]; however, most authors report that when the study is extended at temperatures below T_g, both blocks can crystallize [13–15,17]. In the case of ABA triblock copolymers, it has been found that the B-block remains amorphous when its content is lower than 10%, or its molecular weight is very low. Piao et al. [17] and He et al. [18,19] synthesized either poly(ε-caprolactone)-b-poly(ethylene oxide)-b-poly(ε-caprolactone) ABA triblock copolymers, as well as poly(ethylene oxide)-b-poly(ε-caprolactone) AB diblock copolymers. They used poly(ethylene glycol) (PEG) as precursor and a calcium catalyst. Then, they characterized the materials by using NMR, DSC, WAXS and Polarized Optical Microscopy (POM). Cooling DSC scans carried out by He et al. [18] in AB diblock copolymers of different compositions are presented in Fig. 13.1.

Although almost all the copolymers examined exhibited two crystallization exotherms, in $EO_6{}^2C_{94}{}^{33}$ (the subscripts and superscripts indicate the wt.% and molecular weight, in Kg/mol, of each block respectively) the PEO does not crystallize. $EO_{50}{}^5C_{50}{}^5$ also exhibits a single crystallization exotherm, but in this case, this is due to coincident crystallization of both blocks as demonstrated by WAXS. He et al. previously evaluated the crystallization and melting behaviour of the corresponding homopolymers; they obtained values of $T_c = 31.6°C$ and $T_m = 63.5°C$ for a PCL^8, and $T_c = 42.6°C$ and

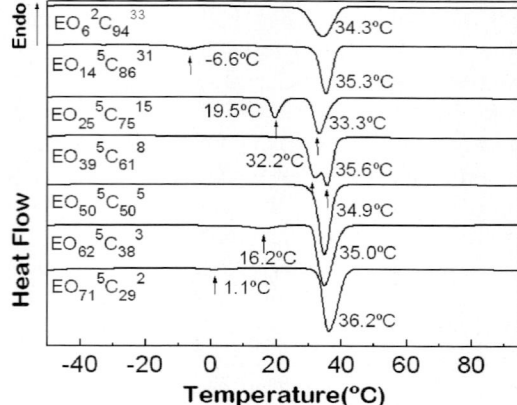

Fig. 13.1. DSC cooling curves (10°C/min) for EOC diblock copolymers (taken from reference [18])

$T_m = 66.6°C$ for a PEO[5] [18]. The crystallization process is more complicated in the double crystalline diblock copolymers than in the homopolymers. In the specific case of PEO-b-PCL diblocks, the crystallization of the first block occurs at supercoolings higher than in the corresponding PEO homopolymer as a consequence of the diluent effect caused by the homogeneous melt. This effect becomes evident when comparing the crystallization temperature of PEO[5] ($T_c = 42.6°C$) with that of the PEO block within $EO_{71}{}^5C_{29}{}^2$ ($T_c = 36.2°C$); i.e., the covalently linked PCL block causes a depression of 7°C in the crystallization temperature of the PEO block.

The influence of one block on the other was evaluated by changing the relative length of the blocks. He et al. [18] maintained the PEO length at 5 Kg/mol, and progressively increased the PCL length from 2 kg/mol up to 31 Kg/mol. The PCL crystallization temperature progressively increased from 1.1°C for $EO_{71}{}^5C_{29}{}^2$ up to 35.3°C for $EO_{14}{}^5C_{86}{}^{31}$. This change of 34.2°C in the crystallization temperature of the PCL block is due to a combined effect of the PCL molecular weight variation, and a diluent effect since the crystallization process takes place from a homogeneous melt. While PCL is the block that crystallizes first or at higher temperatures in $EO_{14}{}^5C_{86}{}^{31}$, in $EO_{71}{}^5C_{29}{}^2$ the PCL block, which has a lower molecular weight, is the second component to crystallize upon cooling; therefore, it is affected by the previously crystallized PEO block that constrains chain mobility. Simultaneously, the authors observed a depression of the PEO crystallization temperature from 36.2°C for $EO_{71}^5 C_{29}^{2.1}$ down to –6.6°C for $EO_{14}^5 C_{86}^{30.6}$ (see Fig. 13.1). It should be noted that all these changes in the order of PEO and PCL crystallization as a function of the molecular weight indicate that no nucleating effects take place in poly(ethylene oxide)-b-poly(ε-caprolactone) diblock copolymers, i.e., none of the blocks nucleates the other.

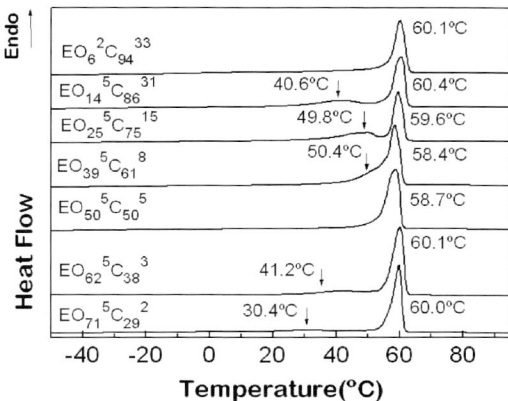

Fig. 13.2. DSC heating curves (10°C/min) for EOC diblock copolymers after the cooling shown in Figure 1, taken from [18]

He et al. [18] also performed subsequent DSC heating curves that are displayed in Fig. 13.2. Again, with the exception of $EO_6{}^2C_{94}{}^{33}$ and $EO_{50}{}^5C_{50}{}^5$, all other copolymers show two melting endotherms. In $EO_6{}^2C_{94}{}^{33}$ the PEO block is amorphous, and in $EO_{50}{}^5C_{50}{}^5$ the melting of both blocks is coincident. An increase of the PCL molecular weight from 2 Kg/mol up to 31 Kg/mol leads to changes of T_{mPCL} from 30.4°C to 60.4°C for $EO_{71}{}^5C_{29}{}^2$ and $EO_{14}{}^5C_{86}{}^{31}$ respectively. On the other hand, the PEO block melts at temperatures between 40.6°C and 60°C due to an increase of the PEO content from 14% to 71%. This means that the T_m of the PEO block is always lower than T_m for the PEO^5 homopolymer (65°C), indicating a diluent effect of the previously melted PCL block in $EO_{71}{}^5C_{29}{}^2$. As can be appreciated in $EO_{14}{}^5C_{86}{}^{31}$, whose PEO block melting point is 40.6°C, when PEO is the second component that crystallizes, the magnitude of the depression is even higher because of the constrains imposed by the previously crystallized PCL block. Thus, the results can then be summarized as follows: the PCL block crystallizes and melts at a higher temperature than the PEO block in $EO_{14}{}^5C_{86}{}^{31}$, $EO_{25}{}^5C_{75}{}^{15}$ and $EO_{39}{}^5C_{61}{}^8$; the blocks crystallize in a coincident fashion in $EO_{50}{}^5C_{50}{}^5$, while the PEO crystallizes and melts at a higher temperature than PCL in $EO_{62}{}^5C_{38}{}^3$ and $EO_{71}{}^5C_{29}{}^2$. WAXS measurements carried out at room temperature evidenced the characteristic reflections of both blocks in the copolymers that crystallizes above 0°C.

As it is usually the case in block copolymers that crystallize from a homogeneous melt (also in the case of weak segregation regime), crystallization dominates the morphology, and superstructures like spherulites can be observed (2-3). POM observations showed that the superstructure in each copolymer was governed by the majority block, showing similarities to that of the corresponding homopolymer; i.e., similar to PEO in $EO_{62}{}^5C_{38}{}^3$ and $EO_{71}{}^5C_{29}{}^2$, and to PCL in $EO_6{}^2C_{94}{}^{33}$, $EO_{14}{}^5C_{86}{}^{31}$ and $EO_{25}{}^5C_{75}{}^{15}$ (18). $EO_{50}{}^5C_{50}{}^5$ and

$EO_{39}{}^5C_{61}{}^8$ exhibit double concentric spherulites (see Fig. 13.3), similar to those found by Shiomi et al. [9] for PEO-b-PCL copolymers with PCL contents of about 60%. $EO_{50}{}^5C_{50}{}^5$ was thoroughly investigated by He et al. [19] by means of DSC, WAXS, SAXS, FTIR and POM in order to find an explanation for the formation of double concentric spherulites and the coincident crystallization of both blocks. PEO and PCL crystallize and melt in a coincident fashion when 10°C/min scans are performed in a DSC. However, FTIR measurements carried out at a cooling rate of 0.2°C/min allowed the observation of the characteristic bands corresponding to the crystalline regions of each block; i.e., 1196 cm^{-1} for PCL and 843 cm^{-1} for PEO. Through these experiments, He et al. [19] were able to distinguish that the PCL block crystallizes first from the melt.

SAXS experiments performed in the $EO_{50}{}^5C_{50}{}^5$ copolymer at room temperature showed four maxima that were attributed to alternated PCL and PEO lamellae (see curve c in Fig. 13.4), in which crystalline and amorphous zones coexist. At 59°C, a temperature where PEO is molten, two maxima are observed and assigned to first and second order reflections due to the alternation of PCL crystalline regions, and mixed PCL and PEO amorphous ones (curve d). At 65°C, where all crystals have melted, the maxima disappear indicating a homogeneous system (curve e).

He et al. [18,19], after considering the above presented results, proposed the following mechanism for the formation of the double concentric spherulites. Upon isothermal crystallization, PCL spherulites appear initially. Then, PEO nucleation takes place within PCL spherulites; nevertheless, PEO crystals grow in the interlamellar regions of the PCL spherulites with a rate that is higher than that of the PCL crystals. This is precisely what induces the concentric growth in the outer zone of the PCL spherulite.

As can be seen from the results presented above, the study of PEO-b-PCL has made evident the complexity of the crystallization and melting processes in double crystalline diblock copolymers that have the following characteristics: their crystallization starts from a homogenous melt, crystallization and melting points of the blocks are very near to each other, and there is an absence of a nucleating effect of each block on the other.

Another very interesting family of double crystalline diblock copolymers is formed by poly(L-lactide) (PLLA) and poly(ε-caprolactone). PLLA has been widely used for biomedical applications, such as absorbable sutures and bond fixation [28] because its excellent biodegradability by hydrolysis. However, the use of PLLA in other kind of products like agricultural films is currently still limited. The main reason for this is its brittleness; nevertheless, this weakness might be improved by means of copolymerization with a lower glass transition temperature component like, for example, PCL, which, as has been previously mentioned, is also biodegradable [29–34]. Biodegradability and mechanical properties are strongly affected by the crystallinity of the blocks, which in turns depends upon microdomain (MD) structures. Dubois et al. have accomplished the synthesis of model PLLA-b-PCL (L_xC_y) diblock copolymers by

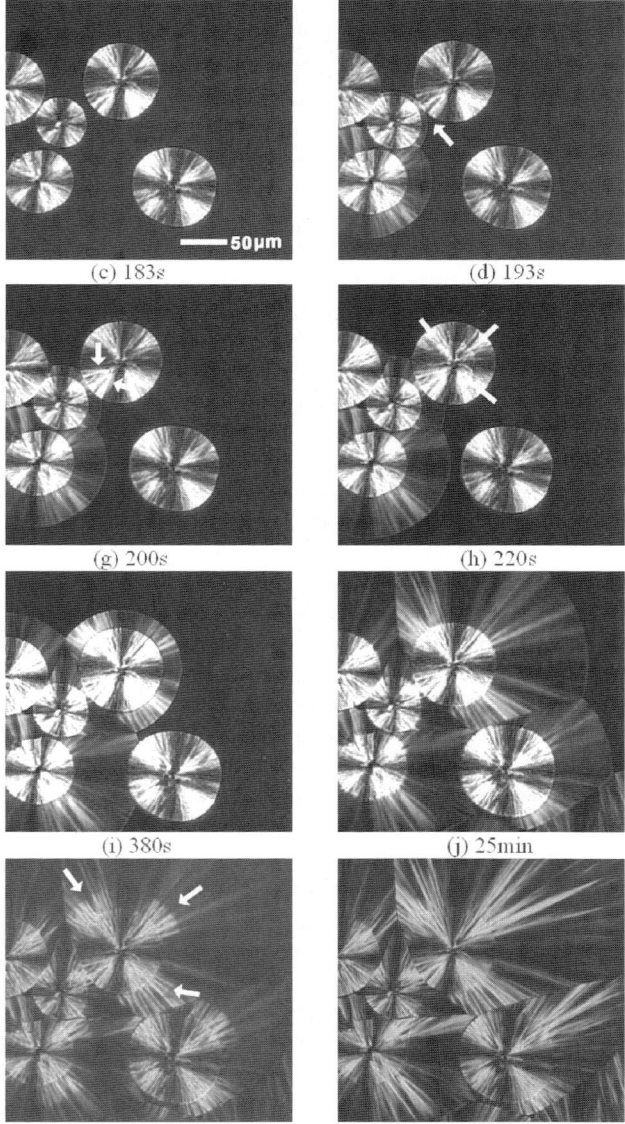

Fig. 13.3. Real-time POM micrographs of $EO_{50}^5C_{50}^5$. The specimen was melted at 80°C for 5 min, quenched down to 38°C at 40°C/min, and finally held at that temperature for the times shown. Taken from [19]

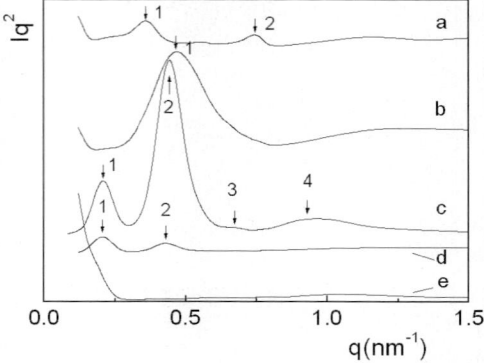

Fig. 13.4. SAXS patterns for PEO (**a**), PCL (**b**) and $\mathrm{EO}_{50}^{5}\mathrm{C}_{50}^{5}$(**c, d, e**). Curves (a,b and c) were obtained at 20°C, while curves (d) and (e) were obtained at 59°C and 65°C respectively. Taken from [19]

controlled "living" sequential block copolymerization initiated by aluminium trialkoxides in solution [35].

A comprehensive study on the melt structure of PLLA-*b*-PCL diblock copolymers and its subsequent morphological transformation upon cooling has been recently reported by Hamley et al. [33,34]. The model diblock copolymers studied by them include the following: $L_{32}^{7}C_{68}^{15}$, $L_{44}^{11}C_{56}^{14}$ and $L_{60}^{12}C_{40}^{9}$, where subscripts and superscripts indicate again composition in wt % and molecular weight of each block in Kg/mol respectively.

Figure 13.5 shows the variation of melting and crystallization temperatures corresponding to each semicrystalline block within PLLA-*b*-PCL diblock copolymers as a function of composition [34]. For comparison purposes, solution blends of PCL and PLLA homopolymers of equivalent molecular weights to those of the diblock copolymers were prepared and their characteristic transition temperatures were also reported in Fig. 13.5. This figure shows that the prepared PLLA and PCL blends are immiscible for the compositions examined as can be gathered by the invariance of the melting points associated with each homopolymer. Instead, the diblock copolymers exhibit signs of miscibility. In particular, the melting temperature of the PLLA block decreases as the content of PCL increases, and in the case of the $L_{32}C_{68}$ sample, the melting point depression of the PLLA block reaches 11°C.

There are no measurements available of χ for the PLLA-*b*-PCL system. However, rough estimates based on solubility parameters of the quantity χN, which determines the segregation strength of the copolymer, vary between 3 and 50 depending on the solubility parameters chosen, temperature range and molecular weight. The uncertainty is rather high; however, the consensus in the literature is that they either form a homogeneous melt or are weakly segregated, but it is considered that they are not in the strong segregation

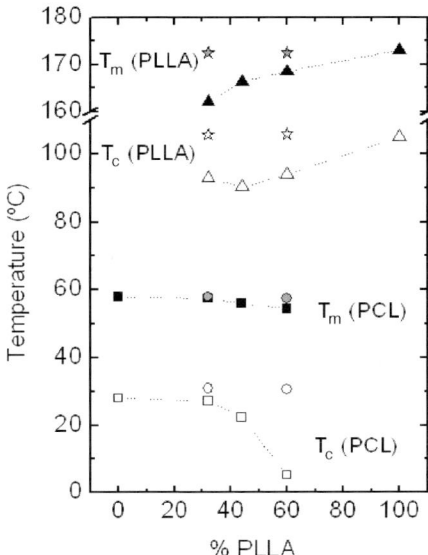

Fig. 13.5. Melting and crystallization temperatures for PLLA and PCL blocks within PLLA-*b*-PCL diblock copolymer versus PLLA composition (data points plus *dashed line*). The data points (without *dashed line*) indicate the transition temperatures for PLLA and PCL components in solution blends. Taken from reference [34]

regime [30, 32–34]. These considerations agree with the above DSC results (Fig. 13.5).

Hamley et al. demonstrated that the copolymer sample $L_{32}C_{68}$ forms a homogenous melt at 190°C (this is the sample that shows a melting point depression of 11°C for the PLLA block in Fig. 13.5), since its SAXS pattern (see Fig. 13.6) is typical of a homogeneous system [4, 34]. Cooling the sample down to 165°C produces a single broad reflection that still indicates the presence of a homogeneous melt. After the sample is quenched to 122°C (a temperature at which PLLA can crystallize while PCL is in the melt) and left to isothermally crystallize at that temperature for 30 min, a PLLA lamellar crystalline structure developed as indicated by the peak positions in a 1:2 ratio. Upon PCL crystallization, when the sample is further cooled to 42°C, the lamellar structure is changed as revealed by the different q values of the maxima and the different intensities of the two characteristic reflections as compared to the SAXS pattern at 122°C. The crystalline structure induced by PCL is also lamellar but of a different periodicity.

Hamley et al. performed simultaneous real time WAXS measurements in the same sample of Fig. 13.6 (i.e., $L_{32}C_{68}$), and some representative results can be seen in Fig. 13.7. The variation of the most intense WAXS reflection of the crystal structure of the PLLA block ($2\theta = 14.8°$), as a function of

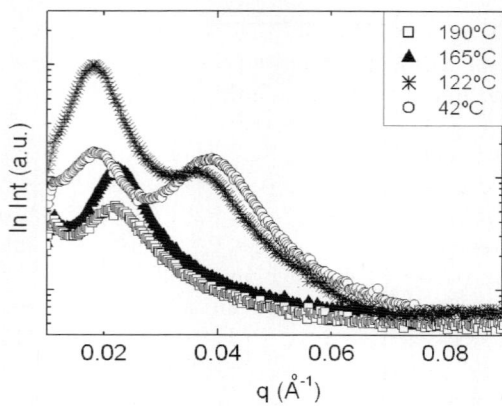

Fig. 13.6. Selected SAXS data for block copolymer $L_{32}{}^{7}C_{68}{}^{15}$. Taken from [34]

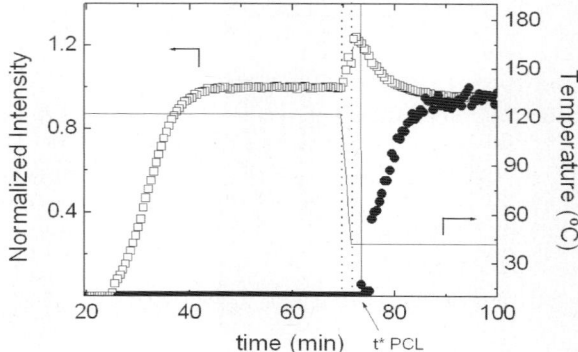

Fig. 13.7. Normalized height of WAXS peaks for block copolymer $L_{32}{}^{7}C_{68}{}^{15}$ from fitted Lorentzian functions: (□) $2\theta = 14.8°$ reflection of PLLA, (●) 110 reflection of PCL. The *solid line* indicates the temperature measured by the DSC instrument (Linkam) employed in the X-ray set-up. The *dashed line* indicated the cooling region from 122 to 42°C. The *vertical solid line* indicated the initial of PCL crystallization (t^*_{PCL}). Taken from [34]

time, describes initially (at 122°C) the crystallization kinetics of the PLLA block until saturation. When the temperature is lowered, as indicated by the temperature ramp shown in the right hand y-axis in Fig. 13.7, the intensity increases as more PLLA is able to crystallize on account of the enhanced supercooling. However, as the PCL block starts to crystallize (as pointed out by the filled circles corresponding to the intensity of the 110 reflection of the crystal structure of the PCL block), the intensity of the PLLA reflection decreases progressively and reaches a plateau when the PCL crystallinity saturates. A similar effect was reported by Hamley et al. for $L_{44}C_{56}$ and $L_{60}C_{40}$ [33]. This effect has been interpreted as a rearrangement of the PLLA lamellar stack to accommodate the crystallization of the previously amorphous PCL lamellar MDs that may even induce a certain level of local melting that could provoke the observed reduction in intensity. Changes in the unit cell of the PLLA were also detected upon PCL crystallization; therefore, a change in the structure factor may also be playing a role in the observed intensity reduction [33, 34].

Polarized optical microscopy (POM) was performed on $L_{32}C_{68}$ yielding remarkable results (34). Even with only 32% PLLA, the sample can crystallize at 122°C (a temperature above the PCL melting point) exhibiting large well developed negative PLLA spherulites that grow until impingement (see Fig. 13.8c). These spherulites contain lamellar stacks that are composed of intercalated semicrystalline lamellar PLLA domains and amorphous PCL domains covalently linked to each other. When the sample is quenched to 42°C, the PCL block crystallizes.

Figure 13.8d shows double crystalline spherulites where PLLA and PCL semicrystalline lamellae coexist. These mixed spherulites are still negative and only the magnitude of the birefringence changed as indicated by the difference

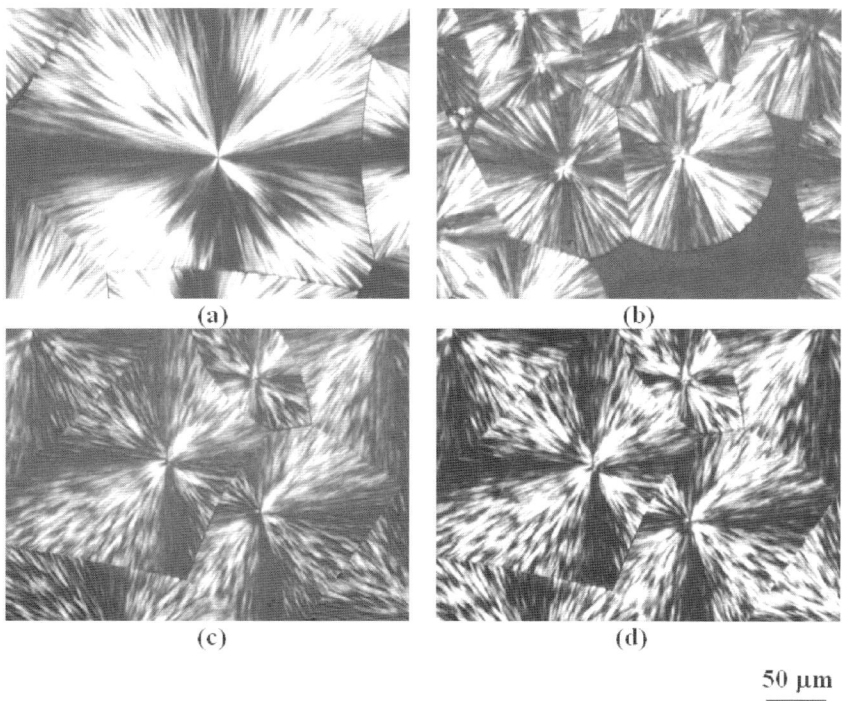

Fig. 13.8. Polarized Optical Micrographs taken during isothermal crystallization: (**a**) homo-PLLA, after 8 min at 122°C; (**b**) homo-PCL, after 10 min at 42°C; (**c**) $L_{32}{}^7C_{68}{}^{15}$ after 30 min at 122°C; (**d**) $L_{32}{}^7C_{68}{}^{15}$ after 15 min at 42°C. Taken from [34]

in the degree of brightness and color of the picture (compare Fig. 13.8c and 13.8d).

Very similar results have been recently obtained by Shin et al. [36] in PLLA-b-PEO-b-PLLA samples (M_n values were in the range of 30-50 kg/mol). They found that PLLA and PEO blocks were miscible in the melt; upon cooling from the melt at 2°C/min, sequential crystallization of PLLA and PEO occurred. The PLLA formed negative spherulites at higher temperatures; then, at lower temperatures, the PEO crystallized within the PLLA spherulites, and the sign of the birefringence was maintained, only the retardation increased in an additive fashion.

The miscibility of the PLLA-b-PCL diblock copolymers studied by Hamley et al. was found to be a function of composition. When compositions closer to symmetric block lengths were employed, a heterogeneous structure in the melt was evidenced by SAXS (see Fig. 13.9). Close inspection of the traces corresponding to 190°C and 165°C in Fig. 13.9 indicated that a lamellar structure was present in the melt (upon close inspection of the data, maxima at q^*, $2q^*$ and $3q^*$ can be seen). Once the sample was allowed to crystallize at 122°C, the spherulitic structure (see [33] for POM micrographs corresponding to this sample) formed is characterized by a lamellar stacking that produces diffuse scattering with a main maximum located at lower q values (plus a shoulder at $2q^*$). An evident change in the lamellar structure is caused by PCL crystallization at 42°C as revealed by SAXS in Fig. 13.9; changes at the unit cell level were also detected by WAXS.

The results obtained by Hamley et al. on PLLA-b-PCL double crystalline diblock copolymers have shown in this case that, regardless of whether the structure of the melt is homogeneous or heterogeneous, crystallization is the dominating driving force leading to extensive rearrangement of the morphol-

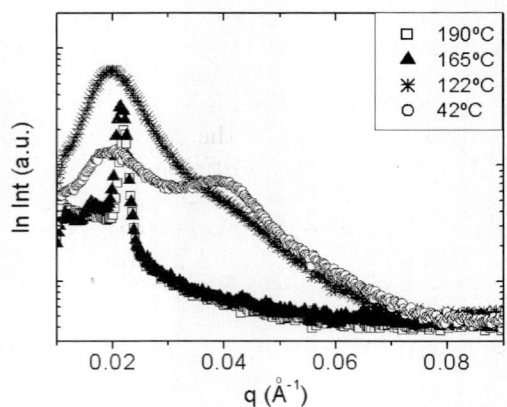

Fig. 13.9. Selected SAXS data for block copolymer $L_{44}^{11}C_{56}^{14}$. Taken from [34]

ogy when each block crystallizes in sequence. This is expected for block copolymers that have low values of χN as in this case [2–4, 33, 34].

Kim et al. [29] also published a report on PLLA-b-PCL diblock copolymers, but they compared different molecular weight materials via X-ray measurements, DSC and POM. They found, for example, that a copolymer with a number average molecular weight of $M_n = 77$ Kg/mol, and a weight fraction of PCL block (w_{PCL}) of 0.32 showed microphase separated structures in the melt up to at least 220°C as determined by rheological measurements. On the other hand, a copolymer with $M_n = 19$ Kg/mol and $w_{PCL} = 0.37$ became homogeneous at a temperature of about 175°C. The melting temperature of the PLLA block, determined through DSC, in the lower molecular weight copolymer was reduced in comparison to that of neat PLLA and the PLLA block in the higher molecular weight block copolymer. In order to determine the MD structures in the melt, they tried to obtain SAXS profiles at temperatures between 180°C and 220°C. Nevertheless, they did not observe any peak corresponding either to a microdomain or to a concentration fluctuation. They attributed this to a very small electron density difference between the PLLA and the PCL block in the melt state. This is not in agreement with the work of Hamley et al. [33, 34], where it was indeed possible to distinguish a microdomain structure in the melt.

The crystallization kinetics of double crystalline diblock copolymers such as PLLA-b-PCL is also interesting. Hamley et al. [33, 34] have reported good agreement between DSC and WAXS isothermal crystallization kinetic data. They have studied isothermal crystallization of the PLLA block at 122°C and found a reduction in the overall crystallization rate and in the Avrami index as the PCL content in the diblock copolymer increases. For instance, the overall crystallization rate of neat PLLA is approximately 1.5 times faster than that of the PLLA block within $L_{60}C_{40}$, and 2.3 times faster than that of the PLLA block within $L_{44}C_{56}$ and $L_{32}C_{68}$. It would seem that the effect of the molten covalently bonded PCL block (at 122°C) on the crystallization kinetics of the PLLA block saturates at a PCL content of 56%. The Avrami index, on the other hand, is progressively reduced from values close to 3 for PLLA to values close to 2 with PCL content in the copolymers, even though PLLA spherulites were observed for all compositions. Similar results were reported by Kim et al. [29].

Müller et al. have performed studies on related double crystalline diblock copolymers that are composed of PCL and poly(p-dioxanone), a biodegradable and polyester-ether whose acronym is PPDX. PPDX is a semicrystalline polymer, commercially employed as bioabsorbable sutures and implantable biomedical devices [37–41]. PPDX-b-PCL diblock copolymers are considered to be in the weak segregation regime [42–45], and they crystallize from a heterogeneous melt with soft confinement for the PPDX block that crystallizes first (at higher temperatures). They found a dramatic increase in the free energy needed for secondary nucleation when the PPDX block was isothermally crystallized at temperatures where the PCL was molten in PPDX-b-PCL di-

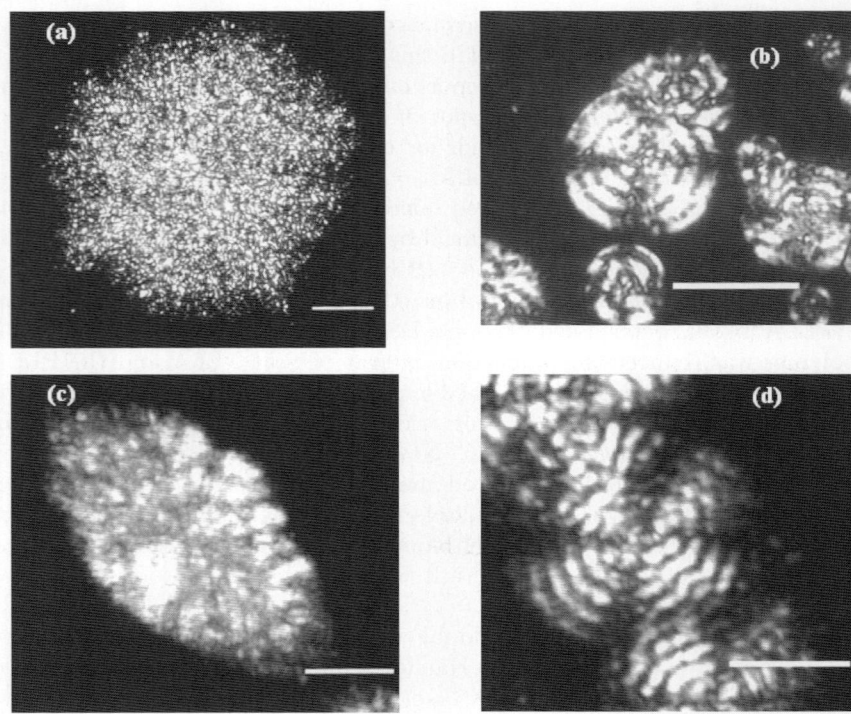

Fig. 13.10. Polarized optical micrographs showing the superstructural morphology of PPDX-*b*-PCL diblock copolymer samples during isothermal crystallization. The bar is equivalent to 30 μm. (**a**) $D_{40}^{5}C_{60}^{7}$, $T_c = 62°C$; (**b**) $D_{40}^{5}C_{60}^{7}$, $T_c = 40°C$; (**c**) $D_{77}^{32}C_{23}^{10}$, $T_c = 60°C$; (**d**) $D_{77}^{32}C_{23}^{10}$, $T_c = 44°C$. Taken from [42]

block copolymers. These block copolymers crystallize from a heterogeneous melt, and even though crystallization dominates the morphological transformation, the energy barrier for a break-out of the spherulitic morphology is much higher than in PLLA-*b*-PCL diblock copolymers [2, 3, 42–44]. Similar mixed spherulites as those formed by PLLA-*b*-PCL diblock copolymers can be formed by PPDX-*b*-PCL diblock copolymers at temperatures were both blocks can crystallize, and the intercalated lamellae of the two semi-crystalline blocks have been visualized by AFM [44]. Figure 13.10 shows examples of the superstructures observed by POM when $D_{40}C_{60}$ is employed. It is interesting to see that if a temperature of 62°C is employed (Fig. 13.10a and [42]), a granular morphology is observed, indicating that even though the PPDX block can crystallize, it can not form well developed spherulites (compare with the spherulites in Fig. 13.8c). Spherulites can be formed by $D_{40}C_{60}$, but only at lower temperatures where the PCL block seems to drive the structure formation (see optical micrographs taken at 46, 44 and 40°C in Fig. 13.10b and [43]). Only when the content of PPDX is as high as 77%, the superstructure formed at 60°C (a temperature at which PCL is molten) resembles a deformed spherulite (see Fig. 13.10c).

A comparison between the polarized optical micrographs taken at temperatures where the PCL block is molten for both PPDX-*b*-PCL and PLLA-*b*-PCL diblock copolymers of similar compositions (compare Fig. 13.8c and Fig. 13.10a), lead to the conclusion that when stronger thermodynamic segregation is present (as in PPDX-*b*-PCL diblock copolymers), the phenomenon of break-out is more difficult. Concurrently, the overall crystallization kinetics is much more strongly depressed at equivalent supercoolings for the PPDX block than for the PLLA block when in both cases they are covalently bonded to molten PCL blocks.

Figure 13.11 shows DSC cooling scans from the melt of selected PPDX-*b*-PCL and PLLA-*b*-PCL diblock copolymers taken from references [34] and [44]. Subsequent heating scans clearly showed the separate melting of each block crystal populations (see refs. [33, 34, 42–44]). It is interesting to note that in the PLLA-*b*-PCL cases, the separate crystallization of each block can be easily identified, i.e., the higher temperature crystallization exotherm corresponds to the crystallization of the PLLA block while that at lower temperatures

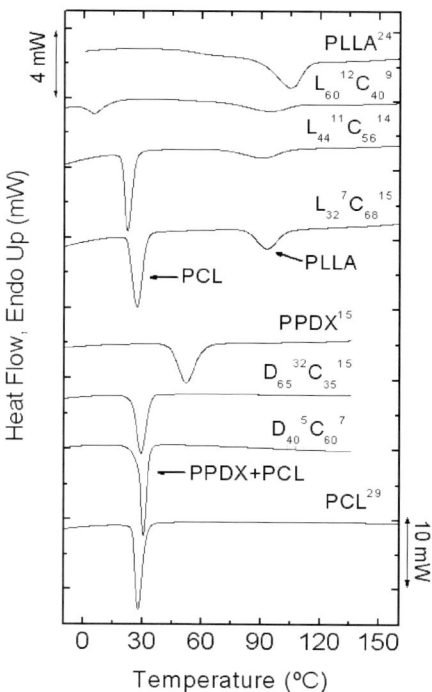

Fig. 13.11. DSC cooling scans (at 10°C/min) for selected PLLA-*b*-PCL and PPDX-*b*-PCL diblock copolymers, and corresponding analogous neat homopolymers [34, 44]

corresponds to the crystallization of the PCL block (see also Fig. 13.5). On the other hand, in the case of the PPDX-*b*-PCL diblock copolymers, only one crystallization exotherm can be seen that corresponds to the crystallization of both blocks as separate evidence by WAXS has indicated [42]. This phenomenon has been studied in detail by Müller et al. [42–45]. The authors have concluded that PPDX crystallization is dramatically retarded by the influence of the covalently bonded PCL block that remains molten at higher temperatures. This in turn causes a depression of the crystallization temperature that can be measured at 10°C/min in the DSC, in such a way, that when the PPDX block finally crystallizes, it does so at the same temperature at which the PCL block crystallizes.

The overall crystallization rate of the PPDX block is depressed to such an extent in PPDX-*b*-PCL diblock copolymers, that isothermal crystallization experiments in the DSC could not be performed in the temperature range where only PPDX crystallizes [3, 44]. Spherulitic growth rates corresponding to PPDX block in $D_{77}C_{23}$ were measured, and the growth rate was an order of magnitude lower than in neat PPDX at equivalent crystallization temperatures. The application of the Lauritzen and Hoffman theory allowed the determination of the energy barrier for secondary nucleation [3, 44]. To increase the overall crystallization rate of the PPDX block, Müller et al. resorted to the self-nucleation of the PPDX block before performing the isothermal crystallization experiment. Only after self-nucleation, the overall crystallization rates of the PPDX block within PPDX-*b*-PCL diblock copolymers were determined by DSC [3, 44].

The above results indicate that a lower thermodynamic segregation may be inducing a smaller effect on the crystallization rate of the block that crystallizes at higher temperatures. This may be connected with the plasticization effect induced by the PCL when it is more miscible with the other block component (i.e., with PLLA). One could envisage that in a system where there is a stronger thermodynamic repulsion, like in PPDX-*b*-PCL (where the Tg's of the block copolymer components are nearly the same as those exhibited by the corresponding homopolymers), PCL can not cause any plasticization; therefore, it can only impair the crystallization of the covalently bonded neighboring block.

Ueda et al. [46] have measured the crystallization rate of the polyethylene block within polyethylene-*b*-(atactic polypropylene), PE-*b*-aPP, with a PE volume fraction of 0.48. The crystallization of the PE block occurs with a molten aPP block covalently bonded to it. The copolymer is reported to be in the strong segregation limit. They also found a substantially lowered crystallization rate as compared to an analog neat PE, which was attributed to a mobility reduction of the chains close to the interphase, and to the presence of non-crystallizable aPP chains close to the growth face which could hinder the growth process.

The isothermal crystallization kinetics of the PLLA block within 50/50 PEO-*b*-PLLA (M_n of the PLLA block was 3.9 Kg/mol) and 10/80/10 PLLA-

b-PEO-b-PLLA (M_n of the PLLA block was 5.2 Kg/mol) was followed by Kim et al. [47] employing real time WAXS at 40°C (a temperature too high for the PEO block to crystallize according to the authors). They have concluded that the crystallization of the PLLA block was retarded as compared to a neat homo-PLLA sample by the presence of the molten covalently bonded PEO.

Bogdanov et al. [15] have studied the isothermal crystallization of a poly(ε-caprolactone)-b-poly(ethylene oxide) diblock copolymer, PCL-b-PEO, with 80/20 weight/weight composition and a total number average molecular weight of 22.5 Kg/mol. The sample exhibited a homogeneous melt at 60°C, and PEO face centered packed spheres at temperatures above the melting point of the PEO block (i.e., 51–58°C), but below the melting point of the PCL block. The PCL block crystallizes first (at higher temperatures) and its crystallization kinetics was measured by DSC and compared to neat PCL. Contrary to the results of Hamley et al. [33, 34] and Müller et al. [42–44] described above, Bogdanov et al. found that both the overall crystallization rate and the Avrami index were similar for the PCL block within the diblock copolymer and for the neat PCL sample. A "somewhat slower" crystallization rate was reported for the PCL block within $(A)_2$-B and ABA copolymers in comparison with neat PCL. In the case of the PEO component, which crystallizes after the PCL block, a significant reduction in the overall crystallization was detected together with a reduction in the Avrami index to values close to 2. These results are consistent with results obtained for the lowest temperature crystallizable block within PPDX-b-PCL and PLLA-b-PCL, as reported below.

In order to study the crystallization kinetics of the PCL block within PLLA-b-PCL or PPDX-b-PCL diblock copolymers without interference from the crystallization of the block that can crystallize first (i.e., at higher temperatures), Müller et al. [44] employed the strategy to crystallize the higher melting temperature block until saturation, and then quench to lower temperatures to follow the isothermal crystallization of the PCL block.

For PLLA-b-PCL copolymers, the PCL component has to crystallize within the neatly arranged interlamellar regions within PLLA spherulites that have previously crystallized at higher temperatures (see Fig. 13.8). WAXS and SAXS results discussed above [33, 34] have indicated extensive rearrangement of the PLLA lamellar stacks once PCL crystallizes, as well as some partial melting (probably at the interphase between the blocks) and unit cell contraction for the PLLA component. Under these conditions of very high topological restrictions, it is not surprising that the crystallization kinetics of the PCL block is slow down as compared to neat PCL of equivalent molecular weight. Furthermore, the overall crystallization rate is strongly reduced as the PLLA content in the copolymer increases. No nucleating effect of PLLA was found on PCL.

When the overall crystallization kinetics of the PCL component within PPDX-b-PCL diblock copolymers is measured after the PPDX has been previously crystallized at higher temperatures, the results are highly dependent on the composition of the block copolymer [3, 42–44]. When the PCL content is equal to 60% or higher, an increase in the overall crystallization rate was obtained as compared to neat PCL. This effect is due to a nucleating effect of PPDX on PCL. Such an effect was demonstrated by detailed DSC studies including self-nucleation (see refs. [42–44]). When the PCL content is lower than 40% (i.e., compositions with 35% and 23% PCL were examined), the topological restrictions outweigh the nucleating effect and the overall crystallization rate was reduced as compared to neat PCL. The Avrami index corresponding to the crystallization of the PCL block was found to progressively decrease from 3–3.5 to 2–2.2 as the content of PCL decreases in the diblock copolymer.

Another important aspect to account for in crystallizable block copolymers is the type of chain folding that takes place during crystallization. It is known that chain folding in amorphous-crystalline block copolymers depends on the segregation strength of the blocks in the melt state [4] and on crystallization temperature [48]. For instance, when the block copolymers are homogeneous or weakly segregated in the melt, there is a trend of the chains to fold in a perpendicular fashion with respect to the block copolymer interphases. On the other hand, when the systems are strongly segregated, chain folding becomes parallel. From DSC, POM images and X-ray experiments carried out on isothermally crystallized samples at 110°C and 140°C, Kim et al. [29] found that the crystallization mechanism in PLLA-b-PCL depends on the molecular weight due to the variation of the segregation level. It should be noted that at 140°C and 110°C only the PLLA can crystallize while the PCL is molten. The authors found that PLLA chain folding of high molecular weight copolymers at both temperatures was parallel to the interphases, while the folding of the low molecular weight copolymer was perpendicular (see Fig. 13.12). These results suggested, in agreement with previous reports in other kind of copolymers, that stem orientation is significantly affected by segregation strength between the blocks in the melt.

The ability of the discussed block copolymers to be selectively degraded by hydrolysis treatment or enzymes could be used for the preparation of nanopatterned templates [30]. Nevertheless, for these copolymers to prove useful in nanotechnology, well-oriented periodic arrays must be reached over a large area. A novel way to create large sized, well-oriented MDs of block copolymers by means of epitaxial crystallization of the crystallizable block onto a crystalline substrate has been introduced recently [49, 50]. Ho et al. [30] successfully applied this technique to induce orientation of PLLA-b-PCL block copolymers on benzoic acid (BA) and hexamethylbenzene (HMB). They were able to obtain well-oriented MDs with flat-on crystalline morphology as shown in Fig. 13.13. Unlike with BA, lattice matching between crystalline PLLA and HMB did not exist, a fact that lead them to conclude that lattice matching does not seem to be a critical issue for the induction of orientation [30].

The coexistence of polyethylene and poly(ε-caprolactone) within block copolymers was at first analyzed by Balsamo et al. in ABC triblock copolymers [51–54]. In this case a greater thermodynamic repulsion is expected be-

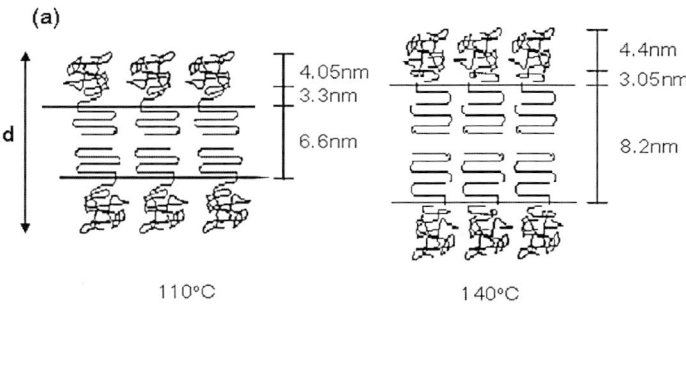

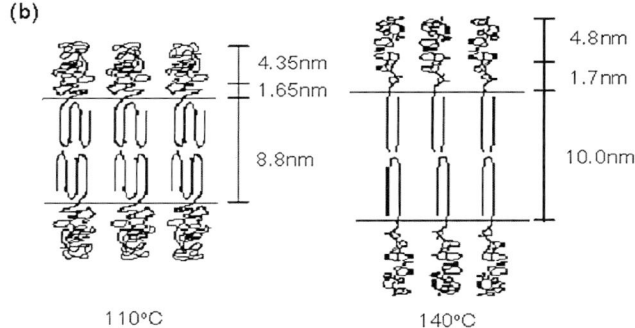

Fig. 13.12. Schematic representation of chain folding and domain spacing of the amorphous PCL block (*top*), amorphous PLLA block (*middle*) and crystalline PLLA block (*bottom*) for (**a**) PLLA-*b*-PCL (high M_n) and (**b**) PLLA-*b*-PCL (low M_n) at 110°C and 140°C. Taken from [29]

tween PE and PCL as compared to PLLA and PCL or even PPDX and PCL. Recently, Nojima et al. published two reports in which the morphology as well as PCL crystallization in melt-quenched PE-*b*-PCL block copolymers of low molecular weight are studied [55,56]. They did not investigate the crystallization of the PE-block because its crystallization rate is extremely fast, so that it crystallizes during quenching from the microphase separated melt. They used three block copolymers, whose M_n varied between 8 and 18 Kg/mol, with a PCL volume fraction of 42–69%. They isothermally crystallized the copolymers at temperatures in the range 5-45°C after they were quenched from 130°C. By means of DSC, SAXS measurements and Transmission Electron Microscopy (TEM) observations they found that the PE block crystallizes during quenching to yield an alternating structure consisting of crystalline lamellar PE and amorphous layers, independently whether the microdomain structure in the melt was cylindrical or lamellar; subsequently, the PCL block crystallizes starting from this PE lamellar morphology after some induction period. It should be noted that the crystallinity of the PE block is rather low,

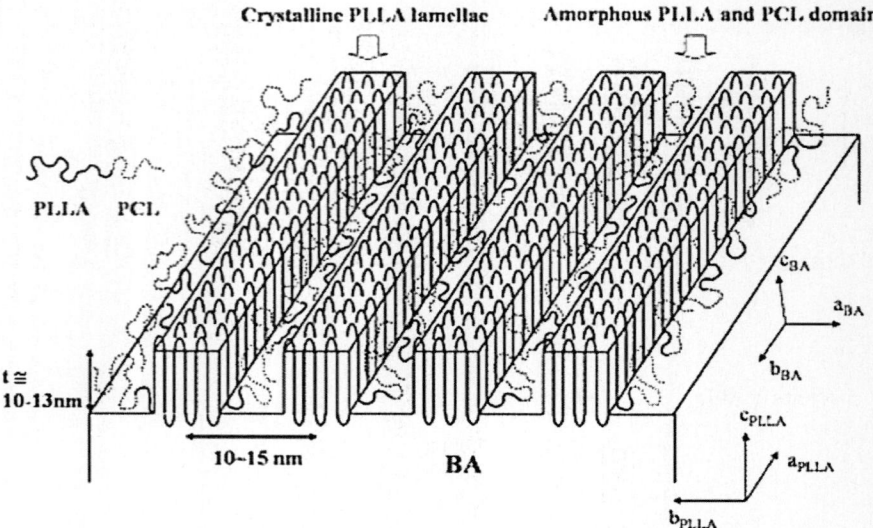

Fig. 13.13. Schematic representation of flat-on PLLA crystalline morphology in PLLA-b-PCL epitaxially crystallized onto BA. Taken from [30]

of about 12%. The PCL melting temperature changes considerably with T_c, as expected, but there is no significant difference between melting temperatures of PCL homopolymer and PCL blocks. Similar results were obtained by Balsamo et al. [51–54] when they analyzed PCL crystallization in PS-b-PE-b-PCL triblock terpolymers. This indicates that the spatial restriction imposed by the PE lamellar morphology does not work effectively against the subsequent crystallization of the PCL blocks.

The final morphology and thermal behavior of PE-b-PCL diblock copolymers upon PCL crystallization depends intimately on the crystallization temperature of the PCL blocks [55]. At low T_c, PCL blocks crystallize within the PE lamellar morphology, and, eventually, this morphology is preserved throughout the crystallization process of the PCL; at high T_c, on the other hand, the crystallization of the PCL blocks destroyed the PE lamellar morphology to result in a new lamellar morphology, as can be appreciated in Fig. 13.14. In this figure, the morphology of an almost symmetric sample of PCL-b-PE is shown; it is possible to observe that the PE lamellar morphology is destroyed and PE crystals are scattered within the PCL lamellar morphology (see Fig. 13.14c). These observations are consistent with the changes of the spacings obtained from SAXS measurements (Fig. 13.15), from which a significant increase of the spacing when the sample is crystallized between 30°C and 45°C is observed. Simultaneously, the increase of the spacing with increasing T_c suggests that the PCL block crystallizes as in the PCL homopolymer with

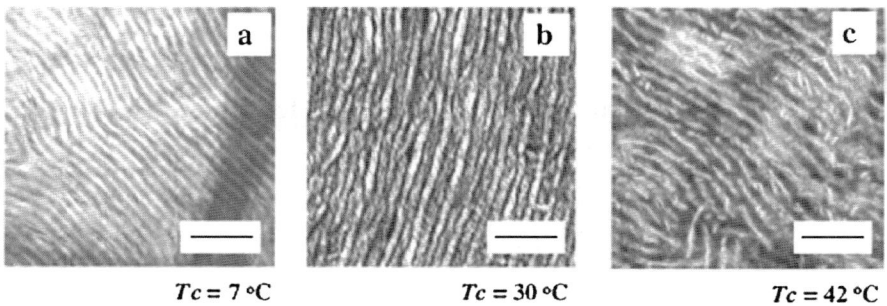

Fig. 13.14. TEM micrographs for a 51:49 (vol%) PCL-b-PE diblock copolymer (total M_n = 18 Kg/mol) isothermally crystallized at the indicated T_c. The bar represents 100 nm. Taken from [55]

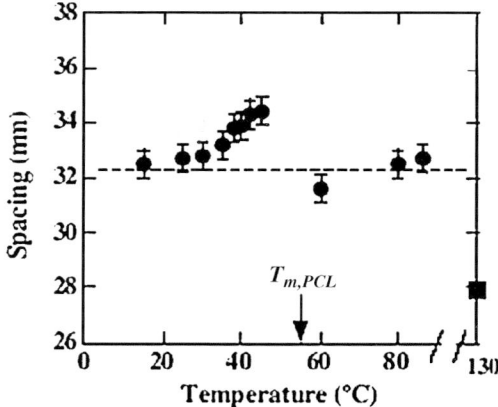

Fig. 13.15. Domain spacings, evaluated from the angular position of the SAXS intensity plot versus temperature for the same PCL-b-PE diblock copolymer of Figure 14. Taken from [55]

no effect of the previous PE lamellar morphology. DSC results were also in agreement with this interpretation.

Later, in 2005, Nojima et al. [56] investigated in greater detail the crystallization behavior of the PCL within PCL-b-PE diblock copolymers and compared it with a PCL-b-PB diblock copolymer. They determined the half-crystallization time by DSC measurements and showed that the crystallization rate of the PCL block within PCL-b-PE diblock copolymers is almost equal to that exhibited by the PCL block within PCL-b-PB at high T_c, but the difference increases significantly with decreasing T_c. This indicates that the occurrence of a morphological transition like in PCL-b-PE at high T_c or in PCL-b-PB at all crystallization temperatures (due to the rubbery nature of PB) disturbs the PCL crystallization process. This is in agreement with

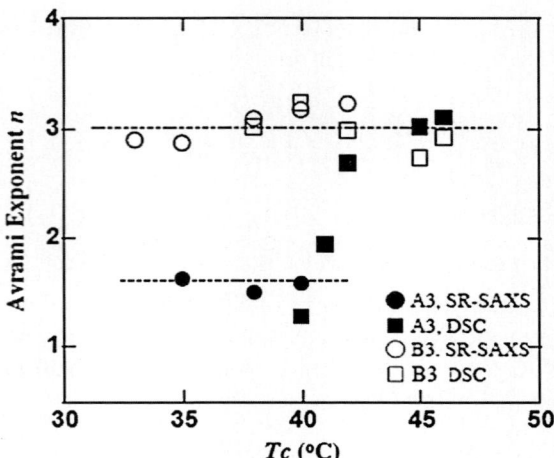

Fig. 13.16. Avrami exponents as a function of crystallization temperature for the following PCL-*b*-PE diblock copolymers: A3 (51:49 PCL:PE, vol%) and B3 (52:48 PCL:PB, vol.%). Taken from [56]

the findings of Balsamo et al. in PS-*b*-PB-*b*-PCL triblock terpolymers [57], when the PCL crystallization kinetics was compared in compositions whose morphological order was changed by annealing the samples in the melt for different times. Macroscopic morphological observations carried out by Nojima et al. also support this interpretation because spherulites could grow only at high T_c, i.e., when the previous microphase separated morphology was destroyed [56]. A similar morphological dependency upon T_c had also been reported by Balsamo et al. in PS-*b*-PB-*b*-PCL triblock terpolymers [58]. Nojima et al. speculate that the large number of nuclei created at low T_c prevented the transformation of MDs into spherulites, an effect that is consistent with the calculated Avrami values [56]. Figure 13.16 shows, for example, that for diblock copolymer A3 (see figure caption) n is about 3.0 at high T_c, but considerably lower, 1.6, at low T_c. These results lead them to postulate that the crystallization of the PCL block produces a morphological transition even though it is occurring within the pre-existing PE lamellae, and a kind of confinement effect may be responsible for the low Avrami index values (2-3).

Several works in the literature have reported that when certain crystallizable AB or ABA double crystalline copolymers polymers are prepared with the A or the B block much shorter than the other component (so that compositions of 80/20 or 90/10 are generated, and the total molecular weight of the copolymer is rather low), the shorter block may remain amorphous [31, 32, 47, 59]. If the block components are weakly segregated, lower molecular weights may induce miscibility, and it is possible that the shorter block may not be able to crystallize. However, another possibility that needs to be checked, especially for immiscible block copolymers, is that of fractionated crystallization [3,60,61]. When a crystallizable phase is segregated in isolated MDs like cylinders or spheres, its nucleation mechanism may change, and as a result, the crystallization can occur at much lower temperatures than for the bulk polymer, and sometimes in several stages (i.e., multiple exotherms; the term fractionated crystallization refers to the subdivision of the crystallization into several exotherms or the fractionation of the crystallization in temperature) A typical block copolymer with a 90/10 composition may be composed of a matrix and spheres of the minor component with a number density of approximately 10^{16} spheres/cm^3. This is typically a much larger value than the number of heterogeneities available in the crystallizable polymer (of the order of maximum 10^9 heterogeneities/cm^3); therefore, some spheres may contain less active heterogeneities that will induce crystallization at larger supercoolings while others may not contain any heterogeneities at all, and their nucleation will be either superficial or homogeneous. The interested reader will find extensive treatments of this phenomenon elsewhere [3,60,61].

Among other studies performed on double crystalline diblock copolymers, a recent report of Janssen et al. [62] on the synthesis of poly(3-hexylthiophene)-b-polyethylene diblock copolymers (P3HT-b-PE) can be mentioned, in which the P3HT weight percentage was varied from 11% to 22%. DSC experiments performed on these materials showed thermal transitions close to that of the neat homopolymers. However, a decrease in the onset of the PE crystallization from 118°C to 107°C accompanied by broadening of the exotherms was observed with increasing P3HT amount P3HT, indicating a hindered crystallization of the PE block.

13.3 ABC Triblock Linear and Star Shaped Terpolymers

The synthesis and characterization of polystyrene-b-poly(ethylene oxide)-b-poly(ε-caprolactone) triblock terpolymers in a linear or star shaped fashion is very interesting because they combine in the same molecule: a glassy amorphous block, polystyrene; a biocompatible crystallizable poly(ethylene oxide) block, and a crystallizable poly(ε-caprolactone) block, which is biodegradable and exhibits miscibility with a variety of polymers.

Floudas et al. [22, 23] carried out the characterization of star shaped PS, PEO and PCL based triblock terpolymers by means of WAXS, SAXS, DSC, POM, Atomic Force Microscopy (AFM) and rheological measurements. The compositions they studied included: $S_9{}^5 EO_{88}{}^{20} C_3{}^2$, $S_5{}^5 EO_{51}{}^{20} C_{44}{}^{45}$ and $S_4{}^5 EO_{36}{}^{20} C_{60}{}^{87}$. SAXS experiments showed that all samples are miscible in the melt, and that the crystallization process produces microphase segregation, indicating that the star architecture induces a reduction of the T_{ODT} in comparison to this transition in sequential copolymers. The miscibility was also favoured by the low molecular weight of the PS block in the stars (i.e., 4.7 Kg/mol).

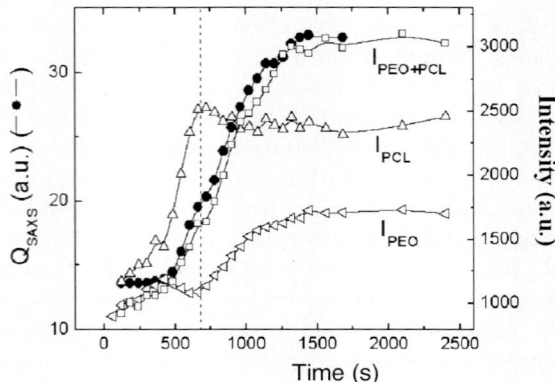

Fig. 13.17. Time evolution of the SAXS invariant, Q_{SAXS}, compared with the intensity of the three most intense WAXS peaks for $S_5{}^5EO_{51}{}^{20}C_{44}{}^{45}$ at 39°C. Taken from [23]

In Fig. 13.17 the evolution of the SAXS invariant is shown; simultaneously, the intensity of WAXS signals due to the PEO monoclinic and PCL orthorhombic structures can be appreciated, together with the intensity of a signal that arise from an overlapping of the reflections of both blocks for the $S_5{}^5EO_{51}{}^{20}C_{44}{}^{45}$ copolymer. At 39°C it is possible to observe the beginning of the PCL crystallization from an increase of the signal attributed to the (110) reflection; once this value stabilized, PEO crystallizes with a consequent small reduction in the intensity of the signal attributed to the (110) plane, characteristic of PCL. This indicates that the PEO crystallization has as a consequence the distortion of the PCL crystals. $S_9{}^5EO_{88}{}^{20}C_3{}^2$ and $S_4{}^5EO_{36}{}^{20}C_{60}{}^{87}$ exhibited, on the other hand, a different behaviour, where only one of the blocks crystallizes; PEO or PCL respectively.

Isothermal crystallizations performed by Floudas et al. [22, 23] between 39°C and 43°C allowed them to conclude that the crystallization of the PEO and PCL blocks only occurs when the two star branches have similar lengths. On the other hand, when one of the branches has three times the length of the other, the crystallization of the shorter block is impeded. Nevertheless, it would be interesting to expand the temperature range in which the crystallization is studied, since other authors have reported crystallization of these blocks at higher supercoolings.

Floudas et al. [22] also studied the crystallization kinetics by means of DSC and applied the Avrami equation, from which n values of about 2 were estimated for all the copolymers. Such a value is interpreted to arise from a crystallization process that takes place with an instantaneous nucleation and a bi-dimensional growth. In addition, they observed that the half crystallization time, $\tau^{1/2}$, depends on composition, and that it is different to that determined in an analogue PS-b-PEO diblock copolymer [22].

In $S_4{}^5EO_{51}{}^{20}C_{44}{}^{45}$, a star where the weight fraction of PEO and PCL is similar, Floudas et al. observed, by using POM, the formation of mixed spherulites, similar to those reported by He et al. (18–19) and Shiomi et al. (16) in PEO-b-PCL diblock copolymers with about 50% of each block. $S_4{}^5EO_{51}{}^{20}C_{44}{}^{45}$ forms axialites in the first stages of crystallization, but then, the superstructure changes and spherulites, similar to those formed by PEO, are developed (22). The formation of concentric superstructures is more evident when the copolymer is partially melted and crystallized at 55°C and 44°C respectively. Similar to the results reported by He et al., Floudas et al. found that PCL is the first block that nucleates; however, the PEO grows faster than PCL. From the above paragraphs is possible to conclude that when the PS content is low, the behaviour of the PS-b-PEO-b-PCL stars is comparable to the behaviour of PEO-b-PCL diblock copolymers.

When linear polystyrene-b-poly(ethylene oxide)-b-poly(ε-caprolactone) (SEOC) triblock terpolymers are considered, a variety of phenomena can be found as a function of the copolymer composition. Figure 13.18 presents the results of Arnal et al. to illustrate two extremes in the diverse complexity of phenomena that these terpolymers can display, i.e., coincident heterogeneously nucleated crystallization and coincident homogeneously nucleated crystallization, see below [3, 24–26]. This behavioural richness is due, as has been previously mentioned, to the competition between microphase separation and crystallization of two crystallizable blocks, PEO and PCL, combined with a glassy amorphous one [3, 24–26].

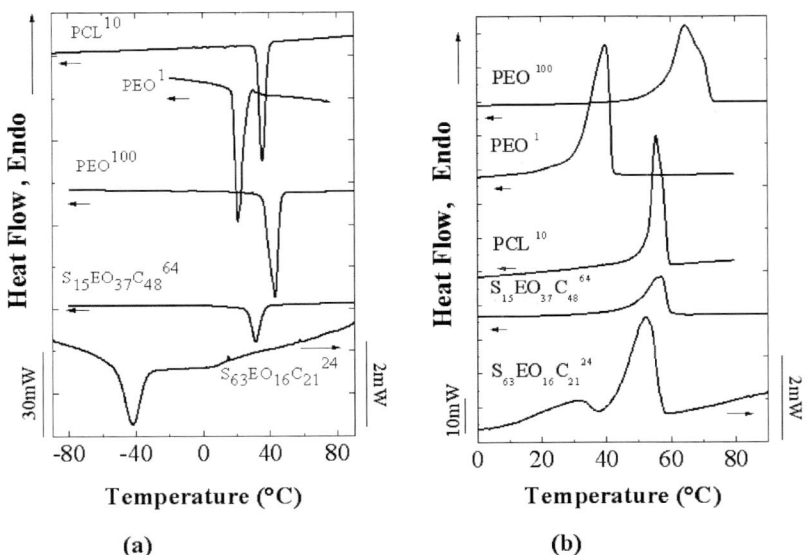

Fig. 13.18. DSC (**a**) cooling and (**b**) heating scans (10°C/min) of PCL and PEO homopolymers and the indicated $S_xEO_yC_z$ copolymers. Taken from [26]

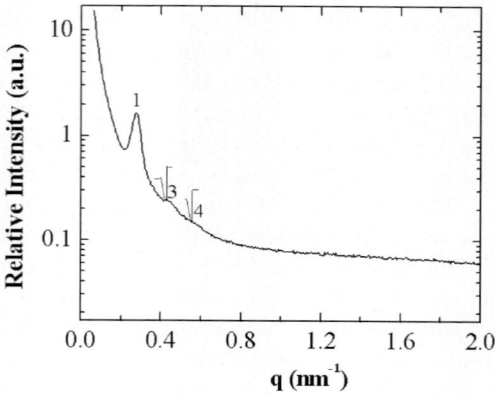

Fig. 13.19. SAXS pattern corresponding to $S_{63}EO_{16}C_{21}^{24}$ triblock copolymer. Taken from [26]

The observation through DSC of two T_g's, one at 85°C, corresponding to the glassy phase, and the other at −62°C, attributed to a PCL/PEO miscible phase, together with the SAXS profiles obtained at room temperature (main signal at $q^* = 0.27$ nm^{-1}, two weaker signals at $\sqrt{3}$ and $\sqrt{4}$ q/q^*, see Fig. 13.19) suggests that the $S_{63}EO_{16}C_{21}$ triblock terpolymer has a PS matrix with PEO and PCL cylinders. The presence of cylindrical microdomains within a glassy matrix is a case of *hard confinement*, in which the crystallization occurs at high supercooling, and in this particular case, in a coincident fashion ($T_c = -42$°C in Fig. 13.18), i.e., at a temperature close to the PEO and PCL T'_gs.

The presence of only one T_g at low temperatures indicates that PEO and PCL blocks form a miscible melt. The extreme supercooling (where temperatures close to T_g are reached) needed for crystallization indicates that homogeneous nucleation is present [3, 60]. As it is known, homogeneous nucleation occurs at higher supercooling than heterogeneous nucleation because it implies the energetically unfavoured creation of new surfaces. A subsequent DSC heating scan of $S_{63}EO_{16}C_{21}^{24}$ exhibits two melting endotherms, attributed to each block (Fig. 13.18), where the middle PEO block exhibits a depression of 4°C with respect to the melting of the same block in the precursor diblock [26]. This clearly implies that the topological restrictions of the PEO, due to the absence of free ends, perturbs the crystallization process. WAXS experiments show characteristic PEO and PCL reflections at 19.3°, 21.3° and 23.5°, corroborating the crystallization of both blocks in monoclinic and orthorhombic unit cells, even though they represent only 16% and 21% of the copolymer.

When PEO and PCL blocks are not confined, like in $S_{15}EO_{37}C_{48}$, a different behaviour is obtained. The observation of the T_{gPS} allowed Arnal et al. to assume that the PS block is segregated from the other two, which are molten; nevertheless, after PEO and PCL have crystallized, SAXS curves do

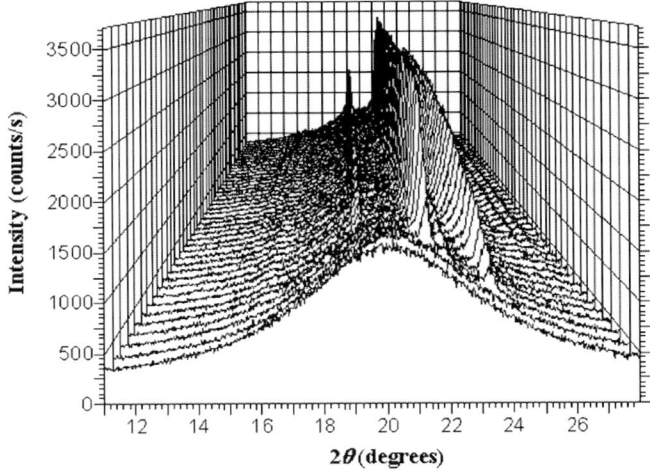

Fig. 13.20. WAXS patterns recorded every 28.5 min during isothermal crystallization for $S_{15}EO_{37}C_{48}^{64}$ triblock copolymer: $Tc = 53.5°C$. Taken from [26]

not show any kind of periodicity [26]. This suggests that the pre-existent order is destroyed by the crystallization. Furthermore, in Fig. 13.18 it is possible to observe that $S_{15}EO_{37}C_{48}$ only has a single melting and crystallization transition for both blocks, but WAXS experiments performed during isothermal crystallization lead the authors to conclude that PCL and PEO crystallize in a sequential fashion (see Figure 13.20).

POM experiments evidenced the formation of mixed PCL and PEO spherulites. At low supercoolings ($T_c = 54°C$), the main crystallizing component is PCL, at high supercoolings, once PCL has crystallized, PEO is incorporated ($T_c = 49°C$) within the superstructures. When PCL is the majority component, like in $S_{10}EO_4C_{86}$, crystallization and melting processes are similar to those observed in PCL homopolymers while PEO remains amorphous [26].

Another type of remarkable double crystalline materials that have been synthesized and characterized by Balsamo et al. are ABC triblock terpolymers composed of polystyrene, polyethylene and poly(ε-caprolactone) (PS-b-PE-b-PCL or SEC) [51–54,57,63]. The morphology, nucleation and crystallization of such copolymers have been recently reviewed [3]. It is interesting to mention that in such terpolymers the PE block induces an antinucleation effect [3,63] on its covalently bonded neighboring PCL block, a remarkable effect that has only been observed in this type of triblock terpolymer. Diblock copolymers or triblock terpolymers with two crystallizing blocks can display all possible effects from the nucleation point of view of one crystallizing block on the other. One block can cause nucleation of the other, or cause no effect, or in the other extreme of behavior induce antinucleation. In addition to the observed antinucleation effect, Balsamo et al. observed through DSC experiments that the crystallization temperature and melting point of the PE block within SEC copolymers decreases as the confinement degree increases in agreement with other kind of double semicrystalline copolymers [54]. Furthermore, the effect of chain tethering influences the behaviour since the melting point and crystallization temperature experience a lower depression when the PE chains are only tethered on the one end, like in polyethylene-b-polystyrene-b-poly(ε-caprolactone) (ESC) triblock terpolymers. Double melting endotherms observed after isothermal crystallizations were interpreted as a result of the melting of two lamellar populations arising from the intrinsic short chain branching distribution within the PE block, and from their location within the copolymer MDs (near or far from the interphases). The isothermal crystallization kinetics showed that the Avrami index decreases as the degree of PE confinement increases in agreement with other authors; indexes as low as 0.5 were interpreted by the authors as a homogeneous nucleation process that is in between sporadic and instantaneous when growth is so fast that nucleation dominates the kinetics of overall transformation [3, 54].

The preparation of ABC triblocks with three crystallizable blocks is still in its infancy [27], and will have to be refined in the future to obtain model materials; however, it is envisaged that having three structurally different blocks may lead to very interesting structured novel terpolymers, albeit complicated.

The formation and properties of inclusion compounds of cyclodextrin with double crystalline PLLA-b-PCL diblock copolymers has been extensively discussed in a series of contributions by Tonelli et al. [64–67], where they highlight the changes that can be introduced in the structure and properties of the copolymers when they form these peculiar inclusion complexes.

13.4 Conclusions

When more than one block is able to crystallize within diblock copolymers or triblock terpolymers, the interplay between phase segregation and crystallization of the blocks can lead to a wide variety of very interesting phenomena. The blocks can crystallize either coincidentally or sequentially, and each block can have a dramatic influence on the nucleation and crystallization kinetics of the other. When the block copolymers are miscible in the melt or in the weak segregation regime, they can display a superstructural morphology whose texture depends on the thermodynamic repulsion between the blocks. Therefore, factors like chemical structure, molecular weight, molecular architecture and number of crystallizable blocks provide a very large number of possibilities to tailor the morphology and properties of these interesting novel materials.

Acknowledgements

We would like to acknowledge financial support from Fonacit (project No. S1-2001000742).

References

[1] I.W. Hamley: *Adv. Polym. Sci.* **148**, 113 (1999).
[2] Y.L. Loo, R.A. Register: Crystallization within block copolymer mesophases. In: *Developments in block copolymer science and technology*, ed by I.W. Hamley (Wiley, New York 2004).
[3] A.J. Müller, V. Balsamo, M.L. Arnal: *Adv. Polym. Sci.*, **190**, 1 (2005).
[4] I.W. Hamley: *The Physics of Block Copolymers* (Oxford University Press, Oxford 1998).
[5] N. Hadjichristidis, S. Pispas, G. Floudas: *Block Copolymers: Synthetic Strategies, Physical Properties, and Applications* (Wiley, New Jersey 2003).
[6] R. Perret, A. Skoulios: *Die Makromol. Chem.*, **156**, 143 (1972).
[7] R. Perret, A. Skoulios: *Die Makromol. Chem.*, **156**, 291 (1972).
[8] P. Cerrai, M. Tricoli, F. Andruzzi, M. Paci, M. Paci: *Polymer* **30**, 338 (1989).
[9] S. Nojima, M. Ono, T. Ashida: *Polym. J.*, **24**, 1271 (1992).
[10] Z. Gan, B. Jiang, J. Zhang: *J. Appl. Polym. Sci.* **59**, 961 (1996).
[11] Z. Gan, J. Zhang. B. Jiang: *J. Appl Polym Sci.* **63**, 1793 (1997).
[12] Z. Zhu, C. Xiong, L. Zhang, X. Deng: *J. Polym. Sci., A: Polym. Chem.* **35**, 709 (1997).
[13] T. Petrova, N. Manolova, I. Rashkov, S. Li, M. Vert: *Polym. Intern.* **45**, 419 (1998).
[14] B. Bogdanov, A. Vidts, A. Van Den Bulcke, R. Verbeeck, E. Schacht: *Polymer* **39**, 1631 (1998).
[15] B. Bogdanov, A. Vidts, E. Schacht, H. Berghmans: *Macromolecules* **32**, 726 (1999).
[16] T. Shiomi, K. Imai, K. Takenaka, H. Takeshita, H. Hayashi, Y. Tezuka: *Polymer* **42**, 3233 (2001).
[17] L. Piao, Z. Dai, M. Deng, X. Chen, X. Jing: *Polymer* **44**, 2025 (2003).
[18] C. He, J. Sun, C. Deng, T. Zhao, M. Deng, X. Chen, X. Jing: *Biomacromol.* **5**, 2042 (2004).
[19] C. He, J. Sun, T. Zhao, Z. Hong, X. Zhuang, X. Chen, X. Jing: *Biomacromol.* **7**, 252 (2006).
[20] O. Lambert, P. Dumas, G. Hurtrez, G. Riess: *Macromol. Rapid Commun.* **18**, 343 (1997).
[21] O. Lambert, S. Reutenauer, G. Hurtrez, G. Riess, P. Dumas: *Polym. Bull.* **40**, 143 (1998).
[22] G. Floudas, G. Reiter, O. Lambert, P. Dumas: *Macromolecules* **31**, 7279 (1998).
[23] G. Floudas, G. Reiter, O. Lambert, P. Dumas, F.J. Yeh, B. Chu, ACS Symposium, Series **739**, Chapter 28, **448** (2000).
[24] A.J. Müller, M.L. Arnal, F. López-Carrasquero: *Macromol. Symp.* **183**, 199 (2002).
[25] M.L. Arnal, V. Balsamo, F. López-Carrasquero, J. Contreras, M. Carrillo, H. Schmalz, V. Abetz, E. Laredo, A.J. Müller: *Macromolecules* **34**, 7973 (2001).

[26] M.L. Arnal, F. López-Carrasquero, E. Laredo, A.J. Müller: *Europ. Polym. J.* **40**, 1461 (2004).
[27] M. Vivas, J. Contreras, F. López-Carrasquero, A.T. Lorenzo, M.L. Arnal, V. Balsamo, A.J. Müller, E. Laredo, H. Schmalz, V. Abetz, *Macromol. Symp.*, **239**, 58 (2006).
[28] S. Vainionpaa, P. Pentti Rokkamen, P. Tormala: *Prog. Polym. Sci.* **14**, 679 (1989).
[29] J.K. Kim, D.J. Park, M.S. Lee, K.J. Ihn: *Polymer* **42**, 7429 (2001).
[30] R.-M. Ho, P.-Y. Hsieh, W.-H. Tseng, C.-C. Lin, B.-H. Huang, B. Lotz: *Macromolecules* **36**, 9085 (2003).
[31] G. Maglio, A. Migliozzi, R. Palumbo: *Polymer* **44**, 369 (2003).
[32] O. Jeon, S.-H. Lee, S.H. Kim, Y.M. Lee, Y.H. Kim: *Macromolecules* **36**, 5585 (2003).
[33] I.W. Hamley, V. Castelleto, R.V. Castillo, A.J. Müller, C.M. Martin, E. Pollet, Ph. Dubois: *Macromolecules* **38**, 463 (2005).
[34] I.W. Hamley, P. Parras, V. Castelleto, R.V. Castillo, A.J. Müller, E. Pollet, Ph. Dubois, C.M. Martin: *Macromol. Chem. Phys.*, **207**, 941 (2006).
[35] C. Jacobs, Ph. Dubois, R. Jerome, Ph. Teyssie: *Macromolecules* **24**, 3027 (1991).
[36] D. Shin, K. Shin, K. Aamer, G.N. Tew, T.P. Russell, J.H. Lee, J.Y. Jho: *Macromolecules* **38**, 104 (2005).
[37] M.A. Sabino, G. Ronca, A.J. Müller: *J. Mater. Sci.* **35**, 5071 (2000).
[38] M.A. Sabino, J.L. Feijoo, A.J. Müller: *Macromol. Chem. Phys.* **201**, 2687 (2000).
[39] M.A. Sabino, J.L. Feijoo, A.J. Müller: *Polym. Degrad. Stab.* **73**, 541 (2001).
[40] M.A. Sabino, L. Sabater, G. Ronca, A.J. Müller: *Polym. Bull.* **48**, 291 (2002).
[41] M.A. Sabino, J. Albuerne, A.J. Müller, J. Brisson, R.E. Prud'homme: *Biomacromolecules*: **5**, 358 (2004).
[42] J. Albuerne, L. Marquez, A.J. Müller, J.M. Raquez, Ph. Degee, Ph. Dubois, V. Castelletto, I.W. Hamley: *Macromolecules*, **36**, 1633 (2003).
[43] A.J. Müller, J. Albuerne, L.M. Esteves, L. Márquez, J.M. Raquez, Ph. Degée, Ph. Dubois, S. Collins, I.W. Hamley: *Macromol. Symp.* **215**, 369 (2004).
[44] A.J. Müller, J. Albuerne, L. Márquez, J.M. Raquez, Ph. Degée, Ph. Dubois, J. Hobbs, I.W. Hamley: *Faraday Discuss.* **128**, 231 (2005).
[45] J. Albuerne, L. Marquez, A.J. Müller, J.M. Raquez, P. Degée, P. Dubois: *Macromol. Chem. Phys.* **206**, 903 (2005).
[46] M. Ueda, K. Sakurai, S. Okamoto, D. Lohse, W.J. MacKnight, S. Shinkai, S. Sakurai, S. Nomura: *Polymer* **44**, 6995 (2003).
[47] K.-S. Kim, S. Chung, I.-J. Chin, M.-N. Kim, J.-S. Yoon: *J. Appl. Polym. Sci.* **72**, 341 (1999).
[48] L. Zhu, B.H. Calhoun, Q. Ge, R.P. Quirk, S.Z.D. Cheng, E.L. Thomas, B.S. Hsiao, F. Yeh, L. Liu, B. Lotz: *Macromolecules* **34**, 1244 (2001).
[49] C. De Rosa, C. Park, B. Lotz, J.C. Wittmann, L. Fetters, E.L. Thomas: *Macromolecules* **33**, 4871 (2000).
[50] C. Park, C. De Rosa, E.L. Thomas: *Macromolecules* **34**, 2602 (2001).
[51] V. Balsamo, A.J. Müller, F. von Gyldenfeldt, R. Stadler: *Macrom. Chem. Phys.* **199**, 1063 (1998).
[52] V. Balsamo, A.J. Müller, R. Stadler: *Macromolecules* **31**, 7756 (1998).
[53] V. Balsamo, Y. Paolini, G. Ronca, A.J. Müller: *Macromol. Chem. Phys.* **201**, 2711 (2000).

[54] V. Balsamo, N. Urdaneta, L. Pérez, P. Carrizales, V. Abetz, A.J. Müller: *Europ. Polym. J.* **40**, 1033 (2004).
[55] S. Nojima, Y. Akutsu, A. Washino, S. Tanimoto: *Polymer* **45**, 7317 (2004).
[56] S. Nojima, Y. Akutsu, M. Akaba, S. Tanimoto: *Polymer* **46**, 4060 (2005).
[57] V. Balsamo, G. Gil, C. Urbina de Navarro, I.W. Hamley, F. von Gyldenfeldt, V. Abetz, E. Cañizales: *Macromolecules* **36**, 4515 (2003).
[58] V. Balsamo, R. Stadler: *Macromolecules* **32**, 3994 (1999).
[59] S.M. Li, I. Rashkov, J.L. Espartero, N. Manolova, M. Vert: *Macromolecules* **29**, 57 (1996).
[60] A.J. Müller, V. Balsamo, M.L. Arnal, T. Jakob, H. Schmalz, V. Abetz: *Macromolecules* **35**, 3048 (2002).
[61] M.L. Arnal, M.E. Matos, R.A. Morales, O.O. Santana, A.J. Müller: *Macromol. Chem. Phys.* **199**, 2275 (1998).
[62] C.P. Radano, O.A. Scherman, N. Stingelin-Stutzmann, C. Müller, D.W. Breiby, P. Smith, R.A.J. Janssen, E.W. Meijer: *J. Am. Chem. Soc.* **127**, 12502 (2005).
[63] V. Balsamo, A.J. Müller, R. Stadler: *Macromolecules* **31**, 7756 (1998).
[64] X. Shuai, F.E. Porbeni, M. Wei, I.D. Shin, A.E. Tonelli, *Macromolecules* **34**, 7355 (2001).
[65] X. Shuai, M. Wei, F.E. Porbeni, T.A. Bullions, A.E. Tonelli, *Biomacromol.* **3**, 201 (2002).
[66] M. Wei, X. Shuai, A.E. Tonelli, *Biomacromol.* **4**, 783 (2003)
[67] F.E. Porbeni, I.D. Shin, X. Shuai, X. Wang, J.L. White, X. Jia, A.E. Tonelli, *Polymer* **43**, 2086 (2005)

14

Monte Carlo Simulations of Semicrystalline Polyethylene: Interlamellar Domain and Crystal-Melt Interface

Markus Hütter[1], Pieter J. in 't Veld[2], and Gregory C. Rutledge[3]

[1] Department of Materials, ETH Zurich, CH-8093 Zurich, Switzerland,
   markus.huetter@mat.ethz.ch
[2] Sandia National Laboratories, Albuquerque, NM 87185, U.S.A.,
   pjintve@sandia.gov
[3] Department of Chemical Engineering, Massachusetts Institute of Technology, Cambridge, MA 02139, U.S.A.
   rutledge@mit.edu

Abstract. The interlamellar domain of semicrystalline polyethylene is studied by means of off-lattice Metropolis Monte Carlo simulations using a realistic united atom force field with inclusion of torsional contributions. Both structural as well as thermal and mechanical properties are discussed for systems with the {201} crystal plane parallel to the interface. In so doing, important data is obtained which is useful for modeling semicrystalline polyethylene in terms of multiphase models. Here, we review the main results published previously by us [P.J. in 't Veld, M. Hütter, G.C. Rutledge: Macromolecules **39**, 439 (2006); M. Hütter, P.J. in 't Veld, G.C. Rutledge: Polymer (in press), (2006)].

On the one hand, the full interlamellar domain was characterized in terms of heat capacity, thermal expansion coefficients, Grüneisen coefficients, and the elastic compliance tensor at atmospheric pressure in the temperature range [350, 450] K. The simulation results corroborate the fact that the properties of the non-crystalline interlamellar phase lie between those of the amorphous melt and the semicrystalline solid, as quantitative comparison with experimental data shows. On the other hand, the interface between polyethylene crystal and melt is characterized in the temperature range [380, 450] K. We invoke the concept of the sharp Gibbs dividing surface to define and quantify the interface internal energy and the interface stresses. We find that the latter are in reasonable agreement with values derived from experimental data. By way of the Herring equation one can also infer that the surface tension of the fold surface is independent of shear strains in the interface.

14.1 Introduction, Motivation

Polymers are typically not fully crystalline below the melting temperature, due to frustration effects. As a result, a significant fraction of the sample consists of non-crystalline material between lamellae. These frustration effects are particularly pronounced when the polymers are considerably longer than the lamella thickness. Chains then form bridges and loops, which are possibly entangled in the region between different lamellae, which in turn has ramifications also for the mechanical properties [1]. The characterization of this interlamellar domain is complicated by the presence and overlap of signals from several different phases and uncertainty in the assignment of a particular response to the interlamellar domain material [2,3]. In addition, in flexible and fast crystallizing polymers such as polyethylene (PE) where crystal phase relaxations are present, relationships between the crystal formation and the structure of the interlamellar material are obscured by fading memory [4]. It is also believed that the constraints imposed by the lamellae influence dramatically the dynamics of the interlamellar chain segments [5].

Two aspects, at least, of semicrystalline polymers must be discussed for arriving at a meaningful description, namely, structure and material properties. The effect of structure on the macroscopic material properties is described in many textbooks, e.g. by Torquato [6]. The importance of characterizing the structure is also realized in continuum modeling approaches, where powerful nonequilibrium thermodynamics techniques are used to incorporate structural information consistently into continuum models, which are then suitable for process modeling [7,8]. In particular, we mention the crystallization rate equations of [8] which not only separates the semicrystalline polymer into crystal and melt, but specifically distinguishes between the fold surfaces and growth surfaces of lamellae. In this way, information gained by microscopic studies can indeed be incorporated into descriptions on a different level. This being said, we address the material properties of the interlamellar domain and of the interface in the following.

Since a substantial amount of material is contained in the interlamellar region, the properties of the latter give significant contributions to the overall material behavior. The properties of the interlamellar material lie between those of the unconstrained amorphous melt and those of the crystalline phase [9–11], and the influence of the crystalline constraints can be addressed experimentally [12–14]. Furthermore, the properties of the crystal-melt interface have various ramifications that can be observed experimentally [15], e.g., interface stresses lead to distortion of the crystal lattice spacing [16–18], and they are possibly responsible for lamella twisting [19]. In addition, the surface tension enters in theoretical models for crystallization rates [20, 21].

To characterize the structure and to quantify the mechanical and thermal properties of the interlamellar, non-crystalline material, Metropolis Monte Carlo simulations have been performed [22–26] on systems kept in metastable equilibrium [27, 28]. Here, we give a summary of our most recent results for a realistic model for polyethylene including torsion interactions. For more details the reader is referred to the original publications [29, 30]. Throughout the manuscript, we concentrate our attention on the {201} crystal surface, because it was found to be energetically favored in simulations [25] and predominant

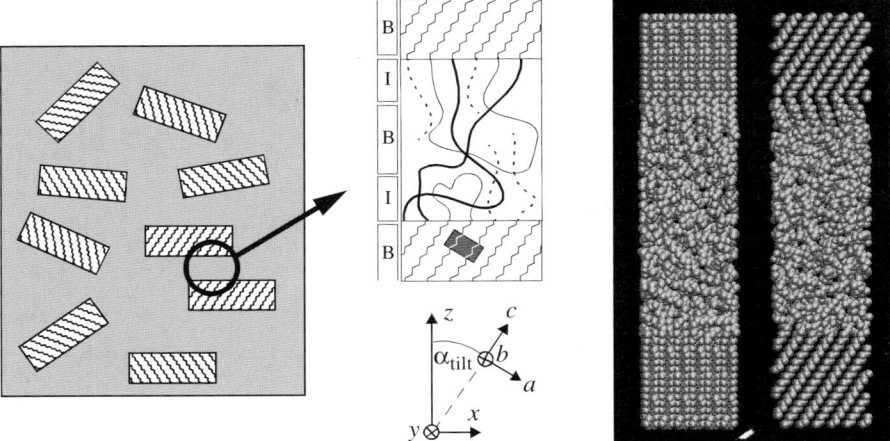

Fig. 14.1. Illustration of semicrystalline polyethylene on the *left*, consisting of crystalline lamellae and non-crystalline material (not necessarily indicative of actual morphology). The simulation box in the middle shows two crystalline lamellae and the interlamellar phase in between, which consists of loops (*thin solid lines*), bridge chains (*thick solid lines*), and tails (*dashed lines*). The polyethylene unit cell (*grey box, middle figure*) with coordinate axes (a, b, c) is tilted by an angle α_{tilt} with respect to the surface normal and the coordinate system (x, y, z). The bars with label "I" indicate the extended interface between crystalline and non-crystalline material, while "B" denotes regions with bulk-like properties, as explained in the text. On the right, snapshots are shown, viewing along the x- and y-direction, respectively

in experimental observations [31]. Figure 14.1 shows an illustration of the simulation box, as well as a snapshot of the simulations.

In the remainder of this contribution we study, on one hand, the entire interlamellar domain as a whole, termed "Study 1" [29]. On the other hand, the crystal-melt interface specifically is examined in "Study 2" [30]. In the course of explaining the results, the benefits of both of these approaches will become evident.

14.2 Methodology

14.2.1 Force Field, Virial Calculation of Stress

Polyethylene is modeled according to the united atom model of Paul et al. [32], as modified subsequently by Bolton et al. [33] and by In 't Veld and Rutledge [26], including the torsion angle terms. Using this force field, kinetic processes in semicrystalline PE have already been modeled accurately [34, 35]. The stable crystal phase, though similar to that for PE, is actually pseudo-hexagonal.

Interactions both between non-bonded united atoms (CH$_2$) on different chain segments and between united atom pairs separated by four or more bonds in the same chain segment were calculated using a Lennard-Jones potential with a cut-off distance $r_c = 2.5\,\sigma_{\rm LJ}$. In addition, three types of bonded interactions are included, accounting for the stiffness in the bond length, bond angle and torsion. The interaction potentials are, in that order,

$$E_{{\rm LJ},ij} = 4\varepsilon_{\rm LJ}\left[\left(\frac{\sigma_{\rm LJ}}{|\boldsymbol{d}_{i,j}|}\right)^{12} - \left(\frac{\sigma_{\rm LJ}}{|\boldsymbol{d}_{i,j}|}\right)^{6}\right], \qquad (14.1)$$

$$E_{l,i} = \frac{1}{2}k_l\left(l_i - l_0\right)^2, \qquad (14.2)$$

$$E_{\theta,i} = \frac{1}{2}\frac{k_\theta}{\sin^2\theta_0}\left(\cos\theta_i - \cos\theta_0\right)^2, \qquad (14.3)$$

$$E_{\phi,i} = \sum_{n=0}^{3} k_n \cos^n \phi_i, \qquad (14.4)$$

with parameters specified in Table 14.1, and $\boldsymbol{d}_{i,j} = \boldsymbol{r}_i - \boldsymbol{r}_j$ the distance between the Cartesian coordinates $\boldsymbol{r}_i$ and $\boldsymbol{r}_j$ of united atoms i and j. Furthermore, l_i is the length of bond i, θ_i is the complement of the bond angle constructed by bond i and $i-1$, and ϕ_i is the bond torsion angle constructed by the angle between the vectors $\boldsymbol{d}_{i,i-1} \times \boldsymbol{d}_{i-1,i-2}$ and $\boldsymbol{d}_{i-1,i-2} \times \boldsymbol{d}_{i-2,i-3}$.

For each of the force field contributions described above, there is a corresponding contribution to the total instantaneous stress tensor $\boldsymbol{\sigma}$, which is expressed in terms of the individual virial contributions. Explicit expressions for these contributions can be found in [26,29]. When using Lennard-Jones interactions for total energy calculations or for virial calculations of the stresses, long range corrections need to be included, as discussed by In 't Veld et al. [26].

14.2.2 Simulation Setup

Monte Carlo Simulation

The simulation box consisted of an immobile crystal phase and a mobile interlamellar phase – the combination of both interfacial material (covered by bars

Table 14.1. Parameters for the Interaction Potentials

Interaction	Parameter	Value
Lennard-Jones	$\varepsilon_{\rm LJ}$ (J/mol)	390.95
	$\sigma_{\rm LJ}$ (nm)	0.4009
Bond length	k_l (J/mol/nm^2)	376.1×10^6
	l_0 (nm)	0.1530
Bond angle	k_θ (kJ/mol)	502.1
	θ_0 (−)	68.0°
Bond torsion	$\{k_0, k_1, k_2, k_3\}$ (kJ/mol)	$\{6.498, -16.99, 3.626, 27.11\}$

"I" in Fig. 14.1) and truly amorphous material (middle bar "B" in Fig. 14.1) – as a model for semicrystalline polyethylene. The goal of studying the interface between crystal and melt also at temperatures different than the melting temperature requires that one imposes certain constraints on the system to keep the interlamellar domain in a metastable state [27]. Crystallization at low temperatures is prevented by keeping the volume constant, while in order to prevent melting at high temperatures the crystal sites are immobile.

Phase space is sampled in Metropolis Monte Carlo fashion by topology altering (end-reptation [36, 37] and end-bridging [38]) and displacement (end-rotation [23], rebridging [39, 40], and single-site displacement) moves [26], and parallel tempering [41] to facilitate more efficient sampling at low temperatures. For Study 1, the temperature profile according to the criteria in [41] is given by $T \in \{350, 359.9, 370.1, 380.6, 391.4, 402.4, 413.8, 425.6, 437.6, 450\}$ K. In Study 2 we used $T \in \{380.6, 391.4, 402.4, 413.8, 425.6, 437.6, 450\}$ K. An illustration of the simulation setup is given in Fig. 14.2.

Once created, the sufficient number of initial configurations generated as described below were randomized and then quenched, via intermediate temperatures, to the desired temperature profile, and equilibrated before any measurements were taken.

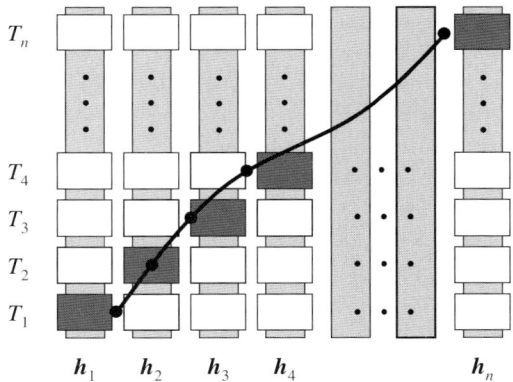

Fig. 14.2. Illustration of parallel tempering scheme as used in this study, with temperatures $\{T_1, \ldots, T_n\}$ and box geometries described by the tensors $\{\bm{h}_1, \ldots, \bm{h}_n\}$. Boxes represent simulated systems, each column denotes a parallel tempering simulation. The solid line represents $P = 1\,\mathrm{atm}$ in the interlamellar phase as obtained by interpolation of the simulated data, and used in Study 1. For Study 2, the box geometries $\{\bm{h}_1, \ldots, \bm{h}_n\}$ are determined by the requirement of atmospheric conditions sufficiently far away from the interface into the crystal and melt, respectively, i.e., $P_\mathrm{c} = P_\mathrm{m} = 1\,\mathrm{atm}$

Crystal Unit Cell

It is widely known that united atom force fields lead to hexagonal symmetry of the crystal structure. However, they are quite accurate and efficient for simulation of amorphous structure. In our case of polyethylene, the united atom force field of Paul et al. leads to pseudo-hexagonal symmetry in the ab-plane, in contrast to the experimentally observed orthorhombic symmetry. Nevertheless, we deliberately accept this, because using a better force field in the crystal brings about the problem of how to transition between the two force fields at the interface. Also, the region of interest is not the crystal phase, but the interlamellar region.

In Study 1, the unit cell is adjusted to satisfy atmospheric pressure conditions at 400 K within the crystal phase. Specifically, the undeformed unit cell had $a = 0.77479$ nm, $b = 0.44626$ nm, and $c = 0.251822$ nm, with all crystallographic angles being 90 degrees. For Study 2, it was necessary to achieve atmospheric pressure conditions in the crystal at all temperatures considered, which can be achieved with unit cells with pseudo-hexagonal symmetry and lattice parameters $a(T) = (0.774053 + 0.0000471 \times (T - 400))$ nm, $b(T) = (0.445817 + 0.0000261 \times (T - 400))$ nm, and $c(T) = (0.252748 + 0.0000014 \times (T - 400))$ nm, where T is the temperature in units of K.

Simulation Box

The simulation box is illustrated in Fig. 14.1. It consists of the interlamellar phase enclosed between two lamellar crystals oriented with the {201} plane normal to the z-direction of the simulation cell.

Further specifications concerning the simulation cells used are summarized in Table 14.2, such as the dimensions of the rectangular simulation box, the thickness of the interlamellar domain $l_{z,\text{il}}$, the number of tails n_{tail}, bridges n_{bridge} and loops n_{loop}. Here, we note that the Monte Carlo moves used keep both n_{tail} and the sum $n_{\text{bridge}} + n_{\text{loop}}$ constant. The loop and bridge populations are not constant individually but rather determined dynamically from

Table 14.2. Parameters and numbers as used in Study 1 and Study 2. The symbols are explained in the text. Quantities with a superscript star are temperature dependent, and only average values are reported in this table

	Study 1	Study 2
box size: (x, y, z) (nm)	(2.77, 1.79, 8.72)	(2.77, 2.67, 12.64)
$l_{z,\text{il}}$ (nm)	7.66	7.22
n_{tail} (−)	12	18
$n_{\text{bridge}} + n_{\text{loop}}$ (−)	18	27
ρ_{il} (g/cc)	0.7947	0.79 *
N_{sites} (−)	1536	3750 *
$N_{\text{sites,il}}$ (−)	1296	1950 *

the simulation. Notice also that the simulated lamellar surface is larger in Study 2 compared to Study 1, but the fraction $n_{\text{tail}}/(n_{\text{tail}} + n_{\text{bridge}} + n_{\text{loop}})$ is identical in both studies, i.e., the same surface is studied. The average mass density of the interlamellar domain, ρ_{il}, is adjusted in Study 1 in order to achieve atmospheric pressure conditions on average over the interlamellar domain at 402 K. In Study 2, we aimed at atmospheric conditions not on average but rather in regions where the influence of the interface is negligible, i.e., in the mid-plane of the simulation box. The total number of sites in the entire simulation cell and in the interlamellar domain are denoted by N_{sites} and $N_{\text{sites,il}}$, respectively.

Keeping in mind that only the relevant surface region of the lamellae in the simulation cell in Fig. 14.1 is shown and needed for the simulation, these simulations are representative of semicrystalline morphologies having thicker crystal lamellae. Assuming realistic lamellae thicknesses, our simulation parameters can be translated into an estimate of the molecular weight of the polyethylene, $M_w \sim 10^4$ g/mol, and a degree of crystallinity, $\phi_c \sim 62\%$, as shown in [27]. Thus, the crystallinity and lamellar spacing studied here are comparable to the values cited by Hoffman [42] and Crist et al. [14].

For purposes of elastic property calculations, tensile and compressive deformations were simulated at ε_i ($i = 1, 2, 3$) in the interval $[-8.75\%, 8.75\%]$ in steps of 1.25%, with a few unimportant exceptions. Combined tensile deformations, used in calculation of the off-diagonal stiffness coefficients, were simulated for values of pairs $\{\varepsilon_i, \varepsilon_j\}$ ($i, j = 1, 2, 3; i \neq j$) with magnitudes $\varepsilon_i = \varepsilon_j$ in the range $[-4.375\%, 4.375\%]$ with increments of 0.625%. Shear deformations were performed under simple shear for ε_i ($i = 4, 5, 6$) in the range $[-5\%, 5\%]$ in steps of 1.25%.

14.2.3 Thermal and Elastic Properties of Interlamellar Domain

The calculation of elastic properties in Study 1 was performed by simulations in the NhT-ensemble for specific values of T and h (see Fig. 14.2). Here, h is the tensor that describes both the size and shape of the system, e.g. $V = \det(h)$ [43]. At each point (T, h), the pressure is calculated as $P = -\text{Tr}\left[\sigma(T, h)/3\right]$. Since our focus is on the interlamellar domain, stress contributions from the rigid crystals are not included in Study 1. Lines of constant pressure can thus be obtained through interpolation. In Fig. 14.2, the solid line is a schematic representation of the condition $P = 1$ atm. The pressure in the interlamellar domain at each temperature can be adjusted by varying only that component of h which describes the interlamellar thickness perpendicular to the interface. In the other two orthogonal directions in the interface plane, the extension of the interlamellar domain is more rigidly constrained due to continuity with the crystal lattice. The condition $P = 1$ atm leads to a relation between cell volume and temperature, $V_0(T)$. Any property X can hence be considered in terms of $X(V, T)$ or $X(P, T)$. In particular, temperature derivatives at constant volume, $(\partial X(V, T)/\partial T)|_V$

(along vertical lines in Fig. 14.2), and at constant pressure $P = 1\,\mathrm{atm}$, $(\partial X(P,T)/\partial T)|_P = (dX(V_0(T),T)/dT)$ (along the solid line in Fig. 14.2), can be calculated.

After appropriate interpolation procedures [29], isochoric and isobaric heat capacities at atmospheric pressure were calculated,

$$C_V = (\partial E/\partial T)|_V , \tag{14.5}$$

$$C_P = (\partial E/\partial T)|_P + P\,(\partial V/\partial T)|_P . \tag{14.6}$$

Furthermore, stresses were calculated as functions of strain and temperature. For each temperature, each component of stress was fit to a second order Taylor series expansion in terms of the strains, about the $P = 1\,\mathrm{atm}$ reference volume $V_0(T)$ at each specific temperature. Based on the stresses, the elastic moduli C_{ij} and the Grüneisen coefficients γ_i ($i = 1, 2, 3$) of the non-crystalline interlamellar phase were calculated using

$$C_{ij} = (\partial \sigma_i/\partial \varepsilon_j)|_{T,\varepsilon_{k\neq j}} , \tag{14.7}$$

$$\gamma_i = -V_0\,C_V^{-1}\,(\partial \sigma_i/\partial T)|_{V=V_0} = C_V^{-1}\,(\partial S/\partial \varepsilon_i)|_{T,\varepsilon_{k\neq i}} , \tag{14.8}$$

where the Voigt notation is used throughout the manuscript. The Grüneisen coefficients provide a measure of entropic contributions to the elastic moduli.

14.2.4 Energy and Stresses in the Crystal-Melt Interface

Chain connectivity between the crystalline and non-crystalline domains and the finite stiffness of polyethylene result in a finite thickness of the transition region that has no counterpart in the crystal-liquid interface for low molecular weight substances. On the molecular scale, this transition region extends well beyond what may be regarded as the surface of the crystal phase, prompting use of the term "interphase" to describe this transition region [44]. For thermodynamic purposes, however, it is convenient to characterize the properties of this interphase as those associated with an interfacial dividing surface. In order to approximate this region (termed "I" in Fig. 14.1) using a sharp interface, a coarse-graining step is involved. The procedure adopted here uses the concept of the Gibbs dividing surface [45–47].

The finite width of the transition region is reflected in the position dependent profiles of mass density $\rho(z)$, internal energy density $e(z)$ and stress tensor $\boldsymbol{\sigma}(z)$, which can all be obtained from Monte Carlo simulations. In accord with these profiles as a function of the coordinate z along the surface normal, one can define the corresponding properties of the interface,

$$\rho_{\mathrm{int}} := \int_{-\infty}^{\infty} \left[\rho(z) - \rho^{\mathrm{step}}(z)\right] dz , \tag{14.9}$$

$$e_{\mathrm{int}} := \int_{-\infty}^{\infty} \left[e(z) - e^{\mathrm{step}}(z)\right] dz , \tag{14.10}$$

$$\pi_{\alpha\beta} := \int_{-\infty}^{\infty} \left[\sigma_{\alpha\beta}(z) - \sigma_{\alpha\beta}^{\mathrm{step}}(z)\right] dz \quad (\alpha, \beta = x, y) . \tag{14.11}$$

Each quantity with superscript "step" denotes a Heaviside step function which on either side of the step takes the values identical to the corresponding bulk values of the crystal and melt domains sufficiently far away from the transition region. Therefore, regions far away from the interface do not contribute to the integrals, i.e., the integrand approaches zero rapidly. The criterion of a massless interface, $\rho_{\text{int}} = 0$, determines the position of the step, i.e., the position of the Gibbs dividing surface, z_{div}. The latter is then used also in the step functions for e_{int} and $\pi_{\alpha\beta}$, see Fig. 14.3. The physical interpretation of the interface properties given by (14.9-14.11) as 'excess quantities' is now obvious: It is the differences between the real profile and the extrapolation of the bulk values up to the sharp interface. In that sense, they isolate the effect of the interface. We note that determination of the surface tension γ, i.e., the Helmholtz free energy per unit area, from the interface stresses is non-trivial, as expressed by the Herring equation [18, 48, 49]

$$\pi_{\alpha\beta} = \gamma \delta_{\alpha\beta} + \left. \frac{\partial \gamma}{\partial \varepsilon_{\alpha\beta}} \right|_T , \qquad (14.12)$$

where $\delta_{\alpha\beta}$ denotes the identity matrix. Because one of the adjoining phases is solid, γ depends on the strain in the interface, $\varepsilon_{\alpha\beta}$ ($\alpha, \beta = x, y$).

Since the crystal-melt interface under consideration has zero curvature, it is natural to assume that atmospheric pressure conditions prevail sufficiently far away from the interface, putting constraints on the lattice parameters of the pseudo-hexagonal unit cell, and on the interlamellar domain. All these parameters of the box geometry are included in the label $\boldsymbol{h}$ in Fig. 14.2. Dark

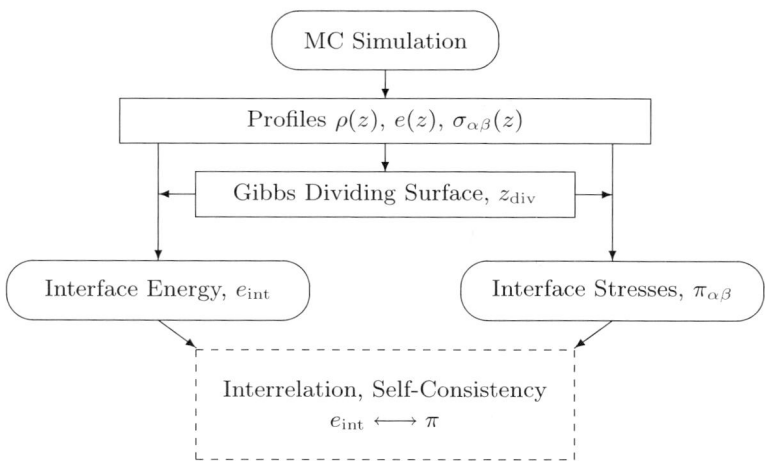

Fig. 14.3. Illustration of the procedure to calculate interface internal energy, e_{int}, and the interface stresses, $\pi_{\alpha\beta}$, as used here for the polyethylene {201} crystal surface

boxes indicate the systems in which atmospheric conditions are established in the bulk crystal and melt domains, respectively.

14.3 Results and Discussion

14.3.1 Conformational Properties

The average lengths of bridge, loop and tail segments in the interlamellar phase as functions of temperature were investigated in Study 1. In the simulation procedure, the probability for attachment and removal of a single CH_2 group is independent of the chain length n, and hence one can anticipate that the chemical potential for a chain segment of n united atoms takes the form $\mu(n,T) = \mu_0(T) + \mu_n(T)n$ with $\mu_n(T) > 0$. As a consequence, the loop, bridge and tail distributions should depend exponentially on n, i.e., $p(n,T) \sim \exp[-\mu(n,T)/k_B T]$ for sufficiently large n, in agreement with our simulation results (not shown).

As far as the temperature dependence of the average segment length is concerned, one finds that it increases with temperature for both bridge and tail populations, from $\langle n_{\text{bridge}} \rangle \approx 139$ and $\langle n_{\text{tail}} \rangle \approx 44$ at 350 K to $\langle n_{\text{bridge}} \rangle \approx 169$ and $\langle n_{\text{tail}} \rangle \approx 55$ at 450 K. These temperature dependences originate primarily from the factor $1/T$ in the exponential of the distribution function. In contrast, for loops one observes a slight decrease in average length with increasing temperature, from $\langle n_{\text{loop}} \rangle \approx 36$ at 350 K to $\langle n_{\text{loop}} \rangle \approx 30$ at 450 K. This reverse trend is attributed to torsional hindrances, which become more stringent the shorter the loops. Hence, we conclude that the temperature dependence of $\mu_n(T)$ is significant in the case of loops, due to torsional contributions.

Next, we consider the equilibrium topology, i.e., the relaxed state long after crystallization has stopped, of the {201} crystal/amorphous interface in polyethylene, which has ramifications for material properties of the interlamellar phase. In particular, we focus on the statistics of loops. We abbreviate with [$mn0$] the reentry vector or end-to-end vector for a loop segment, [$\pm m\, l_x, \pm n\, l_y, 0$], with l_x and l_y representing the projected length of a unit cell vector (a or b) at the crystal surface in the x- and y-directions, respectively. Figure 14.4 shows results at temperatures 350 K and 450 K, where the length of the reentry vector increases from left to right, and n is assumed to be integer. Firstly, one observes that loops with reentry vectors oriented along [0 n 0] are the most common. This population is dominated by the [0 1 0] loops, which are the shortest of all possible loops in the {201} interface for the pseudo-hexagonal unit cell considered here. Secondly, the loop populations decrease with increasing distance between reentry points. Thirdly, the "rest" population is comprised exclusively of loops with reentry vectors longer than that for the [1 1 0] direction. Comparison of the results for PE and the freely-rotating chain (FRC) show that the torsion leads to longer loops [29]. This is

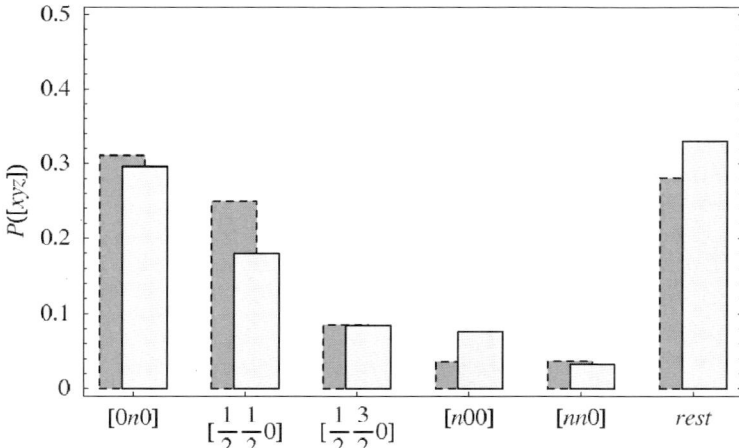

Fig. 14.4. Reentry distribution at the undeformed state of polyethylene as a function of reentry orientation at two temperatures; light bar: 350K, dark bar: 450K. The last entry "rest" lumps together all the remaining loops not explicitly considered in the other sets. The shortest reentry vector length in a particular direction increases from left to right. See text for notation. Reproduced from [29] with written permission from ACS Publications

in accord with our observation above that the loop length distribution is substantially influenced by the torsion potential. Lastly, Fig. 14.4 indicates that reconstruction of reentry topology associated with changes in temperature between 350 K and 450 K has a relatively small effect.

14.3.2 Thermal and Elastic Properties of Interlamellar Domain

In this subsection, the interlamellar domain is characterized in terms of thermal and elastic properties, as obtained from our Study 1 MC simulations. The most important results are summarized in Table 14.3.

Isochoric and Isobaric Heat Capacities

The isochoric and isobaric heat capacities of the interlamellar domain are displayed in Fig. 14.5 as functions of temperature at atmospheric pressure. In terms of the scheme in Fig. 14.2, isochoric temperature derivatives describe changes along vertical columns of constant h evaluated on the solid line representing $P = 1$ atm, while isobaric temperature derivatives capture changes along the line $P = 1$ atm. The isochoric heat capacity $C_V(T)$ at $P = 1$ atm is of the order of 26 J/K/mol CH_2, which was also checked using the fluctuation formula [50] with consistent results. The isobaric heat capacity $C_P(T)$

Table 14.3. Summary of the properties of the interlamellar phase and of the sharp interface, as discussed in the text. Superscripts: values from Study 1 at $P = 1\,\text{atm}$ (a1) in the temperature range $[350, 450]\,\text{K}$, (a1-) in the temperature range $[360, 450]\,\text{K}$, and (b) at $T = 435\,\text{K}$; (a2) denotes values from Study 2 in the temperature range $[380.6, 450]\,\text{K}$ and $P_c = P_m = 1\,\text{atm}$. For ranges of temperature, we give only the values of that specific property at the lowest and at the highest temperature. Note that all properties summarized here have a monotonic temperature dependence in the temperature range given, except for C_V and α_3, which both have a shallow maximum

	Property	Value
Interlamellar Domain	C_V (J/K/mol CH$_2$)	$[25.0 \pm 0.8, 25.5 \pm 0.3]$ [a1]
	C_P (J/K/mol CH$_2$)	$[32.7 \pm 0.2, 31.8 \pm 0.2]$ [a1]
	$\{E_1, E_2, E_3\}$ (GPa)	$\{0.49, 0.77, 0.27\}$ [b]
	$\{G_1, G_2, G_3\}$ (GPa)	$\{-0.17, 0.12, 1.17\}$ [b]
	K (GPa)	0.89 [b]
	γ_1 (−)	$[0.77 \pm 0.04, 0.43 \pm 0.01]$ [a1]
	γ_2 (−)	$[0.74 \pm 0.03, 0.44 \pm 0.02]$ [a1]
	γ_3 (−)	$[0.77 \pm 0.04, 0.43 \pm 0.02]$ [a1]
	α_1 (10^{-4}/K)	$[2.38, 0.41]$ [a1-]
	α_2 (10^{-4}/K)	$[-0.45, 0.27]$ [a1-]
	α_3 (10^{-4}/K)	$[6.54, 6.49]$ [a1-]
Crystal-Melt Interface	e_{int} (J/m^2)	$[0.299 \pm 0.006, 0.340 \pm 0.004]$ [a2]
	$(\partial e_{\text{int}}/\partial T)\vert_h$ (10^{-3} J/m^2/K)	$[-2.5 \pm 0.6, 0.9 \pm 0.2]$ [a2]
	π_{xx} (J/m^2)	$[-0.293 \pm 0.009, -0.244 \pm 0.007]$ [a2]
	π_{yy} (J/m^2)	$[-0.379 \pm 0.008, -0.428 \pm 0.007]$ [a2]
	π_{xy} (J/m^2)	$\sim 0 \pm 0.006$ [a2]

at $P = 1\,\text{atm}$ is of the order $32\,\text{J/K/mol}\,\text{CH}_2$, i.e., approx. 20% larger than $C_V(T)$, but slightly lower compared to extrapolated experimental values of an amorphous polyethylene melt (from $33.1\,\text{J/K/mol}\,\text{CH}_2$ to $37.8\,\text{J/K/mol}\,\text{CH}_2$) at the same pressure and temperatures [51].

Elastic Stiffness and Compliance, Stability

In Fig. 14.6, tensile ($i, j = 1, 2, 3$) and shear ($i, j = 4, 5, 6$) components C_{ij} of the stiffness matrix (Voigt notation) at constant pressure are shown. In

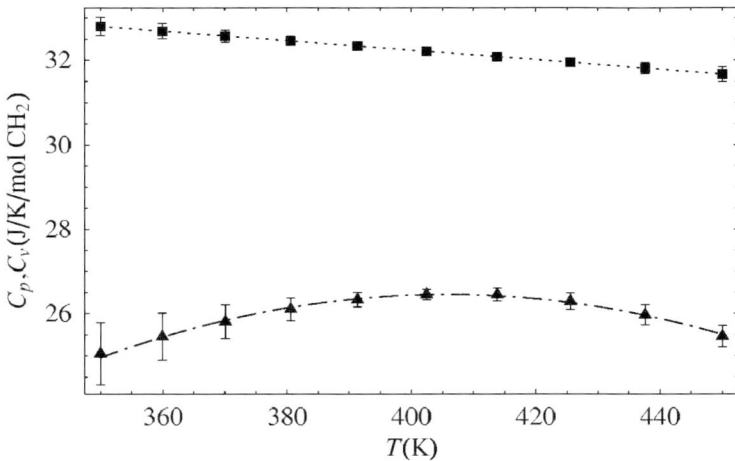

Fig. 14.5. Heat capacities at atmospheric pressure; ■: C_p; ▲: C_V. Reproduced from [29] with written permission from ACS Publications

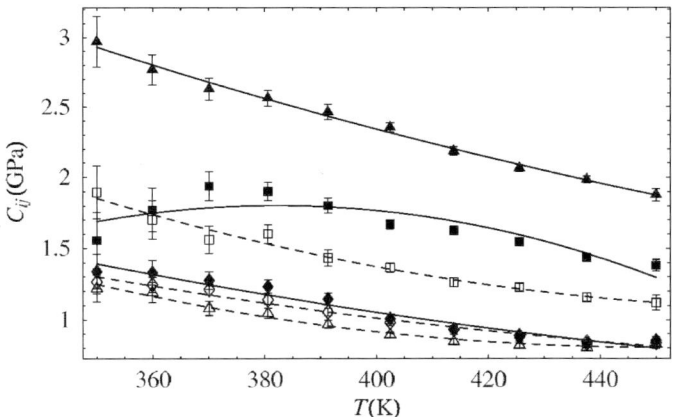

Fig. 14.6. Tensile and shear contributions to the elastic stiffness matrix C at atmospheric pressure; ■: C_{11}; ▲: C_{22}; ◆: C_{33}; □: C_{12}; △: C_{13}; ◇: C_{23}. Lines are drawn as a guide to the eye. Reproduced from [29] with written permission from ACS Publications

general, they decrease with increasing temperature, as expected [52]. The shear components were calculated as functions of temperature, but only for deformation about a single reference cell, $\boldsymbol{h}_0$ at $T = 435$ K. For that temperature, we can report the full stiffness tensor C for the non-crystalline interlamellar material in PE at $P = 1$ atm (in GPa):

$$\mathsf{C} = \begin{pmatrix} 1.54 & 1.21 & 0.83 & 0.00 & -0.18 & 0.00 \\ 1.21 & 2.02 & 0.87 & 0.00 & -0.24 & 0.00 \\ 0.83 & 0.87 & 0.90 & 0.00 & 0.05 & 0.00 \\ 0.00 & 0.00 & 0.00 & \sim 0.00 & 0.00 & -0.20 \\ -0.18 & -0.24 & 0.05 & 0.00 & 0.22 & 0.00 \\ 0.00 & 0.00 & 0.00 & -0.20 & 0.00 & 0.57 \end{pmatrix}. \qquad (14.13)$$

This stiffness tensor has monoclinic symmetry rather than pseudo-hexagonal, since the underlying pseudo-hexagonal crystal is tilted away from the c-axis. The uncertainty in each of the tensile stiffnesses is ± 0.03 GPa; the uncertainty in each of the shear stiffnesses is ± 0.06 GPa, with the exception of C_{44}, where the uncertainty is closer to ± 0.1 GPa. Within the accuracy of our sampling, C_{44} is zero within errors. As a consequence the system is at best only marginally stable, since the determinant of the stiffness matrix in (14.13) is close to zero.

The best estimate of the elastic compliance matrix S at $P = 1$ atm and $T = 435$ K is obtained by inversion of the stiffness matrix and using standard propagation of errors. One obtains for S (in GPa^{-1})

$$\mathsf{S} = \begin{pmatrix} 2.0 & -0.34 & -1.6 & 0. & 1.6 & 0. \\ -0.34 & 1.3 & -1.0 & 0. & 1.3 & 0. \\ -1.6 & -1.0 & 3.7 & 0. & -3.2 & 0. \\ 0. & 0. & 0. & -6.0 & 0. & -2.3 \\ 1.6 & 1.3 & -3.2 & 0.0 & 8.3 & 0. \\ 0. & 0. & 0. & -2.3 & 0. & 0.85 \end{pmatrix}. \qquad (14.14)$$

The uncertainty in C_{44} does not affect most of the compliances, but it does result in unreliable values for S_{44} and S_{66}. From (14.14), we estimate the Young's moduli $E_i = 1/S_{ii}$ ($i = 1, 2, 3$), the shear moduli $G_{i-3} = 1/S_{ii}$ ($i = 4, 5, 6$), and the bulk modulus $K = 1/((S_{11} + S_{22} + S_{33}) + 2(S_{12} + S_{13} + S_{23}))$ [53], with values reported in Table 14.3. Due to the aforementioned uncertainty in C_{44}, the results for G_1 and G_3 should be interpreted with caution. We also mention that Krigas et al. reported shear modulus data for the fully amorphous melt below the detection limit of our simulations, explaining our difficulty in determining an accurate value of G_1 [54]. The results for the Young's moduli are similar to experimental values between 0.02 GPa and 0.4 GPa estimated by Crist et al. [14]. Reported experimental bulk moduli for a PE melt range from 1.38 GPa to 0.87 GPa at 350 K to 450 K, respectively, and from 3.37 GPa to an extrapolated value of 0.84 GPa for semicrystalline PE at the same temperatures [52]. Due to the connectivity between the crystal and the interlamellar material, the latter is expected to be stiffer than the pure melt.

The entropic contributions to the elastic moduli (see Fig. 14.6) is given by the Grüneisen parameters, which show a close to linear temperature dependence in our simulations, ranging from ~ 0.75 at 350 K to 0.4 at 450 K. These results are lower than our previous results [26] which ignored torsion

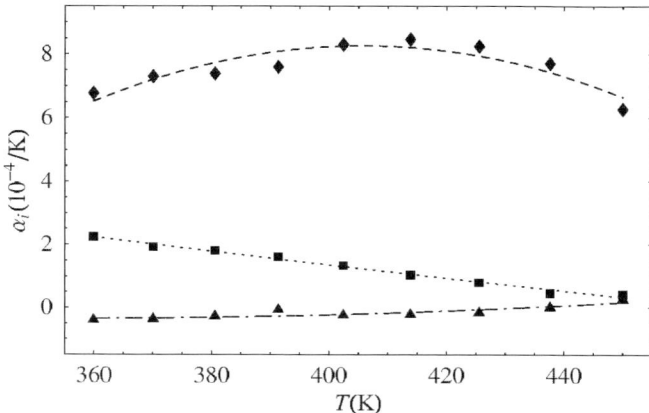

Fig. 14.7. Thermal expansion coefficients as functions of temperature at atmospheric pressure; ■: α_1; ▲: α_2; ♦: α_3. Reproduced from [29] with written permission from ACS Publications

contributions and where the average interlamellar pressures (1300−2500 atm) were substantially different from the one used here (1 atm).

Coefficients of Thermal Expansion

The coefficients of linear thermal expansion are reported in Fig. 14.7 as functions of temperature. The method to their calculation is described elsewhere [26], under the assumption that $\gamma_5 \ll \gamma_{1,2\,\text{or}\,3}$, and interpolated to volumes corresponding to atmospheric pressure conditions. Experimental data for linear thermal expansion coefficients for the amorphous melt range from $7.11 \times 10^{-4}/\text{K}$ to $7.23 \times 10^{-4}/\text{K}$ under comparable conditions [55]. We emphasize that this agrees nicely with the simulated value for α_3 shown in Fig. 14.7. While the thermal expansion in the direction normal to the crystal surface is not immediately constrained by the crystal, it is substantially decreased in the xy-plane due to chain continuity, in accord with the data for α_1 and α_2 reported in Fig. 14.7.

14.3.3 Properties of the Crystal-Melt Interface

Much of the interesting physics in semicrystalline materials is hidden in the transition region between the crystalline domain and the melt-like domain. In particular in polymeric systems with a certain degree of stiffness of the backbone, the chain connectivity between both phases results in a rather wide transition region. In the following, we focus on the characterization of the crystal-melt interface by invoking the Gibbs construction of a sharp

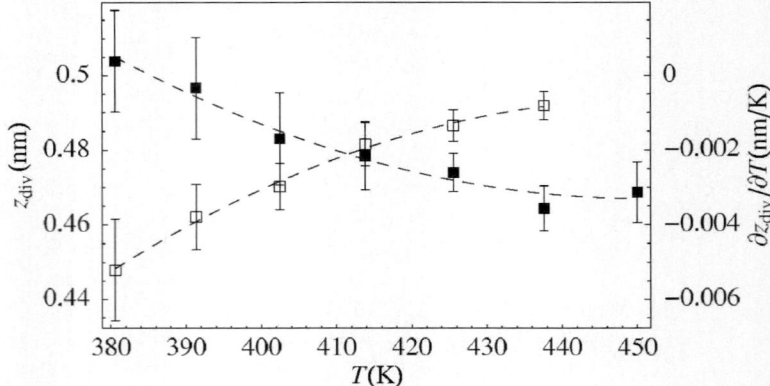

Fig. 14.8. Position of the Gibbs dividing surface z_div plotted versus temperature T at atmospheric conditions in the bulk phases, $P_\mathrm{c} = P_\mathrm{m} = 1\,\mathrm{atm}$ (■), and its temperature derivative at constant box geometry, $(\partial z_\mathrm{div}/\partial T)|_h$ (□). Reprinted from [30] with written permission from Elsevier.

dividing surface. The most important quantitative results are summarized in Table 14.3.

Position of the Gibbs Dividing Surface

The calculation of the interface properties as described in Sect. 14.2.4 requires determination of the position of the Gibbs dividing surface, z_div. We calculated the mass density profiles $\rho(z)$, and then used the criterion $\rho_\mathrm{int} = 0$ with (14.9). The resulting values are reported in Fig. 14.8. There, the position of the Gibbs dividing surface is measured with respect to the real crystal surface, which is defined midway between the top layer of united atoms in the crystal and the first layer of mobile atoms bonded to it. The thickness of the interface (approx. equal to $2z_\mathrm{div}$) decreases for the higher temperatures as a result of weakened chain stiffness and entropic effects, in accord with previous results [26].

The derivative of z_div at constant bulk pressures (i.e., along the diagonal in Fig. 14.2) as obtained from Fig. 14.8 (■) goes from approximately -9×10^{-4} nm/K below $T = 400\,\mathrm{K}$ to zero within error bars above $T = 430\,\mathrm{K}$. On the other hand, the derivative at constant box geometry (i.e., along vertical lines in Fig. 14.2) are about a factor of six larger (□ in Fig. 14.8). We believe that this is explained by the fact that increasing the temperature at constant box geometry leads to increased pressure in the melt phase, which in turn further compresses the interface. Therefore, this additional effect leads to larger values for $(\partial z_\mathrm{div}/\partial T)|_h$ in comparison to $(\partial z_\mathrm{div}/\partial T)|_{P_\mathrm{c}=P_\mathrm{m}=1\,\mathrm{atm}}$.

The error bars in Fig. 14.8 (as also for Figs. 14.9 and 14.10) are calculated by splitting the entire Monte Carlo simulation in ten blocks, from which

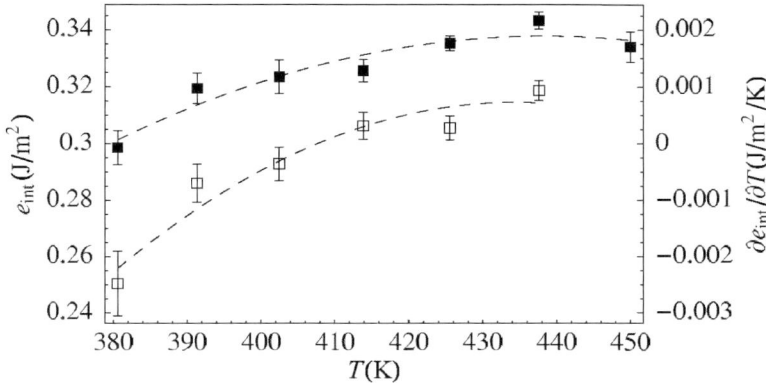

Fig. 14.9. Interface internal energy e_{int} plotted versus temperature T at $P_c = P_m = 1\,\text{atm}$ (■), and its temperature derivative at constant interface strain and constant system volume, $(\partial e_{\text{int}}/\partial T)|_h$ (□). Reprinted from [30] with written permission from Elsevier

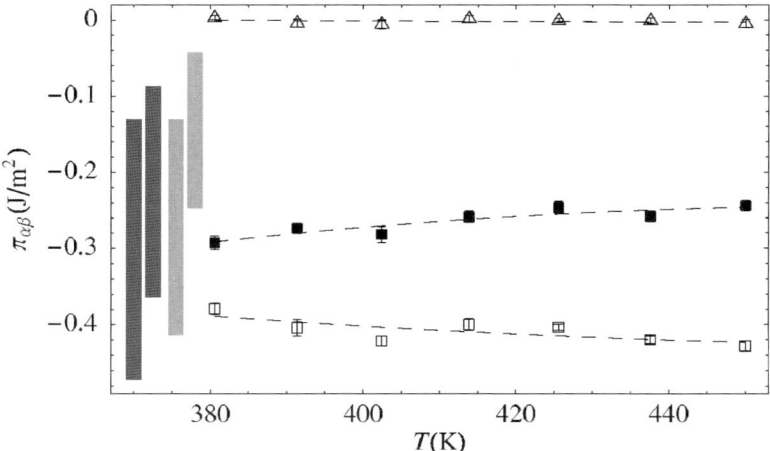

Fig. 14.10. Interface stresses, plotted versus temperature at atmospheric bulk stresses in the adjoining crystal and melt phases: π_{xx} (■), π_{yy} (□), and π_{xy} (△). *Solid bars* indicate the ranges of values given in [16–18] for n-paraffin at $T = 296\,\text{K}$ (*dark grey*), and melt-crystallized polyethylene at $T = 298\,\text{K}$ (*light grey*). Reprinted from [30] with written permission from Elsevier

ten statistically independent block averages are calculated. In turn, these ten averages are used to determine the total average and the associated error. Propagation of errors is employed to obtain the error of the temperature derivatives, based on the three successive data points used for the calculation of the slope.

14.3.4 Internal Energy of the Interface

The internal energy of the interface can be calculated in our MC simulations according to (14.10), with the aid of the position of the Gibbs dividing surface, z_{div} (see Fig. 14.3). The results are shown in Fig. 14.9 (■). The interface energy increases steadily from $0.30\,\text{J/m}^2$ at $380\,\text{K}$ to $\sim 0.335\,\text{J/m}^2$ at the melting temperature ($T_{\text{m}} \simeq 410\,\text{K}$), and remains approximately constant above.

The temperature derivative of the interface energy is particularly interesting. We have mentioned in the previous section the difference between temperature derivatives at constant bulk pressures in contrast to constant box geometry. Although the fact that they are different may not come as a surprise, it is unexpected that these two types of temperature derivatives may even have opposite sign, as illustrated in the following. While the derivative at constant bulk pressure (change along the diagonal in Fig. 14.2) has a positive slope according to the data in Fig. 14.9 (■) at low temperatures, the derivative at constant box geometry (change along columns in Fig. 14.2) is negative below the melting temperature (□ in Fig. 14.9). An explanation for the different signs of the slopes needs to take into account that the temperature derivative of the interface energy depends not only on the profile $e(z)$, but also on z_{div}, which in turn originates from the profile $\rho(z)$. For a detailed explanation the reader is referred to [30].

As shown in detail in [30], the heat capacity at constant total volume of semicrystalline polyethylene contains phase change contributions due to a change in volume fraction (here, $\partial z_{\text{div}}/\partial T|_{\boldsymbol{h}}$) and due to a change in the internal energy of the interface (here, $\partial e_{\text{int}}/\partial T|_{\boldsymbol{h}}$), which is the reason for calling the heat capacity of semicrystalline polymers "apparent". We mention that we refrain from calling $(\partial e_{\text{int}}/\partial T)|_{\boldsymbol{h}}$ a "heat capacity at constant surface area" of the interface, as suggested by analogy to its bulk counterpart "heat capacity at constant volume". We recall that e_{int} is an excess property by definition, and its behavior upon changing temperature is strongly interwoven with the thermodynamic behavior of the two adjoining bulk phases.

14.3.5 Interface Stresses

The values for the interface stresses calculated according to (14.11) for $P_{\text{c}} = P_{\text{m}} = 1\,\text{atm}$ (i.e., on the diagonal of the schematic in Fig. 14.2) are reported in Fig. 14.10.

Two features of the data in Fig. 14.10 are striking, namely the sign of the diagonal stresses and the anisotropy. The fact that the interface stresses are negative means that the interface tries to expand, which can be rationalized with these two arguments. Firstly, the chains are in perfect crystalline registry in the crystal domain, but then the chains exit the crystal and attempt to gain more configurational entropy, i.e., they exert pressure on each other to gain more space. Secondly, (short) folds preferentially try to increase the distance between anchor points due to the stiffness of the backbone. Our results are of the same order of magnitude as the ones reported by Cammarata, Eby and Fisher [16–18], which are between $-0.1\,\mathrm{J/m^2}$ and $-0.3\,\mathrm{J/m^2}$ for the {001} surface.

By virtue of the Herring equation (14.12), measurements of the surface tension of the fold surface are also interesting, which in turn enters in theoretical models for experimentally measurable crystallization rates [20,21]. In view of the interface stress values in Fig. 14.10 it is interesting to note that experimental values for the scalar surface tension are in the range $\gamma = +0.1\,\mathrm{J/m^2}$. The fact that these values are of the same order of magnitude as the interface stresses, but with opposite sign, draws attention to the strain dependence of the surface tension in the Herring equation (14.12). For the diagonal components, the term $(\partial\gamma/\partial\varepsilon_{\alpha\alpha})|_T$ over-compensates the isotropic contribution γ by far, which serves as a measure of the strong dependence of the surface tension on tensile strains. We also observe $|\pi_{xx}|, |\pi_{yy}| > \gamma$, which we suppose to be related to the presence of short folds [15]. This is in agreement with our simulations for which the average loop length is rather short, as reported in Sect. 14.3.1

The off-diagonal stress π_{xy} is zero within error, according to Fig. 14.10. From this one can conclude that the scalar surface tension is independent of small shear deformations in the plane of the interface, as can be inferred from the Herring equation (14.12).

The rigorous definition of the interface stresses relies on mechanical equilibrium, i.e., the shear stresses in the out-of-plane directions being zero and σ_{zz} independent of z. In order to measure such effects in our simulations in relation to the in-plane interface stresses, the definition (14.11) is also applied to the out-of-plane components of the stress profile. One finds that $|\pi_{xz}| \lesssim 0.05\,\mathrm{J/m^2}$, $|\pi_{yz}| \lesssim 0.01\,\mathrm{J/m^2}$, and $|\pi_{zz}| \lesssim 0.05\,\mathrm{J/m^2}$. Hence, the stress integrals involving the z-direction are 20% or less in magnitude compared to π_{xx}, and even smaller when compared to π_{yy}. Nevertheless, they are significant, and we believe that the stresses π_{xz} and π_{zz} are a signature of the tilted chains exiting the crystal with a certain persistence along the backbone.

14.4 Summary and Discussion

The structural, thermal and mechanical characterization of the interlamellar domain and of the {201} crystal-melt interface of semicrystalline PE was performed, and compared with experimental data where available. Monte Carlo simulations complete with three-fold torsional potential were used with a united atom representation of polyethylene. We have employed two different strategies to assess the properties of the interface.

14.4.1 Entire Interlamellar Domain

The morphology was quantified in terms of loop reentry distributions and the average lengths of loops, bridge molecules, and dangling chain ends, i.e., tails. The distribution of reentry vectors on the {201} surface of PE showed that the shorter the reentry vector the more probable its occurrence. In particular, the shortest reentry vector [0 1 0] was the most common. We found that the length distribution of the different types of segments (loops, bridges, tails) over a large range decays exponentially with segment length, i.e., that the chemical potential, μ_n, for addition/removal of united atoms to a segment is independent of the segment length. However, our simulation suggested that μ_n is strongly dependent on temperature for loops, in contrast to bridges and tails, which we attribute to the effect of torsion and hence chain stiffness on short loops.

Thermodynamic and mechanical properties of the interlamellar domain have been determined at average atmospheric conditions in a range of temperatures, $P = 1\,\text{atm}$ and $T \in [350, 450]\,\text{K}$, namely the isobaric and isochoric heat capacity, Grüneisen coefficients, and the anisotropic thermal expansion coefficients. The latter clearly resembled melt-like behavior in the direction normal to the surface, while in-plane one observes low, crystal-like expansion. In large systems consisting of several lamellae as shown on the left in Fig. 14.1, this strong anisotropy in thermal expansion will dominate the behavior close to the crystal surface. When measuring the thermal expansion of the entire system, averaging comes into play which in turn depends most probably on the size, density, and orientation distribution of the lamellae. In that sense, the simulations performed here provide estimates for the ingredients needed in such more complex studies.

A full mechanical characterization of any material for deformations in the linear regime is given by the stiffness or compliance matrix, respectively. For the interlamellar domain at $P = 1\,\text{atm}$ and $T = 435\,\text{K}$, we extracted exactly these matrices, from which Young's moduli and shear moduli were determined. According to the simulations, the bulk modulus of the interlamellar domain lies between between the experimental values reported for a purely amorphous melt and the semicrystalline solid.

14.4.2 Sharp Crystal-Melt Interface

The interface between the lamellar crystals and the non-crystalline, interlamellar region was studied using the technique of the Gibbs dividing surface. In so doing, one is able isolate the effects of the interface alone, irrespective of thickness of the lamellae and, to some degree, of the interlamellar domain. Therefore, the properties attached to the sharp interface can be used in a three-component model with arbitrary composition, which accounts for the interface contribution explicitly, in addition to the crystal and melt bulk contributions.

Our simulations resulted in values for the in-plane diagonal stress components $\pi_{xx} \simeq -0.27\,\mathrm{J/m^2}$ and $\pi_{yy} \simeq -0.4\,\mathrm{J/m^2}$, which compare reasonably well with experimental values. The anisotropy is a signature of the tilted chains exiting the lamella at the {201} surface. The sign of these stresses indicates that the interface is under pressure, due to entropic effects and due to connectivity between the crystal and non-crystalline segments. The Herring equation taught us that the surface tension (i.e., Helmholtz free energy per unit area) depends strongly on the tensile strains, but is independent of shear strains in the interface plane. Furthermore, the interface internal energy, e_{int}, and its change versus temperature was studied and quantified. The latter is of interest because it enters into the "apparent" heat capacity as measured for the entire semicrystalline material.

Throughout the manuscript, we carefully specified the variables under control. In particular when taking temperature derivatives, we observed differences between keeping pressure constant or the geometry of the system. We point out that such book keeping is essential, because it is not only responsible for quantitative differences, e.g., C_V vs. C_P, but it can even result in a alteration of the overall sign, as shown in Study 2 of the interface energy and stresses.

The internal energy and the stresses of the interface were both calculated on the basis of the respective profile obtained from our MC simulations, i.e., based on (14.10, 14.11). The way in which they are defined has clear physical meaning, as explained previously. However, one may ask about the existence of a single thermodynamic potential, namely a Helmholtz free energy of the interface, from which the internal energy and the interface stresses can be derived, in close analogy to the corresponding relations for bulk materials. Whether or not the data reported here have a chance of being derived from a single Helmholtz free energy can be discussed by testing the Maxwell relations, i.e., the relations between mixed second order derivatives of the thermodynamic potential. Doing so requires data on the strain dependence of the internal energy of the interface, which are currently not available to reasonable accuracy (see [30] for more details). Furthermore, one must realize that the thermodynamic state of the interface depends on the conditions in the two adjoining bulk phases, e.g., the pressure. Under the assumption that the difference between the surface tension γ and e_{int} equals the temperature times the entropy of the interface, s_{int}, our results for e_{int} combined with the experimental value $\gamma \sim 0.1\,\mathrm{J/m^2}$ for the fold surface [20, 21] leads to $Ts_{\mathrm{int}} \simeq 0.2\,\mathrm{J/m^2}$. However, due to the difficulties just mentioned this number should be used with caution.

14.4.3 Perspectives

Measurements of the interlamellar domain and interface, as presented here, develop their full strength when put in a wider context and combined with the material properties of the adjoining phases, e.g. the effect of the internal energy of the interface on the heat capacity of the semicrystalline material.

Another such example is the connection between interface stresses and crystal lattice distortion, and lamellar twisting. The latter phenomenon requires two additional ingredients, in addition to the interface stresses: Firstly, the interface stresses must occur asymmetrically at the two opposite lamella surfaces [56, 57], which can not be achieved in our simulations by construction of the simulation cell. Reasons for such asymmetries have to be found on different grounds. Secondly, lamellar twisting can only be predicted if also the material properties of the crystalline lamellae are incorporated, either based on experimental [58, 59] or simulated data [60].

Table 14.3 summarizes the most important results of our MC simulations in terms of thermal and mechanical properties. It is striking that no mechanical response data, e.g., stiffness or compliance data, are available for the sharp interface. In that respect we mention that measuring the change in the interface stresses with respect to small deformations (tensile, shear) in directions of the interface plane with reasonable accuracy is difficult, as discussed in more detail in [30]. However, we have the feeling that such data would be useful for calculating the average mechanical response of semicrystalline polyethylene, in conjunction with similar data for the melt and crystal phases [60] and appropriate multiphase models.

Acknowledgment

This work has been supported by the Swiss National Science Foundation under Grant Number 81EZ-68591, by the ERC program of the National Science Foundation under Grants 0079734 and EEC-9731680 (Center for Advanced Engineering Fibers and Films – CAEFF), and by the MRSEC Program of the National Science Foundation under Award Number DMR-9808941.

References

[1] R. Seguela: J. Polym. Sci. Pol. Phys., **43**, 1729 (2005)
[2] G.R. Strobl, W. Hagedorn: J. Polym. Sci B **16**, 1181 (1978)
[3] C.C. Naylor, R.J. Meier, B.J. Kip, K.P.J. Williams, S.M. Mason, N. Conroy, D.L. Gerrard: Macromolecules **28**, 2969 (1995)
[4] R.H. Boyd: Polymer **26**, 1123 (1985)
[5] C. Alvarez, I. Šics, A. Nogales, Z. Denchev, S.S. Funari, T.A. Ezquerra: Polymer **45**, 3953 (2004)
[6] S. Torquato: *Random Heterogeneous Materials* (Springer, New York, 2002)
[7] M. Hütter: Phys. Rev. E **64**, 011209 (2001)
[8] M. Hütter, G.C. Rutledge, R.C. Armstrong: Phys. Fluids **17**, 014107 (2005)
[9] E.H. Kerner: Proc. Phys. Soc. Lond. **69B**, 808 (1956)
[10] S. Uemura, M. Takayanagi: J. Appl. Pol. Sci. **10**, 113 (1966)
[11] R.M. Christensen: J. Mech. Phys. Solids **38**, 379 (1990)
[12] R.M. Christensen, K.H. Lo: J. Mech. Phys. Solids **27**, 315 (1979)

[13] S.D. Gardner, C.U.Jr. Pittman, R.M. Hackett: Compos. Sci. Technol. **46**, 307 (1993)
[14] B. Crist, C.J. Fisher, P.R. Howard: Macromolecules **22**, 1709 (1989)
[15] J. Rault: J. Macromol. Sci. Phys. **B12**, 335 (1976)
[16] R.C. Cammarata, R.K. Eby: J. Mater. Res. **6**, 888 (1991)
[17] H.P. Fisher, R.K. Eby, R.C. Cammarata: Polymer **35**, 1923 (1994)
[18] R.C. Cammarata, K. Sieradzki: Annu. Rev. Mater. Sci. **24**, 215 (1994)
[19] P.D. Calvert, D.R. Uhlmann: J. Polym. Sci. **11**, 457 (1973)
[20] B. Wunderlich: *Macromolecular Physics* (Academic Press, New York, 1976)
[21] J.D. Hoffman: Polymer **23**, 656 (1982)
[22] S. Balijepalli, G.C. Rutledge: J. Chem. Phys. **109**, 6523 (1998)
[23] S. Balijepalli, G.C. Rutledge: Macromol. Symp. **133**, 71 (1998)
[24] S. Balijepalli, G.C. Rutledge: Comput. Theor. Polym. Sci. **10**, 103 (2000)
[25] S. Gautam, S. Balijepalli, G.C. Rutledge: Macromolecules **33**, 9136 (2000)
[26] P.J. in 't Veld, G.C. Rutledge: Macromolecules **36**, 7358 (2003)
[27] G.C. Rutledge: J. Macromol. Sci. Phys. **B41**, 909 (2002)
[28] P.G. Debenedetti: *Metastable Liquids: Concepts and Principles* (Princeton University Press, Princeton NJ, 1996)
[29] P.J. in 't Veld, M. Hütter, G.C. Rutledge: Macromolecules **39**, 439 (2006)
[30] M. Hütter, P.J. in 't Veld, G.C. Rutledge: Polymer **47**, 5494 (2006)
[31] D.C. Bassett, A.M. Hodge: Proc. R. Soc. Lond. A **377**(1768), 25 (1981)
[32] W. Paul, D.Y. Yoon, G.D. Smith: J. Chem. Phys. **103**, 1702 (1995)
[33] K. Bolton, S.B.M. Bosio, W.L. Hase, W.F. Schneider, K.C. Hass: J. Chem. Phys. B **103**, 3885 (1999)
[34] N. Waheed, M.S. Lavine, G.C. Rutledge: J. Chem. Phys. **116**, 2301 (2002)
[35] M.J. Ko, N. Waheed, M.S. Lavine, G.C. Rutledge: J. Chem. Phys. **121**, 2823 (2004)
[36] M. Vacatello, G. Avitabile, P. Corradini, A. Tuzi: J. Chem. Phys. **73**, 548 (1980)
[37] R.H. Boyd: Macromolecules **22**, 2477 (1989)
[38] K.V. Pant, D.N. Theodorou: Macromolecules **28**, 7224 (1995)
[39] L.R. Dodd, T.D. Boone, D.N. Theodorou: Mol. Phys. **78**, 961 (1993)
[40] V.G. Mavrantzas, T.D. Boone, E. Zervopoulou, D.N. Theodorou: Macromolecules **32**, 5072 (1999)
[41] D.A. Kofke: J. Chem. Phys. **117**, 6911 (2002)
[42] J.D. Hoffman, R.L. Miller: Polymer **38**, 3151 (1997)
[43] M. Parrinello, A. Rahman: J. Appl. Phys. **52**, 7182 (1981)
[44] P.J. Flory, D.Y. Yoon, K.A. Dill: Macromolecules **17**, 862 (1984)
[45] A.W. Neumann, J.K. Spelt (eds.): *Applied Surface Thermodynamics* (Marcel Dekker, New York, 1996)
[46] J.B. Hudson: *Surface Science : An Introduction* (John Wiley, New York, 1998)
[47] A. Zangwill: *Physics at Surfaces* (Cambridge University Press, New York, 1988)
[48] J.W. Gibbs: *The Scientific Papers of J. Willard Gibbs*, vol. 1 (Longmans-Green, London, 1906)
[49] C. Herring: The use of classical macroscopic concepts in surface energy problems. In: *Structure and Properties of Solid Surfaces* ed by R. Gomer and C.S. Smith (The University of Chicago Press, Chicago, 1953) pp. 5–72
[50] M.P. Allen, D.J. Tildesley: *Computer Simulation of Liquids* (Clarendon Press, Oxford, 1986) Ch. 2
[51] U. Gaur, B. Wunderlich: J. Chem. Phys. Ref. Data **10**, 119 (1981)

[52] O. Olabisi, R. Simha: Macromolecules **8**, 206 (1975)
[53] J.F. Nye: *Physical Properties of Crystals* (Clarendon Press, Oxford, 1985)
[54] T. Krigas, J.M. Carella, M.J. Struglinski, B. Crist, W.W. Greassley, F.C. Shilling: J. Polym. Sci., Polym. Phys. Ed. **23**, 509 (1985)
[55] R.A. Orwoll, P.J. Flory: J. Am. Chem. Soc. **89**, 6814 (1967)
[56] H.D. Keith, F.J.Jr. Padden: Polymer **25**, 28 (1984)
[57] B. Lotz, S.Z.D. Cheng: Polymer **46**, 577 (2005)
[58] I.M. Ward: *Mechanical Properties of Solid Polymers* (Wiley, Chichester [etc.], 1983)
[59] D.T. Grubb: Elastic properties of crystalline polymers. In: *Materials Science and Technology*, Volume 12: Structure and Properties of Polymers, ed by R.W. Cahn, P. Haasen and E.J. Kramer (VCH, Weinheim [etc.], 1993) pp. 301–356
[60] G.C. Rutledge: Modeling Polymer Crystals. In *Simulation Methods for Polymers*, ed by M. Kotelyanskii and D.N. Theodorou (Marcel Dekker, New York, 2004), Ch. 9

15

The Role of the Interphase on the Chain Mobility and Melting of Semi-crystalline Polymers; A Study on Polyethylenes

Sanjay Rastogi[1,3], Dirk R. Lippits[1], Ann E. Terry[1,2], and Piet J. Lemstra[1]

[1] Department of Chemical Engineering, Dutch Polymer Institute, Eindhoven University of Technology, Den Dolech 2, P.O. Box 513, 5600MB Eindhoven, The Netherlands
[2] ISIS Facility, Rutherford Appleton Laboratory, Chilton, Didcot, Oxfordshire OX11 0QX, England, UK
[3] IPTME, Loughborough University, Leicestershire LE11 3TU, England, UK
s.rastogi@tue.nl

Abstract. In semi-crystalline polymers, a range of morphologies can be obtained in which a chain may traverse the amorphous region between the crystals or fold back into the crystals leading to adjacent or non-adjacent re-entry, depending on the molecular architecture and crystallization conditions. This causes topological variations on the crystal surface and the occurrence of an interphase between the crystalline and amorphous domains, thus affecting the mechanical properties. In this chapter, we will discuss how the morphology within the interphase plays a prominent role in drawability, lamellar thickening and melting of thus crystallized samples. Normally, for linear polymers it is anticipated that extended chain crystals are thermodynamically most favorable, and ultimately, taking the example of linear polyethylene, it is shown that such chains would form extended chain crystals. However, this condition will not be realized in a range of polymers upon crystallization from the melt, such as those which do not show lamellar thickening or in branched polymers where the side branches cannot be incorporated within the crystal and hence fully extended chains are not possible. From a series of experiments, it is shown that with sufficient time and chain mobility, although extended chain crystals are not achievable, the chains still disentangle and a thermodynamically stable morphology is formed with a disentangled crystallizable interphase. The disentangled interphase has implications in the melting behavior of polymer crystals. It is feasible to melt these crystals by simple consecutive detachment of chain segments from the crystalline substrate. Clear distinction in different melting processes is observed, by the differences in the activation energies required for the consecutive detachment of chain segments or clusters of chain segments. The differences in the melting behavior, revealed during different heating rates, have consequences on the chain dynamics.

15.1 Introduction

The semi-crystalline structures often formed by crystallizable polymers are known to consist of thin crystalline lamellae separated by amorphous regions [1–3]. In the case of crystallization from the melt, the polymer chains must dientangle from the melt in order to form a regular conformation and align parallel to each other to form thin plate-like lamellae whereas the (remaining) entanglements are confined to the amorphous regions, a process often referred to as reeling-in. It is still unclear whether the polymer chains do actually disentangle or merely that during crystallization, entanglements are pushed to the surface. However, independent of the mechanism involved, the molecular structure of the amorphous region strongly depends not just on the chemical nature and inherent shape of the polymer, but also on the crystallization conditions either in the quiescent state or obtained during flow. Experimental efforts to decouple the mechanical properties of polymers from the crystalline and amorphous fractions have not been particularly successful because of the dependence of the molecular organization in the amorphous region upon the crystallization conditions. Moreover, a third structural component, an interphase of intermediate order, could exist between the amorphous and crystalline phases, as has been proposed both theoretically and experimentally, which means that a sharp demarcation line between the amorphous and crystalline phases is unlikely. This added complication is necessary if one considers that the chains emerge at the crystalline surfaces with a high degree of molecular alignment. These chains must either fold back into the crystallite, whether by adjacent re-entry, or must reside in the amorphous matrix. In the case of crystallization from the melt, where the crystallization conditions are often very far from equilibrium, extensive and perfect folding will strongly depend on molecular weight and molecular architecture – and in most cases will be highly improbable. Assuming the crystallites are of an infinite extent in the basal plane, then at small distances away from each crystallite surface, most of the chains present will have originated from the crystallite. The average chain orientation here will not be random as in the bulk amorphous matrix but will be distributed around the normal to the crystallite surface: this is the proposed semi-ordered interfacial component, the degree of ordering strongly depending upon the crystallization conditions. In several cases this may also give rise to a crystallographic register and influence further the physical and mechanical properties of the polymer.

A comprehensive review of the subject, supporting the existence of this third component in the structure of unoriented, semi-crystalline polymers, has been compiled by Mandelkern [4], and has been further supported by WAXD studies by Windle [5]. However, this still remains a topic of controversy. Terms used to describe the suggested third structural component have included "semi-ordered", "intermediate", "rigid amorphous", "interfacial", "interzonal", "interphase" and "transitional zone", and give quite a clear picture of what is envisaged (for this review, we will use the term "interphase").

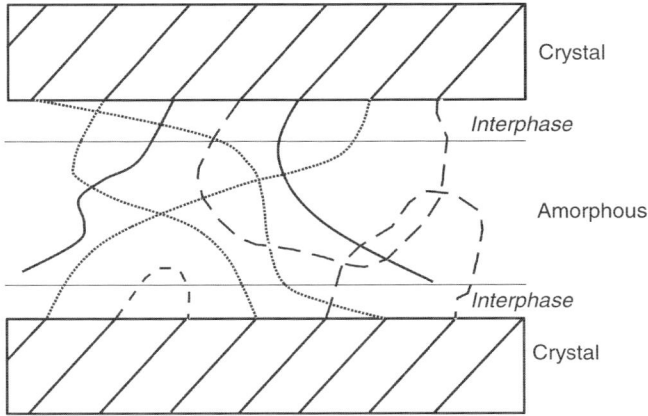

Fig. 15.1. Schematic diagram of the interphase, showing the examples of loops (*dashed*), bridges (*dotted*), and tails (*solid*), reproduced from Gautam et al. [6]

Figure 15.1, reproduced from the work of Rutledge and co-workers [6], illustrates the molecular picture of the interphase, in which three types of chain populations exist: "bridges" that join two crystal lamellae, "loops" that have their entry and exit points on the same crystal lamellae, and "tails" that terminate in the amorphous phase.

At this stage it is essential to recapitulate the existing knowledge on chain tilting and its influence on the interphase. It is well established that polyethylene shows a crystalline lamellar structure and that there is chain folding in the interphase region, one of the important observed features being that the polymer chain stems are not generally orthogonal to the basal plane of the lamellae. Such basal planes are identified by a chain tilt angle, defined as the angle between the lamellar normal and the c-axis of polymer chain stems. The in-situ process of chain tilting during heating has been shown previously and it is known that the initial chain tilt angle is strongly dependent on crystallization conditions. For example, Bassett and Hodge [7] studied polyethylene spherulites using electron microscopy and found a regular texture showing well-defined lamellae having crystal stems inclined at an angle ranging from 19° to 41° to the lamellar normal, with 34° (corresponding to the {201} facet) being the most common. Khoury [8] observed the presence of a predominant chain tilt angle of approximately 34° in polyethylene spherulites grown from the melt at high undercoolings. Under special conditions chain tilt angles higher than 45° could be also observed.

Chain tilting is known to exert a strong influence on the structure and hence the properties of the interphase. Frank suggested that mutual exclusion imposes steric constraints on the structure of the interphase between crystalline and amorphous regions that is only relieved by the presence of chain ends, folding back into the crystal and chain tilt [9]. Yoon and Flory [10] and

Kumar and Yoon [11] also considered that the flux of chains into the amorphous region would be reduced by the chains tilting and the effect of this on chain folding. Mandelkern [12] also suggested that chain tilt angle is an important factor in any detailed description of the mechanism of crystallization and that it also influences the interfacial free energy associated with the basal plane. Bassett et al. [13] and Keith and Padden [14] discussed the role of tilted growth of polymer crystals in allowing better packing and enhanced space for the accommodation of disordered conformations and relatively bulky loops within interfacial layers. Recently, Toda et al. [15] have conclusively shown that the direction of chain tilting in polyethylene single crystals is different for each growth sector – which results in a selected handedness to the spiral terraces during crystal growth.

Using off-lattice Monte Carlo simulations, Rutledge and co-workers [6, 16] have reported a recent, detailed study on the tilted-chain interphase. Their simulations reveal the thermodynamic properties for a metastable interphase with different degrees of chain tilt. The interphase was considered to contain three types of chain populations, as shown in Fig. 15.1. Although the density decreases as one moves from the crystalline to the amorphous phase, an increase or peak in the density is observed around 0.6 nm away from the crystal surface. In the same manner, the order parameter, which starts at unity in the crystal, drops smoothly to zero as one moves into the amorphous except for a marked drop at a distance of $\sim$0.6 nm from the crystal face. The results from density and orientational order were further supported by the occurrence of a transverse structure parallel to the crystal surface. This is explained by a higher number of loops in the interphase, while tails and bridges contribute more to the amorphous region, with a well defined fold surface near 0.6 nm, at which a large number of chains fold back into the crystal. The transverse structure observed indicates a number of chains running parallel to the crystal surface. Since a large areal density (or flux) of chains leave crystal surface, the interphase generates a fold surface in order to reduce the flux to a level sufficient to accommodate disordering. Crystallographic planes could be assigned to the fold surface.

The findings described above form a basis for our current work examining the influence of the interphase on the structure and ultimately on the properties of linear polymers. In this chapter, we will first show that depending on the crystallization conditions, the amount of loops or entanglements in the interphase, and thus the deformation behavior of polymers, can be changed and controlled. We will initially consider the example of Ultra High Molecular Weight Polyethylene (UHMW-PE) and the role of entanglements upon its drawability.

15.2 Control of Entanglement Density Upon Crystallization

As we have outlined above, it is important to consider the role of entanglements within the interphase. Entanglements as such are ill-defined topological constraints, which are usually visualized in textbooks as four strands leading away from a mutual contact, see Fig. 15.2.

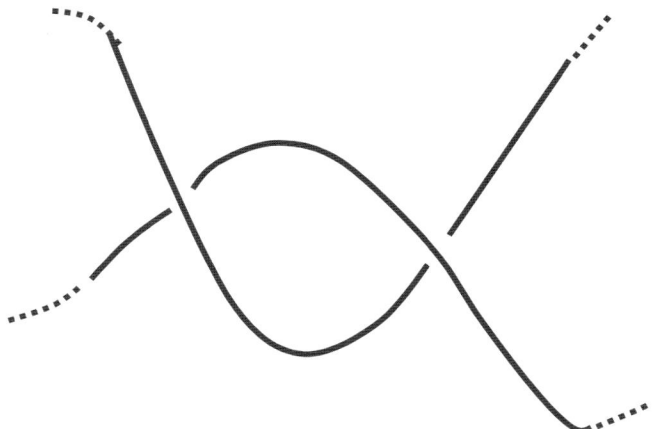

Fig. 15.2. Simple picture of an entanglement

Entanglements can be removed effectively by dissolution in a good solvent or by slow (isothermal) crystallization from the melt or from solution. Slow crystallization promotes disentangling since the process of "reeling-in" of chains from the entangled melt onto the crystal surface is promoted and this in turn enhances the maximum drawability, despite the fact that this increases the crystallinity! However, there is a critical lower limit for the number of entanglements between crystals in order to achieve high drawability and in the extreme limit of very slow isothermal crystallization, linear standard polyethylenes (not the UHMW-PE grades) can loose their drawability completely and become brittle materials due to lack of coherence in between the individual crystals.

15.2.1 Crystallization via Dilute Solution

The way to remove entanglements, viz., the manner in which topological constraints limit the drawability, is seemingly well understood and crystallization from semi-dilute solution is an effective and simple route to make disentangled precursors for subsequent drawing into fibers and tapes [17, 18]. A simple 2D model visualizing the entanglement density is shown in Fig. 15.3. Here φ is

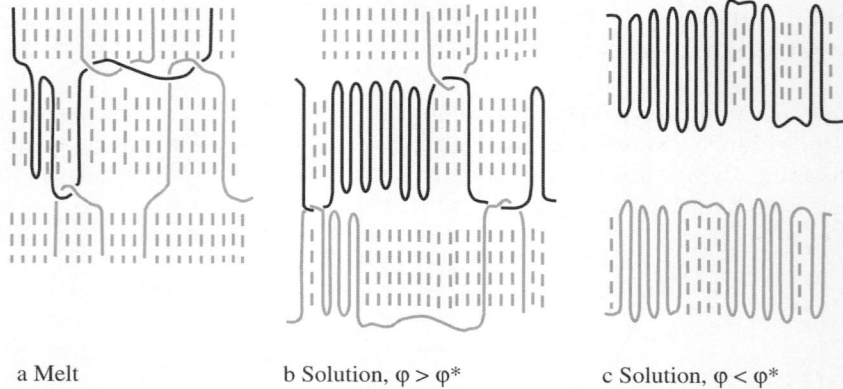

Fig. 15.3. A simple 2D model envisaging how the entanglement density varies upon crystallization at decreasing polymer concentration, φ. φ^* is the critical overlap concentration for polymer chains

the polymer concentration in solution and φ^* is the critical overlap concentration for polymer chains. The entanglement density can be controlled by the (inital) polymer concentration in solution or stated otherwise, the average molecular weight between entanglements in solution, $\langle M_e \rangle_{sol}$ scales with $\langle M_e \rangle / \varphi$. Upon crystallization, the entanglements are trapped in between the crystals as shown in Fig. 15.3.

In a crosslinked system, the maximum drawability, $\lambda_{\max}$, scales with $\langle M_c \rangle^{0.5}$, where $\langle M_c \rangle$ is the average molar mass between two crosslinks. Assuming that trapped entanglements act as physical crosslinks and no chain slippage occurs, the maximum drawability upon crystallization from solution, $\lambda_{\max}$, scales with $\langle M_e \rangle / \varphi$, viz. from a 1% solution, the maximum drawability is enhanced by a factor of 10. Of course, additional dis-entangling can occur during crystallization as well as chain slippage, but this simple model of relating maximum drawability to trapped entanglements holds surprisingly well [17].

The simple picture for an entanglement as shown in Fig. 15.2, in which the chains are hooked together, is a simple model which suffices to explain the strongly enhanced drawability of UHMW-PE, cast or spun from solution as shown in Fig. 15.5 but can not explain, for example, the immediate loss in drawability upon melting and re-crystallization of solution-cast UHMW-PE. The fast decay in drawability can be understood by a local intermolecular re-arrangements of stems as shown in Fig. 15.4. Upon crystallization from notably dilute solutions, adjacent re-entry is promoted and chains are located within one crystal plane. Unfolding along the crystal planes is rather easy in view of the low shear moduli of polyethylene crystals. Upon heating and re-crystallziation, a relatively small displacement of stems can create a situation which is completely different and sharing/unfolding is now hindered by

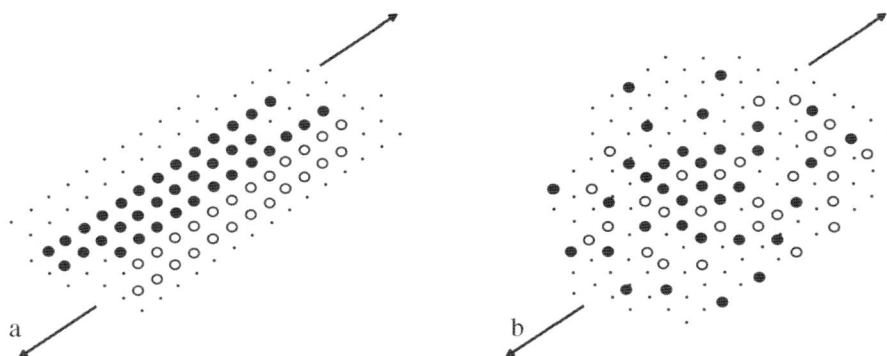

Fig. 15.4. Simplistic representation of a chain fold in a polymer crystal. The dots represent stems of molecules folding along the {110} plane, viewed along the c-direction. No folds are drawn for the sake of simplicity. This figure has been taken from reference Lemstra, PJ, van Aerle NAJM, Bastiaansen CWM (1987) Polymer Journal, 19, 85

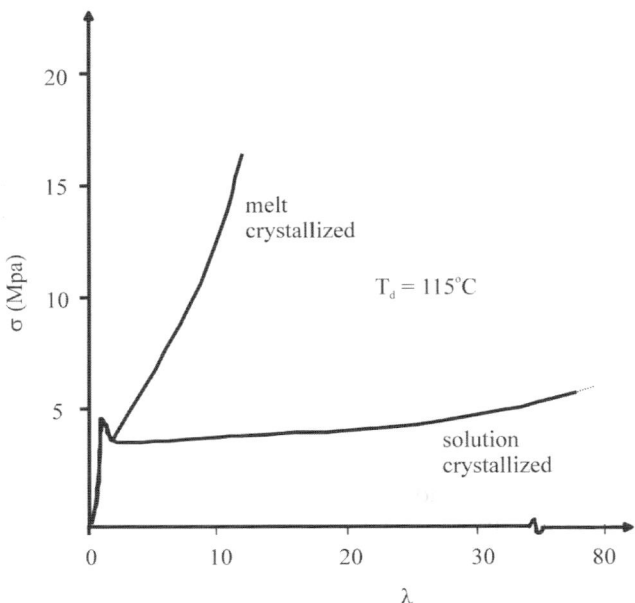

Fig. 15.5. Stress strain curves for solution cast UHMW-PE films compared to melt crystallized films. λ refers to the draw ratio. Films were drawn at drawing temperature $T_d = 115°C$

crossing-over of chains in different crystal planes, see Fig. 15.4b. In the model shown in Fig. 15.4, no entanglements are shown since the drawability and loss of drawability is related to the arrangement of stems within the crystals and not via trapped entanglements in between the crystals.

Independent of the crystallization route taken, the role of entanglements upon the mechanical properties will be the same, i.e. acting as physical crosslinks on the time-scale of drawing experiments of semi-crystalline polymers. This can be demonstrated by comparing the deformation of solution cast UHMW-PE films to melt crystallized films. From the stress-strain curve shown in Fig. 15.5, it is evident that above the α – relaxation temperature and below the melting temperature, solution cast films do not show strain hardening and can be drawn to more than 40 times the original length. In contrast, melt crystallized films show strain hardening and could be drawn only by six to seven times. This difference in the strain-hardening behavior can be attributed directly to the difference in entanglements present as described above.

The schematic representation of stems within the crystals is, of course, an oversimplification to explain the drawing behavior of UHMW-PE films. In actual practice, superfolding and the crossing of stems belonging to the same chain will occur. The presented model, however, serves to demonstrate that adjacent re-entry and the locality of molecules within a crystal will cause a structured interphase to form. This will facilitate the process of ultra-drawing, comprising the breaking of lamellar crystals via shearing, tilting and subsequent unfolding of clusters. The subsequent instantaneous loss in drawability upon melting and re-crystallization is due to the re-arrangement and intermixing of stems involving only local chain motions rather than movement of the complete chains as proposed for self-diffusion in polymer melts.

15.2.2 Exploitation of the Hexagonal Phase in Polyethylene

Another way to disentangle linear polyethylenes, and thus to control the interphase without the need of using a solvent, is to anneal the polymer in the hexagonal phase. Bassett has discussed the role of the hexagonal phase in the crystallization of polyethylene extensively in an earlier chapter in this book. Briefly, polyethylene exhibits a number of different crystal structures, with the hexagonal phase being observed in linear polyethylenes at elevated pressure-temperature in isotropic samples or at ambient pressure in oriented samples. For this reason, we have to distinguish between these two situations, namely isotropic and oriented polyethylene, however, we will focus only on isotropic polyethylene and will refer readers to reference [18, 19] for an overview of oriented polyethylene.

For isotropic polyethylene, the hexagonal phase is usually observed at elevated pressure and temperature, in fact, above the triple point Q located at 3.4 kbar and 220°C according to the pioneering work of Bassett et al. [20] and

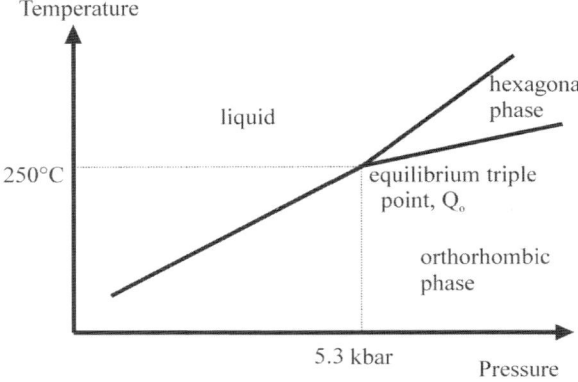

Fig. 15.6. Pressure-Temperature phase diagram for polyethylene

Wunderlich et al. [21]. Later, more detailed studies involving in-situ light microscopy and X-ray studies showed that the equilibrium point, Q_o, is located at even higher temperatures and pressures, 250°C and 5.3 Kbar respectively, see Fig. 15.6 [22–24]. The hexagonal phase is a so-called "mobile phase" with a high degree of chain mobility along the c-axis. Annealing in the hexagonal phase promotes chain re-folding and the formation of extended chain crystals via disentangling of chains and consequently, aids drawability. In the limit of full chain-extension, all entanglements disappear and the material can not be drawn anymore because it becomes brittle due to lack of coherence via trapped entanglements, rather similar to the case of crystallizing from dilute solutions below the overlap concentration, albeit with a completely different morphology, extended vs. folded-chains. For UHMW-PE (according to ASTM definitions $M > 3.10^6$ D) it is therefore possible to disentangle the chains even in the solid state by exploiting the hexagonal phase. The resulting organized molecular structure in the amorphous phase enhances the subsequent drawing operation as is shown by Ward et al. [25].

15.2.3 Via Synthesis

A much more elegant route to obtain disentangled UHMW-PE, and thus to control the interphase, is by direct polymerization in the reactor. In order to make UHMW-PE, a relatively low polymerization temperature is needed and a situation is easily encountered where the polymerization temperature is lower than the crystallization temperature of UHMW-PE in the surrounding medium in which the catalyst is suspended. In this situation, the growing chains on the catalyst surface tend to crystallize during the polymerization process. These UHMW-PE reactor powders, often referred to as "nascent" or "virgin" UHMW-PE, can be remarkably ductile. It was shown by Smith et al. [26] that reactor powders, in the same manner as solution cast UHMW-PE, could be drawn easily into high-modulus structures.

Again the proposed entanglement model can be invoked to explain the ductility of compacted UHMW-PE reactor powders. The growing chains on the crystal surface can crystallize independently of each other and consequently disentangled UHMW-PE is obtained as a direct result of polymerization followed by crystallization. In the limit of low catalyst activity or when using single site homogeneous metallocene based catalyst in solution, one could envisage that one molecule will form its own single crystal, viz monomolecular crystal.

Depending on the polymerization conditions, the crystal size of the nascent morphology can be also controlled. As with the melting point dependence of polyethylene on the crystal thickness (fold length), the triple point, Q, in the pressure-temperature phase diagram of polyethylene is also dependent on the crystal dimensions. In particular, for nascent UHMW-PE reactor powders that consist of crystallites with very small dimensions, it was shown that a metastable hexagonal phase could be observed at pressures and temperatures as low as 1 Kbar and $200°C$, respectively [27, 28] and upon annealing a transformation into the thermodynamically stable orthorhombic phase occurs. The observation that a thermodynamically stable crystal structure is reached via a metastable state of matter is not unique for polyethylene, nor for polymers, indeed it has been invoked as early as in 1897 by Ostwald, commonly expressed as Ostwald's stage rule [29]. For polyethylene, it has been shown that crystals, at elevated pressures, initially grow in the hexagonal phase and after a certain time or once a certain crystal size has been attained, these hexagonal crystals are transformed into thermodynamically stable orthorhombic crystals.

Figure 15.7 shows electron micrographs of two different virgin UHMW-PE's. Electron micrographs clearly show that the lamellae in the laboratory synthesized sample, A, thicken substantially for the same annealing conditions, for example pressure, temperature and time, compared to a commercially synthesized grade, sample B. There is a marked difference in their drawability; the laboratory grade can be drawn in the solid-state, below the melting point, whereas the commercial one cannot be. Details of molecular weight, molar mass distribution, polymerization temperature are given in Table 15.1, however, the difference in drawability arises due to the reduction in the number of entanglements at the interphase of the laboratory grade sample compared to those present in the commercial grade. This highlights once again the influence of entanglements and the interphase in these materials and their effect on the molecular chain mobility even during lamellar thickening. Influence of entanglements on lamellar thickening has been probed further by solid state NMR.

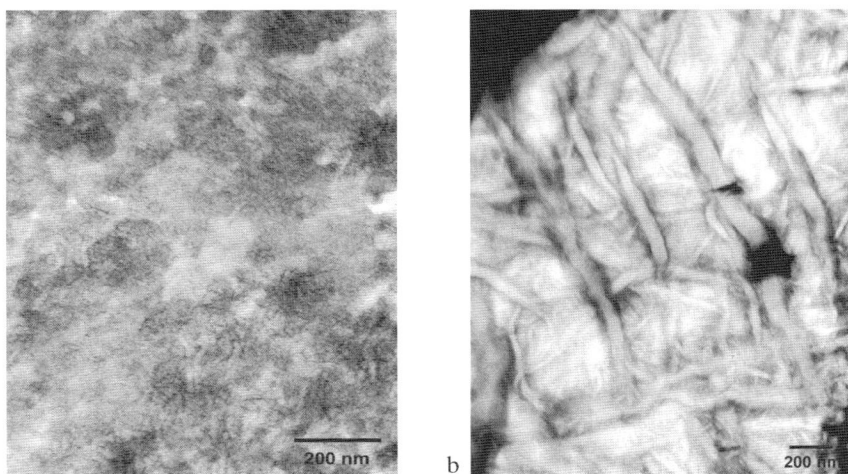

Fig. 15.7. Electron micrographs of two different virgin UHM-PE samples, (**a**) commercially synthesized Z-N (grade B) and (**b**) laboratory synthesized Z-N (grade A). The samples were annealed in the hexagonal phase at 1500 bars, 190°C for 30 minutes. Under these conditions the samples were thermodynamically metastable hexagonal phase (For details see references [27,28])

Table 15.1. Molecular weight, molecular weight distribution, polymerization temperature

	M_w [g mol^{-1}]	M_w/M_n	$T_{synthesis}$
Controlled synthesis, Z-N (*Grade A*)	3.6×10^4	5.6	50°C
Commercial, Z-N (*Grade B*)	4.54×10^4	10.0	80°C

15.3 Influence of the Interphase on Molecular Mobility in Crystalline Domains

Recently, with the help of solid state NMR, Uehara et al. [30] investigated the role of entanglements present at the interphase on molecular chain mobility along the c-axis of the crystal lattice of UHMW-PE. They used solution- and melt-crystallized films and nascent powders and observed regular lamellar stacking in the solution crystallized films, whereas the nascent powders and melt crystallized samples showed conventional non-stacked lamellar morphology. By ^{1}H pulse NMR measurements, they defined three different relaxation processes occurring during heating. In process 1 (heating from room temperature), activation of molecular motion at the boundary between crystal/amorphous regions takes place. During process 2 (above 60-90°C i.e. the α–relaxation temperature), the crystallinity increases with an acceleration of the entire molecular motion caused by sliding of molecular chains in the crystalline region. Raising the temperature further above 130°C (process 3) leads to the start of sample melting. For solution-grown crystals and nascent powder samples, the crystalline relaxation exhibited all three processes as well as constraint of the amorphous chains; however, the transition between processes 1 and 2 occurred at a higher temperature for the nascent morphology. In contrast, melt-grown crystals did not show process 1 and directly led to process 2 upon heating. This therefore suggests that the accelerated molecular motion in the crystal/amorphous boundaries occurs following lamellar thickening via a solid state reorganization, i.e. without melting, for the highly crystalline nascent and solution crystallized samples having fewer entangled chains in the amorphous phase. For melt-crystallized samples, the higher density of entanglements trapped on the lamellar surface precludes such a boundary relaxation; thus, lamellar thickening may not occur. These results imply that the trapped entanglements also play an important role in lamellar rearrangement during annealing such as might occur during welding of semi-crystalline polymers.

15.4 From the Interphase to the Interface: The Welding of Semi-crystalline Polymers

By considering the influence of the interphase upon annealing, it is possible to shed some light on the welding behavior of semi-crystalline polymers which has received much less attention than that of amorphous polymers. Because of the ill defined morphology of the interphase the welding characteristics of semi-crystalline polymers are quite different from amorphous polymers and are far from being understood.

In the case of solution-crystallized UHMW-PE, it is possible to make solution-cast films in which the lamellar crystals are regularly stacked, see Fig. 15.8. Upon heating these (dry) solution cast films above approximately 110°C, it has been observed, by in-situ synchrotron measurements and time-resolved Longitudinal Acoustic Mode Raman spectroscopy [31], that the lamellar thickness increases finally to twice its initial value, from 12.5 nm to 25 nm, with the loss of the well-stacked lamellar arrangement after the doubling process. Figure 15.8b shows the model for the chain re-arrangement during heating. However, the drawability of the annealed solution-crystallized films was maintained after lamellar doubling since no stem intermixing, or entangling, occurs as discussed previously in the stem-rearrangement model, Fig. 15.4.

A well-defined amount of co-crystallization is possible across the interface of two adjacent crystals by annealing two stacked, completely wetted, solution cast films of UHMW-PE [32]. It was found that doubling of the lamellae across the interface enhances the peel energy to such a level that the films could not be separated anymore. By contrast, "pre-annealing" one side of the

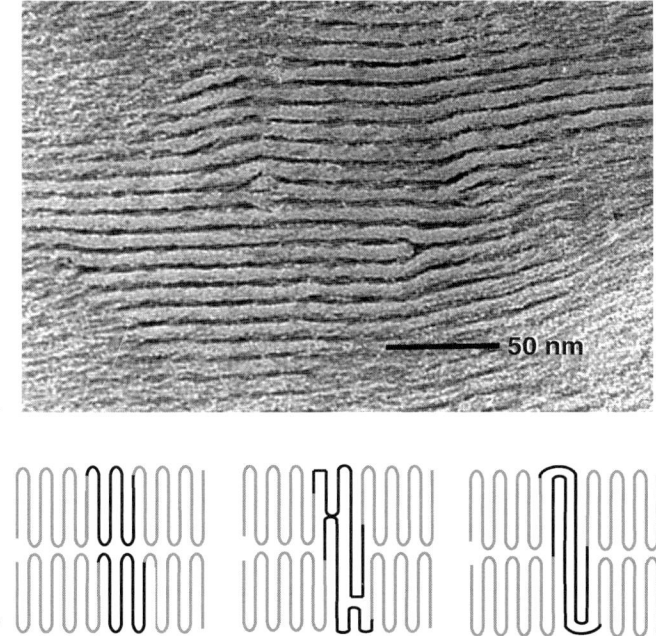

Fig. 15.8. (a) Electron micrograph and (b) schematic representation for the regular lamellar structures formed within solution-cast films of UHMW-PE

film prohibited co-crystallization across the interface and these films could still be separated easily. It was therefore concluded that a limited amount of chain diffusion across the interface occurs during doubling of the lamellae, as facilitated by the well-defined structure of the interphase due to adjacent re-entry that occurs upon crystallization from solution.

Thus far we have considered the influence of the interphase in relation to the mechanical properties of semi-crystalline linear polymers, in particular for the case of UHMW-PE, i.e. its role in the drawability or in the welding of such materials. With the recent advent of synthesis routes to produce highly controlled model systems, the investigation of the interphase can be extended further.

15.5 Influence of Chain Folding on the Unit Cell

Normally, for linear polymers it is anticipated that extended chain crystals are thermodynamically the most favorable, and ultimately given an example of linear polyethylene, it is shown that chains within the folded chain crystals tend to move along the c-axis via chain sliding diffusion to attain a thermodynamically stable morphology. However, the possibility of chain diffusion within the crystal lattice cannot be realized normally when the molecular structure changes from linear to branched, in particular, where the side branches cannot be incorporated within the crystal. Below, we have summarized our recent studies on such branched polymers and will demonstrate a thermodynamic stable morphology that these polymers ultimately tend to achieve. To reach the objective, it is essential to recapitulate first our findings on model systems, for example, linear and branched ultra-long alkanes – which are essential for an understanding of the experimental observations of branched polyethylenes, for example, ethylene – octene copolymers.

It has been known since the early 1970s that the thickness of polyethylene lamellae can have some influence on the lattice parameters of the unit cell, with a tendency towards higher density at larger crystal thickness. Further studies of this effect are hindered in the case of high polymers by the polydispersity of the materials. Polymers are mixtures of chains of different lengths and the lack of purity is likely to lead to a higher level of defects than would occur in a pure system. This gives a very strong dependence of structure on the age of the crystal and the way in which it was crystallized. During the past two decades ultra-long monodisperse alkanes, with chain lengths up to 390 carbons, have become available. These materials, which are model substances for low molecular weight polymers, have provided many new insights into polymer crystallization in general and polyethylene crystallization in particular.

15.5.1 Monodisperse Ultra-long Linear Alkanes

Ungar and Zeng [33] have comprehensively summarized the research on strictly monodisperse materials from their first synthesis in 1985 until 2001. From the earliest studies it became apparent that, due to the monodispersity of the materials, the thickness of lamellar crystals formed are always an integer fraction of the extended chain length (allowing for any chain tilt), such that the polymers always crystallize in the extended chain form or fold exactly in half (once-folded), or in three (twice-folded), etc. This behavior means that, when the alkanes are crystallized at a particular temperature, the entire lamellar population has very closely the same thickness and stability. The use of such an ultra-pure system to study the impact of thickness on lattice parameters removes many of the problems inherent to polymers, whilst maintaining the most important characteristic of chain length.

What follows is a brief overview of wide-angle X-ray diffraction study using the high-resolution time resolved capabilities of the European Synchrotron Radiation Facility to investigate the effect of chain folding and unfolding on lattice parameters in a series of five ultra-long alkanes of different chain lengths, $C_{102}H_{206}$, $C_{122}H_{246}$, $C_{198}H_{398}$, $C_{246}H_{494}$ and $C_{294}H_{590}$ [34]. Both $C_{102}H_{206}$ and $C_{122}H_{246}$ can only be obtained in the extended chain form under typical dilute solution crystallization conditions, all the other alkanes undergoing

Table 15.2. The ideal crystal thickness for differently folded forms of the alkane samples studied, assuming six carbons per fold

Alkane sample	Crystal thickness, nm
$C_{102}H_{206}$ extended	13.2
$C_{122}H_{246}$ extended	15.7
$C_{198}H_{398}$ extended	25.4
$C_{198}H_{398}$ once-folded	12.0
$C_{246}H_{494}$ extended	31.52
$C_{246}H_{494}$ once-folded	15.1
$C_{294}H_{590}$ once-folded	18.2
$C_{294}H_{590}$ twice-folded	11.2

chain folding under some crystallization conditions. Table 15.2 shows the samples used and the crystal thicknesses obtained assuming six carbons per fold (these thicknesses agree well with the thicknesses measured on similarly prepared crystals using Raman LAM and small angle X-ray scattering). The effect of heating at 2°C/min upon the lattice parameters of these alkanes was examined.

Examining the lattice parameters in detail by fitting the (110) and (200) peaks provides three separate but inter-related sets of information. Firstly, as expected, the lattice expands on heating due to the increased thermal motion of the chains. In the case of the extended chain crystals, which are already close to equilibrium (at least with respect to size), this is all that happens, Fig. 15.9. Constant thermal expansion of the a and b lattice parameters can be seen, the coefficients of thermal expansion derived from this data are 4×10^{-3} Å/°C for the a-axis, and -3×10^{-4} Å/°C for the b-axis, these basically remaining constant for each of the chain lengths examined and, as expected, the final melting point is at a higher temperature for the longer chain molecules. Secondly, if the crystals consist of folded chains, the thickening of these crystals that occurs above 120°C is accompanied by a contraction in the crystal lattice. Figure 15.10 shows the contraction in lattice parameters for the once-folded form of $C_{246}H_{494}$, as shown previously for the extended chain form in Fig. 15.9. The contraction during thickening is very striking, and even more so in the case of $C_{294}H_{590}$ where it occurs on the transition both from twice-folded to once-folded crystals, and at the transition from once-folded to extended chain crystals, Fig. 15.11. This is the first time that such a transition has been followed in real-time using WAXD, although the transformation to a less folded form had been measured previously by DSC. The contraction of the lattice that accompanies unfolding is superimposed on top of the thermal expansion. Finally, prior to this contraction, there is an increase in the full width at half maximum (FWHM) of each of the crystal Bragg peaks over a temperature range of several degrees, which can be associated either with an

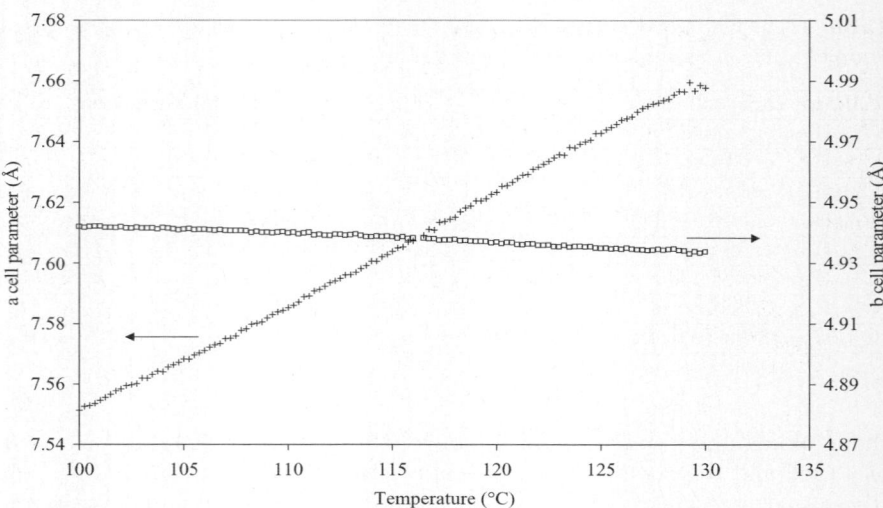

Fig. 15.9. The variation in the a (+) and b (□) cell parameters for $C_{246}H_{494}$ **extended chain** crystals. The *left hand* axis refers to the a cell parameter and the *right hand* axis to the b cell parameter. Just prior to melting the cell parameters drop probably due to an increase in the error as the peak intensity rapidly drops to zero, reducing the accuracy of the peak fitting routine

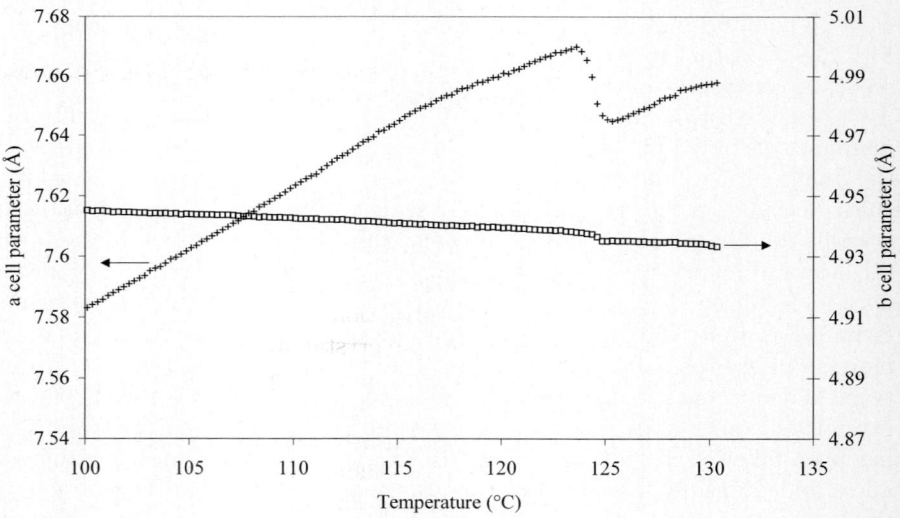

Fig. 15.10. The variation in the a (+) and b (□) cell parameters for $C_{246}H_{494}$ **once-folded chain** crystals. The *left hand* axis refers to the a cell parameter and the *right hand* axis to the b cell parameter

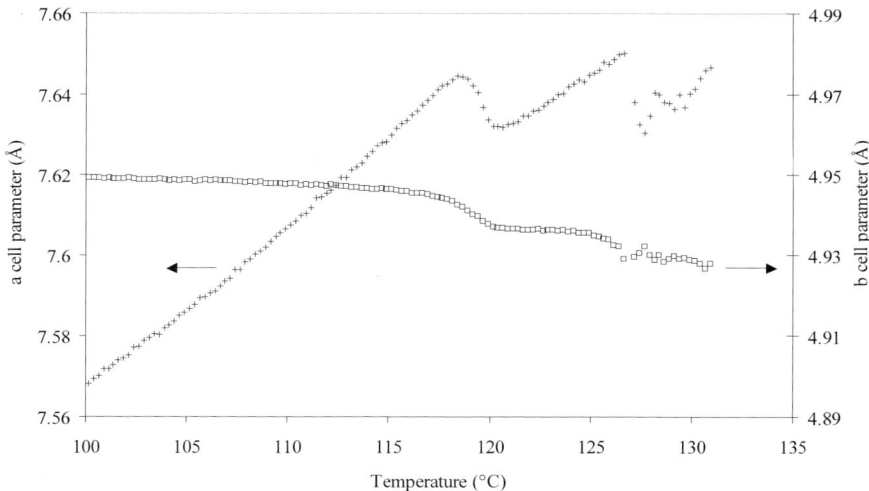

Fig. 15.11. The variation in the a (+) and b (□) cell parameters for $C_{294}H_{590}$ **twice folded chain** crystals. The *left hand* axis refers to the a cell parameter and the *right hand* axis to the b cell parameter.

increase in local strain in the crystals, or a reduction in crystal size in the lateral direction.

The use of different alkanes with different crystal thicknesses and numbers of folds within the crystal enables us to assert with confidence that the contraction in the lattice is due to the thickening process. The fact that the alkanes, due to their strict monodispersity, form crystals with only a few closely defined thicknesses enables this contraction to be seen much more clearly than it is in polyethylene. In polyethylene a range of different crystal thicknesses may be present, and the thickening process generally occurs over a wider range of temperatures, smearing out any step-like effect that may exist.

It may be concluded that a rapid contraction of the crystalline lattice with morphological changes at the surface of the crystal, and hence the interphase, occurs as ultra-long, monodisperse alkanes undergo a transition between different integer folded forms on heating and that this contraction takes place in all the materials studied, despite apparently different routes being taken between achieving the pre- and post-transition crystal forms. The different lattice parameters can be associated with a particular crystal thickness and fold surface density and the observed contraction of the lattice is remarkably clear evidence of the effect of surfaces on the polymer crystal structure.

The extent to which the thickening process occurs by solid-state reorganization of the parent crystals or by melting and recrystallization into thicker crystals is still a matter of discussion and probably depends on the polymer in question and the annealing temperature. Chain mobility within lamellae has been investigated by solid-state NMR [35]. Recent studies have also pointed to the possible role of mobile phases, such as the hexagonal phase in polyethylene, enabling the thickening process to occur. Computer simulations have also been used to address these issues, although this is still hampered by the cooperative nature of the process and therefore the requirement for a large simulation size. In the WAXD study of lamellar thickening by Terry et al. [34], the intensity of the diffraction peaks allows any changes in the degree of crystallinity to be monitored. It was found that in the case of unfolding of $C_{198}H_{398}$ from once-folded to extended, and the unfolding of $C_{294}H_{590}$ from twice-folded to once-folded, there was little change in the intensity during the transition. This implies that the transition must be a solid-state process, or that any melting is very localized. However, other transitions, for example the unfolding of $C_{294}H_{590}$ from once-folded to extended, showed that the material almost melts completely and then recrystallizes into the thickened form. This difference in behaviour must reflect the difference in energetic barrier in the different cases, perhaps due to differences in the degree of thickening required in the transition between the different forms and also perhaps to differences in the initial lamellar thickness.

Thus far, we have only considered studies using simple, highly crystalline linear n-alkanes. Introducing short chain branches into the amorphous region of the crystals further influences the interphase and the corresponding lattice parameters. Two monodisperse branched n-alkanes were also synthesized by Brooke et al. [36], $C_{96}H_{193}CH(R)C_{94}H_{189}$ where $R = CH_3$ and $(CH_2)_3CH_3$, i.e. a methyl and a butyl branched alkane, respectively. The crystallization of polyethylene copolymers, for which these branched n-alkanes serve as an analogue, is highly complex [37]. Some of the key differences of the molecular organization of branched polymers compared to linear alkanes during crystallization are now described.

15.5.2 Monodisperse Ultra-long Branched Alkanes

When examining the crystallization behavior of branched alkanes, consideration should also be given as to whether the branches due to their length will be excluded to the crystal surface, as we have mentioned earlier. It is generally accepted that methyl branches can be incorporated at interstitial sites leading to distorted lamellae [38–40], whereas hexyl branches are definitely rejected from the crystalline core. Whether ethyl branches are included or not depends upon the cooling rate used for crystallization; at higher cooling rates, the possibility of incorporation into the crystal is increased. A further complexity is that the ethylene sequences, if sufficiently long, will be able to fold, however, this is unlikely to be a tight fold, rather a longer fold is envisaged in order to keep the comonomers out of the crystal. As the concentration of comonomers increases, so the lamellar thickness will decrease and overcrowding of the branches in the amorphous region will result [7, 41, 42].

This overcrowding is relieved either by the lamellae curving, the chains in the crystal tilting or even by limiting the lateral size of the crystals. Consequently, thin, imperfect crystals are formed under such conditions.

Detailed SAXS studies using synchrotron radiation on both monodisperse linear and branched n-alkanes have been performed by Ungar et al. [43–46] under isothermal and/or non-isothermal conditions at atmospheric pressure. The investigations unveil and explain the complex chain folding process during crystallization. It was observed that the crystals formed in the early stages of crystallization have lamellae with the thickness of a non-integer fraction (NIF) of the extended form. This transient form soon transforms into an integer form, either via thickening at high annealing temperatures (to the extended form) or thinning at low annealing temperatures (to the once folded form, F2). There is a lower entropic barrier for the random attachment of the chain in the NIF as compared to that for once-folded crystals where all end groups must be positioned at the crystal edge before deposition, and so the chain prefers the fast random attachment. There is evidence to suggest that this disordered layer contains long uncrystallized chain ends (cilia) from molecules which have "half crystallized" and which are subsequently drawn into the crystals by a process of chain translation. This implies that there must be a high degree of chain mobility even within the crystalline lattice to allow such reorganisation of the chains within the growth front, such mobility as was observed during the morphological changes that occur during crystal thickening of linear n-alkanes.

In the past few years, Ungar et al. have elucidated the structure and mechanism of the formation of the NIF form using SAXS (including electron density profiles), Raman longitudinal acoustic modes (LAM) spectroscopy and Differential Scanning Calorimetry (DSC). A comparison of the crystallization mechanism of the linear and branched alkanes [47] indicated that in the sequence, melt $\rightarrow$ NIF $\rightarrow$ F2, the melt $\rightarrow$ NIF step is fast but the NIF $\rightarrow$ F2 step is slow in linear alkanes. The crystallization mechanism is understood in the following manner. In a NIF lamella of a linear alkane, a half-crystallized molecule generally has two cilia. During crystallization, neither of the two cilia is long enough to make a complete adjacent reentry, provided no rearrangements in the already crystallized part occur. This retards the formation of F2 crystals. However, the reverse is true for the branched alkane. An isothermal crystallization study of the methyl branched alkane $C_{96}H_{193}CH(CH_3)C_{94}H_{189}$ revealed that the high rate of NIF $\rightarrow$ F2 transition is attributed to the fact that, here, the only successful deposition mode of the first stem (first half) of the molecule is the one which places the branch at a lamellar basal surface and the chain end at the opposite surface of the same lamella. The other half of the molecule (uncrystallized cilium) is then ideally suited to complete a second traverse of the crystal [47].

To gain a better insight into the crystallization mechanism and resulting crystal structures of the branched alkanes with respect to the branch length, the phase behavior of the butyl branched alkane $C_{96}H_{193}CH(C_4H_9)C_{94}H_{189}$

has been investigated at elevated pressures. A well-defined morphology is expected where the chains are adjacent reentrant, due the enhanced chain mobility along the c-axis at elevated pressures and temperatures. If we envisage the folding of a single molecule then the branch, which occurs exactly after 96 C-atoms and is followed by 94 C-atoms, will lie almost in the middle of the fold. Variations in the crystal structure have been followed in-situ with WAXD while the morphological aspects have been investigated using in-situ SAXS. The high-pressure measurements were done using a piston cylinder type pressure cell similar to the one designed by Hikosaka and Seto [50] capable of attaining a maximum pressure of 5.0 kbar and temperatures from room temperature to 300°C.

At atmospheric pressure, the butyl branched alkanes show a similar packing to linear alkanes and linear polyethylene. Orthorhombic packing (and monoclinic due to shear in the sample) is maintained even though the branches will be excluded to the lamellar surface. For the as-synthesised sample, the lamellar spacing corresponds to chains that are once-folded and perpendicular to the basal plane, i.e. the chains are not tilted. Upon heating, some of the chains will begin to tilt with a chain tilt angle of approximately 35°, and upon cooling this tilted structure is retained. Non-integer folds were not observed. This suggests that during non-isothermal crystallization, the chains tend to resort to the once folded structure and one can assume that the butyl branch is excluded to the surface.

Initially, if one applies a pressure of 3.8 kbar to the sample, no change in behavior is seen compared to that observed in linear alkanes [51]. The intensity of the diffracted pattern will decrease but that is purely to the thickness of the sample decreasing with increased pressure. As the sample is heated, the monoclinic phase will disappear first at approximately 160°C, with a corresponding increase in the intensity of the orthorhombic peaks. The sample finally melts at approximately 258°C.

Interestingly, the effects of the branches on the phase behaviour and the significance of the interphase were clearly shown by cooling from the melt at elevated pressures, Fig. 15.12. Upon crystallizing from the melt, crystal formation occurs directly in the orthorhombic phase. The orthorhombic (110) and (200) reflections gain intensity with increasing supercooling. A weak monoclinic reflection appears at approximately 148°C. With subsequent cooling at $\sim 70°C$, a relatively broad and weak new reflection appears next to the monoclinic reflection. These reflections have been assigned considering the earlier work by Hay and Keller [52]. The appearance of the new reflection is followed by a sudden drop in the intensity of the orthorhombic reflection and a simultaneous increase in the intensity of the monoclinic reflection. The presence of the new reflection becomes more evident with further cooling. The (110) and (200) orthorhombic peaks show a sudden shift to higher angles implying a densification of the orthorhombic crystalline lattice with the appearance of the monoclinic reflection and the new reflection. Figure 15.12b shows that after the appearance of the new reflection at $\sim 70°C$, a dramatic decrease along

15 Role of the Interphase in Chain Mobility and Melting   305

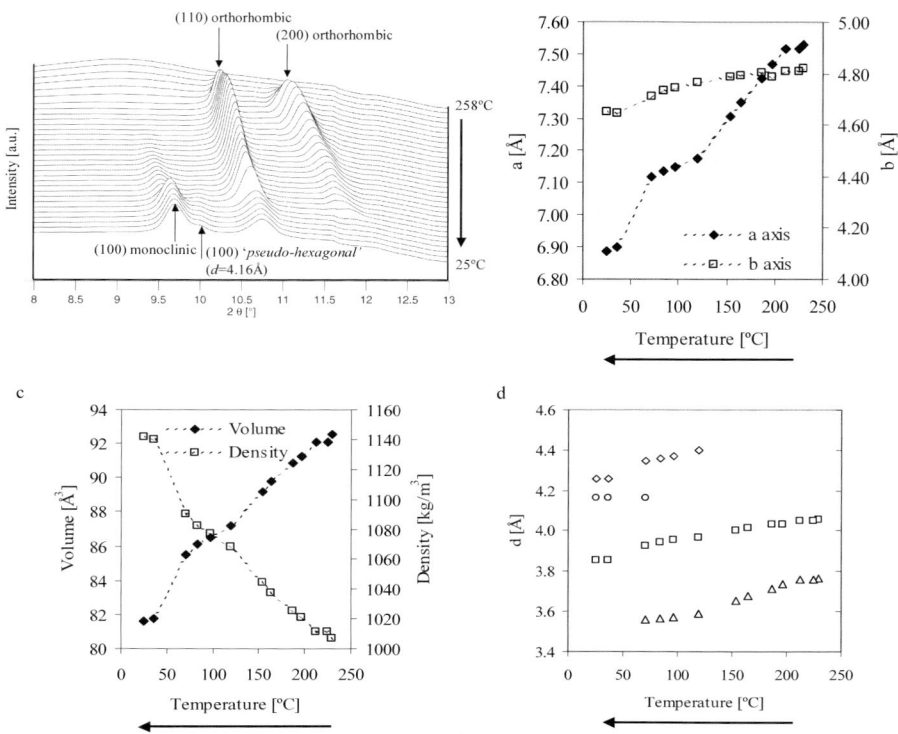

Fig. 15.12. (a) A series of integrated WAXD patterns for the crystallization of butyl branched alkane at elevated pressures of 4.0 kbar while cooling at a rate of 4°C/min. The orthorhombic (110) and (200) reflections appear first which gain intensity with increasing supercooling. A weak (100) monoclinic reflection appears at ca. 148°C. Upon further cooling at ca. 70°C, sudden drop in the intensity of the (110) orthorhombic reflection is observed with the appearance of a new reflection and a simultaneous increase in the intensity of (100) monoclinic reflection. (b) Plot showing contraction of the orthorhombic unit cell parameters during cooling at elevated pressures. (c) A linear decrease and increase in the volume and density of the orthorhombic unit cell is respectively observed during cooling at ca. 4.0 kbar. (d) Plot of Bragg d values of various crystalline reflections versus temperature during cooling at elevated pressure of 4.0 kbar: (100) monoclinic (◇), (100) pseudo-hexagonal (○), (110) orthorhombic (□), (200) orthorhombic (△). X-ray wavelength used for these experiments is 0.744Å

the a axis of the orthorhombic unit cell occurs together with a sharp decrease along the b axis. Densification of the orthorhombic unit cell ($\rho = 1141 \text{kg/m}^3$ at 4.0 kbar, 25°C) is evident from Fig. 15.12c.

Definite assignment of a phase to the new reflection is not straightforward as only one reflection is observed; for the present it has been termed a

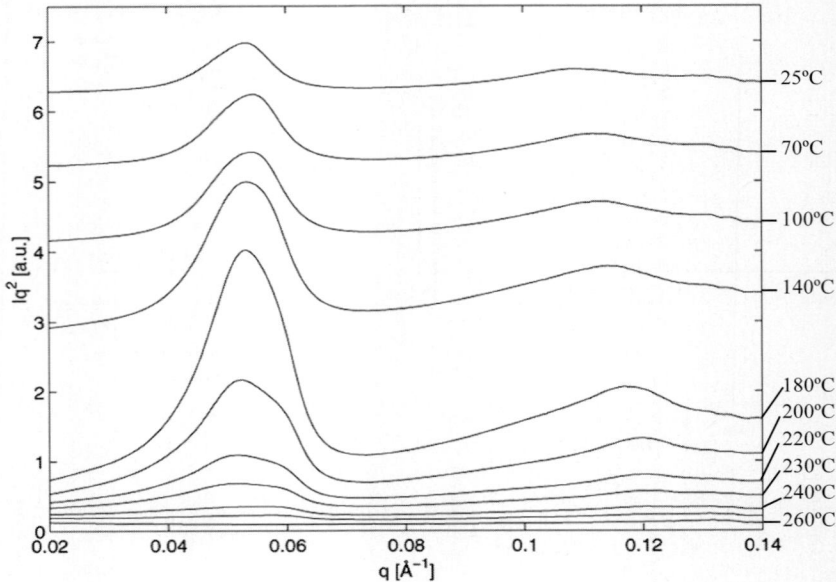

Fig. 15.13. Time-resolved SAXS patterns recorded during cooling at a rate of 4°C/min at 3.8 kbar show a relatively broad first order reflection at 118Å and a second order reflection at 58Å. An overall drop in SAXS intensity follows on cooling

pseudo-hexagonal phase as its spacing is close to that of the hexagonal phase in polyethylene (d = 4.16Å c.f. d_{100hex}(polyethylene) = 4.2Å). We already know that the butyl branches due to their length are rejected from the crystalline lattice and so are segregated to the fold surface of the crystal. It is suggested that with pressure, this new phase appears when these butyl branches crystallize together. The separation of the butyl branches from each other is greater than that of the main chain within the crystal due to the fact that a fold occurs at the surface. The observed densification of the orthorhombic unit cell therefore acts to relieve the strain induced at the surface by the expanded nature of the crystallization of the branches. This is supported by the observations by SAXS, Fig. 15.13, that the d-spacing of the once-folded form increases in value although the intensity is decreased at the same time as this new phase appears in the WAXD patterns. The crystallization of the butyl branches at the interphase of the crystalline and amorphous regions would lead to a decrease in the electron density difference between the crystalline and amorphous regions.

It is remarkable that the new phase, once it appears shows very little expansion or contraction whether caused by temperature or pressure. Again this must be because of the constraints imposed by the folds on the surface, fixing the separation of the branches. Indeed, the orthorhombic reflections can

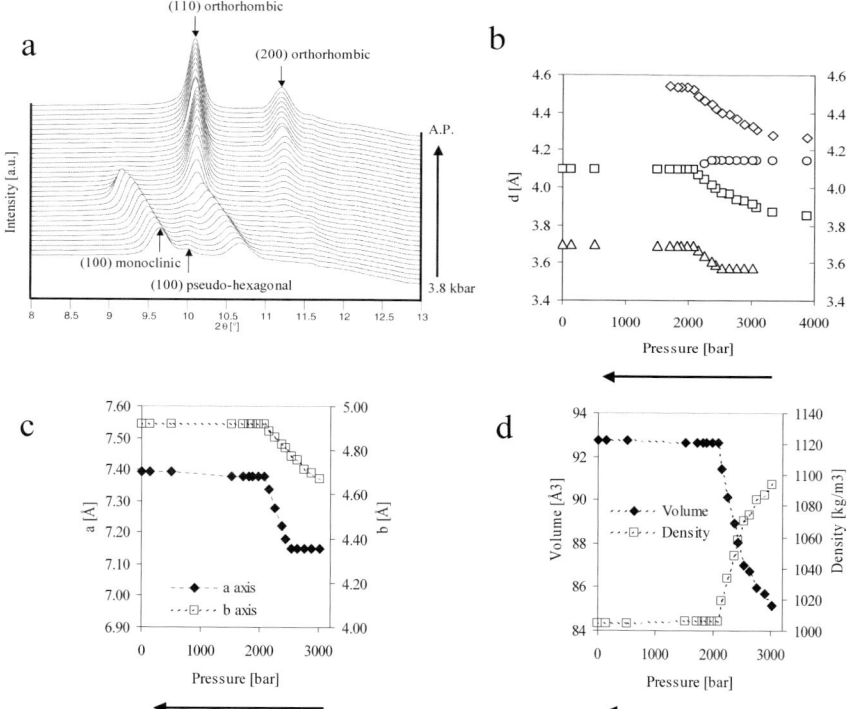

Fig. 15.14. (a) 3-D WAXD plot shows that upon releasing pressure at room temperature, (100) monoclinic and (110) orthorhombic reflections move to lower angles while the (100) pseudo-hexagonal reflection stays at the same position. The latter finally vanishes followed by the disappearance of the monoclinic reflection resulting in an overall increase in the intensity of the orthorhombic reflections. (b) The d values for the various crystalline reflections plotted against pressure. Symbols represent: (100) monoclinic ($\diamond$), (100) pseudo-hexagonal ($\circ$), (110) orthorhombic ($\square$), (200) orthorhombic ($\triangle$). Upon releasing pressure at room temperature, the d value of the (110) orthorhombic reflection increases from 3.85Å to 4.08Å while the $d_{100hex.} = 4.16$Å merges with that of the (110) orthorhombic reflection at ca. 2.0 kbar. At atmospheric pressure the d values for the various crystalline reflections attain values similar to those in linear alkanes. (c) The a and b axis of the orthorhombic unit cell increase and subsequently become constant below 2.0 kbar upon releasing pressure. (d) Volume and density of orthorhombic unit cell plotted as a function of pressure at room temperature. X-ray wavelength used for these experiments is 0.744Å

be seen to expand and contract. On heating or release of pressure, when the d-spacings of the orthorhombic (110) and the pseudo-hexagonal reflections reach the same value, set by the nearest neighbour separation determined by a fold, then no further expansion of the lattice is observed, Fig. 15.14.

15.5.3 Homogeneous Copolymers of Ethylene-1-Octene

In recent years, ethylene-1-octene copolymers with densities between 870 and 910 kg/m^3 have attracted considerable academic and industrial interest. One of the main characteristics of these copolymers is that they are "*homogeneous*" copolymers, that is, these polymers do not display any differences in comonomer distribution along the chains other than the differences related to statistical fluctuations. Copolymers such as LLDPE that do not meet the above definition are considered to be *heterogeneous*. LLDPE shows a superposition of multiple ethylene sequence length distributions(ESLDs). The difference between homogeneous and heterogeneous copolymers lies in the different process of polymerization resulting in quite different chain microstructures and ultimate product properties. The homogeneous copolymers are synthesized with the aid of a single set of reactivity values, corresponding to a single active catalyst site, resulting in a single peaked ESLD. Whereas, synthesis of heterogeneous copolymers involves the presence of two or more active catalyst sites, resulting in an overall multiple distributions of ethylene sequence lengths – causing multiple peaked ESLD. Crystallization kinetics and the subsequent morphology are strongly influenced by the specific chain microstructure – specifically the ESLD, which varies with the mole fraction of octene. Especially in the case of ethylene-1-octene copolymers, the chain microstructure largely affects the properties, since the ethylene units will form crystallites, while the octene units are likely to be excluded from the crystallites. Therefore, the exact distribution of the comonomer units along the chain, and consequently the way of stringing together of the ethylene sequences determines the crystallizability of that chain and hence the morphology and the ultimate properties. As is evident, the chain microstructure will also influence the chain conformation in the melt and in the solution, because the bond angles of the monomers and comonomers will differ.

Ethylene octene copolymers show a very interesting and intriguing crystallization behaviour and morphology upon increasing the comonomer content from zero to very high values. As has been stated above, most importantly for crystallization are the sequences of the crystallizable units present in the copolymer chains – ethylene in the present case. Even for homogeneous copolymers there is always a distribution of ethylene sequences and therefore, distributions corresponding to such copolymer properties as the crystallization temperatures, the resulting crystallite dimensions and melting temperatures. It is conclusively shown that by increasing the comonomer content the morphology changes from a lamellar one, organized in spherulites, to a granular based morphology. Clearly, the term "homogeneous" only applies to the way polymerization is performed. This kind of "heterogeneity" complicates studies of fundamental properties. In this section we will summarize some of our unpublished data, which clearly shows that pressure can be used as a thermodynamic component to segregate variable ESLDs present in the so-called homogeneous copolymers. By pressure-temperature crystallization cycles it is feasible to obtain a regular organization of branches on the crystal surface and adjacent chain-reentry, which may result in the appearance of a new crystalline phase similar to that observed in the case of branched alkanes [51]

In the above section we have shown that it is possible to control number of entanglements per unit chain in a flexible linear chain polymer like polyethylene. This could be achieved either from dilute solution or in melt by high chain mobility along the c-axis of unit cell or by polymer synthesis. Reduction in the number of entanglements, below a critical concentration, promotes drawability of intractable polymer like UHMW-PE having molar mass greater than a million. It is also shown that drawability of UHMW-PE strongly depends on chain re-entry at the interphase. Moreover on crystallization from melt, adjacent chain re-entry resulting into crystallographic registration of folds at the interphase can occur in branched alkanes or homogeneously synthesised ethylene-octene copolymers.

In ethylene-1-octene copolymers, the phase behaviour at elevated pressures is more complicated and has been little studied. High pressures and temperatures lead to extended ethylene sequence crystals (EESCs), i.e. the chain sequence can only extend between branching points, if the polymer chain mobility is high enough, rather than extended chain crystals (EECs). By high pressure DSC, Vanden Eynde et al. [53] showed that the melting and crystallization transitions are shifted to higher temperatures due to a shift in the Gibbs free energy of the melt with increasing pressure. The heating rate at elevated pressures had no significant effect on the thermal behaviour except to reveal that the process by which the ethylene sequences extend is very fast. Similarly, increasing the cooling rate leads to the formation of small and imperfect crystallites since the DSC peaks are broadened. For low 1-octene comononer content, three processes could be resolved: the melting of folded chain crystals, followed by the melting of the relatively large EECs/EESCs superimposed on the orthorhombic to hexagonal transition and a high temperature melting peak of the hexagonal phase. The lack of structural evidence which the DSC data could not provide was addressed by in-situ high pressure SAXS and WAXD measurements.

Recently, we investigated the effect of pressure on the crystallization of ethylene-1-octene copolymers containing 5.2mol% and 8.0mol% 1-octene [54]. Samples that were crystallized at atmospheric pressure from melt showed a disappearance of the wide-angle diffraction peaks and an increase in the amorphous halo, upon increasing pressure to 3.7 kbar at room temperature. However, this "amorphization" (loss of crystallinity) is more likely due to the break up of the large orthorhombic/monoclinic crystallites into small, imperfect crystals at pressures up to 4.0 kbar, as revealed by in-situ high pressure Raman measurements, rather than to be similar to the pressure induced amorphization reported earlier by one of us for the polymer poly-4-methyl pentene-1 [55–57]. In the latter case, the amorphization at room temperature with increasing pressure is a consequence of an inverse density relationship, i.e. the crystalline density is less than the density of the amorphous region.

On crystallizing the ethylene-octene copolymer at approximately 3.8 kbar, similar to branched alkanes, around 200°C crystallization preceded directly via the orthorhombic phase without the intervention of the anticipated hexagonal phase as would be anticipated in linear polyethylenes at these high pressures and temperatures. At $\sim 100°C$, see Fig. 15.15, the d values for (110) and (200) orthorhombic reflections are 4.08Å and 3.71Å. When the sample is cooled below 100°C, a new reflection adjacent to the (110) orthorhombic peak appears at $\sim 80°C$. The position of the new reflection is found to be 4.19 Å and so corresponds to the new phase. No change in the intensity of the existing (110) and (200) reflections is observed, however intensity of the amorphous halo decreases, which suggests that appearance of the new reflection ($d = 4.19$ Å) is solely due to the crystallization of a non-crystalline component. On cooling further as the new reflection intensifies, the (110) and (200) orthorhombic reflections shift gradually. However, at $\sim 50°C$, the (100) monoclinic reflection

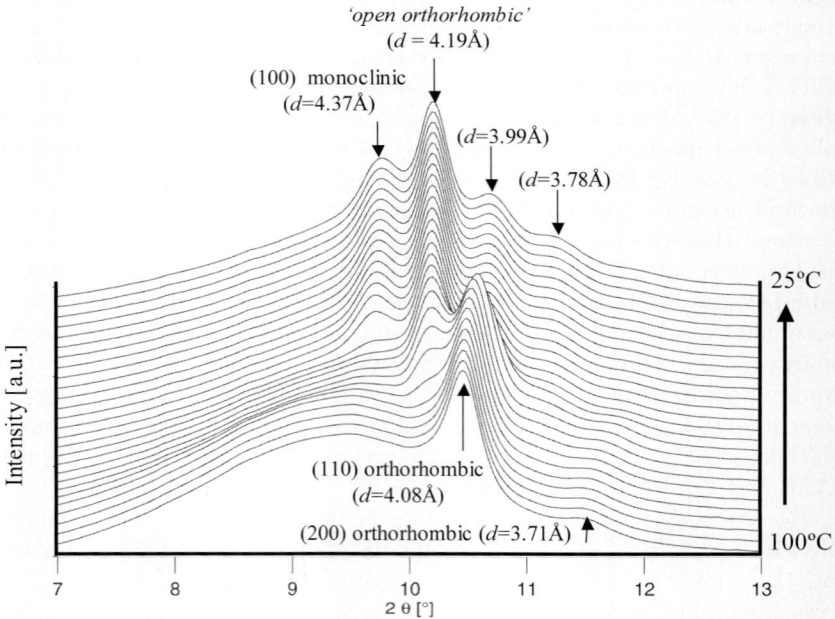

Fig. 15.15. Diffraction patterns of ethylene-1-octene copolymer (5.2 mol%) shown from 100°C to 25°C while cooling at 10°C/min recorded during crystallization from melt at ~3.8 kbar. The open-orthorhombic phase appears at ~80°C, intensity and position of this reflection remains unchanged. The open-orthorhombic phase is followed by the incoming of the (100) monoclinic reflection concomitant with a shift to higher angles and drop in the intensity of the (110) dense-orthorhombic reflection. (X-ray wavelength used for these experiments is 0.744Å)

appears with a concomitant decrease in the intensity of the (110) orthorhombic reflection and a sudden shift in the Bragg d values of the orthorhombic reflections. These results indicate a solid-solid phase transition at 50°C of a large amount of crystals, from the orthorhombic to monoclinic phase. At room temperature it can be seen that the new reflection is much more intense than the (110) orthorhombic reflection and a corresponding secondary new reflection is observed at $d = 3.78$ Å as shown in Fig. 15.15.

If one considers the ratio of the two new reflections (4.19 Å/3.78 Å $\approx$1.108) and compares their intensities, it can be concluded that this phase resembles an orthorhombic phase (for linear polyethylene, $d_{110ortho.}/d_{200ortho} \approx 1.111$). However, the d values for the new reflections are higher when compared to the conventional d values for the orthorhombic phase and do not match exactly with the known triclinic phase in linear polyethylene. Keeping this in mind and assuming no change along the c-axis, the unit cell dimensions for the new orthorhombic phase at 3.8 kbar and room temperature (25°C), were calculated to be $a = 7.56$ Å, $b = 5.03$ Å and $c = 2.55$ Å. The unit cell volume was found to be approximately 96.97 Å^3 and the density 960kg/m^3. This is the first time that this type of orthorhombic phase has been observed in polyethylene. Compared to the conventional orthorhombic phase, the density of the new orthorhombic phase is rather low i.e. the unit cell is more open, for comparison see Table 15.3. The high intensity of the new open-orthorhombic phase at these high pressures means that the new phase is formed by the crystallization of the majority of the amorphous component with the hexyl branches at the surface of the crystal, that is the interphase crystallizes.

The relatively high intensity of the *open-orthorhombic phase*, at these high pressures, means that this phase cannot be attributed to crystallization of the hexyl branches at crystal surface alone – unlike in case of branched alkanes. However, similar to branched alkanes the fold surface incorporating hexyl branches must be a prominent factor in forming the *open-orthorhombic*

Table 15.3. Comparison of the crystalline lattice parameters, volumes and densities for various polyethylenes

	Crystallization Conditions	a Å	b Å	c Å	Volume Å^3	Density Kg/m^3
Linear polyethylene	Atmospheric pressure, 25°C	7.40	4.94	2.55	93.52	996
Branched Polyethylene (ethylene-1-octene 5.2 mol%)	Atmospheric pressure, 25°C	7.52	4.99	2.55	95.63	974
Branched polyethylene (dense-orthorhombic)	3.8 kbar 25°C	7.16	4.81	2.55	87.82	1060
(open-orthorhombic)	3.8 kbar 25°C	7.56	5.03	2.55	96.97	960

phase. In fact one envisages a major part of the orthorhombic lattice, which resorts to a less dense packing even at these high pressures as driven by the hexyl branches [58]. In this manner, the stresses due to crystallization of the hexyl branches residing on the fold surface lead to a further compression of the "original" orthorhombic lattice (termed as *dense orthorhombic phase* at $d_{110} = 3.99$ Å) and a partial conversion of this phase into a monoclinic phase.

On releasing pressure, multiple reflections observed in the sample crystallized at elevated pressure and temperature merge into simple 110 and 200 reflections of the orthorhombic phase. The starting WAXD pattern looks similar to the very first diffraction pattern recorded at 1.4 kbar, as shown in Fig. 15.16. If pressure is once again increased at room temperature, the (110) and (200) orthorhombic reflections split into two distinct reflections. The (110) reflection splits into two distinct d values, $d = 4.19$ Å (open-orthorhombic phase) and $d = 3.99$ Å (dense – orthorhombic phase). The intensity of the dense-orthorhombic reflection ($d = 3.99$ Å) decreases with the appearance of the (100) monoclinic reflection ($d = 4.48$ Å).

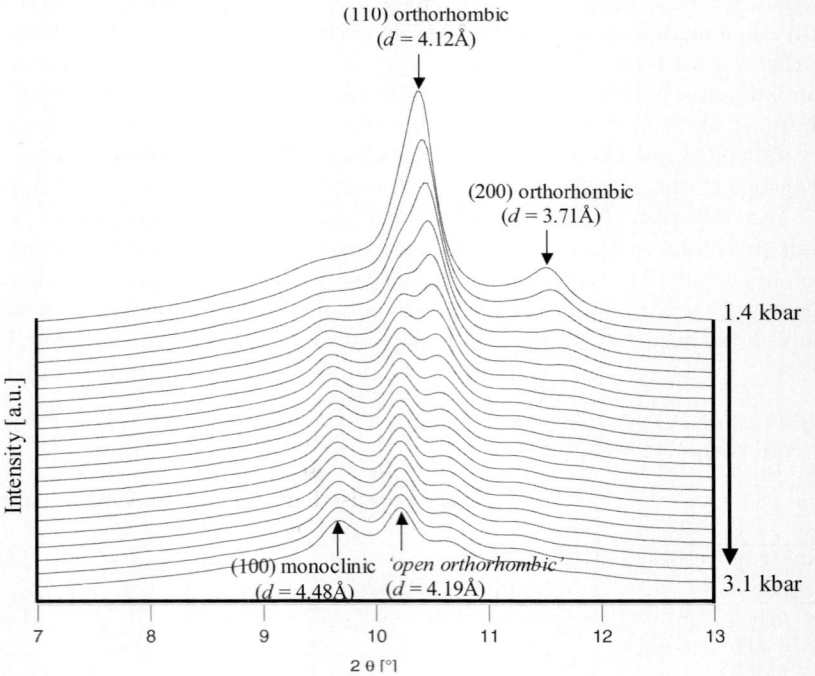

Fig. 15.16. Wide Angle X-ray Diffraction patterns showing the reappearance of the three crystalline phases namely monoclinic, open-orthorhombic and dense-orthorhombic during increase in pressure at room temperature

From Fig. 15.16 it can be further confirmed that with the appearance of the monoclinic phase, the intensity of the open-orthorhombic reflection ($d = 4.19$ Å) remains unchanged and the reflection does not shift with compression. Thus the volume and density of the open-orthorhombic unit cell is invariant with pressure, a feature unusual compared to the changes in volume and density of the dense-orthorhombic phase. These observations strongly suggest that the origin of the open-orthorhombic phase is crystallization of the interphase, where pressure facilitates organization process of the interphase to the extent that it could be detected by conventional X-rays.

Pressure facilitates disentanglement process, phase separation of different molecular weights, reorganization of chains at elevated pressure – temperature and formation of reasonably large crystals along the ab plane in the copolymers and branched alkanes. Thus pressure facilitates in achieving well-defined crystallographic registered fold surfaces at the interphase – which on compression crystallizes and gives rise to specific reflections in WAXD pattern. Concomitant to the appearance of new reflections, shrinkage in the unit cell is observed suggesting lattice contraction with the crystallization of the fold surface – a result in full agreement with recently reported results on lattice contraction with chain unfolding in n-alkanes and varying co-monomer content and size.

15.6 Beyond Flexible Polymers: Rigid Amorphous Fraction

The discussion of the influence of the interphase need not be limited to just linear polyethylenes. Interphases of several nm have been reported in polyesters and poly-hydroxy alkanoates. One major difference between the interphase of a flexible polymer like polyethylene and semi-flexible polymers like PET, PEN, PBT is the near absence of regular chain folding in the latter materials. The interphase in these semi-flexible polymers is often defined as the rigid amorphous phase (or rigid amorphous fraction, RAF) existing between the crystalline and amorphous phases. The presence of the interphase (or rigid amorphous fraction) is more easily realised in these semi-flexible polymers containing phenylene groups such as polyesters.

Similar to polyethylenes the morphology of these polymers is also described as a lamellar stack of crystalline and non-crystalline layers. This so-called "two phase model" is applied for the interpretation of X-ray diffraction data as well as for heat of fusion or density measurements. However, it is well known that several mechanical properties, as well as the relaxation strength at the glass transition temperature, cannot be described by such a simplistic two-phase approach, as discussed by Gupta [59]. From standard DSC measurements [60], dielectric spectroscopy, shear spectroscopy [61], NMR [62], and other techniques probing molecular dynamics at the glass transition (α-relaxation) temperature, the measured relaxation strength is always smaller than expected from the fraction of the non-crystalline phase. The difference in mobility is caused by different conformations of the chains as detected by IR [63] and Raman-spectroscopy, or due to spatial confinement because of the neighboring lamellae. To explain the disagreement between the expected values of relaxation strength and the measured values, Wunderlich and coworkers [60] proposed a three-phase model. In their approach, the non-crystalline phase is subdivided in one part that does contribute, and a second part that does not contribute to the relaxation strength at the glass transition temperature. In addition, Wunderlich and coworkers distinguished between a mobile and a rigid fraction of the polymer. The rigid fraction consists of the crystalline phase and that part of the amorphous phase that does not contribute to the glass transition. This results in a "three phase model" consisting of the crystalline (CRF), the rigid amorphous (RAF) and the mobile amorphous (MAF) fractions.

Although the existence of the RAF seems to have been well demonstrated experimentally, probing of the properties of this part is under investigation. One of the most important questions is: what is the chain packing in the RAF and how does it differ from that in the regular MAF. It is also interesting to know whether the chain packing in RAF is defined by the crystallization conditions (crystallization temperature, nucleation density, pressure etc). Schick et al. [64] demonstrated that the RAF in semicrystalline PET does not exhibit a separate glass transition temperature, in the entire temperature range up to the melting temperature, Tm, while the parameters of sub-Tg relaxation for RAF and MAF are essentially the same. Recently, using temperature-modulated DSC, Schick et al. [65] also showed that similar phenomena take place for several other polymers such as bisphenol-A polycarbonate and poly(3-hydroxybutyrate). No changes in the amount of RAF occurred in the temperature range between crystallization and the glass transition temperatures. Therefore, they suggested that the amorphous chains, constrained between the crystalline lamellae in PET, become effectively vitrified upon crystallization, despite a high temperature, while the remaining amorphous chains located between the lamellae stacks continue to be in the liquid state. Therefore, the crystallization temperature has to be considered as an effective vitrification temperature for the RAF. Devitrification of the RAF then should occur upon melting of the crystalline lamellae, consisting of the lamellae stacks.

If this hypothesis is right, the specific volumes, characterizing the RAF and MAF, have to be essentially different below the crystallization temperature. Figure 15.17 exhibits a sketch to illustrate this point. This sketch basically shows a hypothetical thermal expansion behavior associated with the RAF and MAF for PET, crystallized at some arbitrary crystallization temperature, Tc. Above Tc, in the state of equilibrium melt, only one phase occurs, i.e. the specific volumes for the RAF and MAF are the same. If vitrification of the RAF occurs at Tc, the slope of specific volume versus temperature for this fraction should change at Tc, and become characteristic of the glassy state in

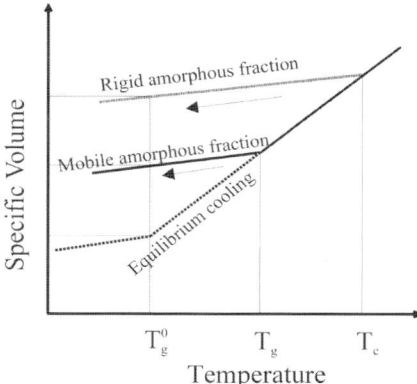

Fig. 15.17. Representation of the volume-temperature relationship for rigid amorphous and mobile amorphous phases of PET

the temperature interval below Tc. In the same manner for the MAF, the slope of specific volume versus temperature, below Tc, should continue to be the same as for the equilibrium melt and change only at the real Tg. Therefore, if room temperature (25°C) is considered as the reference, the specific volume for the RAF at 25°C must be larger than that for the MAF. The same reasoning would lead to the anticipation that the specific volume of the RAF will be a direct function of Tc.

Lin et al. [66] have exploited this variation in specific volume of the RAF to control the barrier properties of polyester films. An attempt to correlate the mechanical deformation of PET with the amount of RAF present in these films has been made recently. Moreover, the observations have been that a sample having a greater amount of RAF, on uniaxial compression, shows considerable loss in crystallinity compared to a sample having lesser amount of RAF. These findings have been reported in a recent publication [67].

15.7 Influence of the Interphase on the Polymer Melt

From the series of experiments reported above it is evident that chain folding at the interphase plays an important role in packing of the chains within the crystalline lattice. In this section we aim to investigate the influence of the interphase on the melting behaviour of crystals and its implications in the polymer melt. Material investigated for the purpose is a solution crystallized UHMW-PE. Salient features on the material have been summarized in the section 4 of this chapter and details have been provided in [31]. Since the solution crystallized UHMW-PE is made from dilute solutions, the number of entanglements between the crystalline and amorphous regions is reduced to an extent that the material can be drawn in the solid state (Fig. 15.5) by more than 100 times.

The reverse route, from such a disentangled solid to the melt, is less well understood. In the past many attempts have been made to use such a disentangled precursor for the melt-processing of intractable UHMW-PE. The basic idea has been that the disentangled molecules will require considerably long time to (re)establish an equilibrium entanglement network. However, against the basic concepts, the observations have been that no memory effect of the disentangled state could be retained within the given experimental time. It was shown that disentangled solution crystallized polymers crystallized from the melt state loses the high drawability (DR > 100) in the solid state, even when the polymer was left in the melt for a few seconds. The crystallized material behaved exactly as a melt crystallized sample [68,69]. This has been explained before by the stem-arrangement model as shown in Fig. 15.4.

The rheological properties of the melt obtained from initially disentangled crystals, such as G', G'' and tan δ were identical to the fully entangled melt state. The experimental observations, suggest that on melting an (equilibrium) entanglement network is restored instantaneously. In view of the long relaxation times, mentioned above, corresponding to the tube renewal time, the absence of any memory effect is rather puzzling.

This problem has been addressed experimentally by Barham and Sadler [70] and theoretically by de Gennes [71]. It was shown by Barham and Sadler using neutron scattering techniques and deuterated polyethylenes that upon melting solution-crystallized polyethylenes the radius of gyration, which is rather low in the case of folded-chain crystals, "jumps" to its equilibrium value corresponding to a Gaussian chain (random coil). The authors introduced the term "coil explosion" for this instantaneous coil expansion upon melting, and suggested that the kinetics of this process is independent of the molecular weight, up to 400.000 g/mol. Considering a single chain forming a single crystal, de Gennes pointed that if a chain starts to melt, the free dangling end of the molten chain will create its own tube and will move much faster than anticipated from reptation theory, rather independent of the molar mass provided that the other end of the chain is still attached to the crystal. Whatever is the cause for chain randomization upon melting, the effect is that the favourable drawability is lost completely, upon re-crystallization from the melt. These observations have been disappointing since no advantage is obtained in processing of the initially disentangled UHMW-PE.

However recently it is shown experimentally in our laboratory that it is possible to retain a disentangled melt state by using disentangled UHMW-PE crystals obtained via controlled synthesis [72]. The obtained melt state proved to be dependent on the melting route. On "slow melting" (heating at a rate of $\sim 0.1°C/min$) it was possible to obtain a heterogeneous melt state where chain ends are distributed heterogeneously in the melt, which on crystallization proved to be still drawable. The heating rate dependence of the melt state suggested that the gain in entropy or "coil explosion" can be controlled.

At this stage we would like to draw differences between the nascent UHMW-PE, obtained on synthesis as described in section 2.3, and the solution crystallized UHMW-PE, described in section 4. Unlike in the solution crystallized sample where crystals of equal thicknesses are regularly stacked (Fig. 15.8a), in nascent UHMW-PE no such regular stacking exists where ultimately using homogeneous synthesis a single chain forming single crystal can be obtained. In the regularly stacked crystals of the solution crystallized polymer thickening on annealing occurs via lamellar doubling [31], whereas in the nascent polymer hardly any lamellar thickening is observed. Such morphological distinctions lead to differences in the first melting temperature of the solution crystallised and the nascent polymer. Details of these distinctions are out of the scope of the present chapter. However, the role of the disentangled chains in melting behaviour of the nascent polymers has been elucidated in a recent publication [73]. These findings are relevant to the disentangled polymers in general and thus are extended to the solution crystallized UHMW-PE in the following section.

Melting Kinetics in Solution-crystallized UHMWPE

Melting of solids can be described using a thermodynamic approach, where the melting temperature is defined as a first-order transition at the intersection of the Gibbs free energy of the solid and liquid state. The transition temperature is then fixed as long as both phases coexist. However, the thermodynamic approach starts from equilibrium conditions and infinite sizes of both phases. For polymers these conditions are not fulfilled; there is no equilibrium, crystallization is not complete and the crystal size is finite. Many semi-crystalline polymers form lamellae crystals, which are 10-30 nm thick and at least one order of magnitude larger in the lateral direction [74, 75] The melting transition is not sharp but covers a certain temperature range that is correlated to the thickness distribution of the lamellae, which has been quantitatively described by using the Gibbs-Thomson equation [76].

$$T_m = T_m^\infty \left[1 - \frac{2\sigma_e}{l \cdot \rho \cdot \Delta H_m} - \frac{2\sigma}{A \cdot \rho \cdot \Delta H_m} - \frac{2\sigma}{B \cdot \rho \cdot \Delta H_m} \right] \quad (15.1)$$

T_m is the experimentally determined melting point, T_m^∞ is the equilibrium melting point for infinite perfect crystals (141.5°C for polyethylene extended chain crystals), σ_e is the surface free energy of the fold planes, σ is the surface free energy of the lateral planes, l is crystal thickness in the chain direction (fold length), ΔH_m is the heat of fusion, A and B are the lateral crystal dimensions and ρ is the crystal density. Normally, the last two terms are ignored due to relatively big lateral dimensions of melt and/or solution crystallized samples (in the order of a few microns) and consequently the melting temperature is only related to the lamellar thickness, l. If l approaches infinity, T_m approaches T_m^∞, which is the equilibrium melting temperature.

The melting aspects involved in UHMW-PE can not be explained by existing thermodynamic concepts alone. Depending on the crystal morphology, different reorganization processes of the amorphous and crystalline regions, which are connected by chains can occur, which can be observed as a shift in the melting region at different heating rates. This shift of the melting region can be either positive or negative.

For polymers where reorganization is feasible, the melting temperature increases with decreasing heating rate [77, 78]. This increase is attributed to crystal thickening and/or crystal perfection. On the other hand, in polymers where no such reorganization occurs (e.g. extended chain crystals), melting temperature decreases with decreasing heating rate. In this case the melting temperature is determined by the extrapolation to zero heating rate. The shift of the melting temperature with increasing heating rate is attributed to superheating [79]. In the case of melt-crystallized polymers, taking the reorganisation processes and super heating into account, the Gibbs-Thomson equation is able to correctly correlate the melting temperature with the crystal dimension. However, the Gibbs-Thomson equation does not always correctly correlate the melting temperature with the crystal dimensions.

In the past we have explained the unusual high melting temperature of nascent UHMW-PE, up to 141°C, by an extensive reorganization process during heating in the DSC [81]. In the case of nascent UHMW-PE, the crystal dimensions are small so the terms A and B can not be ignored in the Gibbs-Thomson equation. However, this view is not correct as discussed elsewhere [73] and below using solution-crystallized UHMW-PE as a model substance.

To recall, the lamellae thickness of UHMW-PE crystallized from solution, double its initial value on annealing below the melting temperature to a maximum of 25 nm. The melting temperature predicted from the Gibbs-Thomson equation for that lamellae thickness is 131°C [80], 5°C lower than the experimentally observed melting point of 136°C. Furthermore, the high melting temperature of 136°C, is lost on second heating where a melting temperature of 131°C is measured [81]. Since this discrepancy of 5°C between the observed first melting point and that predicted by the Gibbs Thomson equation can not be explained by superheating effects alone, we invoke the influence of kinetics in the melting processes.

To get insight in the melt mechanism, annealing experiments below the melting point of 136°C were performed on the solution crystallized polymer. In DSC, the sample was kept for a certain time at annealing temperatures ranging from 127°C to 134°C. Consequently, the sample was cooled to room temperature and reheated (at 10 K/min) to 150°C. Two melting peaks were observed; at 131°C and 136°C, (Fig. 15.18, inset). The peak at 131°C is associated with the melting of material crystallized during cooling from the annealing temperature, and the peak at 136°C to crystal domains in the initial state. The ratio between the areas of the two peaks changes with the annealing time at the given annealing temperature. Figure 15.18 shows an exponential decrease of the high temperature peak area with time having one characteristic

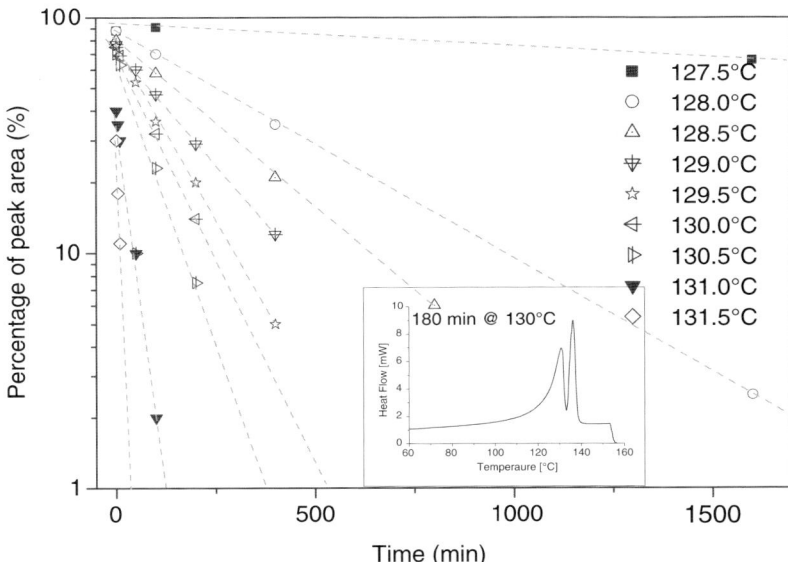

Fig. 15.18. The relative decrease in the area of the 136°C melting peak of the disentangled solution crystallized UHMW-PE with annealing time at different temperatures. In the temperature region (127–132°C) a first order behavior with only one time constant for each annealing temperature is observed. Inset in the figure shows a DSC curve on heating a sample that was annealed for 180 min at 130°C

time constant. Starting from a time law of Debye (Arrhenius) type for the fusion process the enthalpy change reads $H(T,t) = H_0(T)e^{-t/\tau(T)}$, and the time constant (τ) can be related to an activation energy by $\tau = \tau_0 e^{E_A/RT}$.

Figure 15.19 summarizes the relaxation times determined from Fig. 15.18. In the explored temperature region, two different slopes (**a**) and (**b**) of the relaxation times are observed in the log τ versus $1/T$ plot., indicating the involvement of two different activation energies at two different temperature regimes. Slope (**b**) for temperatures above 131°C corresponds to activation energy of 4200 ± 1000 kJ/mol. Slope (**b**) for temperatures below 131°C corresponds to activation energy of 700 ± 50 kJ/mol.

The presence of the two activation energies suggests the involvement of two different melting processes in the solution crystallized polymer. The activation energy of (**b**) 700 ± 50 kJ/mol can be assigned to consecutive detachment of chain segments of ~30 ± 2 nm (with enthalpy of fusion of a single CH_2 group 2.7kJ/mol and C-C distance in the orthorhombic lattice along the c-axis 0.127 nm [82]). This equals roughly one chain (stem) at the lateral surface of the lamellar crystal. These results suggest that, by giving the solution grown crystals enough time at 128°C, the equilibrium melting point predicted by the Gibbs-Thomson equation of alkanes of 25 nm thickness. It

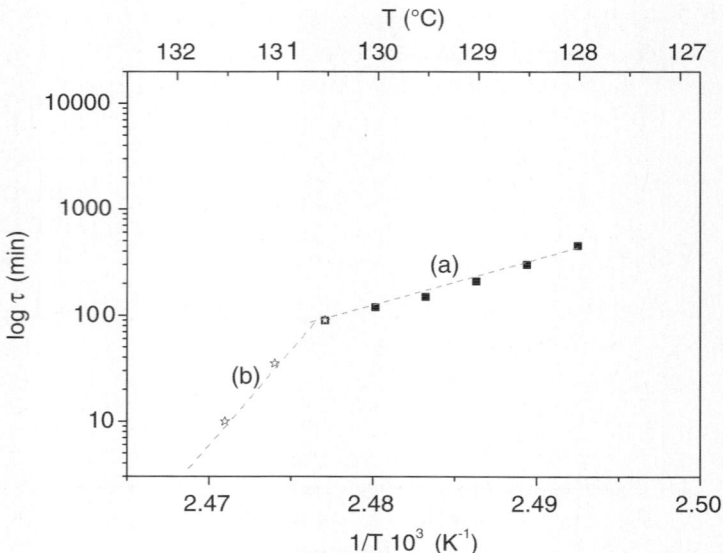

Fig. 15.19. Arrhenius plot for solution crystallized UHMW-PE. The different relaxation times at the given annealing temperatures (determined Fig. 15.18) show two activated processes. The activation energy can be determined from the slopes

is possible to isolate melting of the crystal fold planes, by removal of single chain stems from the crystal substrate. In this way the crystal lattice shrinks from the both sides. At temperatures above 131°C, the activation energy of slope (**b**) corresponds to 4200 ± 1000 kJ/mol. This large activation energy is associated to the breakdown of the crystal lattice by simultaneous randomization of at least 7-8 stems. This result suggests that in the case of solution crystallized UHMW-PE, interphase having adjacent or tight folds, the part of the chain that melts simultaneously corresponds to a length of 7-8 times the lateral thickness, compared to the melt crystallized samples where the random breakdown of the lattice occurs with one stem at a time. If we assume the melting process to be caused by conformational dynamics a greater number of $-CH_2-$ groups are involved in the melting process of solution crystallized UHMW-PE than in the melt crystallized sample with a random distribution of chains in the crystal and large loops in the amorphous phase. Since the number of $-CH_2-$ groups involved in the melting process is greater than the number of CH_2- groups present along the crystal thickness in the solution crystallized UHMW-PE the chain "feels" its length. These results are further supported by solid state NMR studies, which clearly shows that in the amorphous phase of the solution crystallized sample CH_2- groups can adopt restricted conformations compared to the melt crystallized samples. [83]. If we take the number of $-CH_2-$ groups of the respective molecule, incorporated into the crystallite, rather than its thickness, the Gibbs-Thomson equation is able to predict the high melting point of the solution crystallized UHMW-PE.

Therefore, in the case of normal heating rates (10°C/min) the solution crystallized UHMW-PE crystals melt at 136°C and are consequently superheated by 8°C. At 136°C the crystal coil "explodes". If melting occurs from the crystal sides by slow melting this "coil explosion" is circumvented. The heating rate has implications on the chain dynamics and consequently the rate of entanglements formation from the disentangled to the entangled state.

15.7.1 Heating Rate Dependence on the Chain Dynamics – from Disentangled to Entangled Melt

To follow the entanglements formation in the melt, the build-up in the plateau modulus with time is investigated via oscillatory shear rheometry (in the linear visco-elastic regime). This technique is sensitive to follow chain dynamics. The average molecular weight between entanglements, $\langle M_e \rangle$, is inversely proportional to the entanglement density. This is related to the elastic modulus in the rubbery plateau region, G_N^0 by

$$G_N^o = g_N \rho RT / \langle M_e \rangle, \qquad (15.2)$$

where g_N is a numerical factor (1 or 4/5 depending upon convention), ρ is the density, R the gas constant and T the absolute temperature.

The solution-crystallized UHMW-PE, provides a unique opportunity to follow the entanglement process in the initially disentangled material. As shown above the influence of heating rate on the melting kinetics of the crystals is evident. Differences in the chain dynamics, arising due to different heating rates, give rise to differences in the rate of the formation of entanglements which is seen as a build up of plateau modulus [84], Fig. 15.20. Inlay in the figure describes thermal history of the four different samples heated at a rate of 10°C/min, 1°C/min, 0.5°C/min, 0.1°C/min from 125–138°C, respectively. The samples were heated from 138–180°C at a heating rate of approximately 30°C/min. Heating profiles were performed in the ARES rheometer. Once the temperature of 180°C was reached, the samples were subjected to a constant strain of 0.5% at a fixed frequency of 100 rad/s. The frequency was chosen to be in the plateau modulus region of the fully entangled material. The change in modulus, corresponding to the entanglements formation, was followed as a function of time.

From the figure it is evident that with the increasing heating rate the initial modulus increases. Differences in the initial modulus depicting the entanglements present in the initial state is likely to arise with the rate of gain in entropy during phase transformation from crystalline to melt state. Higher the heating rate faster is the gain in entropy, leading to larger amount of entanglements – validating the concept of "coil explosion" proposed by Barham and Sadler. Slower the heating rate, the melting process deviates from the

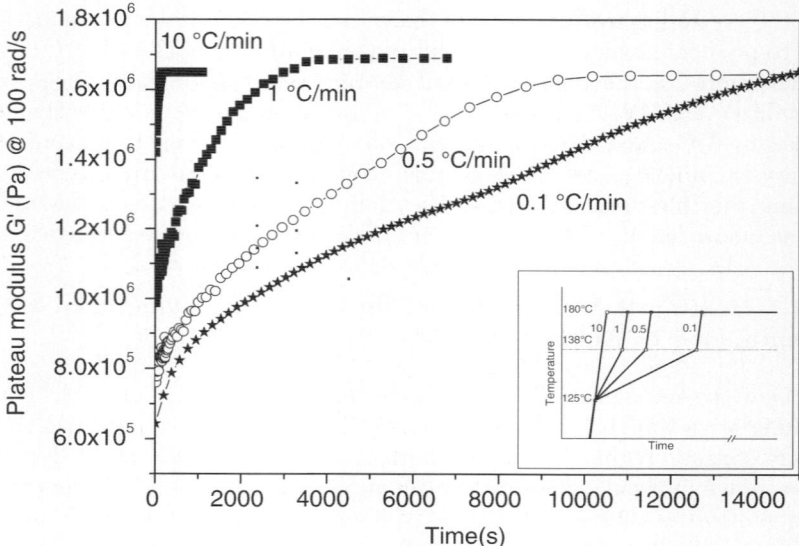

Fig. 15.20. The build-up of modulus in disentangled polymer melts with time. Samples compressed at 50°C, having diameter of 8mm, thickness 1mm, were heated with varing heating rate (10°C/min till 0.1°C/min) from 125°C to 138°C and then heated fast (30°C/min) to 180°C in the rheometer(inlay). A constant strain of 0.5% was applied at a fixed frequency of 100rad/s. The frequency was chosen to be in the plateau region

proposed concept of "coil explosion", suggesting the involvement of different mechanism on melting. The rate of entanglement formation is dependent on the initial state, i.e. lesser the number of entanglements at the initial state larger time required for the build up of the modulus. The time required for the modulus build up to the reach the fully entangled state, even for the material heated at a rate of 0.1°C/min is shorter than the reptation time required for a single chain of molar mass 4.2 million g/mol. These findings are in agreement with the proposed model of de Gennes [71], where he stated that the disentangled chains will reptate faster than the entangled chains. However, the proposed concepts described in the paper by de Gennes cannot be applied fully to the present melting because in his paper de Gennes considered entanglement formation in the single chain forming single crystals during detachment of the chain segments from the crystal lattice. In our case, described in Fig. 15.20 entanglements formation has been followed after melting has occurred.

15.8 Conclusions

In this chapter we have discussed the morphological aspects of the interphase that exists between the ordered three-dimensional crystalline phase and the randomly structured amorphous phase. From a series of experimental observations it is evident that the structure of the interphase strongly depends on the crystallization conditions, for example, a chain from a crystal can be re-entrant to the crystal of origin, with or without forming entanglement(s). The amount of entanglements present between the crystals can be controlled either by crystallization from dilute solution, or by enhancing chain mobility along the c-axis of the crystal through the hexagonal phase or during synthesis. Chain sliding diffusion along the c-axis, which causes lamellar thickening, shows a strong dependence on chain topology. Ultimately, the number of trapped entanglements in between the lamellar crystals controls the deformation characteristics of polyethylenes.

Entanglements at the interphase builds up steric stresses, which are metastable and tend to decrease in number when given sufficient time and chain mobility for reorganization. However, when the chain mobility within the lattice is inhibited with the introduction of side branches, the chains tend to minimize entanglements on the surface during crystal growth. This results into minimization of steric stresses at the crystal basal plane favoring extensive lateral growth. Thus side branches of the homogeneous copolymers and branched alkanes tend to order on the ab plane of the crystal lattice, which crystallizes on compression. Simultaneous to the crystallization of branches, a contraction in the crystal lattice occurs. These observations clearly show that extended chain crystals need not be the thermodynamically stable state; minimization in surface free energy can also be achieved by having disentangled interphase. Figure 15.21 is a schematic drawing showing such a possibility. Intensity of the crystallizable interphase in the open orthorhombic phase strongly depends upon the crystallizable material at the interphase. In our studies, reported in this chapter, content of the crystallizable interphase would be considerably higher in ethylene-octene copolymers compared to sharp folded branched alkanes. The experimental observation that disentangled chain folded crystals in these branched materials are thermodynamically stable is in agreement with theoretical models proposed by Muthukumar and Sommer in this book.

In the final section of this chapter we invoked the unique melting behavior of the partially disentangled solution crystallized UHMW-PE. To summarize the salient findings, in polymers, it is possible to obtain single chain forming single crystals. It is feasible to melt these crystals by simple consecutive detachment of chain segments from the crystalline substrate. Experimentally, clear distinction in different melting processes is observed, by the differences in the activation energies required for the consecutive detachment of chain segments or clusters of chain segments. The consecutive detachment of chain segments occurs at the melting point predicted from the Gibbs-Thomson equation

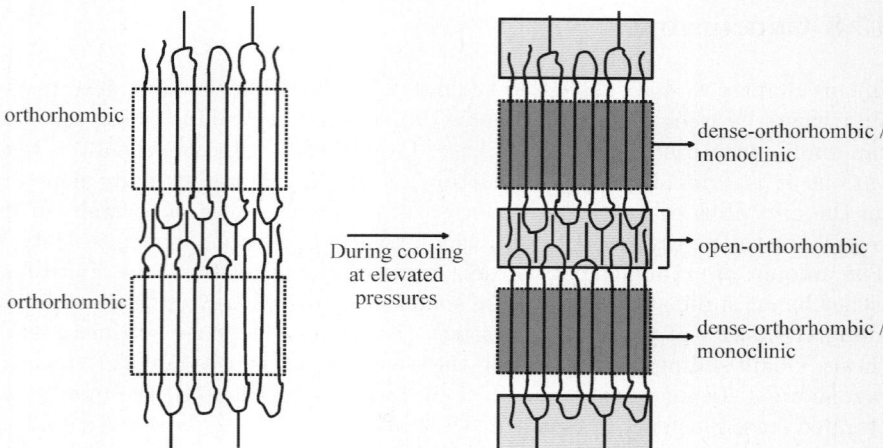

Fig. 15.21. Schematic two dimension presentation for crystallization of the interphase. With crystallization of the interphase in the open-orthorhombic phase contraction in the orthorhombic crystalline lattice occurs. This results into formation of dense orthorhombic or monoclinic phase. (see Figs. 15.15, 15.16). In branched alkanes, because of relatively sharp folds amount of chains contributing to the interphase is considerably less, this attributes to poor intensity of the reflection assigned to the pseudo-hexagonal phase (see Figs. 15.12, 15.14), while the phenomenon of contraction in lattice with crystallization of the interphase is same as in copolymers

whereas, higher temperature or time is required if the chain has to overcome the constraints. The differences in the melting behavior, revealed during different heating rates, have consequences on the chain dynamics. The faster the heating rate, the faster is the gain in entropy during phase transformation from crystal to melt state, resulting into the greater number of entanglements formation on melting the disentangled material. Greater the number of entanglements at the initial state faster is the entanglements formation process.

Acknowledgements

The authors wish to thank Dr. Ankur Rastogi (Dow Chemicals), Dr. Vincent Mathot (DSM Research), Professor dr. Günther W.H. Höhne for their contribution in development of the work. One of the authors (S.R) wishes to thank Max Planck Society and The Dutch Polymer Institute for the financial support. Experimental support provided by the Materials Science beamline ID11 and Ultra Small Angle beamline ID2 (ESRF, Grenoble) is gratefully acknowledged.

References

[1] Flory PJ (1953) Principles of Polymer Chemistry. Cornell University Press, Ithaca, New York
[2] Mandelkern L (1983) An Introduction to Macromolecules. Springer-Verlag, New York
[3] Mandelkern L (1983) Crystallization in Polymers. McGraw-Hill Book Company, New York
[4] Mandelkern L (1992) Chemtracts (Macromol Chem) 3:347
[5] Baker AME, Windle AH (2002) Polymer 42:667
[6] Gautam S, Balijepalli S, Rutledge GC (2000) Macromolecules 33:9136
[7] Bassett DC, Hodge AM (1981) Proc R Soc London A377:25; ibid (1981) A377:39; ibid (1981) A377:61
[8] Khoury F (1979) Faraday Discuss Chem Soc 68:404
[9] Frank FC (1979) Faraday Discuss Chem Soc 68:7
[10] Yoon DY, Flory PJ (1984) Macromolecules 17:868; ibid (1984) 17:862
[11] Kumar SK, Yoon DY (1989) Macromolecules 22:3458
[12] Mandelkern L (1990) Acc Chem Res 23:380
[13] Bassett DC, Hodge AM (1981) Proc R Soc London A377:25
[14] Keith HD, Padden FJ (1996) Macromolecules 24:7776
[15] Toda A, Okamura M, Hikosaka M, Nakagawa Y (2000) Polymer 44:6135
[16] Balijepalli S, Rutledge GC (1998) J Chem Phys 109:6523
[17] Smith P, Lemstra PJ, Booij HC (1982) J Polym Sci Part B: Polym Phys 20:2229
[18] Lemstra PJ, Bastiaansen CWM, Rastogi S (2000) In: Salem DR (ed) Structure formation in polymeric fibers. Hanser, p. 185
[19] Ward IM (1988) Developments in oriented polymers, 2nd Ed. Elsevier, NewYork
[20] Bassett DC (1976) Polymer 17:460
[21] Wunderlich B, Grebowicz J (1984) Adv Polym Sci 60/61:1
[22] Hikosaka M, Rastogi S, Keller A, Kawabata H (1992) J Macromol Sci, Phys Ed B31:87
[23] Rastogi S, Hikosaka M, Kawabata H, Keller A (1991) Macromolecules 24:6384
[24] Hikosaka M, Tsujima K, Rastogi S, Keller A (1992) Polymer 33:2502
[25] Maxwell AS, Unwin AP, Ward IM (1996) Polymer 37:3293
[26] Smith P, Chanzy HD, Rotzinger BP (1985) Polym Comm 26:258
[27] Rastogi S, Kurelec L, Lemstra PJ (1998) Macromolecules 22:5022
[28] Rastogi S, Kurelec L, Cuijpers J, Lippits D, Wimmer M, Lemstra PJ (2003) Macromolecular Materials and Engineering 12:964
[29] Ostwald W (1897) Z Physik Chem 22:286
[30] Uehara H, Yamanobe T, Komoto T (2000) Macromolecules 33:4861
[31] Rastogi S, Spoelstra AB, Goossens JGP, Lemstra PJ (1997) Macromolecules 30:7880
[32] Xue YQ, Tervoort TA, Rastogi S, Lemstra PJ (2000) Macromolecules 33:7084
[33] Ungar G, Zeng X (2001) Chem Rev 101:4157
[34] Terry AE, Phillips TL, Hobbs JK (2003) Macromolecules 36:3240
[35] Schmidt-Rohr K, Spiess HW (1994) in Multidimensional solid-state NMR and polymers Academic Press, London, p. 478
[36] Brooke GM, Burnett S, Mohammed S, Proctor D, Whiting MC (1996) J Chem Soc Perkin Trans 1:1635
[37] Wunderlich B (1980) Macromolecular Physics, Vol 3: Crystal melting. Academic Press, New York

[38] Kortleve G, Tuijnman CA, Vonk CG (1972) J Polym Sci Part B: Polym Phys 10:123
[39] Hosoda S, Nomura H, Gotoh Y, Kihara H (1990) Polymer 31:1999
[40] Vonk CG, Reynaers H (1990) Polymer Commun 31:190
[41] Zachmann HG (1967) Kolloid-Z u Z Polymere 216-217:180
[42] Vonk CG (1986) J Polym Sci: Polym Lett 24:305
[43] Ungar G, Stejny J, Keller A, Bidd I, Whiting MC (1985) Science 229:386
[44] Ungar G, Keller A (1986) Polymer 27:1835
[45] Organ SJ, Keller A, Hikosaka M, Ungar G (1996) Polymer 37:2517
[46] Zeng X, Ungar G (1998) Polymer 39:4523
[47] Ungar G, Zeng X, Brooke GM, Mohammed S (1998) Macromolecules 31:1875
[48] Zeng X, Ungar G (1999) Macromolecules 32:3543
[49] Spells SJ, Zeng X, Ungar G (2000) Polymer 41:8775
[50] Hikosaka M, Seto T (1982) Jpn J Appl Phys 21:L332
[51] Rastogi A, Hobbs JK, Rastogi S (2002) Macromolecules 35:5861
[52] Hay IL, Keller A (1970) J Polym Sci Part C 30:289
[53] Vanden Eynde S, Mathot VBF, Hoehne GWH, Schawe JWK, Reynaers H (2000) Polymer 41:3411
[54] Rastogi A (2002) PhD Thesis, Eindhoven University of Technology; Rastogi A, Terry AE, Rastogi S, Mathot VBF (2004) Macromolecules (under consideration)
[55] Rastogi S, Newman M, Keller A (1991) Nature 353:55
[56] Rastogi S, Newman M, Keller A (1993) J Polym Sci Part B: Polym Phys 31:125
[57] Rastogi S, Hoehne GWH, Keller A (1999) Macromolecules 32:8897
[58] It is to be noted that the reflection assigned to the "new phase" in butyl branched alkanes is relatively weak compared to the reflections observed for the "new phase" in ethylene-1-octene copolymer (5.2 mol%). As explained in this chapter, we attribute the "new phase" to the crystallization of transient layer (butyl branches and fold surface). Considering the anticipated tight folds for butyl branched alkanes, the amount of crystallizable entities in the branched alkanes would be much less compared to ethylene-1-octene copolymers, where the loose folds are expected. We would like to mention that considering the d-value and intensity of the pseudo-hexagonal phase in branched alkanes, this reflection may be referred to as open-orthorhombic phase.
[59] Gupta VB (2002) J Appl Polym Sci 83:586
[60] Suzuki H, Grebowicz J, Wunderlich B (1985) Makromolecular Chemistry 186:1109
[61] Huo P, Cebe P (1992) Macromolecules 25:902
[62] Gabriels W, Gaur HA, Feyen FC, Veeman WS (1994) Macromolecules 27:5811
[63] Cole KC, Aiji A, Pellerin E (2002) Macromolecules 32:770
[64] Schick C, Dobbertin J, Potter M, Dehne H, Hensel A, Wurm A, Ghoneim AN, Weyer S (1997) Thermal Analysis 49:499
[65] Schick C, Wurm A, Mohamed A (2001) Colloid Polymer Science 279:800
[66] Lin J, Shenogin S, Nazarenko S (2002) Polymer 43:4733
[67] Rastogi R, Vellinga WP, Rastogi S, Schick C, Meijer HEH (2004) J Polym Sci Part B: Polym Phys 42:2092
[68] Bastiaansen, CWM., Meijer HEH, Lemstra PJ (1990) Polymer 31, 1435
[69] Lebans, PJR, Bastiaansen, CWM, (1989) macromolecules 3312
[70] Barham P, Sadler DM (1991) Polymer 32, 393

[71] De Gennes PGC (1995) R. Acad. Sci. Paris 321 series II 363
[72] Rastogi S, Lippits DR, Peters GWM, Graf R, Yao Y, Spiess HW (2005) Nature Materials 4, 635
[73] Lippits DR, Rastogi S., Höhne GWH (2006) Physical Review Letters 96, article number 218303
[74] Keller A. (1957) Phil. Mag. 2, 1171
[75] Fischer EW (1957) Nature 12a, 753
[76] Strobl G (1997) The Physics of Polymers, Springer, p. 166
[77] Wunderlich B, Czornyj G (1977), Macromolecules 10, 906
[78] Minakov AA, Mordvintsev DA, Schick C (2004) Polymer, 3755
[79] Toda A, Hikosaka M, Yamada K (2002) Polymer 43, 1667
[80] The authors are aware that depending on the experimental methods used, different numerical Gibbs-Thomson equations exist, see Cho TY, Heck B, Strobl G, (20004) Colloid Polym Sci. 282, 825. A differences arises because of different surface free energy values resulting in a somewhat different melting temperature of 136°C for a crystal thickness of 25 nm. But such discrepancies in the calculated melting temperatures have no implications on our experimental findings
[81] Tervoort-Engelen YMT, Lemstra PJ (1991) Polym. Comm. 32, 345
[82] Considering 1/3rd lesser neighbor interactions on the surface than in bulk, the detachment energy and its diffusion into the melt is likely to be 2.7 kJ/mol since the melting enthalpy of the bulk is 4.11 kJ/molCH$_2$ a value obtained from the ATHAS data bank (http://web.utk.edu/~athas/databank/welcome-db.html).
[83] Yao Y, Graf R, Rastogi S. Lippits DR, Spiess HW (2006), manuscript in preparation
[84] Lippits DR, Rastogi S (2006) manuscript in preparation

16

Polymer Crystallization Under High Cooling Rate and Pressure: A Step Towards Polymer Processing Conditions

Andrea Sorrentino, Felice De Santis, and Giuseppe Titomanlio

Department of Chemical and Food Engineering, University of Salerno, Via Ponte don Melillo, I-84084 Fisciano (SA) – Italy
asorrentino@unisa.it, fedesantis@unisa.it, gtitomanlio@unisa.it

Abstract. Even if many efforts have been spent on the explanation of the mechanisms involved during the polymer crystallization in typical industrial processing conditions, they are still only partially understood. Up to now, due to the remarkable experimental difficulties, in literature only few systematic works have been focused on the effect of high cooling rates and/or solidification pressure on the mechanical and physical properties of the semi-crystalline polymers. In this work, we present two experimental apparatuses, designed and assembled with the aim of obtaining polymer samples under controlled temperature and pressure histories. High cooling rates and pressure, comparable with those experienced by the polymer during industrial processes, were attained in order to produce polymer samples with different morphologies. Exemplar results obtained with Syndiotactic Polystyrene (sPS) show that high cooling rates as well as external pressure are important factors for inducing changes in crystalline polymeric structures.

16.1 Introduction

Crystallization plays an important role in industrial processing of semi-crystalline resins. It strongly affects rheological properties of polymer melts and solutions and influences mechanical and barrier properties of solid objects. Therefore, realistic modeling of technological processes (injection molding, film casting, melt spinning, etc.) involving crystallizable polymers requires that crystallization during processing has to be taken into account.

Unfortunately, during processing operations, crystallization takes place under conditions of cooling rates and pressure much severe than those accessible to available analytical apparatuses (Table 16.1).

An enlargement of the experimental data range is obviously of interest also for a better understanding of basic phenomenon. Indeed, despite the large number of papers concerning polymer crystallization, the role of mesomorphic phase in the formation of the crystalline structures is not completely

Table 16.1. Processing Conditions versus Available experimental range

Analysis	Apparatus	Range Available	Processing Conditions
Calorimetric	DSC – DTA	Cooling rates $<8\,\mathrm{K/s}$ Pressure $<100\,\mathrm{bar}$	Cooling rates $\in 1 \div 1000\,\mathrm{K/s}$
Volumetric	PVT	Cooling rates $<0.03\,\mathrm{K/s}$	Pressure $\in 1 \div 1000\,\mathrm{bar}$
Rheological	Rheometers (rotational, capillary, etc.)	Cooling rates $<0.5\,\mathrm{K/s}$ Shear rates $<1000\,\mathrm{s^{-1}}$	Shear rates $\in 500 \div 10^5\,\mathrm{s^{-1}}$

clear [1, 2]. It is recognized from many experimental evidences, however, that the metastability of morphological entities plays a major role, leading to reorganization, annealing, re-crystallization, super-heating, etc. [3]. One way of avoiding reorganization effects is to increase the scan rate, whereas the amount of reorganization can be established by varying the scan rate.

During crystallization, it is very useful to study the interaction between processing conditions and crystallization-morphology [4]. A great deal of progress can be made by combining different techniques and conducting the various measurements under the same conditions, particularly on samples which experienced the same thermal history. The improvement of simultaneous measurement techniques, such as X-ray, small angle light scattering (SALS), infrared spectroscopy (IR), and dielectric spectroscopy, would be of extraordinary help for providing additional information for the interpretation of solidification/melting transitions, especially for measurements coming from fast crystallization process.

Fast and non-contact methods for the analysis of morphologies evolution during a fast process are highly attractive and, from this point of view, light transmission appears the more promising. In contrast to other methods (calorimetry, X-ray diffraction, densitometry), in fact, measurements of light intensity are very fast, economical, and can be applied in situations (rapid cooling, flow) when other methods are not adequate.

First attempts of monitoring polymer crystallization by light depolarization technique were made in 50s and early 60s [5–7]. Ding and Spruiell [8–10] modified the use of the depolarized light microscopy (DLM) technique so that it could be used to study the overall non-isothermal crystallization kinetics of semi-crystalline polymers under cooling conditions similar to those occurring in the melt-spinning process. On the basis of transmitted light intensity data, they corrected for the scattering that may be present in the transmitted depolarized light intensity data obtained as a result of crystallization in a sample held in a temperature-controlled hot stage. However, the application of the analysis suggested by Ding and Spruiell to some results obtained by Brucato et al. [11] gave rise to unacceptable results as the relative light intensity index shows a maximum during monotonous cooling, which is not an acceptable evolution for a variable that is supposed to represent crystallinity. Lamberti et al. [12] proposed a simple macroscopic model describing the main interactions between a light beam and a semi-crystalline polymer. The proposed model was found to be able to reproduce the observed experimental behavior of light intensities and it was validated by comparison with conventional DSC analysis.

Another important variable affecting the crystallization of a polymer material is the pressure under which it takes place. Basic investigations in this field have been made by Wunderlich and Bassett [13–15]. Their results show that high pressures produce several effects on the properties of polyethylene.

Polyethylene solidified under high pressure usually presents a higher density, a higher melting temperature at atmospheric pressure, a higher crystallinity, and also a peculiar morphology: the formation of a hexagonal phase, intermediate between the stable (orthorhombic) phase and the melt, was evidenced in polyethylene samples crystallized under pressure higher than 3000 bar [14]. Up to now, there are only a limited number of other polymers whose morphology has been studied at elevated pressures [16–18].

However, these investigations were carried out under quasi-isothermal conditions and furthermore pressures are extremely high (typically 2000 bar) with respect to the pressures normally adopted in industrial processes. This implies that the results obtained may not be directly applicable to polymer processing operations, which often involve very high thermal gradients and cooling rates.

Major problems encountered when one tries to apply simultaneous high cooling rates and high pressure rely on the relatively large mass of the sample to ensure the reliability of the data obtained, the hydrostatic character of the stress field applied and the safety of the experimental apparatus.

The effect of pressure on melting temperature represents the largest consequence on kinetics; however, it is not the sole effect on crystallization. Few papers reported some interesting but contradictory observations regarding the actual pressure effect on crystallization at constant super-cooling. Wunderlich [19] reported that crystallization of polyethylene was delayed at elevated pressure (at about 5000 bar). In contrast, it has been reported [20] for a high-density polyethylene that the rate of crystallization was increased with increasing pressure at constant super-cooling. Zoller [21] noted the same tendency as reported by Wunderlich for polyethylene terephthalate, and yet, this effect was not observed for polypropylene and polyamide 6,6.

Thus, also in this case an increase in the quantity and quality of the available experimental data can help to remove ambiguity and aid to understand the polymer solidification in more detail. Interesting examples of the complex morphology that can be achieved in a transformation process come from the structural analysis of injection molded samples in Syndiotactic Polystyrene [22–25].

Exemplar micrographs of injection molded sPS samples (about 2 mm thick) are reported in Fig. 16.1 [25, 26]. These samples show a distribution along thickness direction of transparent, amorphous layers (white layers in

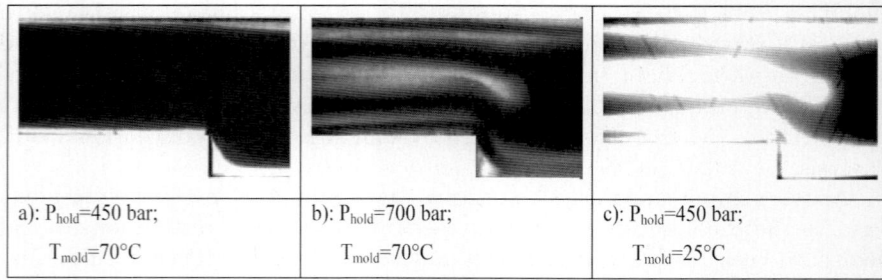

| a): P_{hold}=450 bar; T_{mold}=70°C | b): P_{hold}=700 bar; T_{mold}=70°C | c): P_{hold}=450 bar; T_{mold}=25°C |

Fig. 16.1. Micrographs of injection molded samples in polarized light. The distinctive processing conditions are reported in label (P_{hold} = Packing pressure; T_{mold} = Mold surface temperature)

the micrograph reported in Fig. 16.1) and opaque, crystalline layers (black layer in the micrograph). Thus, under appropriate processing conditions, the amorphous skin-semicrystalline intermediate layer-amorphous core multilayer structures can be found across the thickness direction. This complex multi-layered structure is strongly dependent on the processing variables. In particular, an increase in the packing pressure produces a considerable enlargement of the amorphous core layer (Fig. 16.1 a → b). Equivalently, a reduction of the molding temperature produces a thickening of all amorphous layers with particular effect on the skin layer (Fig. 16.1 a → c). The situation is further influenced by the stress-induced crystallization. The material exposed to the proper levels of stresses, especially at low temperatures, for a sufficient length of time, is induced to crystallize due to the accelerating influence of stress [23].

The final orientation distribution in the mould piece is dependent on the cooling rate, the injection speed and the packing pressure. All these parameters strongly affect the spectrum of relaxation times and the kinetics of crystallization. On one side, a reliable modeling of the overall crystallization kinetic in all processing conditions is the precondition for a correct description of every industrial processing. On the other side, the success of any computer modeling, however, largely depends on the quality of the input information used for describe the material behavior. The super-position of simultaneous cooling rates, packing pressure and molecular orientation is hard to describe without a capable computer simulation. Certainly, additional efforts are needed to overcome experimental difficulties by improving techniques, by combining complementary techniques, or by choosing the optimal material sample. In this work are shown two experimental techniques able to characterize polymer samples in a wide range of cooling rates and pressures. Experimental characterization of Syndiotactic Polystyrene (sPS) help us to illustrate the importance that high cooling rates and pressure have on the solidification process and hence on the final crystalline texture of common polymeric materials.

16.2 Material and Methods

The studied Syndiotactic Polystyrene (Questra QA101) was supplied by the Dow Chemical Company. The molecular weight characteristics of this material were: $M_w = 320000$ g/mol and $M_w/M_n = 3.9$. sPS is a semi-crystalline polymer, which stimulates interest because of its impressive material properties, its unusual polymorphism and its sensibility to processing conditions. Up to now, four different phases were obtained and characterized [27]. In particular, the α and β forms contain chains in planar zigzag conformation and can be obtained either by melt crystallization or by annealing of amorphous samples at proper temperatures [28]. The crystallization of the α form is favored by fast cooling from the melt, by low isothermal temperatures or by cold crystallization from the quenched glass. Crystallization at high temperatures (close to the melting temperature) or under a moderate cooling rate from the melt leads to formation of the β form; otherwise, always a mixture of the two phases (α and β) is obtained. In addition, the sPS presents peculiar relative values of the densities of the different phases. In fact, the crystalline density of the α phase, $1.033\,\text{g/cm}^3$, calculated from the parameters of the unit cell, is smaller than the density of the amorphous phase, $1.048\,\text{g/cm}^3$, whereas the predicted density of the β phase, $1.068\,\text{g/cm}^3$, is larger than the density of the amorphous phase [28, 29].

16.2.1 DTA Experiments

Non-isothermal crystallization kinetic of sPS was investigated using a "Mettler 822 DSC" analyzer equipped with a liquid nitrogen cooling accessory. The heat flow and temperature of DTA were calibrated with standard materials, indium and zinc, prior and after the investigation. Nitrogen gas was purged into DTA furnace during the scans to prevent oxidative degradation at high temperature. Sample weights were chosen between 5 and 10 mg. The as-received material was put in the DTA aluminum pans and heated at 310°C for 15 min to erase any thermal history. Non-isothermal crystallization was carried out at various cooling rates ranging between 0.3 and 100 K/min.

16.2.2 High Cooling Rates Device

An innovative apparatus, which is shown in Fig. 16.2, was adopted for achieving fast cooling crystallization tests [11, 30]. It includes a hot (oven zone) section and a quenching zone section. Sample heating is attained by two radiant electric heaters and the cooling system consists of a couple of gas or gas-liquid (typically air and water) operated nozzles, which spray symmetrically both faces of the sample holder. This cooling system was designed as to determine a large range of cooling rates. As shown in Fig. 16.2 the polymer sample, a thin film (50–100 μm), with an embedded thermocouple is confined between two thin cover glasses that act as sample holders. The sandwich, sample –

 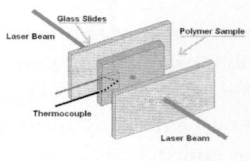

Fig. 16.2. Quenching device and sample assembly scheme

cover glasses, is fastened to a sliding rod, which can be quickly shifted from the hot to the quenching section.

To the purpose of monitoring the crystallinity evolution, an optical set-up was built and it is schematically shown in Fig. 16.3. A laser beam, past the polarizer, crosses the sandwiched polymer film (the sample in Fig. 16.3) while it is subjected to the cooling treatment. The apparatus is able to carry out simultaneous detections of both the depolarized beam intensity and the overall beam intensity, downstream from the film under analysis. The results of these measurements can be related to sample crystallinity content, on the basis of optical properties of each phase [31]. The apparatus can reach very high cooling rates (up to few thousands K/s @ 200°C) by spraying a mixture of air and water on the sample surfaces. Under these conditions, however, the water droplets interact with the laser, strongly reducing the signal intensity detected; it obviously leads to some difficulties in the analysis of results.

16.2.3 Rapid Solidification Under Pressure

Another homemade apparatus was designed and assembled with the aim of obtaining polymer samples solidified under known temperatures and pressure histories [32, 33]. The apparatus, based on the confining fluid techniques, applies hydrostatic pressure on the sample during fast solidification.

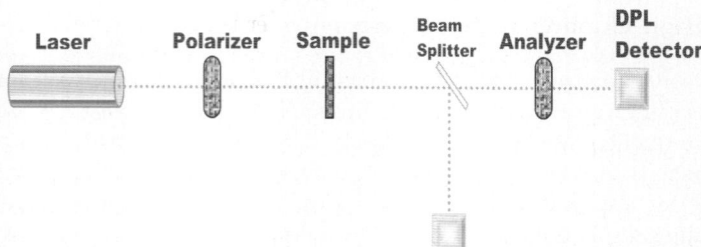

Fig. 16.3. Optical assembly detection scheme

Characterization of samples solidified under known temperature and pressure histories allows correlating temperature and pressure histories to final morphology and properties. The main objective is to attain cooling rates and pressures higher than those achieved by the other available experimental devices which applying hydrostatic pressure on the solid samples. This objective was already achieved: polymeric samples were solidified under simultaneous 1250 bar and 40 K/s (measured at 200°C), see below (Fig. 16.4). The polymeric samples, thin films (100-300 µm), were firstly melted and maintained at the desired temperature and pressure for a suitable time. The samples were then cooled down to ambient temperature under various cooling rates, while the pressure was maintained constant; the values of both temperature and pressure were monitored constantly during the tests.

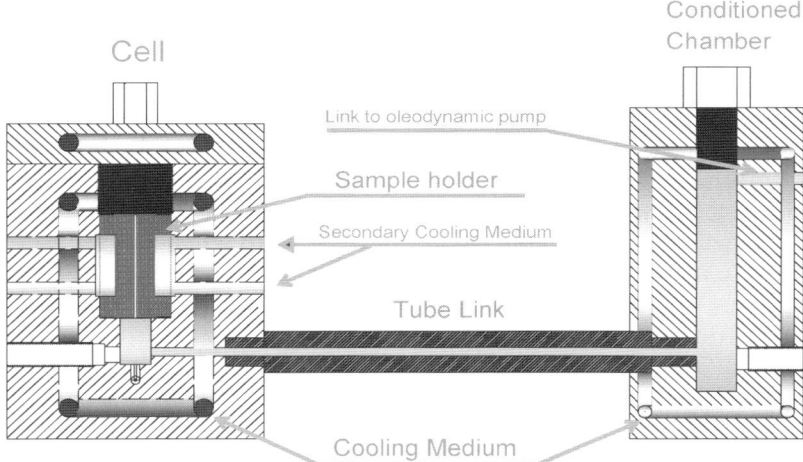

Fig. 16.4. Schematic representation of the apparatus for solidification under pressure

The equipment, schematized in Fig. 16.4, consists of a heated-pressurized steel cylinder, named "cell", where the polymer is confined in mineral oil, and a separate conditioned chamber where the pressure is applied on the pressure-transmitting medium by means of a manual oleodynamic pump. A long thermal conditioned steel tube links the secondary chamber and the pressurized cell. Such a construction avoids any overheating of the pump elements. The insert, that contains the polymer sample, is also cylindrical and can easily be removed from the cell. It is made of a copper-beryllium (98/2) alloy, that has good mechanical properties and elevated thermal conductivity that guarantees a uniform sample cooling. Inserts of different geometry were available: the more excavated allowing the higher cooling rates (Fig. 16.5).

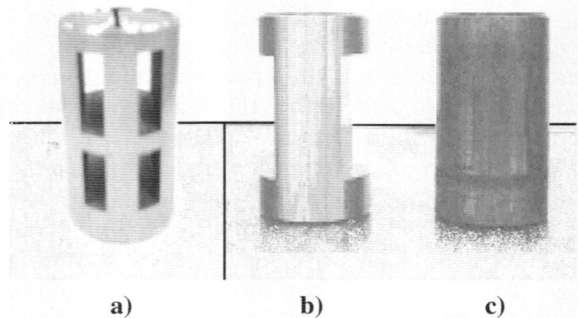

Fig. 16.5. Pictures of inserts with various geometries. Maximum cooling rate accessible @ 200°C: a) 40 K/s; b) 5 K/s; c) 1.5 K/s

16.2.4 Morphological Characterization

All the samples obtained were analyzed by means of X-ray analysis and densitometry.

X-ray diffraction spectra were recorded with a "Philips PW 1830" X-ray generator and a flat camera with a sample-to-film distance of 220 mm (Ni-filtered Cu-Kα radiation) and 1 hour exposure time. A "Fujifilm MS 2025" imaging plate (0.1 mm/pitch) and a "Fuji Bio-imaging Analyzer System", were used to gather and digitalize the diffraction patterns. The degree of crystallinity X_c from the WAXD was evaluated by the spectra according to the Hermans-Weidinger methods [34].

Density was measured by using a gradient column prepared from water and a water solution of sodium chloride. The column was calibrated with glass beads of known density. The samples were placed in the column and allowed to equilibrate for 60 min before the measure were taken. The experimental density of the samples was analyzed with the following model:

$$\rho = \rho_\alpha X_\alpha + \rho_\beta X_\beta + \rho_a(1 - X_\alpha - X_\beta) \qquad (16.1)$$

where ρ, ρ_a, ρ_α and ρ_β are the density of the sample, the amorphous phase, the α phase and the β phase, respectively. X_α and X_β are the volume fraction of α and β phase, respectively. Equation (16.1) allows calculation of the volume fraction of α and β phases once total degree of crystallinity ($X_\alpha + X_\beta$) is evaluate from WAXD measurements, with the proviso that the crystallinity density of single phases are known.

16.3 Experimental Results

Typical results of the quenching experiments, i.e. both overall and depolarized light intensities as well as temperature are reported in Fig. 16.6 as function of

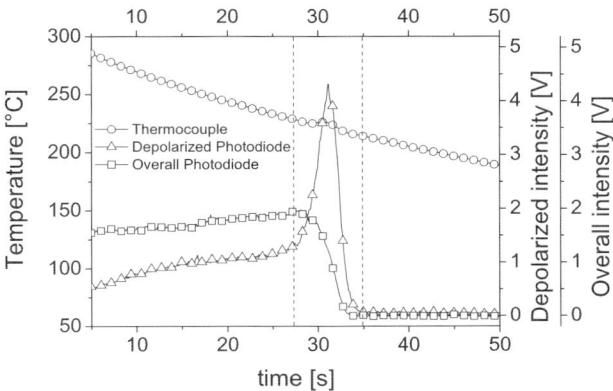

Fig. 16.6. Typical signals output of sPS experiments and values of overall and depolarized intensities adopted for parameters identification

time. The overall light intensity (measured at time t, downstream of a film of thickness S), $I_O(t, S)$ initially shows a quasi constant value. When crystallization starts the light intensity decreases, reaching a new constant level, lower than the initial one. The depolarized light intensity, $I_D(t, S)$, initially shows a constant value, likewise the overall light intensity, then it increases, attains a maximum and then decreases, achieving a new constant level, lower than the initial one. The evolution of the temperature recorded just inside the sample is also reported in Fig. 16.6. A perturbation in the temperature signal is well evident in correspondence of the maximum in the depolarized light intensity. Probably, it is due to the heat generation during the crystallization. From the picture, it is possible to identify the characteristic temperature values of the process (i.e. the starting and the crystallization end).

The temperature at which the crystallization started, were it attained the 50% (for the quenching experiments were the polarized light intensity attained its maximum) and where it finished are reported in the Fig. 16.7, for non-isothermal tests, carried out both in the DTA and in the quenching device. For graphic purpose, the experimental tests carried out at a non-constant cooling rate (quenching experiments) were identified with the cooling rate recorded at 200°C. This temperature was chosen because the large part of the crystallization process takes place close to this temperature. The densities of the solidified samples are reported in Fig. 16.7 where the density levels of α, β and amorphous phases are also shown. As expected, the DTA temperature range where the crystallization takes place increases strongly with the cooling rate and moves to lower temperatures. This behavior is common at both enthalpic and optical results.

The DTA results do not fit with the results coming from optics signal. In particular, the end of the process seems anticipated in the quenching experiments. It can be due to many reasons. First of all, the DTA experiments

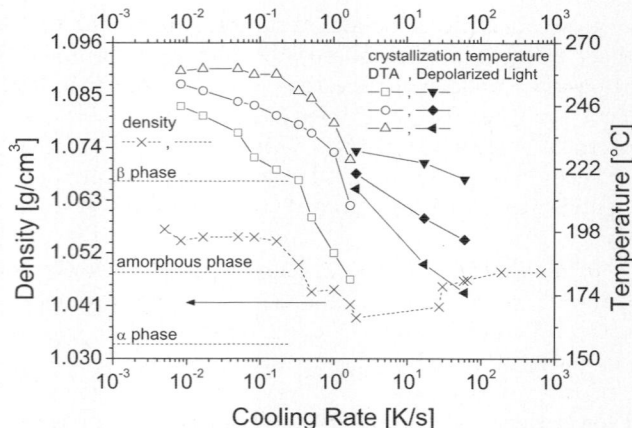

Fig. 16.7. Final Samples Density and Crystallization temperatures as a function of cooling rate @ 200°C. *Open symbols* are data taken from calorimetric (DTA) trace. *Full symbols* are data taken from Depolarized Light measurements

are carried out at constant cooling rates, whereas quenching experiments are obtained with an exponential temperature decrease. In the case of DTA, also the time lag of the instrument must be taken into account. Many corrections were proposed in order to correct enthalpic traces [35,36]. The DTA data proposed in Fig. 16.7 are shown without any type of correction. However, despite the correction proposed, the effect produces generally both a non symmetric shrinkage and a small shift of the crystallization peak versus higher temperature. This effect is quite proportional to the sample mass and cooling rate. The last possibility of misfit between DTA curves and optical signal, is that the last part of the crystallization process, even if it produces a thermal response, does not show appreciable optic effects. It is common in all situations when one tries compare experimental results taken with different techniques. In general, cannot be expected that the same definition of crystallinity holds for all experimental signal. Different techniques present dissimilar sensibility to material properties; it means that for a correct characterization of the polymeric samples the crystallinity degree is not sufficient for a complete characterization.

Densities of final samples are also reported in Fig. 16.7. As the cooling rate increases, the density of the samples clearly decreases from values close to the β phase, to values close to the α phase. This is in good agreement with well known literature results according to an increase of cooling rate produces an increase in α content in the solid sample. The density of the solidified solid samples starts to decrease when the crystallization starts at temperatures lower than 258°C, namely when the cooling rate is higher than 0.2 K/s. It attains a minimum (close to the α phase density) between 2 and 20 K/s and it increases toward the amorphous value for larger densities. Such a dependence of the densities of solidified samples on cooling rate is consistent with their phase composition (α, β and amorphous) determined by X-rays.

These are plotted versus cooling rate recorded at 200°C in Fig. 16.8. Total crystallinity content goes from the value $X_{max} = 60\%$ vol/vol to zero in less than two orders of magnitude of the cooling rate. At cooling rate of about 50 K/s the crystallinity of the solid sample obtained undergoes a sharp decrease; it drops to about 10% of X_{max} at 200 K/s and nearly to zero at cooling rate of 600 K/s.

As also shown in Fig. 16.8, the β phase content decreases continuously with cooling rate, whereas the α phase content shows a monotonic increase up to about 1 K/s (cooling rate transition between DTA and quenching experiments). In particular, the α phase starts to be predominant in the final morphology of the samples when the crystallization process takes place at temperature lower than 250°C or equivalently under cooling rates higher than 0.5 K/s. Indeed, the effect of cooling rates on the final morphology of sPS samples is rather complex and difficult to predict starting from experimental data recorded at low cooling rates. This behavior is still more complex in presence of high pressures.

Final phase composition ($\alpha + \beta$) of sPS samples solidified in the apparatus described in this work at 5 K/s @ 200°C are reported in Fig. 16.9. These were evaluated on the basis of X-ray diffraction characterization [33]. The data clearly show a reduction in the overall final crystallinity by effect of an increase of solidification pressure from 0 to 450 bar. At room pressure (Fig. 16.8) overall final crystallinity degree in the solid samples gradually decreases with cooling rate, and α phase is predominant at 5 K/s. The effect of an increasing pressure

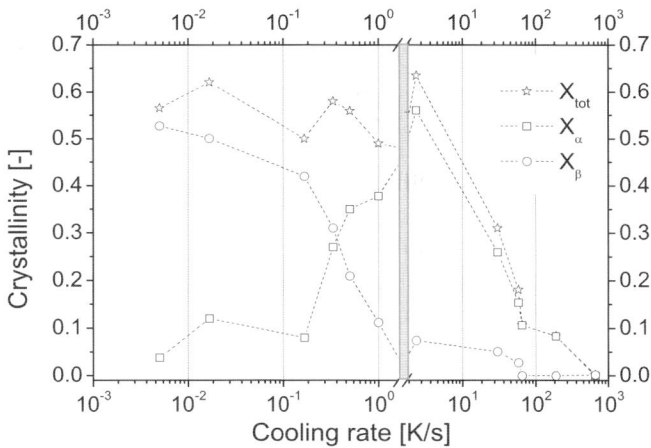

Fig. 16.8. Phase distribution in non-isothermal experiments on sPS samples. The gray band divides DTA experiments from quenching experiments

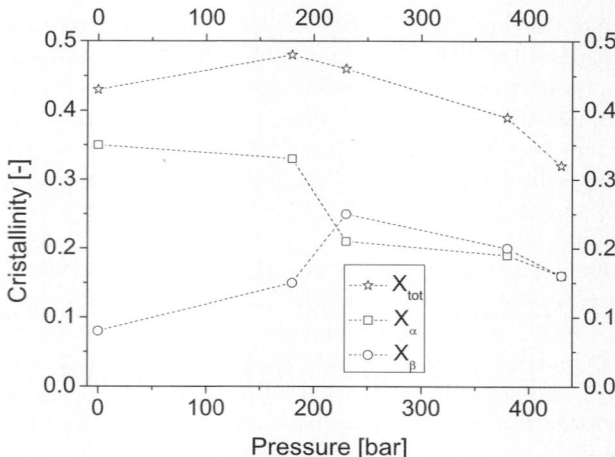

Fig. 16.9. Phase distribution in high pressure non-isothermal crystallization experiments. The solidifications were carried out at a cooling rate of 5 K/s @ 200°C

is toward an increase of the β phase content. Indeed, as expected, an increase in the solidification pressure promotes the content of the higher density phase.

16.4 Discussion

Growth rate data as well as overall crystallization kinetic of polymeric material shows a bell shape as a function of temperature with a maximum located between melting temperature and glass transition temperature. This very general behavior cannot be experimentally observed for many fast crystallizing polymers, like PE, iPP, sPS, etc. For these polymers, in fact, standard calorimetric experiments cannot be performed on the time scale of nucleation and crystallization. In general, for commercial polymers this situation is the rule rather than the exception. For this reason, the majority of analysis of the experimental data in terms of kinetic model is carried at temperatures approaching the melting point, where the crystallization rate is dominated by the thermodynamic driving force. The diffusion term parameters (transport process at the interface between the melt and the crystal surface) are generally used as simple fitting parameters. This not only leads to a poor description of the crystallization at temperature close to the glass transition, but also avoids any theoretical conclusion on the diffusion process. The possibility of achieving high cooling rates during the crystallization from the melt can be important also for elucidating complex polymorphic behavior as in the case of sPS.

An interesting connection between stability and kinetics may be also implied from the stability diagram of the sPS, as displayed in Fig. 16.10. The

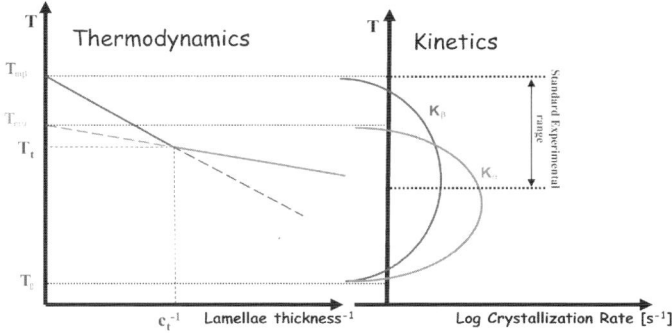

Fig. 16.10. Comparison between Kinetics and stability diagram of the sPS

thermodynamic stability lines of the two phases are reported on the left hand side of the Fig. 16.10 as function of the reciprocal lamellae thickness. At any value of the lamellae thickness, the stable phase is presented as a solid line whereas the meta-stable phase with a broken line. The intersection of the phase lines defines a triple point Q, where all three phases (the melt, α and β crystals) can coexist as stable phases. ¿From the viewpoint of kinetics, the smaller the critical nucleus, the faster the crystallization rate of crystalline phase. The crystallization rate of α form is, thus, expected to be faster than that of β form in the temperature range approaching the glass temperatures and viceversa close to the melting temperature. In the case of sPS, this conclusion has been further approved by the comparison of crystallization kinetics bell shaped curves of the two phases. The maximum crystallization rate of β form was found to be about ten times smaller than that of α form, however at high temperature (close to melting points of the two phases) the relation inverts and crystallization rate of β form become considerably higher than that of the α form [28, 33].

The effect of pressure on the crystallization behavior has been generally attributed to the effect of increasing melting point with pressure, which in turn is equivalent to amplify the degree of super-cooling. However, as reported by Hohne [17,37] this behavior is not true for all the crystal phases. In particular, for the sPS the α phase shows a decrease of the melting point with pressure whereas that of the β phase shows an opposite trends. The opposite behavior of the melting temperature of the two crystal phases with pressure is a consequence of the fact that the density of the amorphous phase is smaller than the density of the β and larger than the density of the α phase. Indeed, the well known Clausius-Clapeyron equation gives for the derivative of the transition temperature with respect to pressure:

$$\frac{dT_m}{dp} = T_m \frac{V_c - V_a}{\lambda_f} \tag{16.2}$$

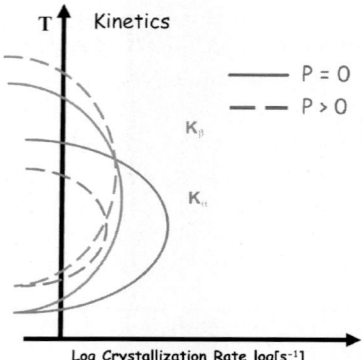

Fig. 16.11. Schematic representation of the crystallization rate as a function of the crystallization temperature and pressure

where T_m is the melting temperature, p the pressure, λ_f the heat of fusion, V_c and V_a the crystalline and amorphous specific volume, respectively.

For the β phase, (16.2) describes a continuous increase of the melting temperature with the pressure and vice versa for the α phase. Also the glass transition temperature was found to increase considerably with pressure. Thus, as sketched in Fig. 16.11, a pressure increase produces a strong reduction in the crystallization range of the α phase, whereas the amplitude of the crystallization range of the β phase is almost unchanged (it is systematically shifted versus higher temperature). The small variation in maximum crystallization value, however, is an indication that crystallization kinetics is even affected by pressure at constant amount of super-cooling [38].

16.5 Conclusions

It is important to grasp properties of polymer systems from the standpoints of the optimal design, the process control, and the savings of energy and resources in polymer industries. In particular, crystallization can dramatically modify dynamics of polymer deformation as well as the properties of the solid material. Even if it seems quite clear for the majority of the researchers in polymer science, for many of them it is surprisingly hard to accept that a correct description of the crystallization in processing conditions requires experimental data covering a wide range of pressures and temperatures, which may change at largely different rates. This work has attempted to analyze the crystallization process of Syndiotactic Polystyrene in a very wide range of experimental conditions. At ambient pressure, the relative degree of α phase increases with the cooling rate up to reaching 100% for cooling rates higher than 1 K/s. For cooling rates higher than 10 K/s, however, the overall final degree of crystallinity in solid samples gradually decreases with increasing cooling rates. The influence of pressure seems to be mainly limited to the α phase. Indeed, solidification tests under different cooling rates and pressures clearly showed that β phase prevails if sPS samples are solidified under high pressure, whereas β phase is not present at all if samples are solidified at ambient pressure in the whole range of cooling rates of interest for common transformation processing. Albeit, all these behaviors can be explained starting from thermodynamic and kinetics considerations, it is difficult, even impossible to extrapolate from quasi-static laboratory conditions. Crystallization under processing conditions reveals new effects, absent under common laboratory conditions.

References

[1] G. Strobl: Eur. Phys. J. E **18**, 295 (2005)
[2] S.Z.D. Cheng, B. Lotz: Polymer **46**, 8662 (2005)
[3] B. Wunderlich: *Macromolecular physics. Crystal structure, morphology, defect*, vol 1 (Academic Press, New York, 1973)
[4] A. Ziabicki, L. Jarecki, A. Sorrentino: e-Polymers **072** (2004)
[5] K. Fischer, A. Schram Angew Chem **68**, 406 (1956)
[6] J.H. Magill: British Journal of Applied Physics **12**, 618 (1961)
[7] J.H. Magill: Polymer **2**, 221 (1961)
[8] Z. Ding, J.E. Spruiell: J Polym Sci B **34**, 2783 (1996)
[9] Z. Ding, J.E. Spruiell: J Polym Sci B **35**, 1077 (1997)
[10] J.E. Spruiell, P. Supaphol: J Appl Polym Sci **86**, 1009 (2002)
[11] V. Brucato, F. De Santis, A. Giannattasio et al: Macromol. Symp. **185**, 181 (2002)
[12] G. Lamberti, F. De Santis, V. Brucato et al: Appl. Phys. A **78**, 895 (2004)
[13] B. Wunderlich, T. Arakawa: J. Polym. Sci. A **2**, 6397 (1964)
[14] D.C. Bassett, B. Turner: Nature Phys. Sci. **240**, 146 (1972)
[15] D.C. Bassett, S. Block, G.J. Piermarini: J. Appl. Phys. **45**, 4146 (1974)
[16] C. Angelloz, R. Fulchiron, A. Douillard et al: Macromolecules **33**, 4138 (2000)
[17] G.W.H. Hohne, Thermochim. Acta **332**, 115 (1999)
[18] Y. Kojima, M. Takahara, T. Matsuoka et al: J Appl Polym Sci **80**, 1046 (2001)
[19] B. Wunderlich, L. Mellilo: Makromol. Chem. **118**, 250 (1968)
[20] T. Hatakeyama, H. Kanetsuna, H. Kaneda et al: J. Macromol. Sci. **B10**, 359 (1974)
[21] J. He, P. Zoller: J. Polym. Sci. B **32**, 1049 (1994)
[22] A.M. Evans, J.C. Kellar, J. Knowles et al: Polym. Eng. Sci. **37**, 153 (1997)
[23] R.C. Lopez, C.L. Cieslinski, R.D. Wesson: Polymer **36**, 2331 (1995)
[24] Y. Ulcer, M. Cakmak, J. Miao et al: J. Appl. Polym. Sci. **60**, 669 (1996)
[25] R. Pantani, A. Sorrentino, V. Speranza et al: In Proc. PPS 2002, (Taipei Taiwan, 2002)
[26] R. Pantani, A. Sorrentino, V. Speranza et al: In Proc. ICHEAP 6, (Pisa Italy, 2003)
[27] C. De Rosa, G. Guerra, V. Petraccone et al: Polym. J. **23**, 1435 (1991)
[28] A. Sorrentino, M. Tortora V. Vittoria: New developments in syndiotactic polystyrene. In: Recent Res. Devel. Appl. Pol. Sci., (2006): ISBN: 81-308-0129-9

[29] Z. Sun, R.J. Morgan, D.N. Lewis: Polymer **33**, 660 (1992)
[30] F. De Santis: Influence of solidification conditions on structural evolution of thermoplastic polymers. PhD Thesis, University of Palermo, ISBN 88-7676-227-2 (2003)
[31] R.J. Samuels, Structured polymer properties: the identification, interpretation, and application of crystalline polymer structure, (New York: John Wiley 1974)
[32] A. Sorrentino, D. Picarella, R. Pantani et al: Review Scientific Instruments **68**, 245 (2005)
[33] A. Sorrentino: Injection Moulding of Syndiotactic Polystyrene. PhD Thesis, University of Salerno, ISBN 88-7897-001-8 (2005)
[34] L.E. Alexander: X-Ray diffraction Methods in Polymer Science, (Krieger Publishing Co., Florida 1985)
[35] G. Eder, H. Janeschitz-Kriegl In: *Transport Phenomena in Processing*, ed by S.I. Guceri (Technomic Publ. Co. 1993) pp. 1031–1042
[36] V.B.F. Mathot: *Calorimetry and Thermal Analysis* (Hanser, Munich 1994)
[37] C.S.J. van Hooy-Corstjens, G.W.H. Hohne, S. Rastogi: Macromolecules **38**, 1814 (2005)
[38] A. Sorrentino, R. Pantani, G. Titomanlio: In Proc. PPS 21, (Leipzig, Germany 2005)

17

Stress-Induced Phase Transitions in Metallocene-Made Isotactic Polypropylene

Claudio De Rosa, Finizia Auriemma

Dipartimento di Chimica, Università di Napoli "Federico II", Complesso Monte S.Angelo, Via Cintia, 80126 Napoli, Italy
claudio.derosa@unina.it

Abstract. The deformation behavior of semicrystalline polymers associated with polymorphic transformations under tensile deformation is discussed in the case of isotactic polypropylene (iPP). The mechanical properties and polymorphic transformations occurring during plastic deformation of iPP samples with variable stereoregularity, containing only rr stereo-defects, are presented. Thermoplastic materials showing high stiffness, or high flexibility, or elastic properties can be produced depending on the concentration of defects. We report a phase diagram of iPP where the regions of stability of the different polymorphic forms are defined as a function of the degree of stereoregularity and deformation. The values of critical strain corresponding to the structural transformations depend on the stereoregularity that affects the relative stability of the involved polymorphic forms and the state of the entangled amorphous phase. In the case of elastomeric iPP, we show that samples of different stereoregularity present different types of elasticity depending on the degree of crystallinity. The more stereoregular samples, with rr content in the range 7–11% show elastic behavior in spite of the high degree of crystallinity (40–50%). Since elasticity is generally a property of the amorphous phase, probably elasticity in these samples is partially due to the enthalpic contribution associated with the crystallization of the mesomorphic form into the α-form occurring upon releasing the tension. In the case of the less stereoregular sample, with rr content of $\approx$17%, the degree of crystallinity is very low, and elasticity has essentially entropic origin, as in conventional elastomers.

17.1 Introduction

Semicrystalline polymers consist of two phases, crystalline and amorphous. The morphology that develops in polymers upon crystallization from the melt may be described in terms of lamellar crystals alternating to amorphous regions, forming a highly entangled network. The lamellar crystals usually have thickness of order of some ten nanometers, whereas the lateral dimensions are much larger. Since the polymer chains in the crystalline phase are oriented perpendicular to the lateral dimensions of the lamellae and the length of the chains is much larger than the lamellar thickness, each chain may run through several crystalline and amorphous regions. Connections between neighboring crystals are thus ensured by chains emanating from one lamella that enter the other and by entanglements involving chains that re-enter into the same crystalline lamella, after passage through a portion of the adjoining amorphous layer [1].

This complex morphology entails that large, irreversible deformations may occur by cold drawing semicrystalline polymers at a temperature higher than the glass transition temperature. During stretching, indeed, the initial isotropic structure, characterized by the random orientation of the polymer segments in the amorphous and crystalline regions, is gradually transformed into a fiber morphology, i.e. highly anisotropic structure, characterized by preferential orientation of the macromolecular chains along the stretching directions [2]. The development of a fibrous morphology starting from the spherulitic structure is an irreversible process, the original spherulitic morphology can be re-obtained only by melting of the fibrous material and successive recrystallization. Therefore, the stretching of crystalline polymers inevitably involves large plastic (irreversible) deformations.

The mechanisms of plastic deformation at microscopic level of amorphous polymers are mainly crazing and shear yielding [3–5]. In semicrystalline polymers, although the glass transition temperature, density, infrared spectrum and other properties of the amorphous phase interdispersed between the crystalline lamellae are close to those of bulk amorphous polymers, the mechanisms of plastic deformation are very different from those of the amorphous materials, since also the crystalline phase plays a key role [1]. However, because of the presence of the entangled amorphous phase, the mechanisms of plastic deformation of semicrystalline polymers are also different from those of other crystalline materials (for instance metals).

The mechanism of plastic deformation in polymers is rather complicated and involves different phenomena, which occur on the same time scale of applied stress and on different length scales. The global deformation behavior of semicrystalline polymers at temperatures higher than the glass transition temperature may be regarded as the stretching of two interpenetrated networks, made by the interlocked crystalline lamellae and the entangled amorphous phase, characterized by a large nonlinerar internal viscosity [6]. Deformation is accompanied by slip processes within the lamellae, and intralamellar mosaic block slips, and, at larger strain, when the stress acting on the crystallites reaches a value at which the crystalline blocks are no longer stable, by stress-induced melting and recrystallization in new oriented crystallites, whose assembly forms fibrils [6]. The principal modes of deformation on the crystallographic length scale may be slips, twinnings, [2, 6] martensitic transformations, [7] stress-induced melting [8] and recrystallization, [1] and formation of nanoblocks in the amorphous phase [9]. Collective intra-lamellar slip processes, collective motions of lamellar stacks [2,3,6] and formation of microvoids [3,7,9] may occur at a larger length scale. Furthermore, transient phenomena may also be involved. For instance, orientation of portion of crystalline lamellae with the chain axes nearly perpendicular to the stretching direction may occur at low deformations, that is, after the initial Hooke's elastic range, close to the yield point, and disappears at higher deformations [6, 10].

The stress-strain behavior of polymeric materials depends on the properties of the material, including molecular weight, polydispersity, the microstructure of the chains (i.e. concentration and distribution of stereodefects and regiodefects, constitutional defects as typically the presence of comonomeric units), packing, chain entanglements, crystallinity, heterogeneity, defects in the crystals (e.g. dislocations, point defects, structural disorder), and several other parameters as temperature, pressure, load rate, the shape of the item under load, etc. At variance with cross-linked amorphous polymer networks, that show large reversible deformations and positive temperature coefficients of stress, semicrystalline polymers generally exhibit negative stress-temperature coefficients and only short-range elasticity when stretched samples are relaxed by releasing the tensile stress.

According to a generalized view the mechanisms that govern the process of tensile deformation of semicrystalline polymers at low and moderate deformations appears to be strain-controlled, rather than stress-controlled [6, 10–13]. This may be evidenced by the fact that along the true stress-strain curves of several polymeric materials different regimes are discernible, corresponding to changes in the differential compliance that take place at defined critical points [6, 10–13]. These critical points have been interpreted as i) the onset of isolated inter and intra-lamellar slip processes after the initial Hooke's elastic range (point A); ii) change into a collective activity of slip motions of crystal blocks at the point of maximum curvature of the true stress-strain curve (point B); iii) the beginning of destruction of crystal blocks followed by re-crystallization with formation of fibrils (point C) and iv) the beginning of disentanglement of the amorphous network or strain hardening due to the stretching of the amorphous entangled network at high deformations (point D).

The values of the strains at critical points A, B and C are constant, for each class of polymer, when varying crystallinity, temperature, strain rate and crystal thickness [6, 10–13]. Opposite to the strain, the stresses at critical points vary, with larger values for higher crystallinities and lower values for higher temperatures. The observation complies with the general assumption that in semicrystalline polymers (and in heterogeneous systems in general e.g. composite materials) whereas the stress is not homogeneously distributed, the strain, instead, is homogeneously distributed [3,6,14]. At low stresses or strains the forces transmitted by the interconnected crystallites dominate, whereas at high strains the rubber-like network forces are superior [14].

The yield point in engineering stretching experiments is always located shortly above point B. The position of the critical strain at the point C, at which the critical stress that starts destructing the crystal blocks is achieved, depends on the interplay between the entanglement density of the amorphous phase and the intrinsic stability of crystals [6]. A higher entanglement density implies a higher stress that is generated when the sample is stretched. The more stable the crystallites, the higher the stress needed for their destruction [6].

The role of cavitation versus plastic deformation of crystals during stretching of crystalline polymers has been recently studied by Galeski [15]. Cavitation corresponds to formation of microvoids in the amorphous layers confined between crystalline lamellae. During stretching of semicrystalline polymers there is a competition between cavitation and activation of crystal plasticity: easiest phenomena occur first, that is cavitation in polymers with crystals of higher plastic resistance, and plastic deformation of crystals in polymers with crystals of lower plastic resistance. The stress level at yield point corresponds to the onset of cavitation and not necessarily to the onset of plastic deformation of crystals. In polymers where cavitation is preponderant over the plastic deformation of crystals, low strain hardening and intense chain disentanglement take place during drawing at high draw ratio [15]. Strain hardening and no significant disentanglement takes place, instead, during plastic deformation of polymers up to high strains, provided that there is no cavitation [15]. Cavitation is negligible either for polymers having crystals with low plastic resistance or for polymers having crystals with high plastic resistance subjected to cavity-free deformation as for instance plane strain compression in a channel die [15].

This complex picture may be further complicated by occurrence of polymorphic transformations during plastic deformation at large degrees of deformation, that is, after the yield point. In fact, in many polymers the crystalline form that develops upon stretching may be different from the stable form present in the melt-crystallized undeformed samples. Moreover, in some cases, stretching may cause the disruption of lamellar crystals through the pulling out of chains from the crystals, leading to the formation of a mesophase, i.e. a solid phase characterized by large amount of structural disorder that may be considered as intermediate between amorphous and crystalline phases [16]. When the crystalline form that develops by stretching is metastable, it may transform back into the more stable form previously present in the unoriented sample, or into another polymorphic form, by removing the tensile stress. In some highly crystalline polymers the polymorphic transition occurring upon releasing the tension is reversible and is associated with a non trivial recovery of the initial dimensions of the sample (long range elasticity) [17–21]. The entity of plastic versus elastic deformation experienced by the material upon releasing the stress, may critically depend on the relative stabilities of the two crystalline phases that develop during successive cycles of stretching and relaxation. The enthalpic gain of the reversible crystal-crystal phase transition occurring upon releasing the tension may play a key role in the elasticity of these materials [20, 21].

The non-trivial role of the crystalline phase in the deformation behavior of semicrystalline polymers is here illustrated in the case metallocene-made isotactic polypropylene (iPP).

The development of single center catalysts for the polymerization of olefins has allowed production of new materials having microstructures that cannot be obtained with conventional Ziegler-Natta catalysts [22–25]. In the case of polypropylene, samples with any type and degree of stereoregularity, from highly isotactic to highly syndiotactic polypropylene can be produced [25]. The fine-tuning of the microstructure is nowadays possible through the rational choice of the catalytic system, and polypropylenes characterized by different kinds and amounts of regio- and stereo-irregularities, different distributions of defects, and different molecular weights, are now available [25]. The mechanical properties of these polymers depend on the crystallization behavior, which, in turn, depends on the chain microstructure and, in particular, on the stereoregularity [17–21].

The mechanical properties and the polymorphic transitions occurring during stretching of metallocene-made iPP samples with different amounts of stereo-irregularities (mainly isolated rr triads) are here discussed. Depending on stereoregularity, thermoplastic materials showing high stiffness, or high flexibility, or elastic properties can be produced [17, 18]. We show that different polymorphic transitions are involved during stretching of these samples [17, 18] and that stress-induced phase transitions are strain controlled rather than stress controlled [26]. A phase diagram of iPP is built up where the regions of stability of the different polymorphic forms of iPP are defined as a function of stereoregularity and degree of deformation. Finally we show that the elastic behavior of less crystalline and stereoregular samples is associated with a reversible polymorphic transition and that elastomeric iPP samples present rubber-like elasticity which originates from different mechanisms, depending on the degree of crystallinity.

17.2 Mechanical Properties of Unoriented Films

Samples of iPP of different stereoregularity containing only one kind of stereo-defect (isolated rr triads), with variable concentration in a wide range and uniformly distributed along the chains, and containing no measurable amount of regioerrors, have been prepared with the different metallocene catalysts of Fig. 17.1, activated with methylalumoxane (MAO) [27–31]. The amount of rr defects depend on the structure of the catalyst, in particular the indenyl substituents, and the conditions of polymerization, and can be varied in the range between 0.5 and 17% [17,18]. Correspondingly the samples show melting temperatures variable between 162 and 45°C (Table 17.1) [17,18].

Samples used for the study of the structural transformations occurring during stretching and for the mechanical tests have been prepared by compression molding. The X-ray powder diffraction profiles of melt-crystallized

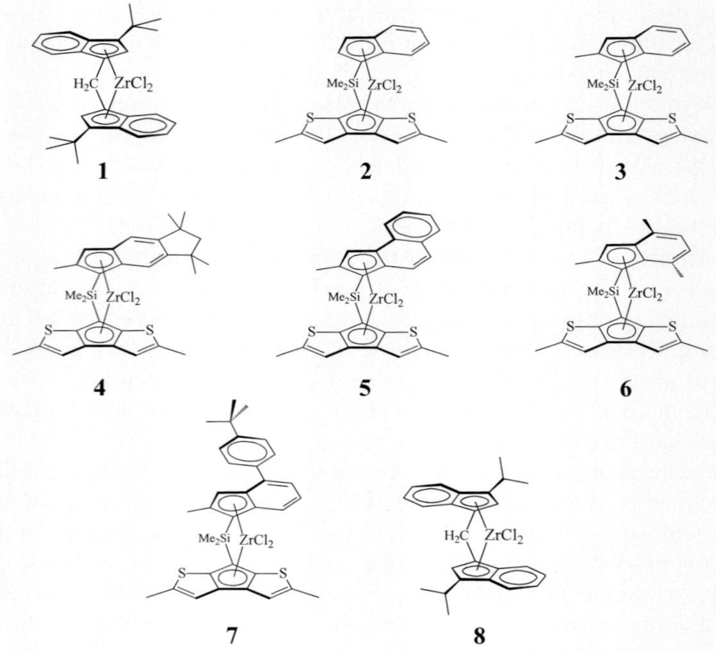

Fig. 17.1. Structure of C_2-symmetric (**1,8**) and C_1-symmetric (**2–7**) pre-catalysts

Table 17.1. Molecular masses (M_v), melting temperatures (T_m) and content of rr triad and $mmmm$ pentad stereosequences of iPP samples prepared with catalysts of Chart 1[a]

Sample	catalyst/cocatalyst	M_v^b	T_m (°C)[c]	mm %	mr %	rr %	mmmm %
iPP1	1/MAO	195,700	162	98.5	1.0	0.49	97.5
iPP2	7/MAO	106,000	140	92.4	5.1	2.54	87.6
iPP3	6/MAO	202,400	133	88.9	7.4	3.70	82.2
iPP4	3/MAO	505,800	119	83.4	11	5.52	73.9
iPP5	5/MAO	210,900	116	82.2	11.8	5.92	72.2
iPP6	2/MAO	166,400	111	76.9	15.4	7.68	64.5
iPP7	4/MAO	123,400	84	66.9	22.0	11.01	51.0
iamPP[d]	8/MAO	143,700	45	54.0	28.9	17.1	35

[a] No or negligible regioerrors (2,1 insertions) could be observed in the ^{13}C NMR spectra of the samples [29,30]. [b] From the intrinsic viscosities. [c] The melting temperatures were obtained with a differential scanning calorimeter Perkin Elmer DSC-7 performing scans in a flowing N_2 atmosphere and heating rate of 10°C/min [18]. [d] iamPP stands for isotactic amorphous polypropylene [31].

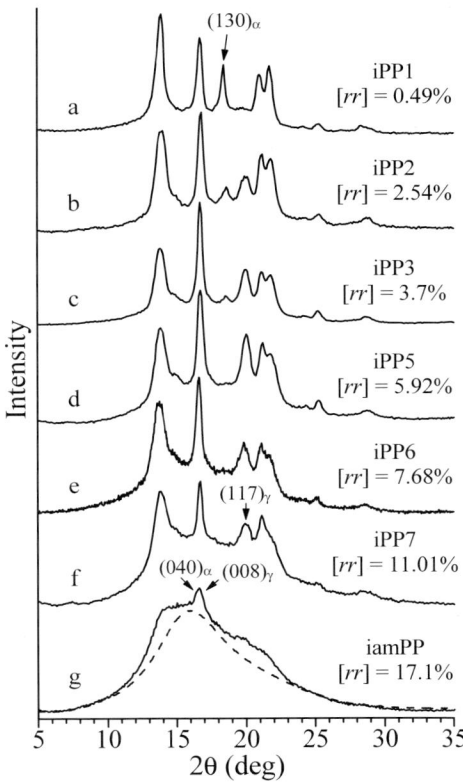

Fig. 17.2. X-ray powder diffraction profiles of iPP samples of Table 17.1 crystallized from the melt by compression molding and cooling the melt to room temperature at 1°C/min [18]. The *dashed line* indicates the diffraction profile of the amorphous phase. The $(130)_\alpha$ reflection of the α form [33] at $2\theta = 18.6°$ and the $(117)_\gamma$ reflections of the γ form [34] at $2\theta = 20.1°$ are indicated. The $(040)_\alpha$ and $(008)_\gamma$ reflections at $2\theta = 17°$ of α and γ forms, respectively, are also indicated

compression molded specimens of iPP samples of Table 17.1 are reported in Fig. 17.2. The samples iPP1-iPP7 crystallize from the melt as mixtures of α and γ forms (Fig. 17.3A and C, respectively), as indicated by the presence of both $(130)_\alpha$ and $(117)_\gamma$ reflections of α and γ forms, respectively, in the diffraction profiles of Fig. 17.2a-f. The intensity of the $(117)_\gamma$ reflection of the γ form at $2\theta = 20.1°$, increases with increasing concentration of rr defects [18].

The iPP sample of lowest stereoregularity (sample iamPP) does not crystallize by cooling the melt to room temperature, but slowly crystallizes in disordered modifications intermediate between α and γ forms (Fig. 17.3B), if the sample, cooled from the melt, is kept at room temperature for several days [32]. In fact, the X-ray diffraction profile of sample iamPP of Fig. 17.2g

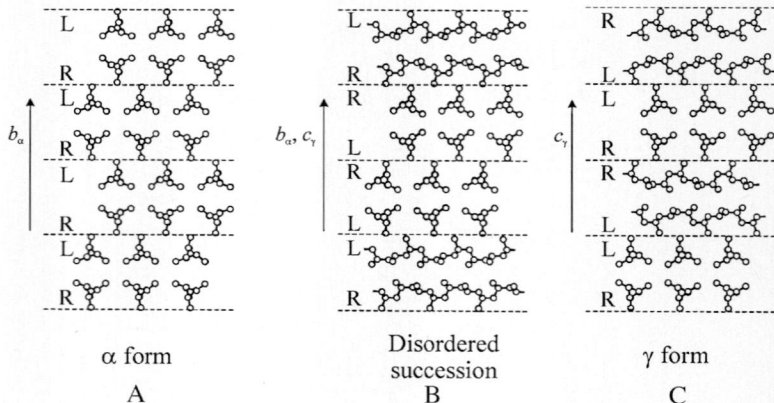

Fig. 17.3. Limit ordered models of packing proposed for α (A) and γ (C) forms of iPP and model of the α/γ disordered modifications intermediate between α and γ forms (B). The *dashed horizontal lines* delimit bilayers of chains. Subscripts α and γ identify unit cell parameters referred to the monoclinic [33] and orthorhombic [34] unit cells of α and γ forms, respectively. In the disordered model (B) consecutive bilayers of chains are stacked along b_α (c_γ) with the chain axes either parallel or nearly perpendicular, making α-like or γ-like arrangements of bilayers [32, 36]

presents only a sharp reflection at $2\theta = 17°$, corresponding to the $(040)_\alpha$ reflection of α form [33] or the $(008)_\gamma$ reflection of γ form [34]. The other sharp Bragg reflections of both α and γ forms, as $(110)_\alpha$ and $(130)_\alpha$ reflections at $2\theta = 14°$ and $18.6°$, respectively, typical of the α form [33], and $(111)_\gamma$ and $(117)_\gamma$ reflections at $2\theta = 14$ and $20.1°$, respectively, typical of the γ form [34] are absent. This indicates that the sample iamPP does not crystallize in the pure α or γ forms, but in a disordered modification intermediate between α and γ forms, [32, 35] containing disorder in the stacking along the b_α or c_γ direction of bilayers of chains with axes either parallel as in the α form or perpendicular as in the γ form (Fig. 17.3B) [35, 36].

The relative amount of γ form, with respect to the α-form in the melt crystallized samples of Fig. 17.2 is reported in Fig. 17.4 as a function of the concentration of rr defects. The most isotactic sample crystallizes basically in the α-form (Fig. 17.2a), with a limit low concentration of γ form of 15–20%. The amount of γ form increases with increasing content of rr defects up to 100% for rr concentrations higher than 6–7% [17, 18, 32].

The degree of crystallinity (Fig. 17.4) decreases only slightly with increasing concentration of rr defects in the range 0–11%, then drops to very low values for the less stereoregular sample iamPP.

The stress–strain curves of compression-molded films of iPP samples of Table 17.1 are shown in Fig. 17.5. The values of the most important mechanical parameters are reported in Fig. 17.6 as a function of the concentration of rr defects. The values of Young modulus decrease with increasing concentration

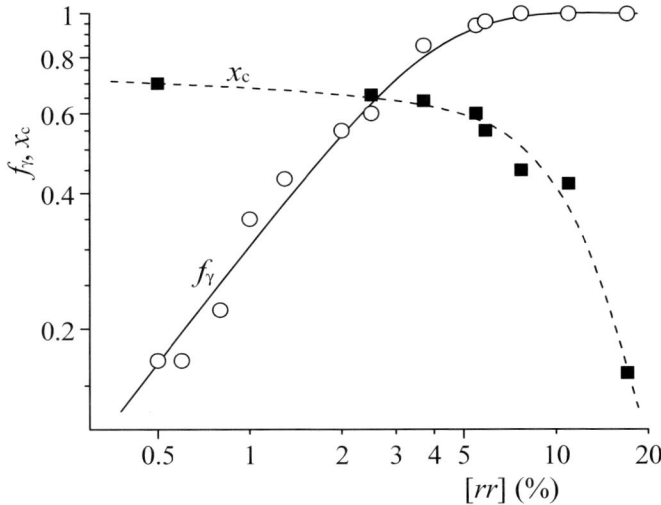

Fig. 17.4. Relative amount of γ form f_γ ($\bigcirc$), and degree of crystallinity x_c ($\blacksquare$), in the iPP samples of Table 17.1 crystallized from the melt by compression molding and cooling the melt to room temperature at cooling rate of $1°C/min$, as a function the concentration of rr defects [18]

of rr defects and decreasing crystallinity, from nearly 200 MPa of the sample iPP1 [18] to nearly 1 MPa of the poorly stereoregular sample iamPP [32] (Fig. 17.6A). The values of deformation at break, instead, increase with increasing concentration of defects (Fig. 17.6C). As expected, parallel to the decrease of the values of the modulus, a decrease of the stress at yielding with increasing concentration of defects is observed (Figure 17.6B) [18, 32].

Samples of iPP containing low concentration of rr defects (up to $[rr] =$ 3.7%, Figure 17.5A) show non uniform stretching behavior and high values of the elastic modulus (Fig. 17.6A) typical of stiff-plastic materials. Nevertheless, they present high ductility (Fig. 17.5A). In fact, these samples can be stretched at room temperature up to remarkable values of the strain (250–350%, Fig. 17.6C) [18].

A strong increase of the ductility and toughness is observed for higher contents of rr defects (Fig. 17.5B). In particular, samples with concentration of rr defects around 5–6% still present behavior of thermoplastic materials with slightly lower strength but much higher deformation at break ($\varepsilon_b \approx 1200\%$, Figs. 17.5B and 17.6C). This indicates that these samples ($[rr] = 5$–6% and melting temperatures around 110–115°C, for instance samples iPP4 and iPP5) behave as highly flexible thermoplastic materials [18].

The less crystalline samples, with concentration of rr defects in the range 7–11% (samples iPP6 and iPP7) show strong strain hardening at high deformation (Fig. 17.5B) with values of the tensile strength (32 MPa) higher than

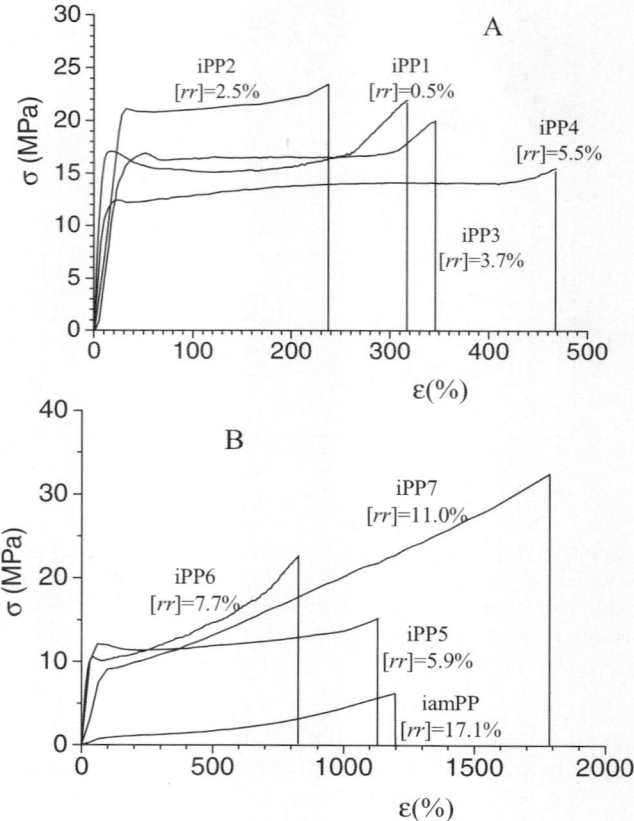

Fig. 17.5. Stress-strain curves of unoriented compression molded films of iPP samples of Table 17.1 [17, 18, 32]

those of the more isotactic and crystalline samples (20–23 MPa, Figures 17.5B and 17.6B) [18]. The strain hardening of these samples may be somehow related to the fact that they present uniform stretching behavior, crystals with low plastic resistance and therefore no cavitation [15]. Once crystals have experienced irreversible plastic deformations at low values of deformation, strain hardening is caused by straightening of the entangled network [15]. Similar strain hardening is experienced also by the lowest isotactic and crystalline sample iamPP [32], even though the achieved values of the tensile strength and the values of the stress at any strain are one order of magnitude lower than those of samples iPP6 and iPP7, because of the much lower level of crystallinity.

Samples with the highest concentration of rr defects ($[rr] = 7$–17%) show elastomeric properties. The stress-strain curves of samples iPP6, iPP7 and

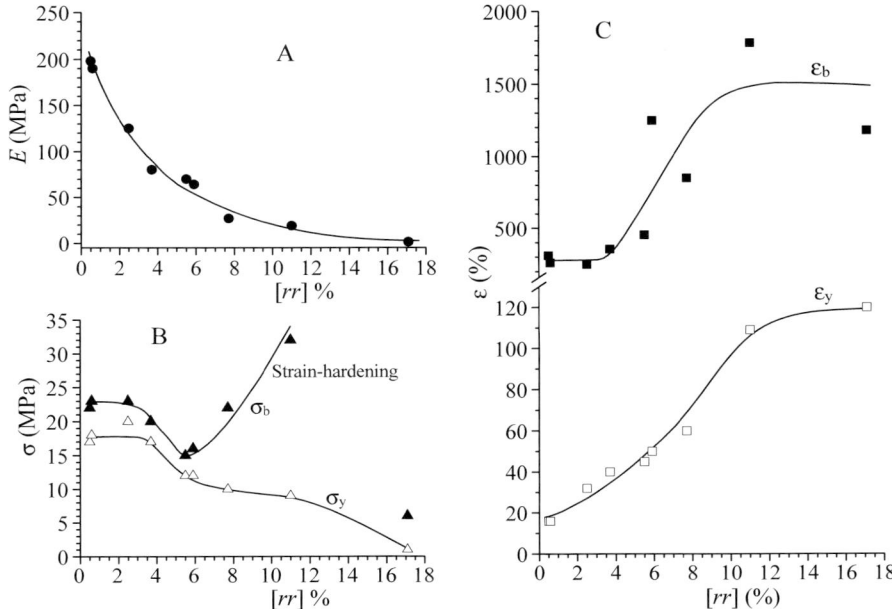

Fig. 17.6. Values of elastic modulus, E, (**A**), stress, σ, (**B**) and deformation, ε, (**C**) at break (*full symbols*) and at yield point (*open symbols*) as a function of concentration of rr defects of iPP samples of Table 17.1 [17, 18, 32]

iamPP present, indeed, typical shape of elastomeric materials (Fig. 17.5B), showing high values of deformation at break and strain-hardening at high deformations. The values of the tension set (t_s), that is the residual deformation achieved upon the release of the tension after stretching, measured at room temperature for unoriented films stretched up to the break or up to a deformation ε, are reported in Fig. 17.7. These values have been obtained as $t_s = 100(L_r - L_o)/L_o$, by stretching unoriented films of initial length L_0 up to the break, as in Fig. 17.5, or up to a deformation ε (final length L_f), then measuring the final length L_r of the relaxed sample ten minutes after breaking or after removing the tension. The low values of the tension set after breaking or after a deformation ε clearly indicate a good elastic behavior of the less stereoregular samples of Table 17.1 [18], especially of the sample iamPP [32].

In the samples iPP6 and iPP7, the elastic properties are associated with remarkable values of the modulus of nearly 20–30 MPa (Fig. 17.6A), and, as discussed before, very high values of tensile strength [18] (Fig. 17.6B). In the case of the poorly stereoregular sample iamPP, low values of the tension set are observed even at high deformation (Fig. 17.7), indicating that the sample iamPP experiences a recovery of the initial dimension after breaking as well as after removing the tension from any deformation [32]. The tension

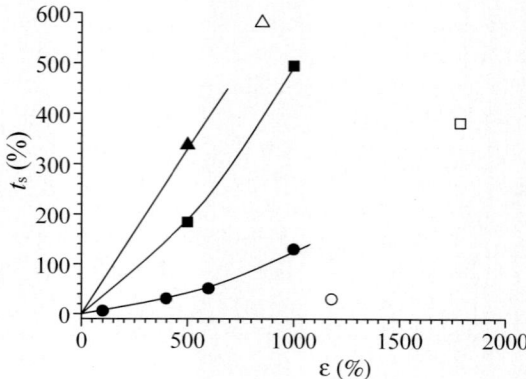

Fig. 17.7. Values of the tension set after breaking (*open symbols*), and after deformation ε (*full symbols*) of unoriented compression molded films of samples iPP6 (*triangles*), iPP7 (*squares*) [18], and iamPP (*circles*) [32].

set increases with increasing deformation, and values of tension set higher than those observed after breaking are obtained for the samples iPP7 and iamPP for values of deformation higher than 400%. This is probably due to the fact that the tension set after a given deformation ε has been measured after keeping the sample in tension for 10 min, allowing relaxation of the sample.

The elastic behavior of samples iPP6, iPP7 and iamPP is due to the fact that these samples are crystalline notwithstanding the low stereoregularity. These samples, indeed crystallize in the γ form of iPP or in α/γ disordered intermediate modifications thanks to the inclusion of most of the rr stereodefects (profiles e-g in Fig. 17.2) [18,32]. The formation of small crystalline domains induces elastomeric properties since crystals act as physical cross-links in the amorphous matrix, producing the elastomeric network. The presence of a high level of crystallinity (40–45%, Fig. 17.4) of samples iPP6 and iPP7, gives high values of the strength (Fig. 17.6), so that interesting thermoplastic elastomers with remarkable values of the modulus and tensile strength are obtained [17, 18]. The small degree of crystallinity of the poorly stereoregular sample iamPP ($\sim$16%), induces good elastic properties in a range of deformation larger than that shown by samples iPP6 and iPP7. The small crystalline domains that develops upon aging, act as physical knots of the elastomeric network, preventing the viscous flow of the amorphous chains and giving a typical thermoplastic elastomeric behavior. The poorly isotactic sample iamPP shows, indeed, poor elastic properties and viscous flow at high deformations in the amorphous state, before crystallization [32].

It has been argued that the outstanding mechanical properties of metallocene-made iPP samples containing only rr defects are related to the easy inclusion of rr defects inside the crystalline phase [17, 18, 32, 36, 37]. The high ductility and good drawability at room temperature of these materials even when the concentration of rr defects is low and the samples basically crystallize in the α form, may be indeed explained by the fact that rr stereodefects are uniformly distributed between crystalline and amorphous phases, which have, therefore, the same composition. In these materials, the presence of defects in the crystals first of all decreases their plastic resistance. Moreover, the similar composition of the crystalline and amorphous phases makes the process of plastic deformation easier, reducing the stress level necessary for the destruction of the preexisting lamellae and re-crystallization of chains in new oriented crystallites with fibrillar morphology.

It is worth noting that metallocene-made iPP samples containing low amount of rr stereodefects (0.1–0.2%) and slightly higher concentration of defects of regioregularity (0.8–0.9% of 2,1 erythro units) are stiff and fragile with high values of the Young's modulus and very low values of deformation at break (ε_b around 6%) [37]. These samples do not undergo plastic deformation at room temperature and break before yielding. This behavior is similar to that of the commercial highly isotactic polypropylene prepared with Ziegler-Natta catalysts. It has been argued that the different effect of stereo- and regio-defects on the mechanical properties of iPP is probably related to their different levels of inclusion inside the crystalline phase. The amount of 2,1 regiodefects included in the crystalline phase, indeed, is much lower than that of rr stereodefects [38,39]. Since crystallization tends to reject the 2,1 regiodefects more strongly than the rr defects [35–37], in regioirregular iPP samples the composition of the crystalline and amorphous phases are not identical [37]. Therefore, regioirregular iPP samples are stiffer and more fragile than iPP samples containing only rr stereodefects, because the crystals show higher plastic resistance due to the low inclusion of regiodefects, and also because the non identical composition of the crystalline and amorphous phases makes the process of plastic deformation much harder, increasing the stress level necessary for the destruction of the crystalline lamellae up to values that produce breaking of the material [37].

17.3 Stress-induced Phase Transitions in Unoriented Fims

The plastic deformations of iPP samples of Table 17.1 are associated with irreversible morphological changes and polymorphic transitions. The structural and morphological transformations occurring during stretching of iPP samples of Table 17.1 have been analyzed by X-ray diffraction. Examples of X-ray fiber diffraction patterns are reported in Figs. 17.8–17.11 for samples having different concentration of rr defects.

The stretching behavior of unoriented films of the more stereoregular samples of Table 17.1 with rr content below 5%, is shown in Fig. 17.8, in the case of

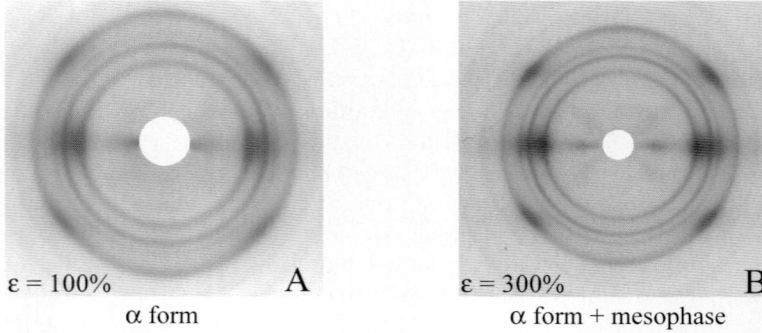

Fig. 17.8. X-ray fiber diffraction patterns of fibers of the sample iPP1 with $[rr] = 0.49\%$ obtained by stretching at room temperature compression molded films at the indicated values of the strain ε [18]

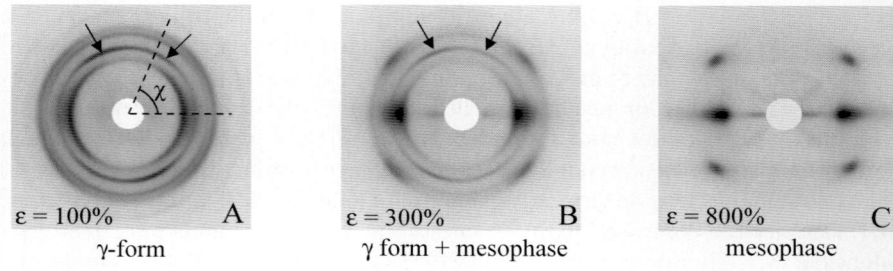

Fig. 17.9. X-ray fiber diffraction patterns of fibers of the sample iPP5 with content of rr defects of 5.9% obtained by stretching at room temperature compression molded films at the indicated values of the strain ε [18]

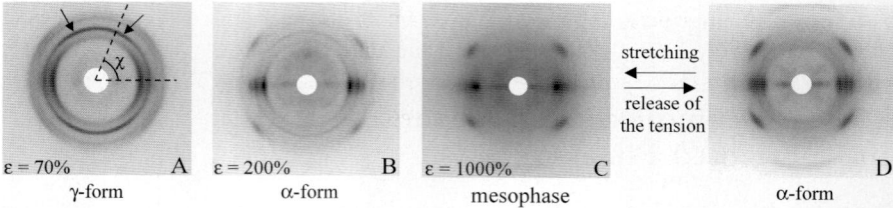

Fig. 17.10. X-ray fiber diffraction patterns of fibers of the sample iPP7 with content of rr defects of 11.01%, obtained by stretching compression molded films at the indicated values of the deformation ε keeping the fiber under tension (**A–C**) and after removing the tension from 1000% deformation (**D**) [18]

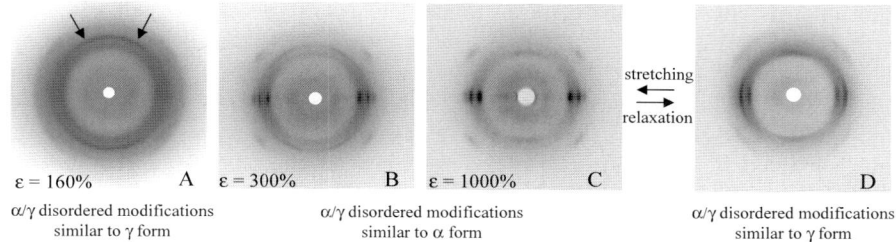

ε = 160% A   ε = 300% B   ε = 1000% C   D
α/γ disordered modifications   α/γ disordered modifications   α/γ disordered modifications
similar to γ form   similar to α form   similar to γ form

Fig. 17.11. X-ray fiber diffraction patterns (**A–C**) of fibers of the sample iamPP obtained by stretching at room temperature compression molded films at the indicated values of strain ε and after releasing the tension from 1000% deformation (**D**) [32]

the most isotactic sample iPP1 ($[rr] = 0.5\%$). For these samples, compression-molded unstretched films are generally crystallized in mixtures of α and γ forms (samples iPP2 and iPP3, Fig. 17.2b and c, respectively), whereas the sample iPP1 is basically in the α form (Fig. 17.2a) [18]. These samples behave as stiff plastic materials even though they can be easily stretched up to 200–300% deformation at room temperature (Fig. 17.5A). The crystalline form initially present in these samples partially transforms into the mesomorphic form already at low draw ratios (Fig. 17.8A) [18]. The formation of the mesomorphic form is indicated in the X-ray fiber diffraction pattern of the sample iPP1 stretched at 100% deformation, by the presence of a broad halo in the range of $2\theta = 14-18°$, typical of mesomorphic form of iPP, subtending non-oriented reflections of the α form (Fig. 17.8A). The stretching at room temperature of the most stereoregular iPP sample, does not produce high orientation of crystals, even at the maximum deformation before breaking (Fig. 17.8B), probably because of the limited possible deformation. A fraction of unoriented crystals of α form is still present even at the maximum possible deformation. Crystals that undergo plastic deformation and achieve orientation rapidly transform into the mesomorphic form, whereas non deformed crystals remains in the crystalline form (α form) of the initial unoriented film [18].

Samples with higher rr content, in the range 4–6%, are highly flexible materials that show very high deformation at break (Fig. 17.5A). For these samples, the γ form, originally present in compression-molded unstretched films (Fig. 17.2d), gradually transforms into the mesomorphic form by stretching. The gradual transformation of the γ form into the mesomorphic form at high draw ratios is shown in the case of sample iPP5 in Fig. 17.9, as an example [18].

Samples with concentration of rr defects in the range 7–11% are thermoplastic elastomers with high strength. For these samples, the γ form present in unstretched films (Fig. 17.2e,f) transforms by stretching into the α form, which, in turn, transforms into the mesomorphic form at very high deformations, as shown as an example in Fig. 17.10 in the case of sample iPP7 ($[rr] = 11\%$) [18].

As discussed above, the poorly isotactic sample iamPP is essentially amorphous and does not crystallize from the melt due to the high concentration of defects. It slowly crystallizes upon aging at room temperature or by stretching in disordered modifications intermediate between the α and γ forms, with a maximum degree of crystallinity of only 16% (Fig. 17.2g) [32,35]. The stretching of this sample, even at high deformation, does not produce formation of the α form or the mesomorphic form, as instead occurs for the samples iPP6 and iPP7, but only α/γ disordered modifications, more similar to the α form are obtained (Fig. 17.11) [32,35]. This different behavior of the sample iamPP is due to the very short length of the regular isotactic sequences, the average value being only five monomeric units [35]. The observed crystallinity in this sample and in iPP samples having very low stereoregularity may be, indeed, explained by the fact that when the concentration of rr defects is high, these defects may be included in the crystals of γ form more easily than in the α form [35,36]. More precisely, isolated rr triad defects can be easily tolerated at low cost of conformational and packing energy in the crystal lattices of α and γ forms of iPP and of the α/γ disordered modifications of Fig. 17.3B, but the inclusion in the γ form or in α/γ disordered modifications of Fig. 17.3B is more probable [32,35,36]. This explains the tendency of the sample iamPP to crystallize in disordered modifications intermediate between α and γ forms (Fig. 17.2g) and the fact that, for this sample, the pure α form is never obtained, even at high deformations [32,35]. Disordered modifications intermediate between the α and γ forms more similar to the α form are instead obtained at high deformations (Fig. 17.11C), where a non-negligible fraction of bilayers of chains arranged with non parallel chain axes as in the γ form is still present [35].

The data of Fig. 17.9–17.11 indicate that the γ form of iPP is mechanically unstable, and tends to transform by stretching into the mesomorhic form in iPP samples with rr content in the range 2–6% [18] (Fig. 17.9), and into the α form (Fig. 17.10) or in disordered modifications intemediate between α and γ forms closer to the α form (Fig. 17.11), in less stereoregular iPP samples with rr content in the range 7–17% [17,18,32,35]. The transformation of γ form into modifications closer to the α form during stretching occurs through a continuum of disordered modifications intermediate between the α and γ forms [35]. More precisely, with increasing the draw ratio, the fraction of consecutive bilayers of chains faced with parallel chain axes as in the α-form increases, whereas the fraction of consecutive bilayers of chains faced with perpendicular chain axes as in the γ-form decreases (Fig. 17.3) [35]. This structural transformation is accompanied by the simultaneous increase of orientation of crystals with chain axes parallel to the stretching direction [32,35].

It is worth noting that during the development of the fibrillar morphology, for deformations below a critical value that presumably coincide with

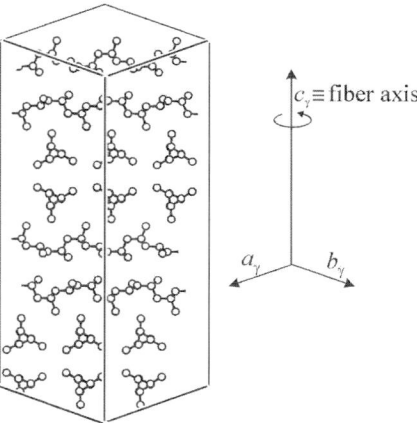

Fig. 17.12. Lamella of γ form oriented with the chain axes directed normal to the fiber axis and therefore with the c_γ axis, the piling direction of bilayers of chains, parallel to the z-axis (fiber axis, cross-β orientation)

the disappearance of γ form, changes of the texture of the sample also occur, consisting in the orientation of portion of the crystalline lamellae with the chain axes nearly perpendicular to the stretching direction instead than parallel, as in the standard fiber morphology (Fig. 17.12) [18, 32]. This non standard crystals orientation is achieved, for instance, in the case of the sample iPP5, iPP7 [18] and iamPP [32] up to values of deformation ε of 500, 400 and 600%, respectively, as indicated by the polarization of the $(040)_\alpha$ reflection of α form, or the $(008)_\gamma$ reflection of γ form at $d = 5.21$Å ($2\theta = 17°$), at oblique angles, indicated with arrows in Figures 17.9A,B, 17.10A and 17.11A. At higher deformations, the diffraction maxima at oblique angles disappear, and this reflection is polarized on the equator, as in the standard fiber morphology (Figs. 17.9C, 17.10C and 17.11C). The nearly meridional polarization of the reflection at $2\theta = 17°$ (($008)_\gamma$ reflection of γ form or $(040)_\alpha$ reflection of α form) in the patterns of Fig. 17.9–17.11A indicates that portion of the crystals of γ form, or in disordered modifications intermediate between the α and γ forms more similar to the γ form, assume an orientation with the c_γ-axes of γ form (b_α axes of α form) nearly parallel to the stretching direction (Figure 17.12) [18,32]. Since the c_γ axes of γ form (b_α axes of α form) are the axes of stacking of bilayers of chains (Fig. 17.3) this non standard mode of orientation of iPP crystals corresponds to lamellae oriented with chain axes nearly perpendicular to the fiber axis (Fig. 17.12).

A similar kind of orientation has been well known for many years in some naturally occurring fibrous proteins such as silks [40]. The perpendicular orientation of chain axes with respect to the fiber axis, described as cross-β, occurs at low draw ratio and has been explained by the fact that in these soft silks the small crystallites are elongated along the hydrogen bond directions, which run perpendicular to chain axes [40]. The cross-β orientation in iPP may be attributed to the simultaneous occurrence of two kinds of slip processes at low deformations, interlamellar and intralamellar [10]. Whereas interlamellar shear leads to a location of the $(008)_\gamma$ reflection of γ form $((040)_\alpha$ reflection of α form) on the meridian, the intralamellar shear pushes the chain axes to align parallel with the stretching direction, and thus shifts the position of the reflection at $2\theta = 17°$ toward the equator.

The fact that the cross-β orientation is apparent up to 500–600% deformation, indicates that not all the crystalline lamellae of γ form originally present in the sample experience simultaneously the uniaxial mechanical stress-field. The non standard mode of orientation of these crystals reflects crystallographic restraints on the slip processes, and topological constrains on the response of crystals to the tensile stress field. In the crystalline domains of γ form, indeed, the chains are oriented along two perpendicular directions, and the crystallites have the shape of elongated entities along the direction normal to the chain axes [18, 32]. Because of the intrinsic structural and morphological characteristics of γ form, at low deformations portion of γ lamellae remains frozen in strained positions of the polymer matrix with the chain axes oriented nearly perpendicular to the stretching direction. By stretching at higher deformations the γ form transforms into the α form. Since in the crystals of α form the chains are all parallel, the deformation also induces orientation of crystals with the chain axes oriented along the stretching direction, as in a standard fiber morphology [18, 32]. This mechanism is confirmed by the fact that this non standard mode of orientation of crystals has been observed only during stretching of iPP samples mainly crystallized in the γ form or in disordered modifications intermediate between the α and γ forms, and has never been observed for iPP samples crystallized in the pure α form.

The structural analysis of fibers of the iPP samples having different stereoregularity, have allowed building the phase diagram of iPP at room temperature reported in Fig. 17.13.

From the phase diagram the regions of stability of the different polymorphic forms of iPP in oriented fibers are defined as a function of stereoregularity and degree of deformation ε. The values of the critical strain at which the polymorphic transitions start and at which the transformation is complete depend on the stereoregularity. The critical values of the stress instead depend also on other parameters as, for instance, the degree of crystallinity of the sample, the amount of structural disorder present in the crystals and on the method of preparation of the test specimens.

It is apparent from Fig. 17.13 that for highly stereoregular iPP samples, with concentration of *rr* defects lower than 1% and content of isotactic pentad *mmmm* higher than 94%, the α form present in the unstretched melt-crystallized sample transforms into the mesomorphic form already at low deformations. For less stereoregular iPP samples, with content of *rr* defects in the range 2–6% and concentration of *mmmm* pentad in the range 68–94%,

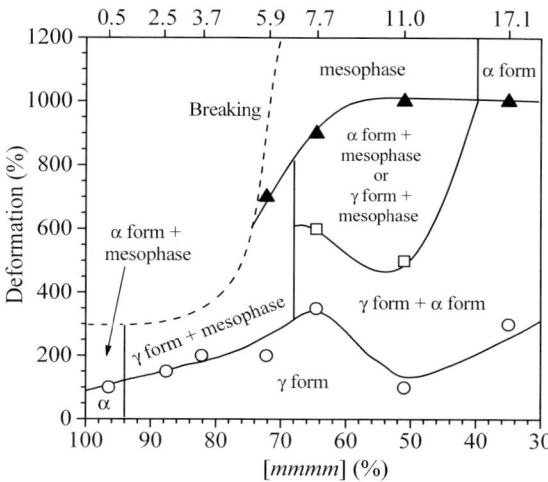

Fig. 17.13. Phase diagram of iPP showing the region of stability of the different polymorphic forms as a function of deformation ε ($\varepsilon = 100(L_f - L_0)/L_0$) and stereoregularity, defined as concentration of the fully isotactic pentads $mmmm$. The values of critical strains corresponding to the boundary lines between the various crystalline forms have been determined from the X-ray fiber diffraction patterns of Figs. 17.8–17.11. The concentration of rr triad defect is indicated in the *upper scale*. The deformation at break is also indicated (*dashed line*)

the unoriented melt-crystallized samples are in the γ form that transforms directly into the mesomorphic form by stretching at values of deformation higher than a critical value. For more stereoirregular samples, with rr content in the range 6–11% and concentration of $mmmm$ pentad in the range 40–68%, the γ form transforms at low deformations into the α form which, in turn, transforms into the mesomorphic form with increasing deformation.

The transformation of α form into the mesomorphic form by stretching at values of deformation higher than a critical value has been already observed in the case of stretching at room temperature of iPP samples prepared with Ziegler–Natta catalysts [41]. These studies have indicated that there was no lamellar structure in the mesomorphic form of the iPP fibers. It has been suggested that the formation of the mesophase occurs through the destruction of the lamellar crystalline phase, probably by pulling chains out from crystals, and the dominant constituent of the mesomorphic form may be oriented bundles of helical chains [41]. Also for our metallocene-made iPP samples we can assume that the formation of the mesomorphic form, from fibers of α form, in the case of the elastomeric samples iPP6 and iPP7 (Fig. 17.10), or directly from the γ form in the case of the more stereoregular samples iPP2–iPP5, occurs via the pulling out of the chains from the lamellae of pre-existing crystalline form and successive re-organization of the chains in crystalline mesomorphic aggregates. The latters are characterized by chains in 3/1 helical conformation, where the parallelism of the chain axes is maintained and only a poor correlation in the lateral positioning of the chain axes is present [42].

The transformation of γ form into the α form and/or into disordered modifications intermediate between the α and γ forms, closer to the α form, that occurs at low deformations in elastomeric stereoirregular iPP samples (with rr content higher than 7%, $[mmmm] < 60\%$), is also not direct. This transition is gradual, and occurs through a continuum of disordered modifications intermediate between the α- and γ-forms, and probably corresponds to the progressive breaking of the pre-existing lamellae with formation of the new ones during stretching. Direct transformation of γ form into α form is, in fact, prevented for steric reasons by the fact that whereas in the γ form the bilayers of chains are stacked along the c_γ-axis direction according to the sequence: [34] ...LRRLLR... (Fig. 17.3C, L and R standing for righ- and left-handed helical chains), in the α form the bilayers are stacked along the b_α-axis direction with a strict alternation of helical hands: [33] ...LRLRLR... (Fig. 17.3A). This transformation would imply simultaneous inversion of helical hand of the chains belonging to every bilayer, making a direct mechanism very unlikely.

It is worth noting that the presence of large biphasic regions in the phase diagram of Fig. 17.13 indicates that not all crystals undergo simultaneously phase transition during stretching, consistent with the fact that in semicrystalline polymers, whereas the strain is homogeneously distributed, the stress is not homogeneously distributed [3, 6, 14] and, at any strain only the crystals of a given form that experience a stress higher than a critical value undergo reorientation and/or transform into another polymorphic form. In other terms, the stress-induced phase transitions during plastic deformations are regulated by the same factors that govern the textural and morphological changes (transformations of spherultic morphology into fibrillar morphology and development of cross-β orientation of crystals), and reflects crystallographic restraints on the slip processes, and topological constraints on the response of crystals to the tensile stress field transmitted by the interconnected crystallites.

The above consideration support the hypothesis that when polymorphic transformations occur during plastic deformation, in iPP but also in many other polymers, the phase transitions are strain controlled rather than stress controlled [26]. As discussed above, the critical values of deformation at which the polymorphic transitions start always correspond to the destruction of the original lamellae of a given crystalline form, and re-crystallization with formation of fibrils in a new crystalline form. These critical values of the strain are indeed higher than the deformation values at point C of the stress-strain curve, that is the point at which the destruction of crystal blocks starts, followed by re-crystallization with formation of fibrils [3, 10–13]. Moreover these critical values depend on the stereoregularity of the sample (Fig. 17.13). This suggests that the two factors that govern the location of the critical strain corresponding to the formation of fibrils, that is the modulus of the entangled amorphous and the stability of the crystal blocks [6], depend on the stereoregularity. Chains of different stereoregularity possess, indeed, different flexibility. In fact, the relative configuration of consecutive stereoisomeric centers along the chain affects the space correlation among skeletal bonds and the rotational energy barriers around the C-C bonds [43]. Since the dynamics of macromolecular chains is largely controlled by these parameters, which can be defined as "the internal viscosity" [44], different degrees of stereoregularity produce a different entanglement density of the amorphous phase. The stereoregularity also affects the stability of crystals (besides the degree of crystallinity) and may influence the relative stability of the different polymorphic forms involved in the structural transformations [26]. In the case of iPP, it has been shown that the presence of rr defects induces crystallization from the melt of γ form and of disordered modifications intermediate between α and γ forms (Fig. 17.2), and affects the polymorphic transitions occurring during stretching of unoriented samples (Fig. 17.13). In particular, the easy inclusion of isolated rr triads defects in the crystal lattices of the α and γ forms of iPP [35, 36], and the fact that when the concentration of rr defects is very high their inclusion in the γ form or in α/γ disordered modifications of Fig. 17.3B is more probable [37], explain the observed tendency of iPP samples of very low stereoregularity to crystallize achieving non negligible values of crystallinity. Moreover this also explains the fact that these samples crystallize in disordered modifications intermediate between α and γ forms [32, 35], where the fraction of consecutive bilayers arranged as in the α-form increases with increasing deformation (Fig. 17.11 and 17.13).

17.4 Oriented Fibers of Elastomeric Samples

The analysis of the mechanical properties (Figs. 17.5–17.7) and the structural characterization of the stereoirregular elastomeric samples iPP6 and iPP7 have shown that, because of the presence of a significant level of crystallinity, unoriented films undergo irreversible plastic deformation that involves structural and morphological transformations. This explains the relatively high values of tension set measured after breaking or after a given deformation (Fig. 17.7), which indicate a non-complete elastic recovery after the first stretching [18]. In particular the more crystalline sample iPP6 presents an elastic recovery lower than that of the sample iPP7 (Fig. 17.7). Unoriented films of the less stereoregular sample iamPP, instead, show tension set values at any deformation and at break lower than those of samples iPP6 and iPP7, because of the lower degree of crystallinity [18, 32].

For these samples a perfect elastic recovery is instead observed in successive cycles of stretching and relaxation of oriented fibers, regardless of stereoregularity [18, 32], as shown by the stress-strain hysteresis curves of fibers of samples iPP7 and iamPP reported in Fig. 17.14. These fibers have been prepared by stretching unoriented films obtained by compression molding up to 1000% and 600% deformation, for samples iPP7 and iamPP, re-

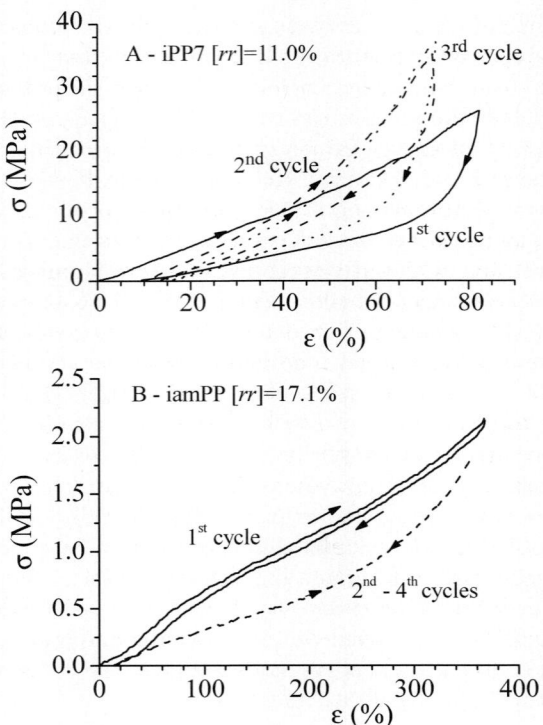

Fig. 17.14. Stress-strain hysteresis cycles recorded at room temperature, composed of stretching and relaxation (at controlled rate) steps according to the direction of the arrows, for stress-relaxed fibers of the samples iPP7 [17,18] (**A**) and iamPP [32] (**B**). The stress-relaxed fibers have been prepared by stretching compression-molded films, of initial length L_0, up to 1000% or 600% elongation, for the samples iPP7 and iamPP, respectively, (final lengths $L_f = 11L_0$ or $7L_0$, respectively), and, then, removing the tension. In the hysteresis cycles the stretching steps are performed stretching the fibers up to final length $L_f = 11L_0$ for the sample iPP7 and $L_f = 7L_0$ for the sample iamPP. (**A**) Continuous lines: first cycle; dashed lines: second cycle; dotted lines: third cycle and successive cycles. (**B**) Continuous lines: first cycle; dashed lines: curves averaged for at least four cycles successive to the first one

spectively, keeping the fibers under tension for 10 min at room temperature, then removing the tension, allowing the specimens to relax. The hysteresis cycles of Fig. 17.14 have been obtained by stretching the so prepared stress-relaxed oriented fibers, having the new initial length L_r, up to the final lengths $L_f = 11L_0$ and $7L_0$ for the samples iPP7 and iamPP, respectively, with L_0 the initial length of the unoriented film. It is apparent that successive hysteresis cycles, measured after the first one, are in all cases nearly coincident, indicating a tension set close to zero and a perfect elastic recovery [18,32].

Fibers of crystalline iPP6 and iPP7 samples show elastic behavior in a non trivial deformation range, which, however is much lower than the maximum deformation achieved during the preparation of the fibers [18]. This is due to the fact that unoriented films, when stretched at high deformation, do not experience total recovery of the initial dimension upon removing the tension (the tension set observed after the first stretching being higher than 100%, Fig. 17.7) [18]. Fibers of the less crystalline sample iamPP, instead, show elastic behavior in a much larger deformation range, nearly coincident with the maximum deformation achieved during the first stretching of unoriented films. In fact, since the unoriented films can be stretched up to very high deformation (up to 1200%) and experience a nearly total recovery of the initial dimension upon removing the tension (the tension set observed after the first stretching being very low even for large deformation, Fig. 17.7), the oriented fibers give elastic response in a large range of deformation, up to the maximum deformation achieved during the preparation of the fibers [32].

The stereoregularity also influences the polymorphic behavior of elastomeric iPP samples upon relaxation of fibers by releasing the tension. In the case of more isotactic and crystalline iPP6 and iPP7 samples, the mesomorphic form obtained by stretching at high draw ratios (Fig. 17.10C), transforms into the α form by releasing the tension (Fig. 17.10D) [18]. This transformation is reversible upon successive stretching and relaxing cycles and is associated with the fully elastic recovery of the sample (Figure 17.14A). The crystalline α form transforms by stretching into the disordered mesomorphic form, which, in turn, transforms back into the α form by releasing the tension (Fig. 17.10C and D) [18]. It is worth noting that the transformation of the disordered mesomorphic form into the α form corresponds to an increase of crystalline order. The crystallization upon removing the tension in stretched fibers is not common in polymers and is opposite to what is generally observed in a common elastomer as the natural rubber, for which crystallization occurs during stretching, whereas melting occurs upon releasing the tension [45, 46]. The crystallization of the mesomorphic form into the α form upon releasing the tension is not observed in the case of flexible or stiff-plastic samples (Figs. 17.8 and 17.9), which do not show elastic behavior. This indicates that in the elastomeric samples elasticity is probably partially due to the enthalpic contribution associated with the crystallization of the mesomorphic form into the α form [18].

Also in the case of the poorly isotactic sample iamPP the elastic recovery observed upon releasing the tension (Fig. 17.14B) is associated with a polymorphic transformation occurring inside the crystalline domains. Disordered α/γ modifications of iPP very close to the α form, having a high fraction of consecutive bilayers facing as in the α form with parallel chain axes, formed during stretching (Fig. 17.11C), transform back into more disordered modifications closer to the γ form upon releasing the tension (Fig. 17.11D), with a simultaneous decrease of the degree of orientation of the crystals [32]. These small crystalline domains act as physical knots in an amorphous matrix. The chains belonging to the amorphous phase, connecting the crystalline regions, undergo a reversible conformational transition between the entropically favored disordered random coil conformation in the unstretched state and the extended conformation in the stretched state. Therefore, the entropic effect due to this conformational transition is responsible for the elasticity. These amorphous chains are entangled and connect, as tie-chains, the small crystalline domains. They act as springs between the crystals being well-oriented and in extended conformation in the stretched state and return in the disordered coil conformation when the tension is removed [32].

It is worth noting that after the first hysteresis cycle, whereas in the case of elastomeric samples iPP6 and iPP7 a remarkable increase of the strength occurs (Fig. 17.14A) (hardening), in the case of the sample iamPP the stress at any strain decreases (Fig. 17.14B) (softening). The hardening in the case of more crystalline elastomeric samples iPP6 and iPP7 maybe, somehow, related to the increase of crystallinity and the structural transitions occurring during stretching. The formation of the metastable mesomorphic form at high deformation and the successive crystallization into the α form upon relaxation may play an important role. In the case of the sample iamPP instead, softening may be somehow related to a decrease of entanglement density due to viscous flow of amorphous chains.

17.5 Conclusions

The non trivial role of crystalline phase during plastic deformation has been examined in the case of samples of isotactic polypropylene prepared with metallocene catalysts. Samples with high molecular mass and variable stereoregularity, containing only rr stereo-defects and no regio-defects, have been analyzed. These samples show crystallization and mechanical properties that depend on the degree of stereoregularity. The continuous change of mechanical properties of metallocene-made iPP samples as a function of concentration of rr defects of stereoregularity is shown in Fig. 17.15. Samples with low concentration of rr defects, up to 3–4%, present high melting temperatures, in the range 162–130°C, and behave as stiff plastic materials. Sample with higher rr content, in the range 4–6%, and melting temperatures around 115–120°C are highly flexible thermoplastic materials, showing very high deformation at break. Samples with concentration of rr defects in the range 7–11% and melting temperature in the range 80–110°C are thermoplastic elastomers with high strength. Samples with concentration of rr defects higher than 11% and melting temperature lower than 50°C are soft materials with elastomeric properties more similar to those of conventional thermoplastic elastomers.

The samples show a complex polymorphism during tensile deformation. The relationships between the different mechanical behavior and the stress-induced phase transitions are discussed in terms of a general view, outlining the concept that stress-induced phase transitions during plastic deformation of

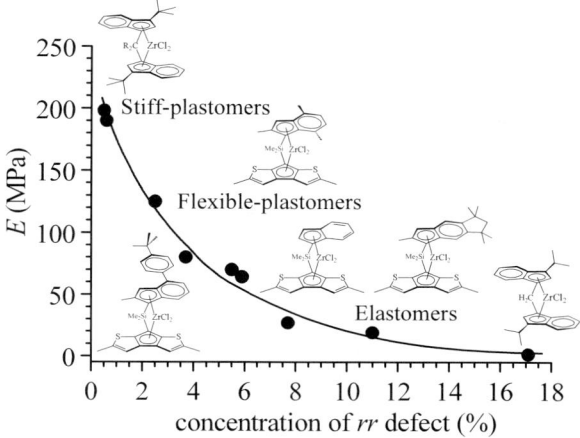

Fig. 17.15. Classification of i-PP samples prepared with different catalysts, as stiff-plastic materials, flexible-plastic materials, and thermoplastic elastomers depending on concentration of rr defects of stereoregularity and Young's modulus (E)

semicrystalline polymers are governed by the same rules that govern their deformation behavior. Polymorphic transformations occur through breaking of preexisting lamellae of the original crystalline form and formation of fibrils of the new crystalline form. These transitions, for a given sample, appear strain controlled rather than stress controlled. The values of the critical strain linked to the polymorphic transitions are namely affected by the chain microstructure, whereas the corresponding values of the stress depend on the degree of crystallinity, the amount of structural disorder present in the crystals and on the relative stability of the two involved crystalline forms. A phase diagram of iPP where the regions of stability of the different polymorphic forms are defined as a function of stereoregularity and degree of deformation has been built.

Stress-induced phase transitions may play a key role also in determining elastic properties in polymeric materials. In fact semicrystalline polymers generally show only short-range elasticity when stretched samples are relaxed by releasing the tensile stress. Long range elasticity, when present, is, instead, generally associated with the occurrence of polymorphic transitions. In the case of elastomeric iPP, we have shown that samples of different stereoregularity present different types of elasticity depending on the degree of crystallinity. The more stereoregular samples, with rr content in the range 7–11% show elastic behavior in spite of the high degree of crystallinity (40–50%). Since elasticity is generally a property of the amorphous phase, probably elasticity in these samples is partially due to the enthalpic contribution associated with the crystallization of the mesomorphic form into the α form occurring upon releasing the tension. In the case of the poorly stereoregular sample iamPP, instead, the degree of crystallinity is low, and elasticity has essentially entropic origin as in conventional elastomers.

Acknowledgment

Financial supports from Basell Ferrara, Italy and from the "Ministero dell' Istruzione, dell'Università e della Ricerca" (PRIN 2004) are gratefully acknowledged. We thank Dr. Luigi Resconi of Basell for providing the polymer samples and for having stimulated this study.

References

[1] Flory, P. J.; Yoon, D. Y. *Nature (London)* **1978**, 272, 226.
[2] Peterlin, A. *J. Mater. Sci.* **1971**, 6, 490.
[3] Galeski, A. *Progr. Polym. Sci.* **2003**, 28, 1643.
[4] Argon A. S.; Cohen, R. E. *Polymer* **2003**, 44, 6013.
[5] Oleinik, E. F. *Polymer Sci. Ser. C.* **2003**, 45, 17.
[6] Men, Y.; Rieger, J.; Strobl, G. *Phys. Rev. Lett.* **2003**, 91, 95502.
[7] Hughes, D. J.; Mahendrasingam, A.; Oatway, W. B.; Heeley, E. L.; Martin, C.; Fuller, W. *Polymer* **1997,** 38, 6427; Yamada, M.; Miyasaka, K.; Ishikawa, K. *J. Polym. Sci.* **1971**, A29, 1083; Takahashi, Y.; Ishida T. *J. Polym. Sci., Polym. Phys.* **1988**, 26, 2267.
[8] Liu, Y.; Kennard, C. H. L.; Truss R. W.; Carlos, N. J. *Polymer* **1997**, 38, 2797.
[9] Ferreiro, V.; Coulon, G. *J. Polym. Sci., Polym. Phys.* **2004**, 42, 687.
[10] (a) Hiss, R.; Hobeika, S.; Lynn, C.; Strobl, G. *Macromolecules* **1999**, 32, 4390; (b) Men Y.; Strobl, G. *J. Macromol. Sci. Physics* **2001**, B40, 775.
[11] Al-Hussein, M.; Strobl, G. *Macromolecules* **2002**, 35, 8515.
[12] Men Y.; Strobl, G. *Macromolecules* **2003**, 36, 1889.
[13] Hobeika, S.; Men Y.; Strobl, G. *Macromolecules* **2000**, 33, 1827.
[14] Hong, K.; Rastogi, A.; Strobl, G. *Macromolecules* **2004**, 37, 10165; *Macromolecules* **2004**, 37, 10174.
[15] Pawlak, A.; Galeski, A. *Macromolecules* **2005**, 38, 9688.
[16] Auriemma, F.; De Rosa C.; Corradini, P. *Adv. Polym. Sci.* **2005**, 181, 1.
[17] De Rosa, C.; Auriemma, F.; Di Capua, A.; Resconi, L.; Guidotti, S.; Camurati, I.; Nifant'ev, I. E.; Laishevtsev, I. P. *J. Am. Chem. Soc.* **2004**, 126, 17040.
[18] De Rosa, C.; Auriemma, F.; De Lucia, G.; Resconi, L. *Polymer* **2005**, 46, 9461.
[19] Auriemma, F.; Ruiz de Ballesteros, O.; De Rosa, C. *Macromolecules* **2001**, 34, 4485.
[20] Auriemma, F.; De Rosa, C. *J. Am. Chem. Soc.* **2003**, 125, 13143.
[21] F. Auriemma and C. De Rosa, *Macromolecules* **2003**, 36, 9396.
[22] Ewen, J. A.; Elder, M. J.; Jones, R. L.; Haspeslagh, L.; Atwood, J. L.; Bott, S. G.; Robinson, K. *Makromol. Chem. Macromol. Symp.* **1991**, 48/49, 253.
[23] Brintzinger, H.H.; Fisher, D.; Mulhaupt, R.; Rieger, B.; Waymouth, R. M. *Angew. Chem. Int. Ed. Engl.* **1995**, 34, 1143.
[24] Kaminsky, W. *Macromol. Chem. Phys.* **1996**, 197, 3907.

[25] Resconi, L.; Cavallo, L.; Fait, A.; Piemontesi, F. *Chem. Rev.* **2000**, 100, 1253.
[26] De Rosa, C.; Auriemma, F.; Ruiz de Ballesteros, O. *Phys. Rev. Lett.* **2006**, 96, 167801.
[27] Nifant'ev, I.E.; Guidotti, S.; Resconi, L.; Laishevtsev, I.. PCT Int Appl WO 01/47939. Basell, Italy; 2001.
[28] Resconi, L.; Guidotti, S.; Camerati, I.; Nifant'ev, I.E.; Laishevtsev, I. *Polym. Mater. Sci. Eng.* **2002**, 87, 76.
[29] Fritze, C.; Resconi, L.; Sculte, J.; Guidotti, S. PCT Int Appl WO 03/00706. Basell, Italy; 2003.
[30] Nifant'ev, I. E.; Laishevtsev, I. P.; Ivchenko, P. V.; Kashulin, I. A.; Guidotti, S.; Piemontesi, F.; Camurati, I.; Resconi, L.; Klusener, P. A. A.; Rijsemus, J. J. H.; de Kloe, K. P.; Korndorffer, F. M. *Macromol. Chem. Phys.* **2004**, 205, 2275. Resconi, L.; Guidotti, S.; Camurati, I.; Frabetti, R.; Focante, F.; Nifant'ev, I. E.; Laishevtsev, I. P. *Macromol. Chem. Phys.* **2005**, 206, 1405.
[31] Balboni, D.; Moscardi, G.; Baruzzi, G.; Braga, V.; Camurati, I.; Piemontesi, F.; Resconi, L.; Nifant'ev, I. E.; Venditto, V.; Antinucci, S. *Macromol. Chem. Phys.* **2001**, 202, 2010.
[32] De Rosa, C.; Auriemma, F.; Perretta, C. *Macromolecules* **2004**, 37, 6843.
[33] Natta, G.; Corradini, P. *Nuovo Cimento* **1960**, 15, 40–51.
[34] Bruckner, S.; Meille, S. V. *Nature* **1989**, 340, 455–457.
[35] Auriemma, F.; De Rosa, C.; Boscato, T.; Corradini, P. *Macromolecules* **2001**, 34, 4815.
[36] Auriemma F.; De Rosa C. *Macromolecules* **2002**, 35, 9057.
[37] De Rosa, C.; Auriemma, F.; Paolillo, M.; Resconi, L.; Camurati, I. *Macromolecules* **2005**, 38, 9143.
[38] VanderHart, D. L.; Alamo, R. G.; Nyden, M. R.; Kim, M. H.; Mandelkern, L. *Macromolecules* **2000**, 33, 6078.
[39] Nyden, M. R.; VanderHart, D. L.; Alamo, R. G. *Comput. Theor. Polym. Sci.* **2001**, 11, 175.
[40] Geddes, A. J.; Parker, K. D.; Atkins, E. D. T.; Beighton, E. *J. Mol. Biol.* **1968**, 32, 343.
[41] Ran, S.; Zong, X.; Fang, D.; Hsiao, B. S.; Chu, B.; Phillips, R.A.; *Macromolecules* **2001**, 34, 2569.
[42] Corradini, P.; Petraccone, V.; De Rosa, C.; Guerra, G. *Macromolecules* **1986**, 19, 2699. Corradini, P.; De Rosa, C.; Guerra. G.; Petraccone, V. *Polymer Commun.* **1989**, 30, 281.
[43] Flory, P. J *Statistical Mechanics of Chain Molecules* John Wiley &Sons, New York: 1969.
[44] Allegra, G. *J. Chem. Phys.* **1974**, 61, 4910.
[45] Treolar, L. R. G. The physics of rubber elasticity. Oxford: Claderon Press; 1975.
[46] Tosaka, M.; Murakami, S.; Poompradub, S.; Kohjiya, S.; Ikeda, Y.; Toki, S.; Sics, I.; Hsiao, B. *Macromolecules* **2004**, 37, 3299.

18

Insights into Polymer Crystallization from In-situ Atomic Force Microscopy

Jamie K. Hobbs

Department of Chemistry, University of Sheffield, Dainton Building, Brook Hill, Sheffield. S3 7HF. UK
Jamie.hobbs@sheffield.ac.uk

Abstract. In-situ observation of polymer crystallization with atomic force microscopy is rapidly becoming a standard method, providing an increasing wealth of real-space information on the growth process at the molecular scale. Here, in-situ studies of dendritic thin film growth are extended to polyethylene, and the conditions for the onset of flat-on crystal growth and dendritic growth are given. Crystallization of oriented films is studied to provide accurate measurements of lamellar growth rates and their spatial and temporal variation. The use of AFM as a tool for the observation of intermediate phases is discussed in light of recent observations on model systems, and it is concluded that AFM under standard conditions is unlikely to discern between two crystal-like phases, but should discern between two melt-like phases. Finally, the recent development of rapid scanning AFM (VideoAFM™) is outlined, and an application which exemplifies the necessity of such high speed techniques is given.

18.1 Introduction

Despite more than fifty years of intensive study, some aspects of the process of crystallization in polymers remain poorly understood and hotly debated. Arguably, this is not surprising – as we delve deeper into any system, and attempt to understand in more detail how it is affected by ever increasing levels of complexity of initial conditions, it is hard, or maybe impossible, to maintain a grasp of simplifying and unifying principals. At the same time, the development of new techniques, and improvements in older methods, open up new areas of study and pose new questions that have rarely been predicted by existing theories.

In a previous article [1] I considered from a personal perspective how in-situ real-space observation of the process of polymer crystal growth and melting can provide insights into the fundamentals of polymer crystallization at the molecular scale. There the aim was to concentrate on universal themes, and on how atomic force microscopy (AFM) could inform our understanding of free, unperturbed, crystal growth. At the same time, the impact of AFM was considered in the context of other imaging techniques, and some developments in the technology that allowed application to polymer crystal growth discussed. In the four years since that article was written, there have been developments both in our understanding of polymer crystal growth, and in AFM technology, that are, potentially, revolutionary. The former considers the formation of polymer structure at the microscale, and shows the (limited) set of parameters necessary to predict polymer microstructure in the absence of flow [2,3]. The latter is the development of very rapid AFM techniques with the potential to probe crystal growth at the molecular scale with millisecond time resolution [4,5], which I will discuss towards the end of the current work. Also, during this time, further work has been carried out on the possible role of intermediate phases during crystal growth [6]. This has helped to fix limits on the expected material properties of any such phases, and hence the ability of the AFM to discern between these phases and the surrounding melt and/or stable crystal phase. It should now be possible to say whether or not the lack of direct observation by AFM is, or is not, evidence that these phases do not exist.

The aim of the current article is to present an up-date on the application of AFM to some of the topics covered in this book. In the time since the previous article [1], in-situ AFM has arguably moved from being a new technique to a standard tool, available in many polymer laboratories, with high temperature measurements now supported by most commercially available instruments. This has, rightly, led to a change in emphasis with quantitative analysis of in-situ data becoming progressively more sophisticated [7]. At the same time, the technique itself has developed so as to facilitate its routine use, with better quantification of the impact of imaging on the crystallization process, and further work to broaden the temperature range accessible, and the ultimate stability of the instrument when imaging at high temperatures. However, there are still developments in instrumentation both recent and potential in the near future, that could have some impact on the study of polymer crystal growth, and these will be discussed briefly in the final section of this work.

18.2 In-situ Observation of Thin Film Growth

The study of growth in very thin films where crystal morphology is strongly influenced by diffusion has become increasingly popular in recent years. The reasons for this are probably both scientific and opportunistic. Since the 1980s there has been considerable work on understanding diffusion controlled and dendritic growth from a theoretical standpoint [?, e.g.]]JH8,JH9, as well as some accompanying experimental work on classical dendrite forming (small molecule) materials [?, e.g.]]JH10,JH11. Recently, the combination of spin-coating – giving very thin, and controllable, film thicknesses – and AFM – allowing direct visualisation of structures on the nanometre scale on a solid substrate – has led to an increasing study of dendrites and related morphologies in synthetic polymers [?, e.g.]]JH12,JH13. Much of this work has concentrated on polyethylene oxide, although other materials are starting to be studied. To-date the potential for real-time, in situ, observation has hardly been exploited, with one notable exception [14]. Here I will briefly discuss some new observations on dendritic crystallization in polyethylene, the standard exemplar for flexible chain polymer crystallization, and the system used in one of the earliest thin film growth studies [15]. Some comparisons will also be made with a data set obtained with polyethylene oxide, the material that is becoming the exemplar for thin film growth.

18.2.1 In-situ Observation of Polyethylene Thin-film Crystallization

Figure 18.1 shows a series of images obtained during the gradual, stepwise cooling of a sample of polyethylene (Mw 120000 g/mol, Mw/Mn 1.1). All images were collected in TappingMode™, with active quality factor damping to reduce the effective quality factor from its natural value of ~260 to ~100, using the Infinitesima ActivResonance Controller, as detailed in [16]. This control over the effective response time of the cantilever allows faster scanning, giving access to the rapid dynamics of this process in polyethylene. Figures 18.1a-h are phase images, in which hard areas appear bright (the crystal and glass substrate) and sticky/soft areas appear dark (the molten polymer). There are some artefacts apparent in the images due to the high scan rates utilised – slight ringing on the right hand side of high features, and brightening on the left hand side of high features as the feedback loop struggles to maintain the correct tapping amplitude. However, these fast scan rates mean that images are obtained rapidly relative to the growth rate of the polymer crystals, avoiding significant distortion of the images due to the serial (line by line) nature of the data collection.

The film thickness is 11–14 nm over the whole of the uncrystallized area in Fig. 18.1a-h. Growth prior to, and including, Fig. 18.1a is of edge-on lamellae (i.e. perpendicular to the plane of the substrate, crystalline polymer chains lying parallel to the substrate) growing steadily at constant temperature. In between Figs. 18.1a and 18.1b the temperature is rapidly dropped by 2°C to 127°C, and there is a marked change in growth morphology with a transition from edge-on to flat-on growth. It is unclear how this transition occurs as the flat-on growth requires a change in chain orientation relative to the edge-on growth of at least 55° (assuming a 35° chain tilt), so simple branching seems unlikely. From 1b-c it is apparent that the newly formed edge-on growth consists of several individual lamellae, spaced by 200–400 nm along the edges of the existing lamellae. From existing data it is not possible to tell definitively if flat-on growth occurs once a sufficiently low temperature is reached (i.e. 127°C in polyethylene), or if the temperature jump causes the branching process (perhaps by the initiation of nuclei due to the (small) stress imparted on the crystal by the temperature jump and the anisotropy of the expan-

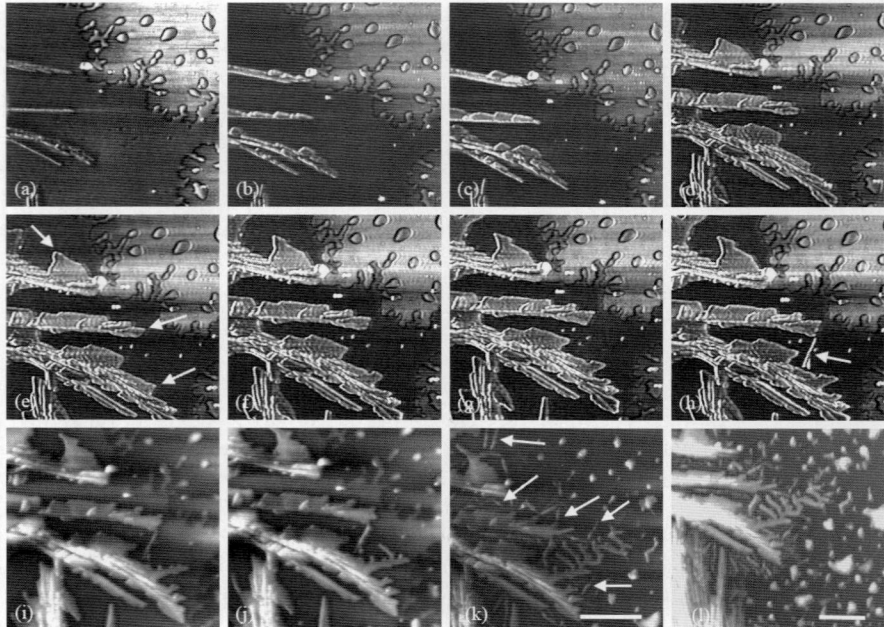

Fig. 18.1. A series of tapping mode AFM images showing the crystallization of a thin film of polyethylene during gradual cooling. The brighter regions to the right of the image are the glass substrate. (**a**) to (**h**) are phase images where black to white represents a change in phase of 40°. (**i**) to (**k**) are topographic images where black to white represents a change in height of 70 nm. In (**l**) black to white represents a change in height of 100 nm. Scale bars represent 1μm, and the bar in (k) applies to images (a) to (k). (a) taken at 129°C, t = 0s, (b) at 127°C, t = 120s, (c) at 127°C, t = 162s, (d) at 122.5°C, t = 674s, (e) at 121°C, t = 760s. The arrows indicate crystal tips that are starting to grow at an accelerated rate relative to the surrounding crystal. (f) at 119.6°C, t = 880s, (g) at 118.1°C, t = 970s, (h) at 117.6°C, t = 998s. The *arrows* indicate crystals that have been nucleated by the AFM tip. (i) at 116.8°C, t = 1149s, (j) at 116.8°C, t = 1329s (k) at 103.4°C, t = 3739s. The arrows indicate crystals that have been nucleated by the AFM tip. (l) at room temperature, t = 1day

sion coefficients of the crystalline lattice). What is clear is that film thickness alone is not sufficient to lead to a transition to flat-on growth, and neither is this transition induced by a particular relationship between film thickness and crystal thickness, as this would imply more likelihood of flat-on growth at high temperatures where the crystal thickness is greater.

Once the transition to flat-on growth has occurred, the lamellae continue to grow maintaining an approximately rectangular shape, until a temperature of 122.5°C is reached in Fig. 18.1d. Here another transition in behaviour is apparent, again induced by temperature rather than film thickness (even after very long growth times in similar films at higher temperatures this transition was not seen). The corners of the rectangular crystals start to grow faster than the adjoining planar surfaces (examples arrowed in figure 18.1e). This transition is most likely related to the Mullins-Sekerka instability [17] – the corners project furthest into the surrounding melt, and, if diffusion has a strong influence on growth rate, will grow faster than neighbouring regions. Further cooling shows a clear onset of dendritic growth, with rapid growth of the crystal tips leading to the formation of a central trunk which in turn breaks up through further surface instabilities into side branches. Figures 18.1i-k show the progression of this dendritic growth during cooling (note these are topographic images).

AFM involves mechanical contact between a sharp probe and the sample surface, and, although careful control over imaging conditions usually prevents damage to the sample, there is still the possibility of interacting with the growth process. In Fig. 18.1h-k there are numerous small crystals independently nucleated in between the growing dendrites (examples arrowed in 1h and 1k). In areas that were not being continuously imaged such infilling nucleation is not seen at these temperatures – it is clear that the tip is having a nucleating effect at these high supercoolings, albeit a weak one. This is not surprising – the size of the critical nucleus reduces with increasing supercooling, and small fluctuations in tapping force (due to external vibration and/or errors in the feedback control) will lead to local variations in density that will, on reducing temperature, have an increasing probability of leading to nucleation. Similarly, transient shear forces will induce chain orientation that will have an increasing probability of inducing nucleation as temperature is dropped. On cooling to room temperature (Fig. 18.1l) there is considerable infilling growth that consumes much of the remaining crystallisable material. Unfortunately this image is complicated by the presence of a 'double-tip' artefact (i.e. contamination is attached to the tip so there are two asperities sufficiently close to the surface to allow imaging, giving two offset images of high features). However, there do appear to be numerous small crystallites in between the larger dendritic structures that have not grown by branching from these structures, and must have nucleated during the rapid cooling to room temperature.

18.2.2 In-situ Observation of Polyethylene Oxide Dendritic Crystallization

For the sake of comparison, Fig. 18.2 shows a series of in-situ topographic images showing the gradual growth of a polyethylene oxide (PEO) dendritic structure crystallized at room temperature (PEO obtained from Polymer Laboratories Ltd., Mw 50000 g/mol, Mw/Mn 1.1). In this case the sample was prepared by solution casting (from chloroform) onto a clean glass slide. This gives a near monolayer coverage with a thickness of 1–2 nm with a few larger

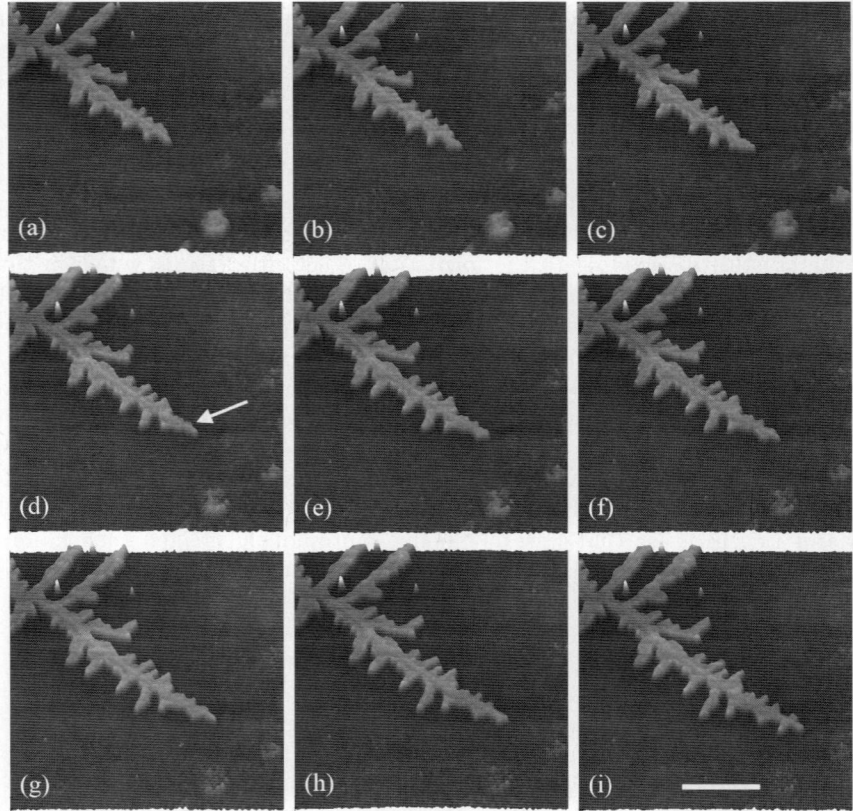

Fig. 18.2. A series of topographic images of thin film crystallization in polyethylene oxide, presented in pseudo 3D. The scale bar represents 1μm. Black to white represents a change in height of 20 nm. (**a**) taken at t = 0s, (**b**) taken at t = 85s, (**c**) taken at t = 361s, (**d**) taken at t = 578s. The *arrow* indicates a side branch that started from a subtle change in direction of the growth tip. (**e**) taken at t = 794s, (**f**) taken at t = 938s, (**g**) taken at t = 1110s, (**h**) taken at t = 1327s, (**i**) taken at t = 1543s

droplets, the biggest of which provide nucleation sites from which the dendritic structures grow on cooling.

Under these conditions the crystal growth is occurring close to a transition between dendritic and tip-splitting, or seaweed, growth [14]. Branching occurs primarily following subtle changes in direction of the growth tip, so the branch position is defined during the passing of the primary growth front. An example of such a branch is arrowed in Fig. 18.2d, the previous image showing the growth tip as it was just prior to the change in direction that apparently resulted in the branch point. The long range diffusion process is

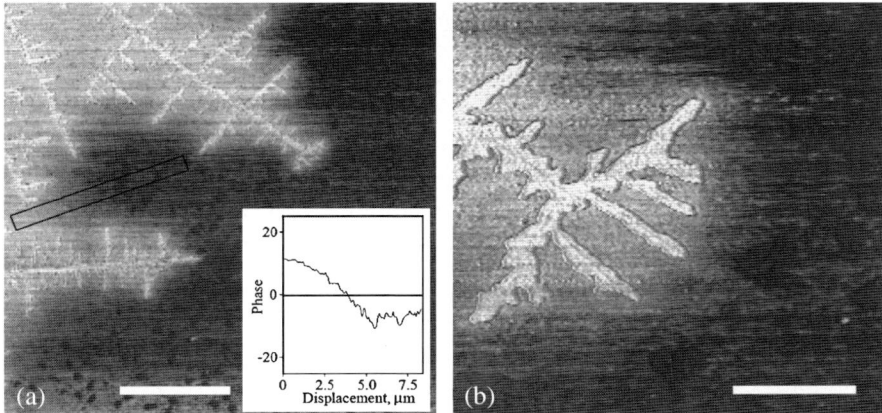

Fig. 18.3. A pair of phase images showing the growth of a polyethylene oxide crystal in a thin film. In both images black to white represents a change in phase of 50°. In (a) the scale bar represents 5μm. In (b) the scale bar represents 1μm. The inset in (a) shows the average phase profile taken through the region indicated with a *box* – i.e. it is the average variation in phase with displacement along the *long axis* of the box, the average taken over the width (*short axis*) of the box

clearly visible through the gradual depletion of the droplets of molten polymer present in the bottom right of Fig. 18.2a. Close inspection shows that this process occurs in a surprising manner. Instead of maintaining an equilibrium droplet shape, the droplet is apparently pulled apart gradually with remarkably little relaxation, maintaining its original footprint but eventually breaking up into a large number of smaller domains. What is also clear are the long distances over which material is diffusing to reach the growth front, and hence the very large extent of the depletion zone caused by the growth process. Figure 18.3 shows a pair of phase images of such a structure in which the diffusion field is strikingly apparent. Here the glass surface and the crystalline polymer appear bright (hard, non-adhesive surfaces), while the glass that maintains an adsorbed layer of molten polymer, and the molten polymer droplets, appear dark (soft, adhesive surfaces). The step in thickness measured between these two regions (i.e. with and without molten PEO on the glass surface) is 1-2 nm. This causes a step in phase, shown in the inset on Fig. 18.3a, of approximately 20° (note phase degrees are approximately the cosine of the true phase difference between the cantilever drive and response). The depletion zone, forming a smooth envelope around the growing crystallites, is graphically shown. This opens up the intriguing possibility of directly imaging the extent of the depletion zone and how it changes during growth, providing real-space information on the time dependence of surface polymer density as it varies due to the diffusion field set up by the growing crystal. This will be the subject of a future publication.

18.3 Observations on Growth of Shish Kebabs

AFM is unlikely to provide a tool for direct observation of the growth of extended chain 'shish' crystals that are frequently found following the imposition of extensional flow on polymer melts – the growth is likely to be too fast, and the process of extension is unlikely to provide a sufficiently stable surface to allow imaging. However, following an initial period of extension, it is possible to use AFM to follow, in real-time, the over-growth of lamellar structures – i.e. the 'kebab'. This was the subject of a previous study [18], in which this morphology was particularly chosen as it gave an unambiguous lamellar orientation and hence removed some of the problems associated with a surface technique. The key observations to come from that work were the variation in growth rate of individual lamellae from crystal to crystal and for each crystal with time, and the changing of growth direction as two opposing lamellae approach within a few tens of nanometres of each other, leading to the eventual inter-digitation of the kebab structures (important for the material properties). A later study on melting [19] has shown directly the expected behaviour with the overgrown lamellae melting first to leave the underlying shish, initially with small thickened vestiges of the lamellae that then melt at a temperature several degrees below the final melting temperature of the extended chain backbone.

In [18] we had carried out some rudimentary initial evaluation of the growth rates, all that was possible considering the quality of the data. Since then, improvements both in instrument stability and scan speeds (through use of active control of the cantilever resonance) have allowed higher quality data to be obtained. Figure 18.4 shows part of such a data set, showing the slow growth of polyethylene lamellae from two neighbouring oriented structures, and the ensuing inter-digitation of the growth fronts.

Figure 18.5 shows the analysis of the growth rates of 26 of the lamellae on the left hand side of the images. These images were taken over a period of 308 seconds in which 12 AFM images were collected. Growth rates were measured between every other image, so that the slow scan direction was always in the same direction (i.e. over a period of approximately 52 seconds although each image was collected in 25.6 seconds). The variation in average rate with time shown in Fig. 18.5b shows that the temperature is fairly stable – the change in rate is initially small, and increases for the last couple of points, the opposite behaviour from that which would be seen if the sample was slowly approaching an equilibrium temperature. However, there is a gradual reduction in rate with increasing time and there are several possible explanations for this. It could be due to depletion of crystallisable material compared to non-crystallisable impurities, or thinning of the film due to depletion caused by crystallization, or possibly due to the influence of the opposing crystal population on the growth rate as growing crystals compete for the same material. Considering the film itself is between 70 and 120 nm thick, the reduction in film thickness that occurs due to the volume reduction associated with crystallization is the

18 Insights into Polymer Crystallization 381

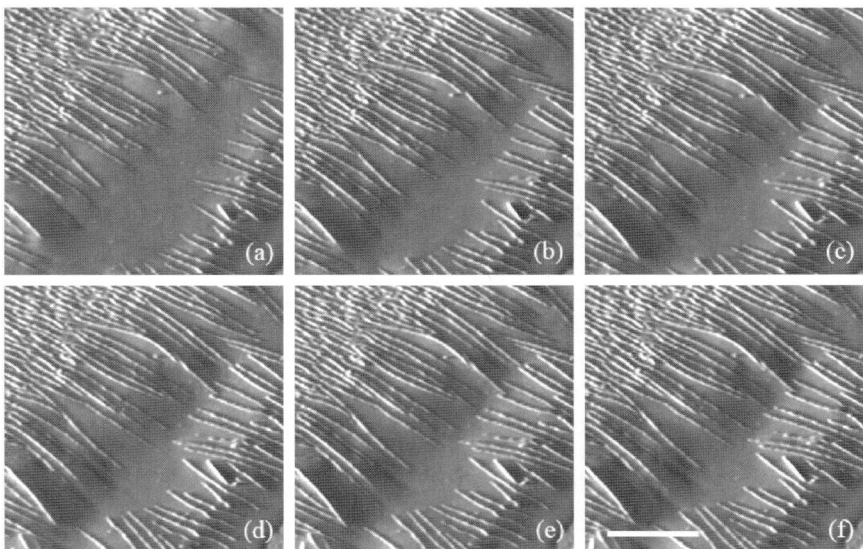

Fig. 18.4. A series of AFM amplitude images showing the growth of lamellae from two oriented nuclei. Black to white represents a change in amplitude of 0.5V compared to a set-point amplitude of approximately 1.3V. The scale bar represents 1μm. (**a**) taken at t = 0s, (**b**) taken at t = 102s, (**c**) taken at t = 153s, (**d**) taken at t = 233s, (**e**) taken at t = 258s, (**f**) taken at t = 309s

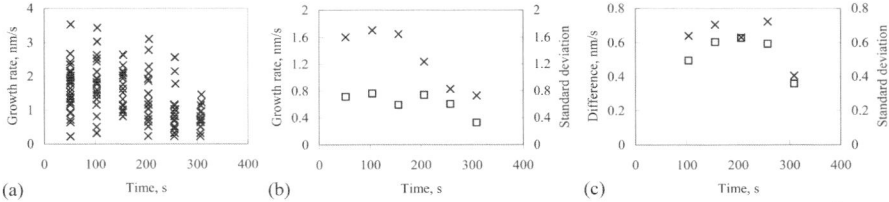

Fig. 18.5. A series of graphs showing analysis of the growth rates of the first 26 lamellae starting from the top left of the first image – i.e. lamellae growing from the left hand oriented nucleus. (**a**) shows the growth rates of all the lamellae measured over approximately 52s periods ending at the time shown on the x-axis. (**b**) shows the average growth rates taken from the data in (a) – × – and the standard deviation of those data sets – □. (**c**) shows the average magnitude of the difference in growth rate of each lamella measured at one time and the subsequent time (e.g. < |growth rate at 104s – growth rate at 52s| >) – × – and the standard deviation in that growth rate difference data set – □

most likely source of slowing in rate – a reduction in thickness of a few tens of nanometres due to depletion would certainly be sufficient to reduce the growth rate. Changes in the relative concentration of impurities are unlikely to have such a marked impact at these very slow growth rates. The possibility that growth slows because of the impact of competition from the encroaching lamellae growing from the neighbouring shish is not borne out by a closer inspection of the rates – there is no correlation between variations in growth rate of opposing crystals as they grow towards each other.

The standard deviations of these growth rates are very large, in line with inspection of the very broad scatter of the actual rates. This re-confirms the results given in [18], where the rate of growth of individual lamellae was found not to be a temperature dependent constant, but rather to vary from lamella to lamella. Figure 18.5c shows the average difference in the growth rate between successive measurements of the rate of each individual lamella, and the standard deviation of that data set. Again, it is clear that the growth rate of a lamella does not remain constant but rather varies considerably with time, in line with our previous observations. What comes from this new analysis is that the variation in growth rate with time for each lamella is as large as the variation between lamellae at any point in time. There is no simple correlation between the growth rates of neighbouring crystals (or of crystals growing towards each other), and the current analysis implies the variation in rate is a random fluctuation.

18.4 Imaging of Precursors to Crystal Growth?

In the previous article [1] I discussed the extent to which AFM could shed light on recent suggestions that crystallization occurs through intermediate, partially ordered, phases [20–23]. To date in-situ observations have not shown direct evidence of an intermediate state between the growing crystal and the melt – images have not been obtained showing a molten polymer transforming into a partially ordered crystal-like phase which then transforms, some time later (i.e. further back along the growing 'crystal') into a highly ordered crystal phase. Similarly, images have not been obtained that imply an intermediate spinodal type densification of the melt prior to nucleation [20–23]. As pointed out previously, to say whether or not this is evidence against the existence of such a gradual transition process, more information on the suggested properties of these new phases relative to the initial melt and the final crystal is needed. Recently such clarification has become available in the case of the intermediate state suggested by Prof. Strobl [6, 20], agreeing with previous suggestions from the Keller group [21] that it will be similar to the hexagonal form of polyethylene – i.e. its properties are more similar to the final crystal than to the melt.

The AFM imaging mode that holds out greatest hope for seeing such differences in local properties is phase imaging, in which the change in phase of the tapping cantilever's response relative to the drive phase is monitored. Unfortunately, as the cantilever is driven at resonance, the contact with the sample surface is intermittent, the shape of the AFM tip and hence the contact area with the sample surface is unknown, and, perhaps most importantly, the contact area changes with the material properties of the surface, a simple analysis of material properties is not possible from the phase image (regardless of some claims that have been made). However, it is generally agreed that phase images reflect primarily variations in dissipative interactions (i.e. energy loss) over the sample surface [24, 25]. Clearly changes in stiffness can also be involved in any dissipation process, as a change in stiffness will lead to a change in tip-sample contact area, and hence to a change in the adhesive interaction between the tip and the sample (and thus energy loss). We have recently [26], following previous work elsewhere [27], shown that the phase image is sufficiently sensitive to differentiate between such subtly different materials as the phase separated components of a block copolymer in which both phases are more than 100°C above their glass transition temperature, at the same time as obtaining an additional contrast between both molten phases and the crystallization of one of the blocks. In that work it was suggested that this contrast was due to the 5–6× modulus difference between the two liquids, which compares to an approximately 1000× contrast between the modulus of the liquid and the crystal. The phase contrast between the most dissipative liquid and the crystal was found to be approximately 4°, compared to a contrast of 0.3–1° between the two liquids (note degrees here cannot be simply interpreted as the AFM used – an Extended Veeco Nanoscope IIIa with a Dimension 3100 – did not utilise a lock-in amplifier to measure phase, but rather the cruder 'phase extender' module that measures something approximating to the cosine of phase at small angles). Clearly (and not surprisingly) phase contrast does not respond linearly with sample modulus. What is apparent from our (and others') work is that phase is generally more sensitive to subtle changes between relatively soft materials. It might be possible to optimise imaging conditions to allow discrimination between two different 'crystal' phases, but to-date this has not been attempted during an in-situ crystallization experiment.

In summary, it is perhaps not surprising that intermediate phases between melt and crystal that are more similar in properties to the crystal than to the melt have not been observed by in-situ AFM measurement. As used in standard operation, the AFM phase image is not very sensitive to changes in properties in such relatively stiff materials. Clearly, if there were any changes in density that would lead to distortion of the crystalline material, or other changes in morphology associated with the transition from metastable to stable crystalline phase, these should be observable by in-situ AFM – that they have not been observed implies that such changes are minimal, if the metastable phases exist. However, the changes suggested by the spinodal decomposition model would be more likely to be visible as these do involve changes in relative stiffness of different regions of the melt over lengthscales comparable to those observed in the phase separated block copolymers. That such changes have not been seen must be taken as evidence against this model for a precursor state, or at least as an indication of the changes in local properties that could be happening – i.e. they must be less than the variations seen in a 'hard segregated' block copolymer.

18.5 Developments in Atomic Force Microscopy – High Speed AFM

Both of the example applications given above use active control of the response time of the tapping AFM cantilever so as to reduce the time necessary to take individual AFM images. At the extreme [16] this approach can approximately half the time required to take a single image – a significant improvement but one that is unlikely to give access to much new physics. With conventional technology the problem remains that the rates of growth accessible with AFM are so slow that other techniques almost never probe this regime. In the case of optical microscopy and X-ray scattering, this is simply a matter of patience, but for other techniques the reasons are more fundamental. In differential methods, such as DSC, the signal-to-noise depends on the rate of the transition, and such slow transitions are practically inaccessible. In all techniques where an average over a large sample volume is obtained, the danger of studying temperatures where rates are so slow is that an average that is assumed to have been taken over many similar events occurring simultaneously may in fact be an average of a few events occurring in series, giving a misleading result.

Whatever the reason, there is little available data from other techniques with which to compare in-situ AFM data, which is a problem in itself if AFM is to be used to build on an existing bed of knowledge. However, more importantly, polymer crystallization is a kinetically controlled process, and kinetics are important at many different levels. The initial lamellar thickness is controlled by kinetic considerations rather than thermodynamic. The crystal morphology (e.g. isolated lamellae, 'axialite' aggregates, or spherulites) depends on the interplay of surface tension and diffusion [3], and hence the actual rate of growth relative to the translational and rotational diffusion coefficients of the polymer melt influences the observed structure. Finally, Ostwald's stage rule has been invoked by several groups [6, 21, 28] in the discussion of the role of intermediate (metastable) phases in the crystallization process, and the actual extent of any metastable state will depend on the relative rates of growth and transformation into more stable forms – so the structure at a sub-molecular level is also, arguably, not well described by thermodynamic considerations. Although it is possible to image many polymers as they crystallize with AFM, it is usually necessary to look either close to the glass transition temperature or the melting temperature, where diffusion, and the small driving force respectively, reduce the growth rate considerably. Only in exceptional cases, usually through copolymerisation, are growth rates sufficiently slow that the whole temperature range can be accessed [16]. The question inevitably arises whether the information obtained by in-situ AFM studies of crystal growth actually applies under the circumstances that the vast majority of published data is exploring.

In the case of crystal melting the above arguments are just as applicable. Indeed, the rate of melting cannot be easily slowed down, as secondary processes such as crystal perfecting intervene and can prevent melting from occurring. There are very few examples of real-space in-situ studies of melting available [19, 29–31], although crystal thickening has been more extensively examined [?, e.g.]]JH31,JH32,JH33. How heating rate, crystal thickness, crystal history, and crystal shape/morphology interact on an individual crystal basis is almost completely unexplored, largely because of the lack of availability of suitable tools.

Recently a new type of AFM, the VideoAFM™, has been developed which by-passes many of the problems that have previously limited the AFM scan speed [5]. Images can be obtained in tens of milliseconds, compared to tens of seconds. This opens up the possibility of in-situ observation of crystallization with sub-lamellar resolution at timescales more than 1000 times shorter than those previously obtained. To-date this has been applied to the crystallization of polyethylene oxide at temperatures where growth rates of 100s of nm/s are observed [34].

Details of this instrument are given in [5, 34, 35], but the main issues that affect image interpretation will be recapitulated here. The VideoAFM™ uses an adaptation of contact mode imaging, in which the cantilever maintains constant contact with the sample surface through the action of a direct force applied to the cantilever tip, forcing the cantilever to respond at a frequency considerably greater than it's first bending mode. The image consists of the raw cantilever deflection signal, so image contrast is a combination of both the relative 'height' of different areas of the sample and the relative slope, although a simple deconstruction of the image to obtain the sample topography is not possible from data obtained using the standard microscope configuration [35]. For the pixels collected at the highest rate (i.e. those in the middle of the image, considering the sinusoidal tip velocity), the pixel frequency is approximately 8 MHz, so it is not possible to carry out oscillating measurements such as sample tapping that give mechanical contrast. So, with the caveats given above, the images reflect sample topography rather than material properties, and, in the case of crystallization and melting, the motion of melt-crystal interfaces is all that can be seen, rather than direct determination of the location of melt or crystal as is possible in conventional TappingMode™ AFM.

Figure 18.6 shows an example set of data taken during a heating experiment carried out on polyethylene oxide (Polymer Laboratories, Mw 50000 g/mol, Mn ~ 1.1). The sample was first crystallized at 56°C and then heated at 10°C/minute until it started to melt. The heating process was stopped while the images were collected. Each image was collected in 35 milliseconds at a rate of 14 frames/second (only each alternate image is collected so the

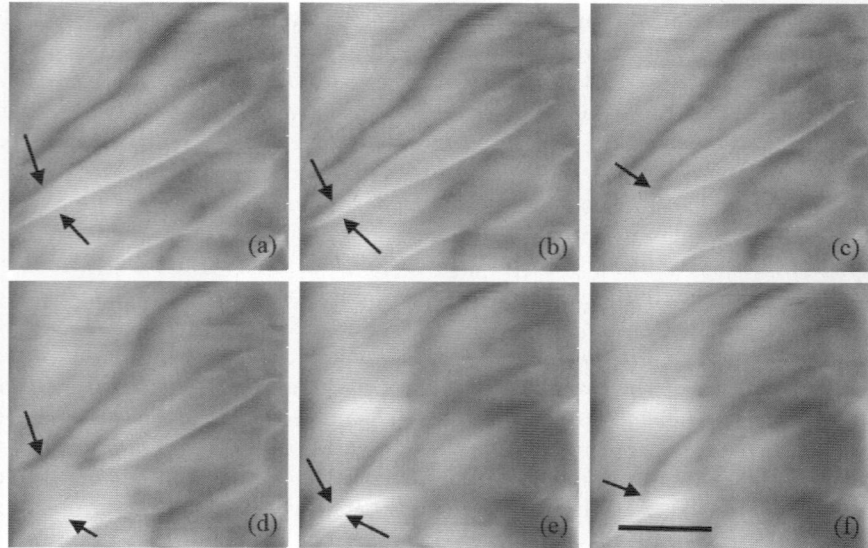

Fig. 18.6. A series of VideoAFM™ images showing the rapid melting of a sample of polyethylene oxide. The scale bar represents 1μm. The *greyscale* shows the raw deflection of the cantilever. (**a**) collected at t = 0s, (**b**) collected at t = 0.35s, (**c**) collected at t = 0.42s, (**d**) collected at t = 0.7s, (**e**) collected at t = 1.82s, (**f**) collected at t = 1.89s. The *arrows* indicate the positions of lamellae edges referred to in the text

scan direction is always the same). Other experimental details are the same as those given in [34].

As expected, the melting process occurs back from the lateral edges of the crystals. Here melting is considerably more rapid than has been previously followed in-situ at this resolution, with the entire sequence taking only 1.89 seconds. The arrows indicate the receding edges of two crystals, one in the first three images and the crystal below in the next three images. In both cases the crystal edges melt back at approximately constant rate to leave a narrow spur of crystal (in Figs. 18.6b and 18.6e) with high aspect ratio that then melts completely by the subsequent frame. In previous studies of melting with AFM [19] such high aspect ratios are not seen. It would be expected that such morphologies would be short lived because of their very high surface area to volume ration. It is likely that in this case it is the rapid melting process which reduces the opportunity for re-organisations driven by surface tension, so allowing this unexpected morphology to be observed. Here is a clear example where high speed scanning allows access to behaviour that is not observed at more conventional imaging rates, but which is most likely typical of the behaviour that is occurring when other techniques are used to follow a polymer process.

An additional advantage of high speed scanning, as discussed in [35], is the extra stability it affords. As images are collected in tens of milliseconds, vibrations and motion of the microscope cause motion of entire images, rather than distortion within images – i.e. the movie that is collected 'shakes' but each frame is still coherent. This means that it should be possible to use the microscope under harsher conditions than has been previously possible, maybe even providing access to 'real' processing conditions including sample deformation during crystallization.

Clearly there are many issues still to be addressed with the instrumentation of the VideoAFM to allow it to become a routine tool for the study of polymer crystallization. Not least is the need for image contrast mechanisms based on material properties, so that molten and crystalline areas can be clearly discerned. Also, there is a need to control the interaction force between the tip and the sample so as to allow more delicate samples to be imaged in a reproducible and non-destructive manner. However, the technique is already starting to provide new insights into the processes of crystallization and melting, and provides a route for considerable advances in the future.

18.6 Conclusions

I have given a brief overview of some recent work on the application of in-situ AFM to polymer crystallization. The ability to directly visualise the process continues to show unexpected behaviours and provide impetus for new directions in our exploration of this intriguing area of polymer science.

Crystallization of thin films, where diffusion strongly influences morphology, provides a particularly attractive subject for in-situ observation. Here I have barely scraped the surface, with the most exciting possibility of directly visualising the diffusion field with nanometre resolution still to be explored. What is arguably yet to be shown is whether these thin film growth behaviours have any contribution to make to our understanding of crystal growth under more conventional conditions?

Oriented crystallization is of particular importance industrially, both through its use to specifically obtain improved material properties, and through its accidental presence in many products due to the necessary imposition of extensional flow during rapid processing conditions. Here this morphology has been used to allow a more straight forward interpretation of quiescent crystallization, through the analysis of a large number of individual growing lamellae. This has helped to quantify previous observations on lamellar growth rates, and in particular to show that the fluctuations in growth rate that are frequently observed when rates are measured over short length and timescales are random.

One of my initial expectations in starting this work was that it would be possible to observe directly whether crystallization did indeed occur through intermediate states. Here this issue has been discussed in the context of recent work on phase separated block copolymer crystallization, and the conclusion reached that, if the intermediate states are all crystal-like (rather than melt-like), the AFM will not see them under normal imaging conditions.

The recent development of video rate AFM has also been discussed and its first application to crystal melting has been shown. The possibility of obtaining images 1000 times faster than conventional AFM, with time resolution of milliseconds, should provide exciting new information in many areas of polymer crystal growth in the future.

Acknowledgements

I would like to thank Dr Andy Humphris, Infinitesima Ltd., and Dr Cvetelin Vasilev, University of Sheffield, for their input into the VideoAFM™ work. Thanks also to the Engineering and Physical Sciences Research Council, UK, for funding.

References

[1] J.K. Hobbs in "Polymer Crystallization: Observations, Concepts and Interpretations" Springer-Verlag, Berlin Eds J.U. Sommer and G. Reiter, 82–95 (2003)
[2] L. Granasy, T. Pusztai, T. Borzsonyi, J.A. Warren, J.F. Douglas: Nature Materials, **3(9)**, 645 (2004)
[3] L. Granasy, T. Pusztai, G. Tegze, J.A. Warren, J.F. Douglas: Phys. Rev. E. **72**, 011605 (2005)
[4] A.D.L. Humphris, J.K. Hobbs, M.J. Miles: Appl. Phys. Lett. **83(1)**, 6 (2003)
[5] A.D.L. Humphris, M.J. Miles, J.K. Hobbs: Appl. Phys. Letts. **86**, 034106 (2005)
[6] G. Strobl: Eur. Phys. J. E. **18(3)**, 295 (2005)
[7] H. Xu, D. Shirvanyants, K.L. Beers, K. Matyjaszewski, A.V. Dobrynin, M. Runinstein, S.S. Sheiko: Phys. Rev. Letts. **94**, 237801 (2005)
[8] E. Ben-Jacob: Contemp. Phys. **34**, 247 (1993)
[9] J.S. Langer: Science **243**, 1150 (1989)
[10] B. Utter, E. Bodenschatz: Phys. Rev. E. **66**, 051604 (2002)
[11] B. Utter, R. Ragnarsson, E. Bodenschatz: Phys. Rev. Letts. **86(20)**, 4604 (2001)
[12] G. Reiter, J-U. Sommer: J. Chem. Phys. **112(9)**, 4376 (2000)
[13] K. Taguchi, H. Miyaji, K. Izumi, A. Hoshino, Y. Miyamoto, R. Kokawa: Polymer **42(17)**, 7443 (2001)
[14] V. Ferreiro, J.F. Douglas, J. Warren, A. Karim: Phys. Rev. E **65(5)**, 051606 (2002)
[15] H.D. Keith, F.J. Padden, B. Lotz, J.C. Wittmann: Macromolecules **22**, 2230 (1989)
[16] J.K. Hobbs, A.D.L. Humphris, M.J. Miles in "Applications of scanned probe microscopes to polymers" ACS symposium series 897, eds J.D. Batteas, C.A. Michaels, G.C. Walker, ACS, Washington, DC (2005)
[17] W.W. Mullins, R.W. Sekerka: J. Appl. Phys. **34**, 323 (1963)

[18] J.K. Hobbs, A.D.L. Humphris, M.J. Miles: Macromolecules **34**, 5508 (2001)
[19] J.K. Hobbs: Polymer, *in press*
[20] G. Strobl: Eur. Phys. J. E, **3**, 165 (2000)
[21] A. Keller, M. Hikosaka, S. Rastogi, A. Toda, P.J. Barham, G. Goldbeck-wood: J. Mater. Sci., **29(10)**, 2579 (1994)
[22] P.D. Olmsted, W.C.K. Poon, T.C.B. McLeish, N.J. Terrill, A.J. Ryan: Phys. Rev. Lett. **81**, 373 (1998)
[23] M. Imai, K. Mori, T. Mizukami, K. Kaji, T. Kanaya: Polymer **33(21)**, 4451 (1992)
[24] J.P. Cleveland, B. Anczykowski, A.E. Schmid, V.B. Elings: Appl. Phys. Lett. **72(20)**, 2613 (1998)
[25] J. Tamayo, R. Garcia: Appl. Phys. Lett. **73(20)**, 2926 (1998)
[26] J.K. Hobbs, R. Register: Macromolecules, **39**, 703 (2006)
[27] G. Reiter, G. Castelein, J-U. Sommer, A. Rottele, T. Thurn-Albrecht: Phys. Rev. Lett. 87(22), 226101-1 (2001)
[28] E.L. Heeley, C.K. Poh, W. Li, A. Maidens, W. Bras, I.P. Dolbnya, A.J. Gleeson, N.J. Terrill, J.P.A. Fairclough, P.D. Olmsted, R.I. Ristic, M.J. Hounslow, A.J. Ryan: Faraday Discussions **122**, 343 (2003)
[29] R. Pearce, G.J. Vancso: Macromolecules **30(19)**, 5843 (1997)
[30] L.G.M. Beekmans, D.W. van der Meer, G.J. Vancso: Polymer **43(6)**, 1887 (2002)
[31] A.K. Winkel, J.K. Hobbs, M.J. Miles: Polymer, **41(25)**, 8791 (2000)
[32] M.W. Tian, J. Loos: E-Polymers, 51 (2003)
[33] N. Dubreuil, S. Hocquet, M. Dosiere, D.A. Ivanov: Macromolecules **37(1)**, 1 (2004)
[34] J.K. Hobbs, C. Vasilev, A.D.L. Humphris: Polymer **46**, 10226 (2005)
[35] J.K. Hobbs, C. Vasilev, A.D.L. Humphris: The Analyst **131**, 251 (2006)

19
Temperature and Molecular Weight Dependencies of Polymer Crystallization

Norimasa Okui, Susumu Umemoto, Ryuichiro Kawano, Al Mamun

Tokyo Institute of Technology, Department of Organic and Polymeric Materials
Ookayama 2-12-1-S8-37, Meguroku, Tokyo, Japan
nokui@o.cc.titech.ac.jp

Abstract. Crystallization behaviours are characterized with the nucleation and crystal growth rates, which are strongly dependent on temperature and molecular weight in polymeric materials. Their rates show the bell-shape temperature dependence with maximum rates. The maximum rates are characteristic intrinsic values in polymer crystallization mechanism. Temperature and molecular weight dependencies are discussed.

19.1 Introduction

An unequivocal discussion on temperature dependences of nucleation and crystal growth rates requires data measured sufficiently in a wide range of temperatures. One of the historic data of temperature dependence of crystallization in an organic substance is the nucleation of glycerin reported by Tammann [1]. There are many nucleation [2–7] and growth [8–12] data in inorganic compounds. In polymeric materials, a few data can be found out for the temperature dependence of nucleation and crystal growth rates in an identical polymer [13, 14]. Figure 19.1 shows nucleation rate (I) and growth rate (G) in a wide range of temperature for poly(ethylene succinate). Both I and G show a maximum rate. In most of crystalline materials, the crystal growth rate is observed in the temperature range between the melting temperature (T_m) and the glass transition temperature (T_g). On the other hand, nucleation is often found below T_g in inorganic compounds but, in general, crystal growth could not be found below T_g. However, in the case of a thin film, crystal growth can be observed even below T_g. This might be caused by a decrease of T_g with film thickness. The crystal growth is affected mainly by crystallization temperature for a given material. On the other hand, the nucleation is influenced by many experimental conditions. These influences on polymer crystallization will be discussed in this report.

Molecular weight (M) dependence of physical properties is one of the most common characteristics in polymeric materials. For examples, melt vis-

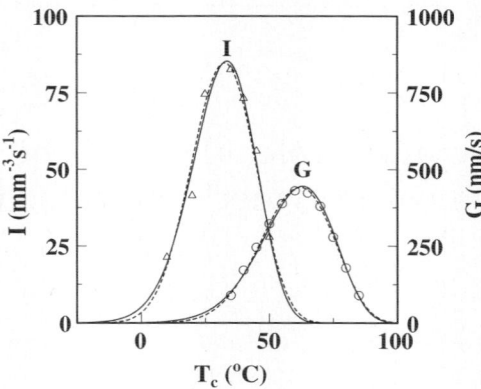

Fig. 19.1. Temperature dependencies of the primary nucleation rate (I) (Δ) and the linear crystal growth rate (G) ($\bigcirc$) for poly(ethylene succinate) (PESU) [14] with a molecular weight (M) of 8,770. The *solid* and *broken lines* are results from the best fitting procedure for G based on Eq. (19.2) and for I based on Eq. (19.11) by the Arrhenius and the WLF expressions of the molecular transport term, respectively

cosity shows remarkable molecular weight dependence, and it can be scaled and expressed as a 3.4 power of molecular weight for molecular chains with entanglements. The influence of molecular weight on polymer crystallization rate has been the most interesting subject of various papers [15–39]. Figure 19.2 shows molecular weight dependence of crystallization rate constant (k) in Avrami equation for poly(ethylene) (PE) at constant crystallization temperature [30]. These dependencies are very complicated and seem to follow no regular rule. The Avrami constant (k) includes I and G. These two rates might be shown in different molecular weight dependences. In many literatures, it is assumed that G is proportional to k in a case of that the Avrami exponent of n remains unchanged. This is not true [31] that G at a given temperature decreases with the molecular weight, however k shows the bell-shape molecular weight dependence. At high molecular weight regions, k increases with an increase in G, at low molecular weight regions k decrease with an increase in G. These evidences indicate that k is influenced by nucleation behaviors. We can assume the proportionality between G and k when crystallization is controlled only by the instantaneous (predetermined) nucleation. These results suggest that molecular weight dependence must be studied separately by G and I and their products of G and I should be compared to k. This report will be discussed separately on I, G and k.

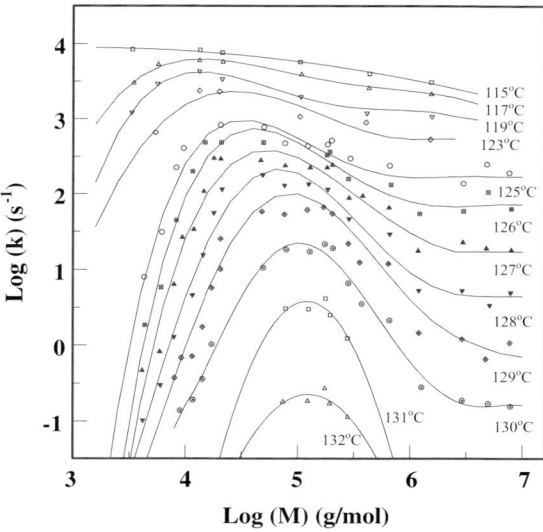

Fig. 19.2. Double common logarithmic plot of the overall crystallization rate (k) against molecular weight for indicated crystallization temperatures for poly(ethylene) (PE) [30]

19.2 Temperature Dependence of Crystallization

19.2.1 Temperature Dependence of Linear Crystal Growth Rate

Most of crystalline materials such as polymeric materials, organic substances and inorganic compounds, show spherulitic growth from the melt and they show often the maximum growth rate as shown in Fig. 19.3. According to a current crystallization theory for the temperature dependence of linear crystal growth rate (G), the following exponential equation is generally employed,

$$G = G_o \exp\left[-\frac{\Delta E_g}{RT} - \frac{\Delta F_g}{RT}\right] \quad (19.1)$$

where G_o is assumed to be a constant without temperature dependence but strongly depended on molecular weight. ΔE_g is the activation energy for the transport process at the interface between the melt and the crystal surface. ΔF_g is the work required to form critical size of secondary nucleus on the crystal growing surface. ΔE_g and ΔF_g terms have opposing temperature dependence; thereby bring about a maximum (G_{max}) in the growth rate. Here, ΔF_g commonly expressed as $\Delta F_g = K_g T_m^o / (T_m^o - T)$ yielding the following equation [40].

$$G = G_o \exp\left[-\frac{\Delta E_g}{RT} - \frac{K_g T_m^o}{RT \Delta T}\right] \quad (19.2)$$

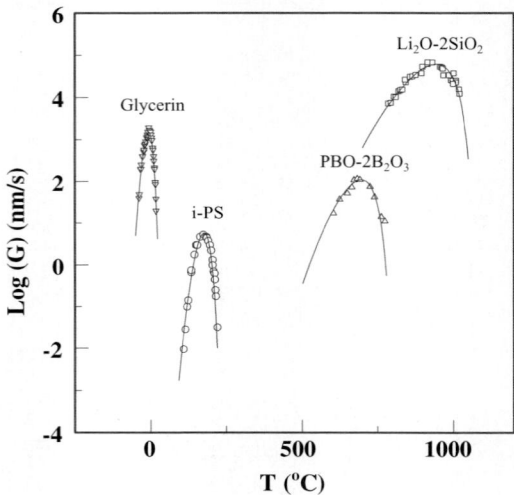

Fig. 19.3. Temperature dependence of common logarithm of crystal growth rate from the melt for a variety of crystalline materials; Glycerine [1], Li_2O-$2SiO_2$ [9], PBO-$2B_2O_3$ [12] and isotactic poly(styrene) [19] (i-PS). *Solid lines* are best fitting by Arrhenius expression of the molecular transport term

K_g is generally expressed as $K_g = nb_o\sigma_e\sigma_u/\Delta H_m$, where n is a function of mode of secondary nucleation, b_o is the thickness of the depositing growth layer, ΔH_m is the heat of fusion and σ_e and σ_u are the end- and the lateral-surface free energies, respectively. ΔT is the degree of super-cooling (T_m^o-T), where T_m^o is an equilibrium melting temperature and T is crystallization temperature.

An application to polymer crystallization leads to that the molecular transport term is considerably important in the lower temperature ranges. The transport term can be expressed in terms of the equation of either Arrhenius type ($\Delta E_g/RT$) or WLF type ($\Delta Q_g/R(T-T_o)$) where ΔQ_g and T_o are adjustable parameters. In analyzing the crystallization data in bulk polymers, the WLF expression has been used much familiar than the Arrhenius-type, since it has been believed that the former expression fits the data better than the later one. The activation energy in the transport term can be associated with an activation process of molecular transport from the melt to the crystal surface. Such activation energy is commonly used as an expression of $C_1/R(T - T_g + C_2)$ [27, 41]. C_1 and C_2 are adjustable parameters but these values are often used as a constant value in many reported references. In the WLF expression, polymer crystallization is forbidden at T_o, which is usually about 50K below the glass transition temperature (T_g). T_o is a hypothetical temperature at which the macro-Brownian (segmental) motion of polymer molecules ceases. It is interesting to note that T_o shows linear relationship to the β relaxation temperature in mechanical relaxation properties [42]. The

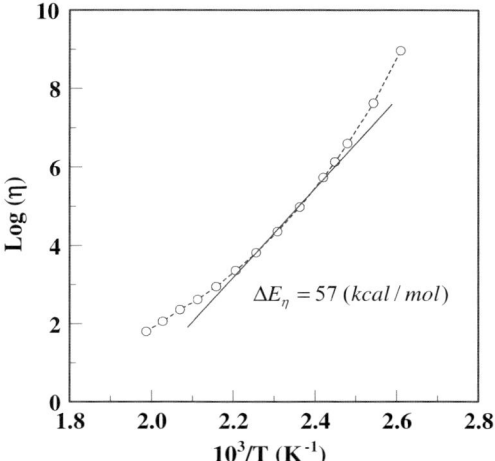

Fig. 19.4. Temperature dependence of melt viscosity for atactic-poly(styrene) [47]. The activation energy between T_g and T_m is approximately about 57 (kcal/mol)

β relaxation is associated with the local molecular motion in polymeric systems. According to the free volume concept, the free volume fraction (f) is often expressed as follow,

$$f = f_g + \Delta\alpha(T - T_g) = \Delta\alpha(T - T_o) \tag{19.3}$$

where f_g is free volume fraction at T_g and $\Delta\alpha$ is the difference between thermal expansion coefficients for glassy and rubbery states. In general, the mean value of f_g is approximately to be 0.025 (=1/40) and $\Delta\alpha$ is about $5*10^{-4}(1/K)$, so that the ratio of $f_g/\Delta\alpha$ (=C_2 in WLF equation) equals to about 50K. The mean value of $\Delta\alpha T_g$ is about $0.1(\approx 4f_g)$ based on reference data [43]. These results indicate as follows,

$$T_o = T_g(1 - f_g/\Delta\alpha T_g) \approx (3/4)T_g \tag{19.4}$$

The ratio of T_g and T_m^o is well known to be 2/3 for the most of polymers [44–46]. That is, T_o comes near to $(1/2)T_m^o$. This result is much used for the best fitting procedure for crystal growth data. Anyhow, the activation energy for molecular transport term can be expressed by either WLF or Arrhenius type in a wide crystallization temperature encompassed through T_{cmax} as seen in Fig. 19.1 and 19.21.

It is worth to check a value of activation energy for the molecular transport term on the base of Arrhenius expression. The best fitting result for i-PS is 57 Kcal/mol for the activation energy of the molecular transport term. Figure 19.4 shows the temperature dependence of melt viscosity of PS [47]. It is interesting to note that the activation energy for crystal growth rate nearly equals to the activation energy in the rubbery state between T_g and T_m. The

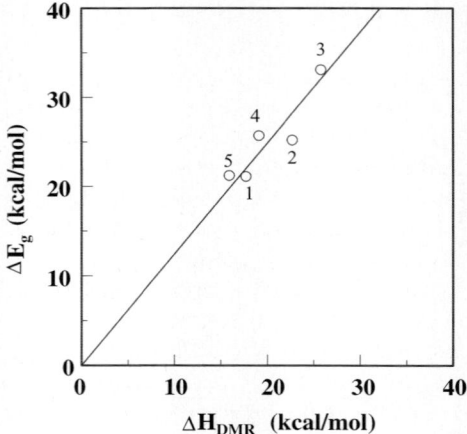

Fig. 19.5. Relationship between the activation energies for viscoelastic relaxation (ΔH_{DMR}) measured by dynamic mechanical relaxation spectra and those for molecular transport (ΔE_g) in crystallization for poly(ethylene tetrephthalate) (PET) / poly(1-4-cyclohexylene dimethylene terephthalate) (PCT) copolymers [48]. Copolymer compositions; (PCT/PET) = 1 ; 100/0, 2 ; 80/20, 3 ; 66/34, 4 ; 5/95, 5 ; 0/100

similar relationship is found in crystallization of poly(ethylene terephthalate) copolymer [48] as shown in Fig. 19.5. The activation energy for the molecular transport term (ΔE_g) in crystallization process shows a linear relation to that for relaxation spectra (ΔH_{DMR}) in mechanical property above T_g. Values for ΔH_{DMR} were measured by a Rheovibron (Toyo Seiki) at a heating rate of 2°C/min and a frequency of 10H_z over the temperature range from –150°C to 200°C. These phenomenological results suggest that the activation energy for the molecular transport term in polymer crystallization is associated with molecular diffusion in the super-cooled melt above T_g. This indicates that the activation process in molecular transport in polymer crystallization could be similar to that in the confined molecular motion in the rubbery state above T_g.

Polymeric materials often show spherulite crystal morphology at a high super-cooling and also show axialite or polyhedral crystal at a relatively high crystallization temperature. According to a regime theory [49–51], the temperature dependence of crystal growth rate will change at a certain temperature range where the crystal morphology is transformed in its shape. In fact, there are many papers for regime changes for various polymers [30, 35, 51–53]. According to the regime theory, growth rates are proposed to three different regimes as follows.

$$\text{Regime} - \text{I} \qquad G \propto b_o i H \qquad (19.5)$$

Regime − II $\quad\quad\quad G \propto b_o(2ig)^{1/2}$ \quad\quad\quad (19.6)

Regime − III $\quad\quad\quad G \propto b_o iH$ \quad\quad\quad (19.7)

H is a terrace length between defect points on depositing crystal-growing surface and i is the secondary nucleation rate on the terrace H producing a step and kink of crystal. Small character of g is the propagation rate of steps on H and assumed to be independent on the degree of super-cooling. For example in crystallization of a thin polyethylene film from the melt [53], the truncated lozenge crystal is observed in the temperature ranges below about 124°C and the lenticular crystal is formed at above 124°C. Such morphological transformations bring a change in the crystal growth rate clearly at about 124°C as shown in Fig. 19.6 [53]. The regime transition from regime I to regime II gives the difference in their slope as seen in Fig. 19.6. However, many questions and arguments have arisen in recent years for applying the regime theory to crystal growth data. It has been doubted that H and g are constant. They might be dependent on the degree of super cooling. In fact, it is expected that the step propagation rate depends strongly on super-cooling [54]. In addition, the morphology will change gradually with temperature and then the regime transition will change gradually with temperature. Such gradual change might not give the slope change in the regime plots. In the case of PESU and i-PS, the regime transitions could not be found in temperature dependences of crystal growth rate regardless of the clear morphological change from spherulite to polyhedral crystals (lozenge for PESU and hexagonal for i-PS). The morphology of PESU changes from a spherulite to a lozenge through an axialite crystals as an increase in crystallization temperature. Figure 19.7 shows temperature dependence of crystal growth rate of PESU with morphological changes. According to the best fitting to the data by a least squared method based on Eq. (19.2), the straight line is observed without any changes in slopes as shown in Fig. 19.7. As another example, morphology of i-PS is changed from spherulite to hexagonal with increasing in crystallization temperature and the growth rate shows bell-shape temperature dependence as shown in Fig. 19.8. The linear plot is also observed for i-PS wherein triangles and circles indicate the hexagonal and spherulite morphologies [55]. This is also true for other polymers accompanying with morphological changes [55]. The molecular transport term is important in regime plots because the appearance of slope change is strongly dependent on the value employed in the molecular transport term. When the molecular transport term has selected a proper value, no regime transition is often found. On the other hand, an inadequate value or disregard of molecular transport term gives a phantom regime transition. We should be very careful for regime plots based on these assumptions.

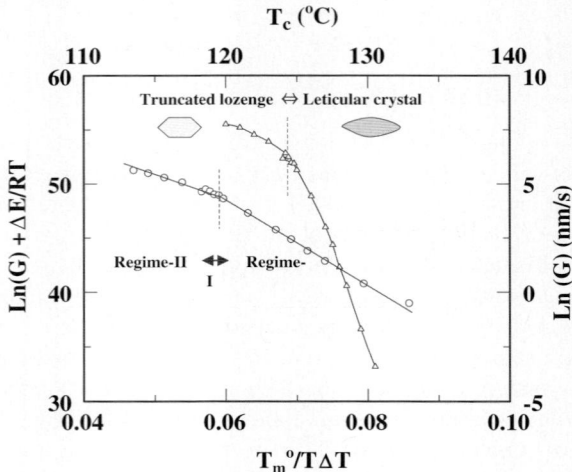

Fig. 19.6. Natural logarithm of G vs. crystallization temperature (T_c) ($\triangle$) and super-cooling ($T_m^o/T\Delta T$) dependence of natural logarithm of G plus the molecular transport term of $\Delta E/RT$ ($\bigcirc$) for PE [53]. *Broken lines* indicate morphological transition from truncated lozenge to leticular crystal

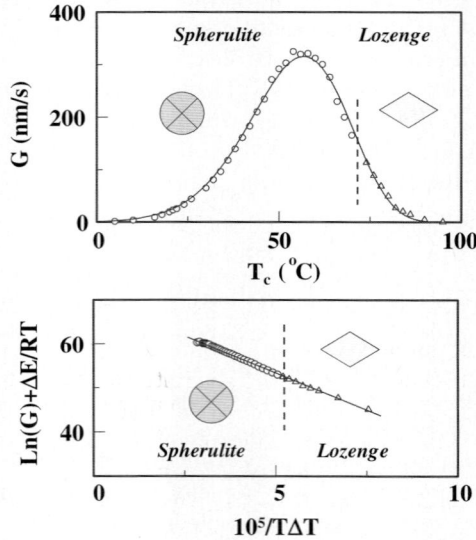

Fig. 19.7. Temperature dependence of G and super-cooling dependence of natural logarithm of G + $\Delta E/RT$ for PESU with M=3,850. *Broken line* indicates morphological transition from spherulite ($\bigcirc$) to lozenge ($\triangle$). *Solid lines* are the best fitting by Arrhenius expression

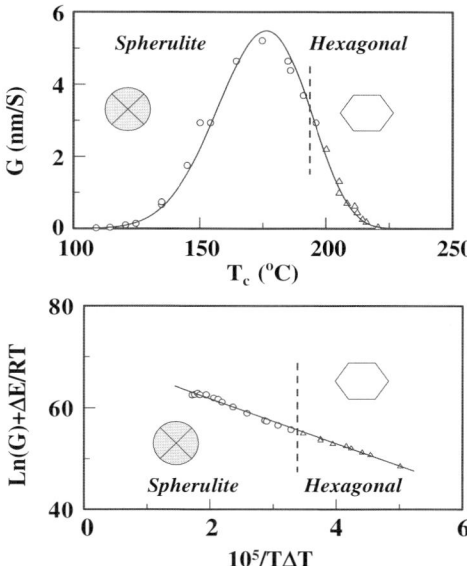

Fig. 19.8. Temperature dependence of G and super-cooling dependence of natural logarithm of G $+\Delta E/RT$ for i-PS [55]. *Broken line* indicates morphological transition from spherulite (○) to hexagonal (△). *Solid lines* are the best fitting by the Arrhenius expression

19.2.2 Temperature Dependence of Nucleation Rate

There are two types of crystal nucleation, such as homogenous and heterogeneous nucleation. Homogeneous nucleation can be defined by spontaneous aggregation of polymer molecules to form a three-dimensional nucleus, which must be beyond the certain critical size below the melting point. Beyond this size, the nucleation occurs sporadically. On the other hand, in heterogeneous nucleation, a limited number of sites become activated instantaneously or sporadically. The heterogeneous nucleation is initiated by the sites, which may be distributed on a surface of impurities or a wall of vessel. Number of impurities could be proportional to volume of sample used in an experimental setup. In fact, the sites, which are generated by impurities in the sample, decrease drastically with a decrease in the volume and then the nucleation changes from heterogeneous to homogeneous. In small droplets of crystallizable materials only homogeneous nucleation can take place [56,57]. The picture of such sites in impurities is not clear but the nucleation rate is strongly influenced by their surface energy (wetting mechanism) and surface morphology (epitaxial mechanism). In addition, heterogeneity in polymer melt structure is also important for nucleation as generally called by memory effect in the molten state prior to crystallization.

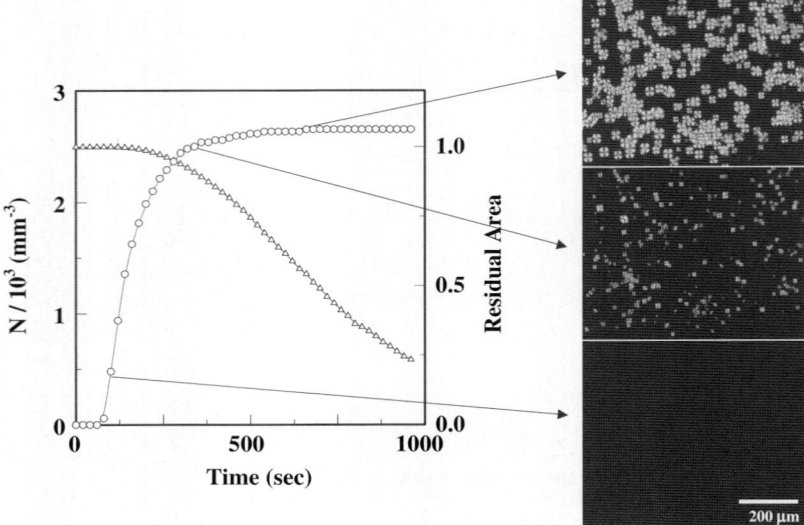

Fig. 19.9. Nucleation density (N) (○) and fraction of the residual non-crystallized area (△) as a function of time with three typical optical photographs for indicated time for PESU with M = 9,150 crystallized at 20°C [59]

Figure 19.9 and 19.10 show nucleation behavior for poly(ethylene succinate) (PESU) and isotactic-poly(styrene) (i-PS) as a function of time at a constant crystallization temperature. Small spherulites, which are assumed to have been started by active individual sites, are observed after a certain induction time whereas the aggregation of polymer molecules is reversible up to the critical size. Beyond that time, an embryo with a size greater than the critical size comes up steady and the nucleus increases linearly with time. The total number of nuclei saturates to a limiting constant value before the crystallization is completed (residual melt region remains still about 80–90% as seen on Fig. 19.9 and 19.10). These results indicate that the total nucleation sites are predetermined by a limited number of active sites in the polymer melt. These nucleation sites might be associated with an amount of impurities and/or heterogeneity in polymer melt. However, the saturation density is strongly dependent on experimental conditions such as crystallization temperature, previous temperature of melting and its holding time of melting prior to crystallization, cooling process to a given crystallization temperature from the melt and other experimental conditions. These nucleation behaviors can be classified as heterogeneous nucleation involving sporadic appearance of nuclei and with a limited number of nuclei. The saturation nucleus density can be expressed as the limited number of effective sites in impurity of foreign particles like a catalytic action in nucleation process. The effective nucleus number (N) in the heterogeneous nucleation with the limited number of ac-

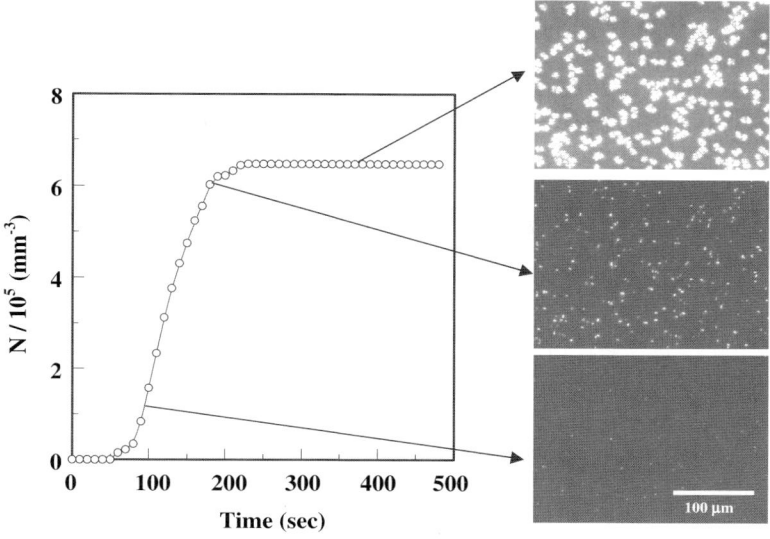

Fig. 19.10. Nucleation density (N) as a function of time with three typical optical photographs for indicated time for i-PS with $M_n=10{,}600$ crystallized at 175°C from the melt at 230°C [61]

tive sites (N_s) is assumed proportional to the residual site ($N_s - N$) with the frequency of nucleation per active site (J_o) and thus can be expressed by the following equation.

$$\frac{dN}{dt} = J_o(N_s - N) \qquad (19.8)$$

Here, J_o and N_s are strongly dependent on the distribution of active sites and their thermally stability. The integration of Eq. (19.8) gives the following equation with the initial condition $N = 0$ at $t = \tau_o$, where τ_o is the mean time to build up a critical nucleus (induction time).

$$N = N_s[1 - \exp(-J_o(t - \tau_o))] \qquad (19.9)$$

Figure 19.11 shows the natural logarithm of the fraction of residual nucleus sites of $(N_s-N)/N_s$ based on Eq. (19.9) as a function of time. The residual fraction decreases linearly with time giving the rate constant of J_o. The straight line can be applied up to 90–95% of the active sites, whereas the non-crystallized area remains about 90%. These results clearly indicate that almost all the active sites with the limited number are controlled by the heterogeneous nucleation mechanism. These characteristics of the active sites are not known. In the other explanation for the saturation density, there is a nucleation exclusion zone [58]. The nucleation will stop and reaches to the saturation density when the exclusion zone is overlap on the whole substrate surface. The zone might be associated with the density fluctuation around the

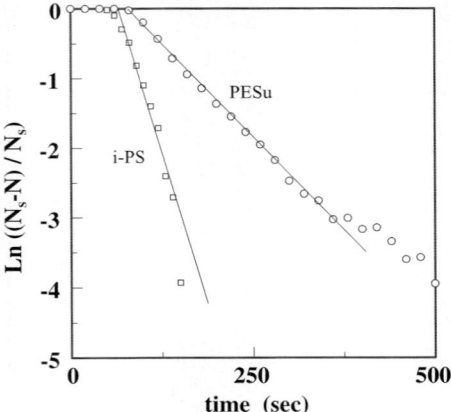

Fig. 19.11. Natural logarithm of the fraction of residual active sites as a function of time for PESU in Fig. 19.9 and for i-PS in Fig. 19.10

active sites or the density fluctuation of polymer molecules caused by their diffusion and the conformational changes in the molten state. Such fluctuation might be correlated with spinodal decomposition in polymeric systems.

Nucleation rate is usually estimated by the number of nuclei per a unit volume and time. Here, the unit volume is employed as an initial volume under an optical microscope as determined by nominal nucleation rate. In practical, the volume of non-crystalline volume decreases as crystallization proceeds and a real nucleation rate must be determined with correction of the actual residual volume. Figure 19.12 shows the temperature dependence of the nominal nucleation rate and the real nucleation rate for PESU [59]. Both rates coincide well within an experimental error since the residual melt region is large enough. However, the real nucleation rate will differ from the nominal nucleation rate when the residual melt region goes small. Figure 19.13 shows the relationship between the nominal nucleation rate I and J_o based on an Eq. (19.9) for i-PS. There are good linear relationships between I and J_o for nucleation from the melt. So, the nominal nucleation rate I can be used in general.

Homogeneous nucleation (sporadic) is often expressed by the following equation proposed by Turnbull and Fisher [60].

$$I = I_o \exp\left[-\frac{\Delta E_i}{RT} - \frac{\Delta F_i}{RT}\right] \quad (19.10)$$

I_o is assumed to be a constant without temperature dependence but strongly dependent on molecular weight. ΔE_i is the activation energy for the molecular transport process. ΔF_i is the work required to form a critical size of nucleus. ΔE_i and ΔF_i terms have opposing temperature dependence leading to a maximum (I_{max}). Here, ΔF_i commonly expressed as $\Delta F_i = K_i T_m^{o2}/(T_m^o-$

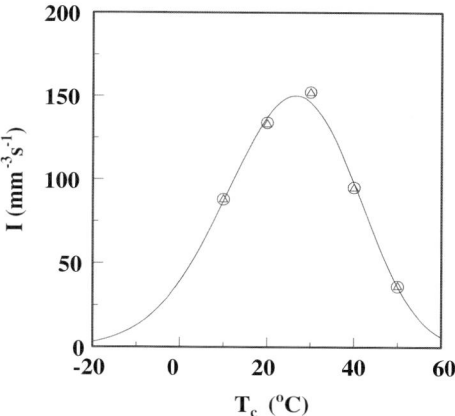

Fig. 19.12. Temperature dependence of nucleation rate for PESU with M = 9,150 [59]. *Open circle* is the nominal nucleation rate reduced by the initial view area and *open triangle* is the real nucleation rate reduced by the real effective area during crystallization. *Solid line* is the best fitting by Arrhenius expression of the molecular transport term, respectively

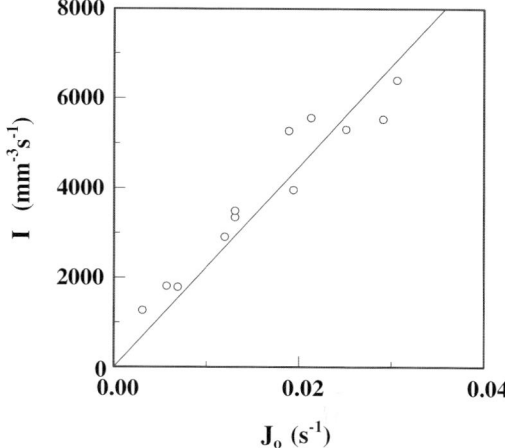

Fig. 19.13. Relationship between the nominal nucleation rate (I) and the rate constant (J_o) for i-PS crystallized at various temperatures from the melt at 230°C

T)^2 yielding the following equation [40].

$$I = I_o \exp\left[-\frac{\Delta E_i}{RT} - \frac{K_i T_m^{o2}}{RT\Delta T^2}\right] \quad (19.11)$$

K_i is generally expressed as $K_i = n\sigma_e \sigma_u^2/\Delta H_m^2$. An application to nucleation rate based on Eq. (19.11) causes that a molecular transport term is considerably important in the lower temperature ranges. Both expressions for the transport term expressed in terms of the equation of either Arrhenius type or WLF type can be fitted very well with experimental data as seen in Fig. 19.1 and 19.31.

Experimental data for number of nuclei, nucleation rate and induction time are generally widely dispersed and sometimes they seem not to be reproducibile. These dispersed data might be associated with different experimental setups, impurities contained in the system and heterogeneity in the molten state. For example, nucleation density is influenced by the previous temperature of melting at which the previously crystallized sample is melted and cooled down to a given crystallization and then heated to melt. The effect of melting condition on nucleation is often called a memory effect in the melt. Figure 19.14 shows nucleation density as a function of the previous temperature of melting for i-PS [61] and PEO [62]. The nucleation density decreases rapidly a thousand-fold (three orders) within 10–20 degree of temperature change. This result indicates that the heterogeneous nucleation sites decrease rapidly with the previous temperature of melting. The pre-existing active sites are strongly dependent on a type of polymeric materials. There are many active sites but the crystal growth rate is very slow in a case of i-PS. In the opposite situation for PEO, the nucleation rate is slow but the crystal growth rate is fast. It is well known that the nucleation rate is much different between crystallization from the melt and from the glass. Figure 19.15 shows the nucleation densities of PESU as function of time for crystallization from the melt and from the glass. The nucleation rate and the saturation density

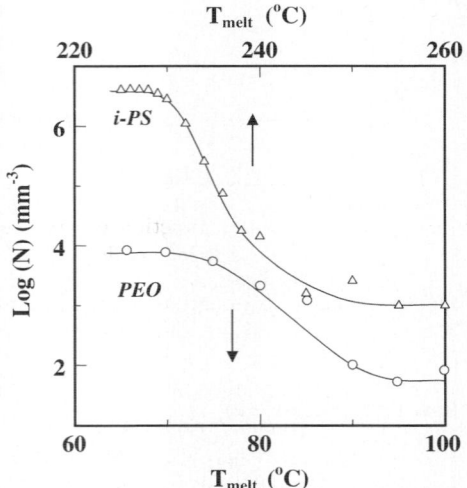

Fig. 19.14. Common logarithm of nucleation density (N) as a function of the melt temperature prior to crystallization for i-PS [61] (Δ) and poly(ethylene oxide) (PEO) [62]($\bigcirc$)

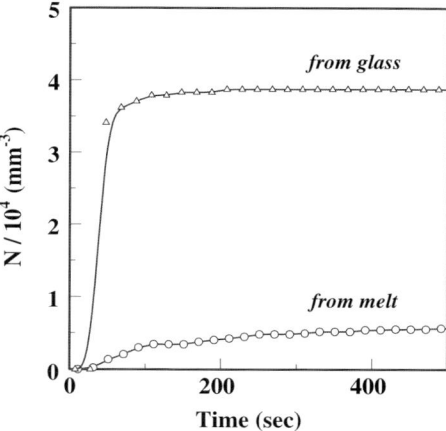

Fig. 19.15. Time dependence of nucleation density (N) for PESU with M = 9,150 crystallized at 30°C from the molten state (○) and the glassy state (△) [59]

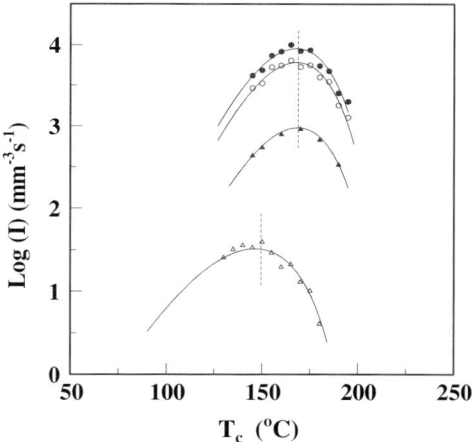

Fig. 19.16. Common logarithm of I as a function of crystallization temperature (T_c) for i-PS [61]. *Circle* and *triangle symbols* indicate as previous temperature of melting at 230°C and 250°C, respectively. *Open* and *solid symbols* indicate as crystallization from the molten state and from the glassy state, respectively

from the glass are much higher than those from the melt. This is also true for the nucleation density of i-PS crystallized from the melt and the glass [61,63]. Figure 19.16 shows the temperature dependence of nucleation rate for i-PS at two different previous temperature of melting (230°C and 250°C). Each crystallization temperature is set up from the molten state with cooling speed of 30°C/min and from the glassy state with the following procedure; samples are quenched to below the glass transition temperature (90°C) with cooling

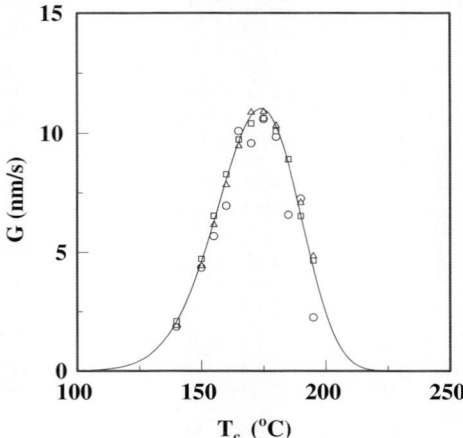

Fig. 19.17. Temperature dependence of the crystal growth rate from the melt at three melt temperatures of 230°C (*circle*), 240°C (*triangle*) and 250°C (*square*). *Solid line* results from the best fitting by Eq. (19.2)

rate of 130°C/min and kept there for 5 minutes and then subsequently heated to a given temperature with heating rate of 130°C/min. Starting from a melt at 230°C leads to the heterogeneous nucleation both for nucleation from the molten and glassy states. The nucleation rate from the glass is higher than that from the melt. Starting from a melt at 250°C, the nucleation rate from the glass is faster than that from the melt and the maximum nucleation temperature from the glass is higher than that from the melt. It is interesting to note that the maximum nucleation temperature is almost the same for the nucleation at 230°C both from the melt and the glass. However, at 250°C the nucleation from the glass is much higher than that from the melt. Here, it is worth to note, nucleation rate is strongly influenced not only by the previous temperature of melting but also by the quenching rate, during the cooling and heating process. Nucleation behavior is highly complicated in comparison to crystal growth behavior. In fact, growth rate is mainly governed by crystallization temperature without any effect by the previous temperature of melting (as shown in Fig. 19.17), cooling rate and other experimental conditions.

In addition to the above melting conditions, nucleation rate is also influenced by cooling program such as cooling rate and cooling history to the setup crystallization. Figure 19.18 shows a schematic cooling program from the molten state to the setup crystallization temperature as an example for PESU. At the first stage of cooling process, a hot stage under optical microscope is cooled down to a certain temperature (T_{set}) from 135°C with the cooling rate of 130°C/min. At the second stage, the cooling rate is changed at T_{set} to 10°C/min from 130°C/min and then the sample holds at the crystallization temperature (T_c). Here, a step cooling temperature is defined as tem-

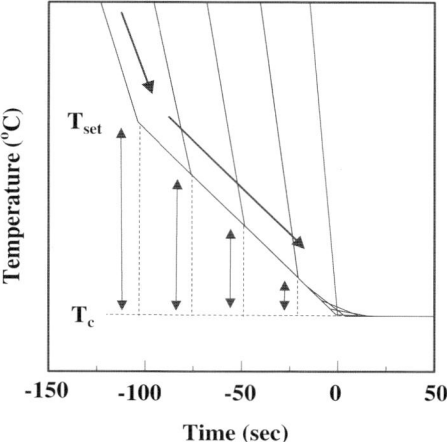

Fig. 19.18. Schematic illustration for cooling process in crystallization of PESU. At the first stage of cooling process, the molten state of polymer sample at 135°C is cooling down to a certain setup temperature (T_{set}) with the cooling rate of 130°C/min. And successively, the cooling rate is changed at T_{set} to 10°C/min from 130°C/min then holds at a given crystallization temperature (T_c)

perature difference between T_{set} and T_c. The melting temperature of 135°C is higher than the equilibrium melting temperature (131°C) for PESU where the molten state could be homogenous without the memory effect in the melt. During cooling down from the molten state with different cooling process, the heterogeneity in the amorphous structure in the super-cooled melt will vary in accordance with the wide distribution of molecular relaxation time in the super-cooled melt. The amorphous structure in the super-cooled melt might give rise to change in the nucleation sites. Figure 19.19 shows a relationship between the nucleation rate and the setup cooling temperature for PESU [59]. The small setup cooling temperature gives the large nucleation rate. This result clearly indicates the molecular relaxation during cooling process affects the nucleation mechanism. The nucleation rate from the glassy state is also influenced by aging effect below glass transition temperature. Quenched amorphous polymer often shows an enthalpy relaxation based on molecular relaxation in the super-cooled melt. Such relaxation mechanism could affect on the nucleation mechanism. In fact, Fig. 19.20 shows that nucleation rate increases with an increase in the enthalpy relaxation [59]. Various experimental conditions as discussed above seem to produce data with no reproducibility. This is not true, reproducible results can be obtained only when experimental conditions are setup in a rigorous manner.

It is interesting to note that the above results might be corresponding to spinodal decomposition during the melt crystallization. When poly(ethylene terephthalate) is quenched down quickly to the setup crystallization temper-

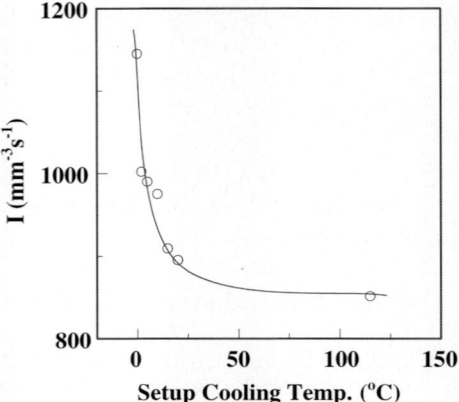

Fig. 19.19. Relationship between nucleation rate and setup cooling temperature [59]. Setup cooling temperature is defined as the temperature difference between T_{set} and T_c

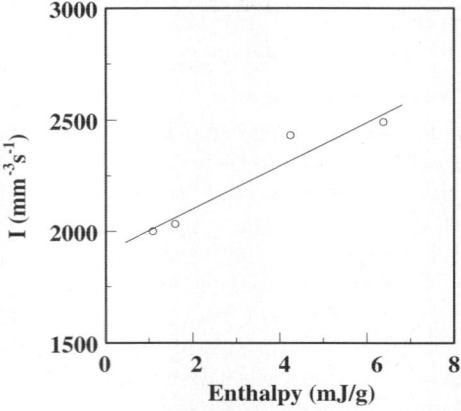

Fig. 19.20. Relationship between the nucleation rate and the enthalpy of relaxation aging at below the glass transition temperatures for PESU

ature from the molten state, the nucleation starts in the high density phase, which might be generated by spinodal decomposition even in a homo-polymer system [64]. It is also found by IR measurement that the sequence length of helix in i-PS increases with a decrease in temperature, especially at the temperature just before crystallization start [65]. In addition to these experimental results, a theoretical model for spinodal decomposition has been proposed in a single homo-polymer system based on density fluctuation with conformational change [66]. In particular, changes in amorphous structure prior to crystallization are often observed in molecular orientation crystallization of polymeric systems. For example, a drawn amorphous PET shows a nematic phase and the nematic PET changes to smectic phase and to crystalline phase when it is annealed [67]. In melt spinning of i-PP, a smectic phase appears first in the spine line forming a shish-structure and lamella crystals are over-grown on the shish yielding shish kebab structure [68].

19.3 Molecular Weight Dependence of Nucleation and Crystal Growth Rates

19.3.1 Molecular Weight Dependence of Crystal Growth Rate

Data existing in the literature for spherulite growth rate of several crystalline polymers have been analyzed as a function of molecular weight expressed as a power law of M^{α_g}. For example, the exponent α_g for poly(ethylene) lies in the range of –1.3 [23] to –1.8 [26] at relatively small super-cooling. On the other hand, for relatively large super-cooling, α_g is nearly –0.5 [15–22, 25]. The value of α_g depends strongly on the degree of super-cooling [29]. The differences are attributed to the reference crystallization temperature such as a constant super cooling or a constant crystallization temperature. The characteristic reference value for the molecular weight dependence of crystal growth rate should be employed just as to zero shear viscosity for the molecular weight dependence of melt viscosity. Figure 19.21 shows the temperature dependence of the linear crystal growth rate as a function of molecular weight

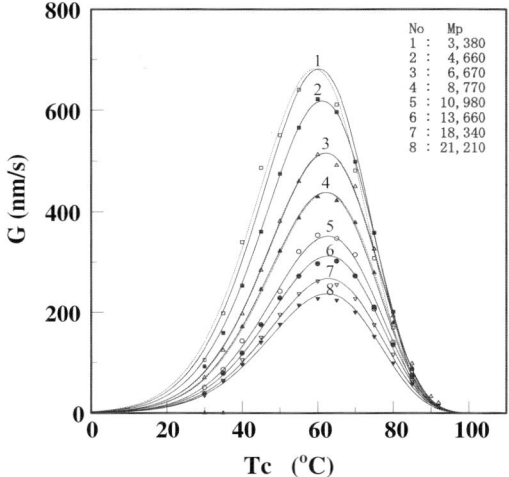

Fig. 19.21. Temperature dependence of the growth rate for PESU with various molecular weights indicated on the graph [29]. *Solid* and *broken lines* are the best fitting by Arrhenius and WLF expressions of the molecular transport term, respectively

for fractionated PESU samples [29]. Each molecular weight fraction shows crystal growth rate with a bell-shape curve. Solid and broken lines are results for the best fitting by Arrhenius and WLF expressions of the molecular transport term, respectively. It is clear that both expressions can fit the data sufficiently. The crystal growth rate (G) remarkably decreases with the molecular weight. The maximum crystal growth rate (G_{max}) and its temperature (T_{cmax}) vary with molecular weight. Molecular weight dependence of T_{cmax} showed similar molecular weight dependence of T_m^o. The G_{max} decreases with increasing molecular weight. The logarithm of G_{max} increases linearly with a decrease in the logarithm of M, yielding a slope of –0.5.

Molecular weight dependence of crystal growth rate can be expressed as a function of adsorption of polymer molecules (A_g) on to the crystal-growing surface, diffusion of the adsorbed molecules (D_g) on the crystal surface and secondary (surface) nucleation based on the adsorbed molecules as schematically illustrated in Fig. 19.22. Accordingly, the growth rate is expressed as follows [29],

$$G \propto A_g \cdot D_g \exp\left[-\frac{K_g T_m^o}{RT \Delta T}\right] \tag{19.12}$$

Lauritzen and Hoffmann have introduced the probability parameter of admolecules on the crystal-growing surface prior to surface nucleation but not considered the molecular weight dependence on it [51]. Adsorption of polymer molecules must be a function of molecular conformation on the crystal surface (substrate) as given by the following equation.

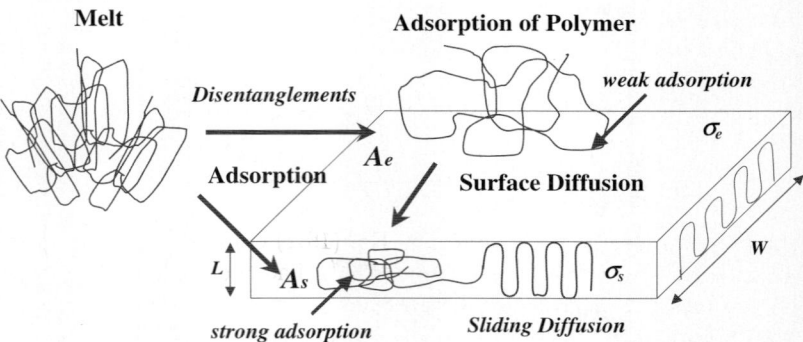

Fig. 19.22. Schematic illustration for secondary nucleation process during crystallization. Random coil chains with entanglements will adsorb on chain folding plane with weak adsorption and lateral crystal growing plane with strong adsorption with loop and train conformations. The weak adsorption is not important for the lateral crystal growth but probably generates the stacked lamellae structure. The strong adsorption will migrate and rotates on the crystal growing front and attach to the crystallographic lattice. These adsorbed segments will resemble a two-dimensional random walk with a number of contacts on the crystal surface

$$A_g \propto M^a \exp\left[-\frac{\Delta E_a}{RT}\right] \quad (19.13)$$

The exponent a lies between zero and one, depending on the molecular conformation on the substrate surface [69]. For example, when polymer molecules are attached with a number of contacts on the surface (loop-train adsorption), a is 0.5. The probability of admolecules (P_{ad}) can be related to the substrate area (S) and the substrate surface energy (σ). A chain folded lamella crystal is constructed with a thickness of L (side area is A_s and its surface energy is σ_s) and a lateral width of W (fold surface is A_e and its surface energy is σ_e) as seen in Fig. 19.22. The value of σ_e is, in general, about ten times larger than that of σ_s. The polymer molecules can be adsorbed weakly on the fold surface area of A_e but strongly on the side surface of A_s. ΔE_a is activation energy for the molecular adsorption on the crystal surface. When the size of lamella is about 10 nm thickness (L) and less than 0.5 µm for W, P_{ad} on the A_s is larger than that on A_e, because ΔE_a on A_s is larger than that on A_e. This means that most of secondary nucleation event occurs on A_s. When the lamella size is bigger (W increases whereas L is constant), the surface nucleation will be unable to disregard on A_e, yielding a multi-layered (stacked) lamellar structure. Molecular diffusion constant of the adsorbed molecules is a function of molecular weight based on reptation or sliding diffusion mechanism on the surface as given by the following equation.

$$D_g \propto M^d \exp\left[-\frac{\Delta E_d}{RT}\right] \quad (19.14)$$

According to reptation mechanism, d equals to -1 for molecules without entanglements and to -2 for entangled molecules [70]. ΔE_d is the activation energy of molecular diffusion for the adsorbed molecule on the crystal surface. Taking these factors into account, the crystal growth rate is expressed as follows:

$$G \propto M^{a+d} \exp\left[-\frac{\Delta E_g}{RT} - \frac{K_g T_m^o}{RT\Delta T}\right] \quad (19.15)$$

The pre-exponential factor of G_o in Eq. (19.2) can be expressed as a function of molecular weight as given by M^{a+d}. The molecular transport term (ΔE_g) involves two activation energies for the molecular adsorption (ΔE_a) and the molecular diffusion (ΔE_d) on the crystal surface. These activation energies might be dependent on molecular weight as well. In addition, the parameter K_g and T_m^o in the nucleation term should be dependent on molecular weight. These factors of G_o, ΔE_g and K_g can be evaluated as a function of molecular weight by the following experimental procedure.

Figure 19.23 shows plots of natural logarithm of G plus $\Delta E_g/RT$ against $T_m^o/T\Delta T$ according to Eq. (19.2) for various molecular weights of PESU. The straight lines were calculated to obtain the best fit to the data by a linear least square procedure, yielding G_o, ΔE_g and K_g. G_o depends remarkably on molecular weight and can be expressed as a power law of $G_o \propto M^{-0.5}$,

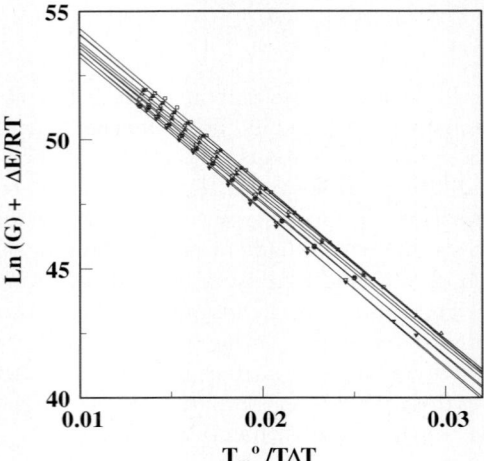

Fig. 19.23. Relationship between super-cooling and natural logarithm of G according to Eq. (19.2) for PESU with various molecular weight [71]

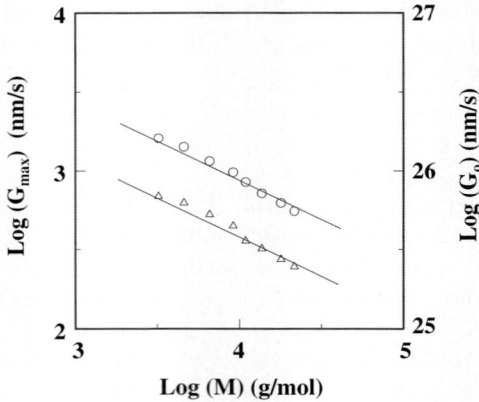

Fig. 19.24. Common logarithm of the pre-exponential factor of G_o (circle) and the maximum crystal growth rate of G_{max} (triangle) as a function of common logarithm of molecular weight for PESU [71]

which shows the same power law of G_{max} as seen in Fig. 19.24. The exponent value of -0.5 can be explained in terms of $a + d$ given in Eq. (19.2). The loop-train adsorption of polymer molecules occurs in polymers with relatively high molecular weight that forms chain folding crystallization. The molecular weight dependence of the loop-train adsorption results in an exponent a equal to 0.5. The adsorbed molecules could migrate on the surface based on the reptation mechanism without chain entanglements, which gives d to be –1. Thus, the estimated value for $a + d$ is –0.5.

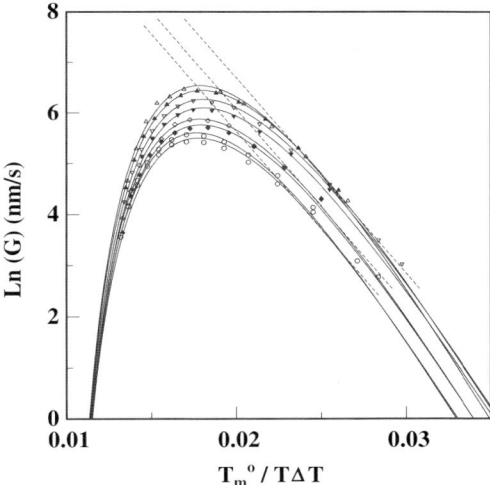

Fig. 19.25. Natural logarithm of G as a function of super-cooling function according to Eq. (19.2). *Broken lines* are drawn by the best fitting in the limited area

On the other hand, ΔE_g and K_g show a slight molecular weight dependence [71]. The ratio of $\Delta E_g/K_g$ gives an almost constant value of 24 [72, 73] yielding no molecular weight dependence. These molecular weight dependencies were also true for the estimated values based on the WLF expression of the molecular transport term [74]. The constant ratio of $\Delta E_g/K_g$ gives rise to the constant ratio of T_{cmax}/T_{mo} [72, 73]. Here, it is better to indicate that the value of α_g is influenced not only by the degree of super-cooling but also by the plotting method based on Eq. (19.2). When the molecular transport is omitted, the extrapolated value of G_o can be estimated by the linear line part in the plots of Ln(G) against $T_m^o/T\Delta T$ as seen in Fig. 19.25 for various molecular weights of PESU. Thus obtained molecular weight dependence of G_o gives -0.9 to α_g. It is worth to note that the molecular transport term gives an important rule to determine the molecular weight dependence of crystal growth rate.

G_{max} is presented by only one rate at a given molecular weight and can be formulated by equating to zero the derivative of Eq. (19.2), either Arrhenius or WLF expression of molecular transport term [75]. The crystal growth rate (G) can be formulated as a function of the maximum crystal growth rate and the reduced super-cooling (Z) based on Eq. (19.2) with the Arrhenius expressions in the molecular transport term, as follows:

$$G = G_{\max} \exp\left[W_g \frac{(1-Z-A_z)^2}{Z(1-Z)}\right] \tag{19.16}$$

where $W_g = \mathrm{Ln}(G_{\max}/G_o)$, $Z = \Delta T/T_m^o$ and $A_z = T_{cmax}/T_m^o$. Figure 19.26 shows plots of natural logarithm of G against the whole term of the right hand

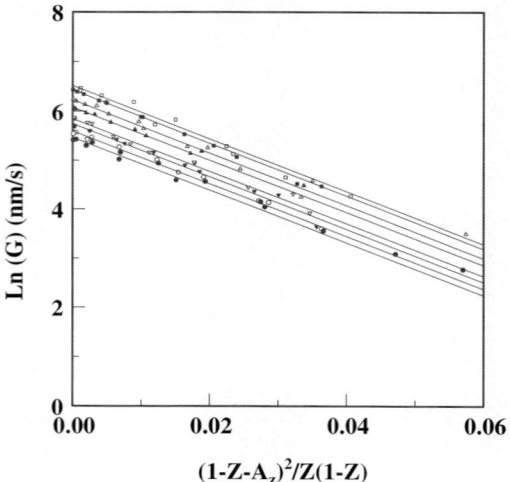

Fig. 19.26. Relationship between natural logarithm of G and reduced super-cooling function of Z according to Eq. (19.16) for PESU with various molecular weight [71]

side of Eq. (19.2). Straight lines are observed for each molecular weight sample, giving $\text{Ln}(G_{\text{max}})$ at the intercept and the ratio of $\text{Ln}(G_{\text{max}}/G_o)$ for the slope. The ratio of $\text{Ln}(G_{\text{max}}/G_o)$ gives an almost constant value of -53.8. In other words, the molecular weight dependence of G_{max} is mainly a consequence of the molecular weight dependence of G_o. It is much more advantageous to use G_{max} rather than G_o for studying molecular weight dependence, because G_{max} can be observed experimentally but G_o cannot.

The whole exponential term in Eq. (19.2), $(\exp[-\Delta E_g/RT - K_g T_m^o / RT \Delta T] = Y_g)$, can be evaluated as a function of molecular weight and crystallization temperature on the bases of the molecular weight dependence of ΔE_g, K_g and T_m^o. The molecular weight dependence of T_m^o was expressed by $T_m^o = 404.4 - 12,587/M$ in the case of fractionated PESU. Figure 19.27 shows three dimensional plots for Y_g as function of molecular weight and temperature [76]. At a given constant temperature above T_{cmax}, the molecular weight dependences of Y_g show upward and leveling-off tendencies. At a given constant temperature below T_{cmax}, they show downward and leveling-off tendencies. At T_{cmax}, the maximum point of Y_g gives rise to the constant value without the molecular weight dependence. These results indicate that it is not required to take the molecular weight dependence of Y_g at T_{cmax} into account. The molecular weight dependence of G depends only on G_o or G_{max}. Figure 19.28 shows the molecular weight dependence of G obtained from the combination of Fig. 19.24 and 19.27. At the higher temperature regions above T_{cmax}, the molecular weight dependence shows the convex curve and at lower temperature regions below T_{cmax} it shows the downward curve. This is true for growth data of PEO [77] as shown in Fig. 19.29. The growth data of PEO

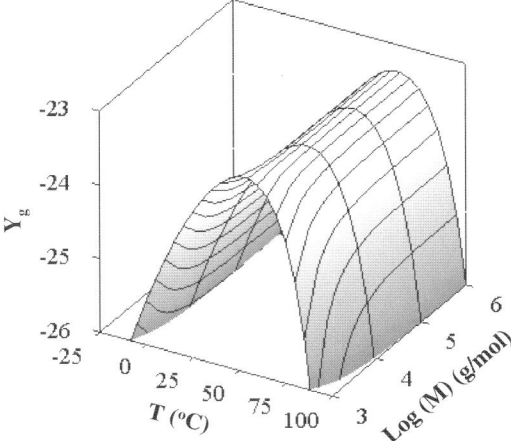

Fig. 19.27. Three-dimensional plots of the whole exponential term in Eq. (19.2) (Y_g) as a function of molecular weight and crystallization temperature for PESU

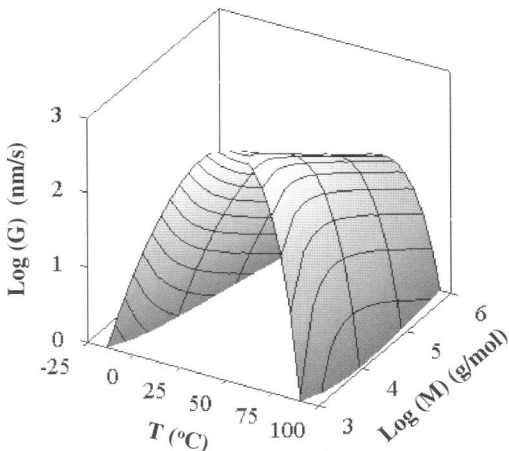

Fig. 19.28. Three-dimensional plots for the growth rate as a function of molecular weight and temperature for PESU

were observed only in the higher temperature regions above T_{cmax}. The molecular weight dependence of growth rate of PEO at constant temperature shows a maximum in a similar way as in Fig. 19.28 in the higher temperature regions. At T_{cmax}, the maximum growth rate shows the linear relationship with the molecular weight, yielding a power law of $G_{max} \propto M^{-0.5}$. That is, the maximum crystal growth can be defined as an intrinsic (or reference) growth rate for crystallization behavior, just like molecular weight dependence of melt

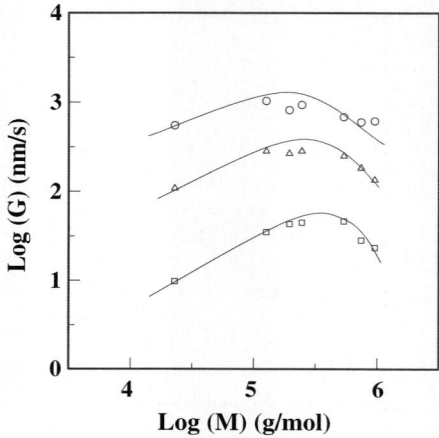

Fig. 19.29. Common logarithm of the growth rate as a function of common logarithm of molecular weight for PEO [77] at three different temperatures of 54.5°C (*circle*), 56.9°C (*triangle*) and 59.1°C (*square*)

viscosity based on zero shear viscosity as an intrinsic property. The zero shear viscosity is expressed as a power law of $M^{3.4}$ for entangled polymers. We can thus conclude that the molecular weight dependence of the linear crystal growth rate must be evaluated by G_{max}, otherwise α_g depends strongly on the degree of super-cooling and molecular weight. Figure 19.30 shows the molecular weight dependence of the maximum crystal growth rate for various polymers. All polymers show a good linear relationship with a slope of –0.5. These results indicate that the molecular weight dependence of maximum growth rate can be scaled and expressed as a –0.5 power of molecular weight to all crystalline polymers.

19.3.2 Molecular Weight Dependence of Nucleation Rate

Figure 19.31 shows the temperature dependence of the primary nucleation rate as a function of molecular weight for the fractionated PESU samples. Each molecular weight fraction shows the bell-shape nucleation rate curve. Solid and broken lines in Fig. 19.31 are results for the best fitting by the Arrhenius and the WLF expressions of the molecular transport term, respectively. The maximum nucleation rate (I_{max}) increases with molecular weight. The molecular weight dependence of I_{max} can be expressed as $I_{max} \propto M^{1.5}$. The exponent value in the power law has been reported to be unity in previous papers [14,78]. According to further extensive studies in the molecular weight dependence of I_{max}, the exponent value can be evaluated to 1.5. It is interesting to note that the molecular weight dependence of the primary nucleation rate shows the opposite dependence compared to the linear crystal growth rate as discussed above section.

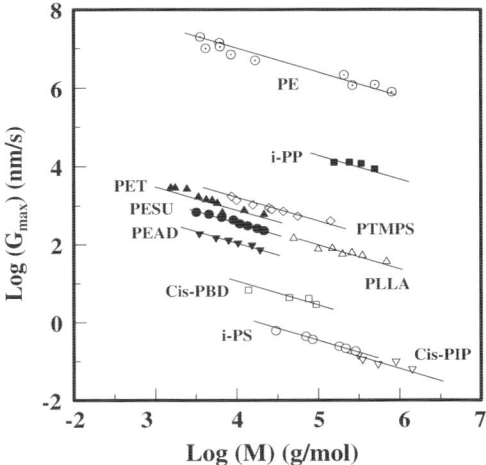

Fig. 19.30. Plots of common logarithm of the maximum crystal growth rate (G_{max}) against common logarithm of molecular weight (M) for various polymers for chain folding crystallization: PTMPS [27], i-PS [28], PLLA [21,25], PESU [29], PET [37], i-PP [36], PEAD [38], cis-PIP [38], cis-PBD [39]. Maximum growth rates for PESU, PET, i-PP and PEAD are observed by our laboratory and those for PTMPS, i-PS, PLLA, cis-PIP and cis-1,4-polybutadiene (cis-PBD) are reported in literatures. PE maximum growth rates are estimated based on reference data [23]

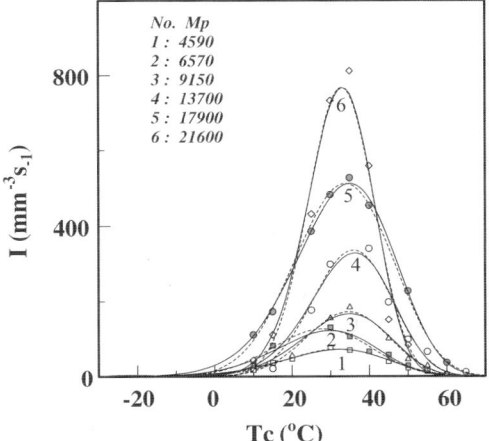

Fig. 19.31. Temperature dependence of I for PESU with various molecular weight indicated on the graph [14]. *Solid* and *broken lines* are the best fitting by Arrhenius and WLF expressions of the molecular transport term, respectively

Molecular weight dependence of primary nucleation rate can be expressed by number of molecules (N_i), diffusion of molecules (D_i) and formation of critical nucleus (ΔF_i) according to the following equation:

$$I \propto N_i \cdot D_i \exp\left[-\frac{\Delta F_i}{RT}\right] \qquad (19.17)$$

Polymer molecules are transformed into chain folded crystal by a process of intra-molecular nucleation. A certain sequence length in a single polymer chain is required to form a critical nucleus. Number of the sequence length is proportional to the molecular weight. However, the probability of finding the same sequence length in the same polymer chain could be higher than that in the other molecules (inter-molecular). Nucleation within the same polymer chain accelerates in velocity with an increase in molecular weight as expressed with M^b. Molecular diffusion of long chain molecules with molecular entanglement is expressed by the inverse of molecular weight squared [70]. Therefore, molecular weight dependence of intra-molecular nucleation is given by $I \propto M^{b-2}$. The value of b is not known at this moment. The contribution of the self-assembly in the intra-molecular chains can be larger than that of the molecular diffusion to the molecular weight dependence.

In the similar way for the molecular weight dependence of growth rate as discussed above, the whole exponential term ($\exp[-\Delta E_i/RT - K_i T_m^{o2}/RT\Delta T^2] = Y_i$) in Eq. (19.2) can be estimated on the base of the molecular weight dependence of each component, such as ΔE_i, K_i and T_m^o. Figure 19.32 shows three-dimensional plots for Y_i as a function of molecular weight and temperature [76]. At a given constant temperature above T_{imax}, the molecular weight dependences of Y_i show upward and leveling-off but at a constant

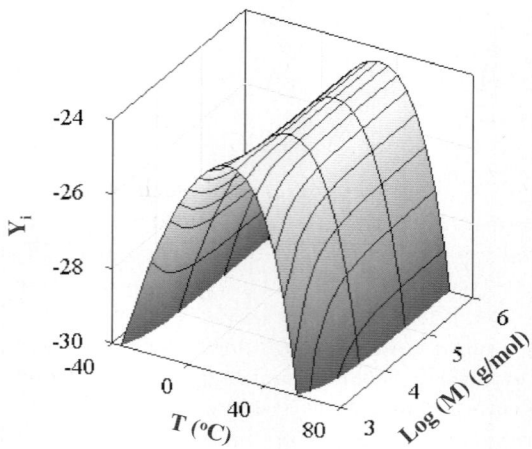

Fig. 19.32. Three-dimensional plots of the whole exponential term in Eq. (19.11) (Y_i) as a function of molecular weight and crystallization temperature for PESU

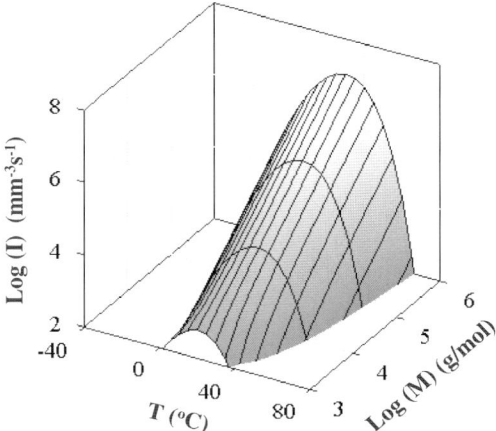

Fig. 19.33. Three-dimensional plots for the nucleation rate as a function of molecular weight and temperature for PESU

temperature below T_{imax}, they show downward and leveling-off. At T_{imax}, the maximum points of Y_i give rise to the constant value without the molecular weight dependence. These results indicate that the molecular weight dependence depends only on I_o or I_{max}. Figure 19.33 shows the total molecular weight dependence of I (together with I_o and Y_i). At T_{imax}, the molecular weight dependence of I_{max} shows a power law as expressed by $I_{max} \propto M^{1.5}$.

19.3.3 Molecular Weight Dependence of Overall Crystallization

Molecular weight dependence of overall crystallization rate (k) seems to be very complicated as seen in Fig. 19.2 for PE. The Avrami constant of k is a function of nucleation rate of (I) and crystal growth rate of (G) as expressed by following equation,

$$k \propto I\, G^m \tag{19.18}$$

where m is a dimension of forming crystal, such as 3 for three-dimensional growth, 2 for two dimensional growth. The nucleation (I) and growth (G) rates show opposite molecular weight dependence as discussed above. When the primary nucleation event is controlled by homogeneous (sporadic) nucleation and the crystal grows three-dimensional (m = 3), the molecular weight dependence of k can be expressed as shown in Fig. 19.34. In higher temperature regions above T_{kmax} (where the overall crystallization shows a maximum), k shows upward and leveling-off with the molecular weight. In the higher molecular weight regions or at the maximum rate of k, the molecular weight dependence of k disappeared. This is true for PEO data [79] as shown in Fig. 19.35. At each crystallization temperature in Fig. 19.35, the molecular weight dependence of k for PEO shows almost the same dependence for

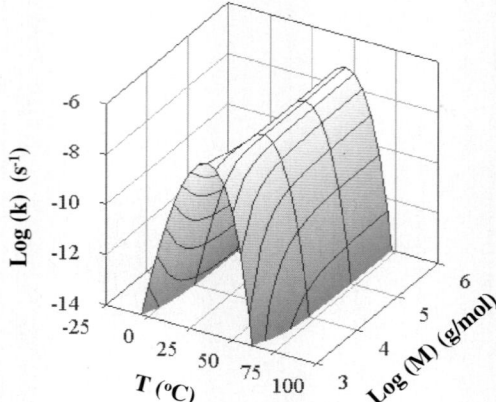

Fig. 19.34. Common logarithm of the overall crystallization rate constant for sporadic nucleation system as a function of common logarithm of molecular weight and temperature. Curves are calculated on the bases of PESU data [76]

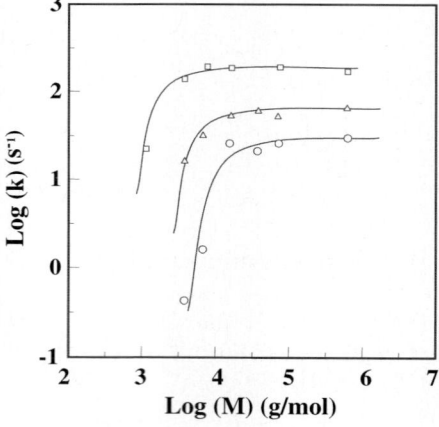

Fig. 19.35. Common logarithm of the overall crystallization rate constant as a function of common logarithm of molecular weight for PEO [79] crystallized at three different temperatures of 35°C (*square*), 45°C (*triangle*) and 50°C (*circle*)

the predicted data as shown in Fig. 19.34. On the other hand when the primary nucleation event occurred in a heterogeneous fashion (instantaneously), the molecular weight dependence of k shows a maximum formation as shown in Fig. 19.36 in higher temperature regions. This is true for PTMPS [31] and PE [30] as seen in Fig. 19.2. The molecular weight dependence of k in Fig. 19.36 is the similar to that of G in Fig. 19.28, because of no contribution of I on k. In a general crystallization mechanism, the primary nucleation can not be controlled by neither 100% sporadically (homogenously) nor 100% instanta-

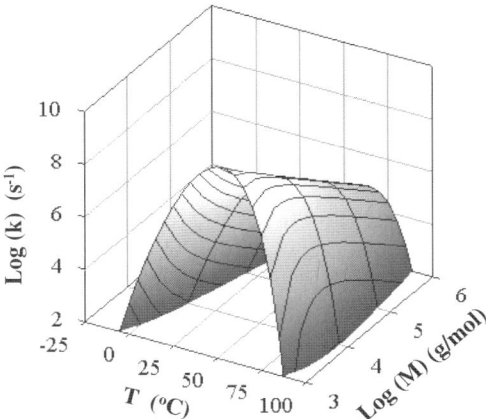

Fig. 19.36. Common logarithm of the overall crystallization rate constant for instantaneous nucleation system as a function of common logarithm of molecular weight and temperature. Curves are calculated on the bases of PESU data [76]

neously (heterogeneously). In fact, nucleus appears sporadically after a certain induction time as seen in Fig. 19.9 and 19.10. However, the total number of nuclei saturates at a limiting constant value much before the crystallization is completed (about 10–20% degree of crystallinity). These nucleation behaviors will affect the overall crystallization rate. That is, the homogenous crystallization based on Avrami concept will take place at the early crystallization time before the saturation of the number of nuclei. However, after the saturated nucleation density, the heterogeneous nucleation will be dominated. In other words, the homogeneous or heterogeneous nucleation based on Avrami can be determined by the nucleation profile as seen in Fig. 19.9 or 19.10. Here, it could be assumed that the contribution of I on k can be expressed by the power-law as follows:

$$k \propto I^c G^m \tag{19.19}$$

where c = 0 for 100% heterogeneous nucleation and c = 1 for 100% homogeneous nucleation. Figure 19.37 shows the molecular weight dependence of k as a function of the contribution factor of c ranging from zero to one. The maximum overall crystallization rate can be expressed as $k_{max} \propto (M^{1.5})^c (M^{-0.5})^m \propto M^\gamma$. In the case of the value of m being 3 for three-dimensional growth, the exponent value of γ changes from –1.5 to zero depending on the contribution factor of c in Eq. (19.19). It is worth to evaluate the degree of nucleation contributed to the overall crystallization mechanism according to the molecular weight dependence of k.

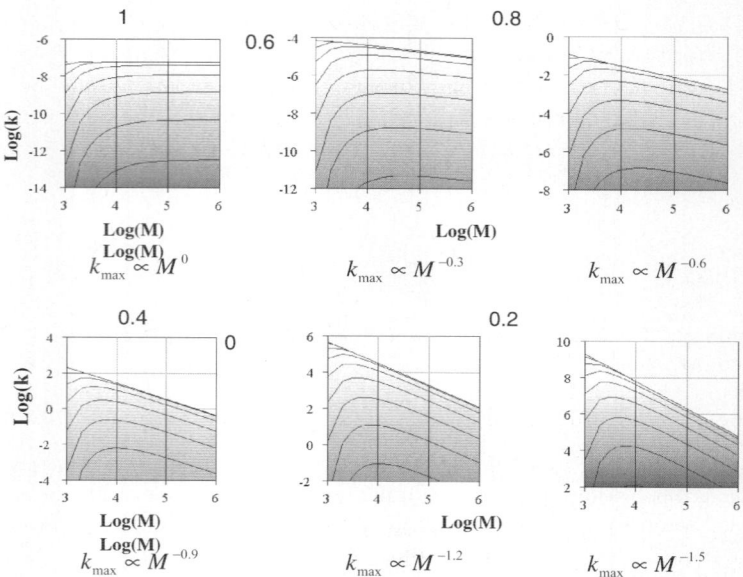

Fig. 19.37. Molecular weight (g/mol) dependence of common logarithm of the overall crystallization rate constant (1/sec) as a function of the contribution factor c ranging from zero to one. Curves are calculated on the bases of PESU data [76]

19.4 Conclusions

Nucleation rate is strongly dependent on experimental conditions such as previous temperature of melting, cooling process from the molten state, heating process from the glassy state and other various conditions. On the other hand, crystal growth rate is dependent on only crystallization temperature but is not influenced by other conditions. Heterogeneous nucleation is mainly controlled by the limiting number of active sites which become effective sporadically. The active site for nucleation might be associated to density fluctuation of polymer molecules caused by their diffusion and their conformational changes in the molten state.

Nucleation and crystal growth rates show the bell-shape temperature dependence showing the maximum rates of I_{max} at T_{imax} and G_{max} at T_{cmax}. These I_{max} and G_{max} are characteristic references to crystallization mechanism. The molecular weight dependence of I_{max} and G_{max} were scaled and expressed by power laws as $I_{\text{max}} \propto M^{1.5}$ and $G_{\text{max}} \propto M^{-0.5}$ for chain folding crystallization for polymeric materials. Data existing in reported references for G_{max} are scaled and expressed as a -0.5 power of molecular weight for all crystalline polymers. The molecular weight dependence of I_{max} shows the opposite tendency to the molecular weight dependence of G_{max}. The molecular weight dependence of the overall crystallization rate (k) is also expressed by $k_{max} \propto M^\gamma$, where the value of γ depended on the nucleation behavior such as sporadically (homogenous) ($\gamma = 0$) or instantaneously (heterogeneous) ($\gamma = -1.5$) nucleation in the case of three-dimensional growth. There is no molecular weight dependence of k in the case of homogeneous nucleation systems but strong molecular weight dependence of k in the case of heterogeneous nucleation systems.

References

[1] Tammann, G., Kristallisieren und Schmelzen, Leipzig, (1903)
[2] Chen, M., James, P.F., Lee, W.E., J. Sol-Gel Sci. Tech., 2, 233 (1994)
[3] Zanatoo, E.D., J. Non-Cryst. Solids, 89, 361 (1987)
[4] Potapov, O.V., Fokin, V.M., Filipovich, V.N., J. Non-Cryst. Solids, 247, 74 (1999)
[5] Fokin, V.M., Kalinia, A.M., Filopovich, V.N., J. Cryst. Growth, 52, 115 (1981)
[6] Burnett, D.G., Douglas, R.W., Physics Chem. Glasses, 12, 117 (1971)
[7] James, P.F., Physics Chem. Glasses, 15, 95 (1974)
[8] Vergamo, P.J., Uhlmann, D.R., Phys. Chem. Glasses, 11, 30 (1970)
[9] Matushita, K., Tashiro, M., Yogyo-Kyokai-Shi, 81, 60 (1973)
[10] Leedecke, C.J., Bergeron, C.G., J. Cryst. Growth, 32, 327 (1976)
[11] Scott, W.D., Pask, J.A., J. Am. Ceramic Soc., 44, 181 (1961)
[12] Luca, J.P., Eagan, R.J., Bergeron, C.G., J. Am. Ceramic Soc., 52, 322 (1969)
[13] Organ, S.J., Barham, P.J., J. Materials Sci., 26, 1368 (1991)
[14] Umemoto, S., Hayashi, R., Kawano, R., Kikutani, T., Okui, N., J. Macromol. Sci., $B42$, 421 (2003)
[15] Takayanagi, M., Kusumoto, N., Kogyo Kagaku Zasshi, 62, 587 (1959)
[16] Boon, J., Challa, G., van Krevelen, D.W., J. Polym. Sci., $A-2, 6$, 1791 (1968)
[17] Lovering, E.G., J. Polym. Sci., $C-30$, 329 (1970)
[18] Godovsky, Y.K., Slonimsky, G.L., Garbar, N.M., J. Polym. Sci., $C-38$, 1 (1972)
[19] van Antwerpen, F., van Krevelen, D.W., J. Polym. Sci., $B-10$, 2423 (1972)
[20] Cortazar, M., Guzman, G.M., Makromol. Chem., 183, 721 (1982)
[21] Vasanthakumari, R., Pennings, A.J., Polymer, 24, 175 (1983)
[22] Gomez, M.A., Fatou, J.G., Bello, A., Eur. Polym. J., 22, 661(1986)
[23] Hoffman, J.D., Miller. R.L., Macromolecules, 21, 3038 (1988)
[24] Chen. H,L., Li, L.J., Ouyang, W.C., Hwang, J.C., Wong, W.Y., Macromolecules, 30,1718 (1987)
[25] Miyata, T., Masuko, T., Polymer, 39, 5515 (1998)
[26] Okada, M., Nishi, M., Takahashi, M., Matsuda, H., Toda, A, Hikosaka, M., Polymer, 39, 4535 (1998)
[27] Magill, J.H., J. Appl. Phys., 35, 3249 (1964)
[28] Lemstra, P.J., Postma, J., Challa, G., Polymer, 15, 757 (1974)
[29] Umemoto, S., Kobayashi, N., Okui, N., J. Macromol. Sci., $B41$, 923 (2002)
[30] Fatou, J.G., Marco, C., Mandelkern, L., Polymer,31, 890 (1990)
[31] Magill, J.H., Polym. Lett., 6, 853 (1968)
[32] Burnett, B.B., McDevit, W.F., J. Appl. Phys., 28, 1101 (1957)
[33] Suzuki, T., Kovacs, A.J., Polym. J., 1, 82 (1970)
[34] Vasanthakumari, R., Pennings, A.J., Polymer, 24, 175 (1983)

[35] Rensch, G.J., Phillip, P.J., Vatanseyer, N., J. Polym. Sci.,*B*24, 1943 (1986)
[36] Tamura, K., Shimizu, S., Umemoto, S., Okui, N., Polym. Prep. Japan, 53, 686 (2004)
[37] Harada, K., Sawada, N., Umemoto, S., Okui, N., Polymer Preprints Japan, 53, 697 (2004)
[38] Shibata, K., Murayama, E., Kawano, R., Umemoto, S., Kikutani, T., Okui, N., Polymer Preprints Japan, 50, 1992 (2001).
[39] Cheng T.L., Su, A.C., Polymer, 36, 73 (1995)
[40] Hoffman, J.D., Lauritzen J., J.I., J. Res. Nat. Bureau Stand., 65*A*, 297 (1961)
[41] Hoffman, J.D., Weeks, J.J., J. Chem. Phys., 37, 1723 (1962)
[42] Bershtein, V.A., Egorov, V.M., "Differential Scanning Calorimetry of Polymers", Ellis Horwood, p-37 (1994)
[43] van Krevelen, D.W., "Properties of Polymers", Elsevier, Amsterdam, p-92 (1990)
[44] Lee, W.A., Knight, G.J., Br. Polym. J., 2, 73 (1970)
[45] Boyer, R.F., "Polymer Yearbook-2", ed. Pethrick, R.A., Harwood Academic Pub., p-233 (1985)
[46] Okui, N., Polymer, 31, 92 (1990)
[47] Boyer, R.F., Eur. Polym. J., 17, 661 (1981)
[48] Yoo, H.Y., Okui, N., to be published
[49] Seto, T., Rep. Prog. Polym. Phy. Jpn., 7, 67 (1964)
[50] Frank, F.C., J. Cryst. Growth, 22, 233 (1974)
[51] Hoffman, J.D., Davis, G.T., Lauritzen Jr., J.I., "Treaties on Solid State Chemistry", ed. Hannay, N.B., Plenum Press, New York vol.3, p-555 (1976)
[52] Point, J.I., Dosiere, M., Polymer, 30, 2292 (1989)
[53] Toda, A., Faraday Dis., 95, 129 (1993)
[54] Toda, A., Colloid & Polym. Sci., 270, 997 (1992)
[55] Miyaji H., Miyamoto, Y., Taguchi, K., Hoshino, A., Yamashita, M., Sawanobori, O., Toda, A., J. Macromol. Sci., *B*42, 867 (2003)
[56] Turnbull, D., J. Chem. Phys., 20, 411 (1952)
[57] Massa, M.V., Dalnoki-Veress, K., PMSE Preprints, 91, 847 (2004)
[58] Markov, I.V., "Crystal Growth for Beginners" World Scientific p-130 (1995)
[59] Kawano, R., Umemoto, S., Okui, N., Fiber Preprints Japan, 59, 267 (2004)
[60] Turnbull, D., Fisher, J.C., J. Chem. Phys., 17, 71 (1949)
[61] Mumun, A., Okui, N., to be published in Polymer
[62] Banks, W., Sharples, A., Makromol. Chem., 67, 42 (1963)
[63] Boon, J., Challa, G., van Krevelen, D.W., J. Polym. Sci., A-2, 6, 1835 (1968)
[64] Imai, M., Kaji, K., Kanaya, T., Macromolecules, 27, 7103 (1994)
[65] Matsuba, G., Kaji, K., Nishida, K., Kanaya, T., Imai, M., Macromolecules, 32, 8932 (1999)
[66] Olmsted, P.D., Poon, W.C.K., McLeish, T.C.B., Terrill, N.J., Ryan, A.J., Phys. Review Let., 81, 373 (1998)
[67] Fukao, K., Koyama, A., Tahara, D., Kozono, Y., Miyamoto, Y, Tsurutani, N., J. Macromol. Sci., B42, 717 (2003)
[68] Somani, R.H., Yang, L., Hsiao, B.S., Fruitwala, H., J. Macromol. Sci., B42, 515 (2003)
[69] Sato T, Richard R. "Stabilization of Colloidal Dispersions by Polymer Adsorption", Surface Science series 9, Marcel Dekker Inc. p-8 (1980)
[70] de Gennes, P.G., J. Chem. Phys., 55, 572 (1971)

[71] Umemoto, S., Okui, N., Polymer, 46, 8790 (2005)
[72] Okui, N., Polymer J. 19, 1309 (1987)
[73] Okui, N., J. Materils. Sci., 25, 1623 (1990)
[74] Okui, N., Polym. Bulletin, 23, 111 (1990)
[75] Umemoto, S., Okui, N., Polymer, 43, 1423 (2002)
[76] Okui, N., et.al to be published
[77] Maclaine, J.Q.G., Booth, C., Polymer, 16, 191 (1975)
[78] Okui, N., Umemoto, S., "Polymer Crystallization", ed. Sommer, J.U., Reiter, G., Springer, Berlin, chapter-19 (2003)
[79] Fatou, J.G., Marco, C., Polymer, 31, 1685 (1977)

20

Step-scan Alternating Differential Scanning Calorimetry Studies on the Crystallisation Behaviour of Low Molecular Weight Polyethylene

Kinga Pielichowska and Krzysztof Pielichowski

Department of Chemistry and Technology of Polymers, Cracow University of Technology, ul. Warszawska 24, 31-155 Kraków, Poland
kpielich@usk.pk.edu.pl

Abstract. Differential scanning calorimetry (DSC) and step-scan alternating (SSA) DSC were applied to investigate the crystallisation behaviour of polyethylene (PE) with molecular weight of 4000, 15000 and 35000. It has been found that PE 15000 is characterised by the highest degree of crystallinity and by the highest crystallisation temperature, as compared with other PE samples studied in the course of this work. The non-reversing component of the crystallisation process depends strongly on the PE molecular weight; parallelly, the reversing component shows minor fluctuations only, confirming thus the irreversibility of the PE crystallisation process.

20.1 Introduction

Crystallisation behaviour of polyethylene (PE) has been intensively studied from both fundamental and application reasons [1–3]. One of the most commonly used techniques is differential scanning calorimetry (DSC). New information on PE crystallisation can be gained if modulated temperature DSC (MT-DSC), which uses a periodical temperature modulation over a traditional linear heating or cooling ramp, is applied. It makes it possible to separate reversing and non-reversing components of the total heat flow; because of this characteristics of the MT-DSC technique, it can be regarded as a useful tool to study polymer crystallisation behaviour.

Since properties of the PE products strongly depend on the polymer molecular weight we report in this study on crystallisation behaviour of well-characterised PE of low molecular weight (4000, 15000 and 35000), studied by MT-DSC technique.

20.2 Experimental

20.2.1 Materials

Polyethylene (PE) with molecular weight 4000, 15000 or 35000 was purchased from Aldrich (Steinheim, Germany).

20.2.2 Techniques

Differential Scanning Calorimetry (DSC)

For the DSC measurements a Netzsch DSC 200, operating in dynamic mode, was employed. Samples of ca. 4 mg weight were placed in sealed aluminium pans. The heating/cooling rate of 10 K/min was applied. Argon was used as an inert gas with flow rate 30 cm^3/min. Prior to use, the calorimeter was calibrated with mercury and indium standards; an empty aluminium pan was used as reference. Liquid nitrogen was used as a cooling medium.

Step-scan Alternating DSC (SSA-DSC)

Step-scan alternating DSC investigations were performed by using a Perkin-Elmer Pyris Diamond DSC. Measurements were done in closed aluminium pans with sample mass of ca. 8 mg under argon flow of 20 cm^3/min. Prior to use the calorimeter was calibrated with indium standard. After a series of optimisation measurements, the following parameters have been chosen: length of the isothermal segment (t_{iso}) = 48 s; linear heating rate in dynamic segments (β) = 2 K/min and temperature jump between two subsequent isothermal segments (step) = 1 deg.

In general, modulated DSC offers extended temperature profile capabilities by e.g. sinusoidal wave superimposed to the normal linear temperature ramp

$$T = T_0 + \beta t + B \cdot \sin(\omega t) \tag{20.1}$$

where T is the program temperature, T_0 is the starting temperature, β is the underlying average heating rate, B is the amplitude of the temperature modulation, and $\omega = 2\pi/\text{p}$ [1/s], is the modulation angular frequency.

The superimposition may be also in form of oscillations, dynamic – isothermal heating and cooling segments, "saw – tooth" profile, etc.

The equation to describe heat flow is derived from a simple equation based on thermodynamic theory in which

$$dQ/dt = C_{pt} \cdot dT/dt + f(t, T) \tag{20.2}$$

where Q is the amount of heat absorbed by the sample, C_{pt} is the thermodynamic heat capacity, $f(t, T)$ is some function of time and temperature that governs the kinetic response of any physical or chemical transformation [4].

By assuming that the temperature modulation is small and that over the interval of the modulation the response of the rate of the kinetic process to temperature can be approximated as linear, one can rewrite Eq. (20.2) as

$$\frac{dQ}{dt} = C_{pt}\left(\beta + B\omega \cos(\omega t)\right) + f'(t,T) + C\sin(\omega t) \quad (20.3)$$

where $f'(t,T)$ is the average underlying kinetic function once the effect of the sine wave modulation has been subtracted, C is the amplitude of the kinetic response to the sine wave modulation and $(\beta + B\omega\cos(\omega t))$ is the measured quantity dT/dt.

The kinetic approach is based on differentiating between fast responses (equilibrium behaviour) and slower kinetically hindered processes including irreversible processes.

Equation (20.3) can be rewritten as

$$\Phi(T(t)) = \Phi_{dc}(T(t)) + \Phi_a(T(t))\cos(\omega_0 t - \varphi) \quad (20.4)$$

where φ is the phase shift, Φ_a the cyclic component, Φ_{dc} the underlying heat flow.

We obtain the reversing component of the heat flow from

$$\Phi_{rev}(T(t)) = \frac{\Phi_a(T(t))}{B\omega_0}\beta_0 \quad (20.5)$$

The kinetic component (the non-reversing component) is then

$$\Phi_{non}(T(t)) = \Phi_{dc}(T(t)) - \Phi_{rev}(T(t)) \quad (20.6)$$

The modulated temperature and resultant modulated heat flow can be deconvoluted using a Fourier transform to give reversing and non-reversing components. The reversing component is evaluated from the periodic part of the heat flow. The non-reversing component is the difference between the underlying heat flow (static heat flow) and the reversing component. The static heat flow is evaluated by an averaging method. Amplitude of the cyclic response and the phase lag are divided into in and out of phase responses by use of the phase lag unless the phase shift is small in which case the out of phase component can be neglected [5].

An alternative evaluation method has been reported based on the linear response approach [6]. In this method, the heat flow rate into the sample is described as

$$\Phi(t) = \int_0^t \dot{C}(t-t')\beta(t')\,dt' \quad (20.7)$$

with

$$\beta(t) = \beta + \omega_0 B\cos(\omega_0 t) \quad (20.8)$$

Insertion of eq. (20.2) into eq. (20.1) leads to the relationship

$$\Phi(T(t)) = C_\beta(T)\beta + \omega B \, |C(T,\omega)| \cos(\omega t - \varphi) \qquad (20.9)$$

with

$$|C| = \sqrt{C'^2 + C''^2} \qquad (20.10)$$

where C' is the storage heat capacity (associated with mobility); C'' is the loss heat capacity (associated with dissipation).

Both the phase shift and the amplitude of the dynamic component are used for the calculation of a complex (frequency-dependent) heat capacity. These quantities can be interpreted in the context of the relaxation theory or irreversible thermodynamics.

In the step-scan alternating DSC (SSA-DSC), the temperature program comprises a periodic succession of short, linear heating and isothermal phases; the measured heat flow contain thus fractions which arise from the heat capacity and those due to physical transformations or chemical reactions [7].

Since there is no complex Fourier transform involved in data deconvolution and no phase lag component to the analysis, the reported advantages include reliable and direct heat capacity measurement with low mass samples over a much shorter time than experiments performed in equivalent large mass heat flux modulated temperature DSC [8].

20.3 Results and Discussion

DSC results are presented in Table 20.1.

Table 20.1. DSC results of the melting and crystallisation of PE with different molecular weight

Sample	Melting		Crystallisation	
	Melting temperature [°C]	Heat of melting [J/g]	Crystallisation temperature [°C]	Heat of crystallisation [J/g]
PE 4000	104.5	114	91.5	106
PE 15000	113.9	136	98.9	129
PE 35000	92.5	77	74.0	69

The degree of crystallinity (X_c) was calculated, assuming the enthalpy of melting of 100% crystalline polymer equal to 293.6 J/g) – Table 20.2.

It can be seen that PE with molecular weight of 15000 has the highest degree of crystallinity – it has formed the most regular crystal structure.

At the next stage, SSA-DSC investigations were performed – Fig. 20.1.

The non-reversing component of crystallisation is very distinct (Fig. 20.2) whereby there are practically no reversing signals – Fig. 20.3.

Table 20.2. Degree of crystallinity (X_c) of PE

PE	X_c [%]
PE 4000	38.8
PE 15000	46.3
PE 35000	26.2

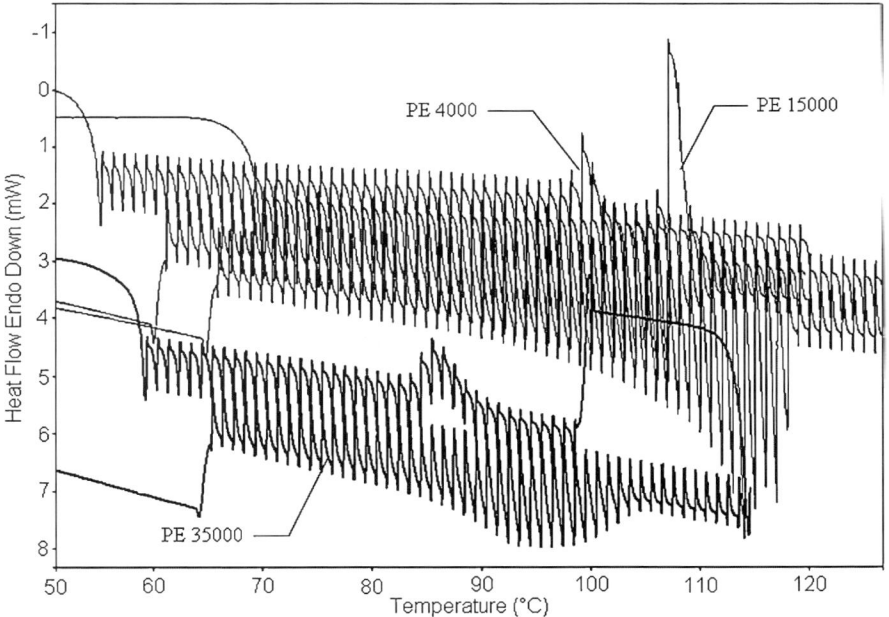

Fig. 20.1. SSA-DSC profiles of curves of PE with different molecular weight

In the fast cooling process from the melt state short-chain PE tends to form metastable folded-chain crystals, while the most stable extended-chain crystals can form only in a slow cooling process. The initial folding in the metastable crystals is rather irregular; the lamella will perform thickening to the extended chain crystals as well as thinning to the once folded-chain crystals during the isothermal annealing [9,10]. Additionally, during a dynamic experiment on cooling from the melt, PE exhibits a crystallisation process whose transition temperature is a function of molecular weight, its branching content and the intramolecular distribution of branches [11]. The internal microstructure or morphology of semicrystalline polymers is central to their properties. The physical properties of semicrystalline polymers are largely dominated by constraints imposed through the extensive fold surfaces of their constituent lamellae and the presence of inter-lamellae tie molecules [12]. Different lamellar habits of defined lamellar thickness are conferred by crystallisation.

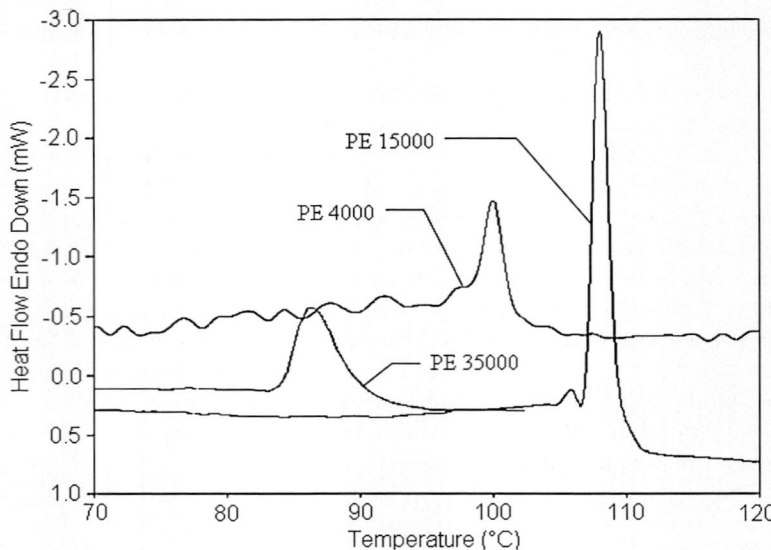

Fig. 20.2. SSA-DSC non-reversing component of the crystallisation process of PE with different molecular weight

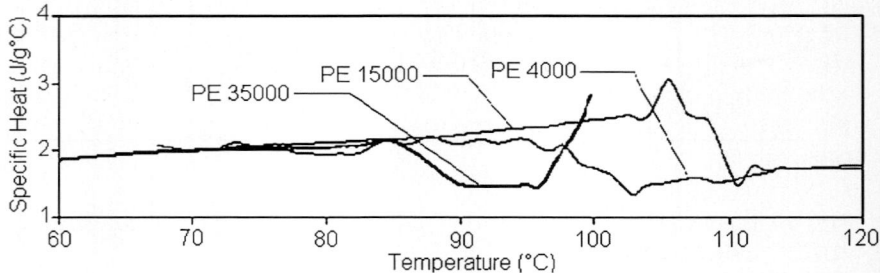

Fig. 20.3. SSA-DSC reversing component of the crystallisation process of PE with different molecular weight

However, they are inherently metastable with respect to crystal thickness, because of the greater reduction in fold surface per unit mass that occurs on thickening [13]. It has been also found that the trans–gauche equilibrium influences also some properties of PE. DMA, indeed, identifies three relaxation regions in semicrystalline PE. They were labeled as α-, β-, and γ-relaxations. The following descriptions were given for these relaxations: The α relaxation is linked to local mobility in the crystalline lamellae, which mobilizes the chain translationally, and affects the amorphous regions tied to the crystals. The β relaxation marks the glass transition, and the γ-transition occurs mainly in the amorphous phase and is a broad relaxation in the temperature- or frequency-domain, interpreted as a localized crankshaft-like motion of the backbone of the chain [14].

These results show that the crystallisation process of low molecular weight PE is a non-reversible process.

Polymer lamellae form because this is the fastest route for macromolecules to crystallize but, once formed, they are metastable with respect to increased thickness which reduces the surface to volume ratio and thereby the specific Gibbs function or chemical potential. Frank and Tosi showed that, nevertheless, secondary nucleation on an infinite substrate could be expected to yield a well-defined thickness, ζ_g, with individual fold lengths fluctuating around a mean and probably subject to later evening out; a lesser length, ζ_g, would result for a substrate of finite height [15]. Solution-grown lamellae do have a well-defined thickness, which does not increase unless and until they are heated well above their growth temperature, commonly by partly melting then recrystallizing at a higher thickness [16].

20.4 Conclusions

In the course of the work it was found that PE 15000 is characterised by the highest degree of crystallinity and by the highest crystallisation temperature, as compared with other PE samples studied in the course of this work. The non-reversing component of the crystallisation process depends strongly on the PE molecular weight; parallelly, the reversing component shows minor fluctuations only, confirming thus the irreversibility of the PE crystallisation process.

Acknowledgement

The authors are grateful for helpful discussions on the subject of PE crystallisation with other participants of the join COST P12 project "Structuring of Polymers".

References

[1] P.J. Barham, M.J. Hill, A. Keller, C.A. Rosney, J. Mater. Sci. Lett., 7 (1998) 1271.
[2] R.L. Morgan, M.J. Hill, P.J. Barham, A. van der Pol, B. Kip, J. van Ruiten, L. Markwort, J. Macromol. Sci. Phys., B38 (1999) 419.
[3] M. Agamalian, R.G. Alamo, M.H. Kim, J.D. Londono, L. Mandelkern, G.D. Wignall, Macromolecules, 32 (1999) 3093.
[4] M. Reading, A. Luget, R. Wilson, Thermochim. Acta, 238 (1994) 295.
[5] M. Reading, Trends in Polym. Sci., 248 (1993) 1.

[6] J.E.K. Schawe, Thermochim. Acta., 260 (1995) 1.
[7] M. Sandor, N.A. Bailey, E. Mathiowitz, Polymer, 43 (2002) 279.
[8] K. Pielichowski, K. Flejtuch, J. Appl. Polym. Sci., 90 (2003) 861.
[9] W. Hu, T. Albrecht, G. Strobl, Macromolecules, 32 (1999) 7548.
[10] S.Z.D. Cheng, A. Zhang, J.S. Barley, A. Habenschuss, P.R. Zschack, Macromolecules, 24 (1991) 3937.
[11] C.C. Puig, Polymer, 42 (2001) 6579.
[12] D.C. Bassett, Principles of Polymer Morphology, Cambridge University Press, Cambridge, 1981.
[13] J.J. Janimak, G.C. Stevens, Polymer, 41 (2000) 4233.
[14] B. Wunderlich, Prog. Polym. Sci., 28 (2003) 383.
[15] F.C. Frank, M. Tosi, Proc. R. Soc. A., 263 (1961) 263.
[16] M.I. Abo el Maaty, D.C. Bassett, Polymer (2005) 8682.

21

Order and Segmental Mobility in Crystallizing Polymers

Aurora Nogales[1], Alejandro Sanz[1], Igors Šics[2], Mari-Cruz García-Gutiérrez[1], and Tiberio A. Ezquerra[1]

[1] Instituto de Estructura de la Materia, C.S.I.C. Serrano 119, Madrid 28006, Spain
[2] Department of Chemistry, State University of New York at Stony Brook, Stony Brook, NY 11794-3400, USA
imte155@iem.cfmac.csic.es

Abstract. The simultaneous combination of scattering techniques, probing structure, with relaxation techniques, detecting modifications of the amorphous phase dynamics, can be helpful in order to obtain complementary information about crystallization processes in polar polymers. The objective of this contribution is to review the improvements in the combination of real time (wide and small angle) X-ray scattering and dielectric spectroscopy aiming at a better understanding of polymer crystallization.

21.1 Introduction

Upon cooling liquid systems either crystallize or vitrify or both. From the liquid state, as temperature decreases, the specific volume of a material decreases linearly with temperature (Fig. 21.1). Below the equilibrium melting temperature, T_m^0 the liquid is in the supercooled liquid state (SCL). If temperature is further decreased the specific volume of the supercooled liquid decreases in the same fashion as in the liquid phase. At the glass transition temperature, T_g, a change in the slope of the specific volume versus temperature dependence is observed marking the glass transition [1]. An important thermodynamical characteristic of the SCL in the temperature window defined by T_g and T_m^0 is that it is unstable due its higher free energy as compared with that of the crystal. Therefore there exists the probability that the supercooled liquid tends to reduce its free energy undergoing a first order phase transition as schematically represented in Fig. 21.1 by the arrow.

This transition by which molecules self-assemble forming crystals is referred to as crystallization [2, 3]. In this case a discontinuous change in the specific volume is expected (Fig. 21.1). A classical experimental approach to characterize the nature of the crystals and the overall fraction of crystalline phase mainly involves scattering and diffraction measurements [4]. Precise

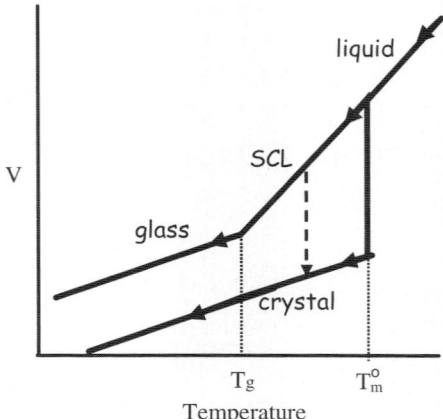

Fig. 21.1. Schematic dependence of the specific volume (V) of a given material as a function of temperature indicating the liquid, super cooled liquid (SCL), glassy and crystalline state

information about the changes occurring in supercooled liquids upon crystallization can be obtained when a real time experimental set up is used [5–7]. Nowadays, as far as crystalline phase development is concerned, both synchrotron and neutron sources offer the possibility to perform real time diffraction experiments [6, 8].

As an example, Fig. 21.2a shows the evolution with temperature of the neutron diffraction patterns of isopropanol upon heating. Here, isopropanol was quenched from the liquid state to avoid crystallization upon cooling [6, 7]. As temperature increases the initial SCL exhibits the diffraction pattern characteristics of an amorphous system. Upon heating from the SCL state isopropanol crystallizes and the diffraction pattern presents narrow Bragg peaks characteristic of a crystalline phase. Further heating above the melting temperature produces a crystal destruction and the characteristic diffraction pattern of an amorphous system, now the liquid phase, is recovered. Scattering techniques can also be used to extract structural information in amorphous materials [9]. However, due to the fact that crystals provoke strong diffraction phenomena, superimposed over a relatively weak contribution of the amorphous phase, mainly information about the crystalline phase is obtained. Thus, processes occuring in the amorphous fraction during crystallization of the supercooled liquid are almost non-detectable for these techniques due to the absence of order. An improvement in the understanding of crystallization in terms of interrelation between crystalline and amorphous development is obtained when diffraction experiments are simultaneously accompanied by dielectric spectroscopy (DS) [10, 11]. At $T > T_g$ molecular mobility in liquids and segmental mobility in polymers is revealed by the α relaxation which appears as a maximum in frequency of the imaginary part, ϵ'', of the complex

21 Order and Segmental Mobility in Crystallizing Polymers 437

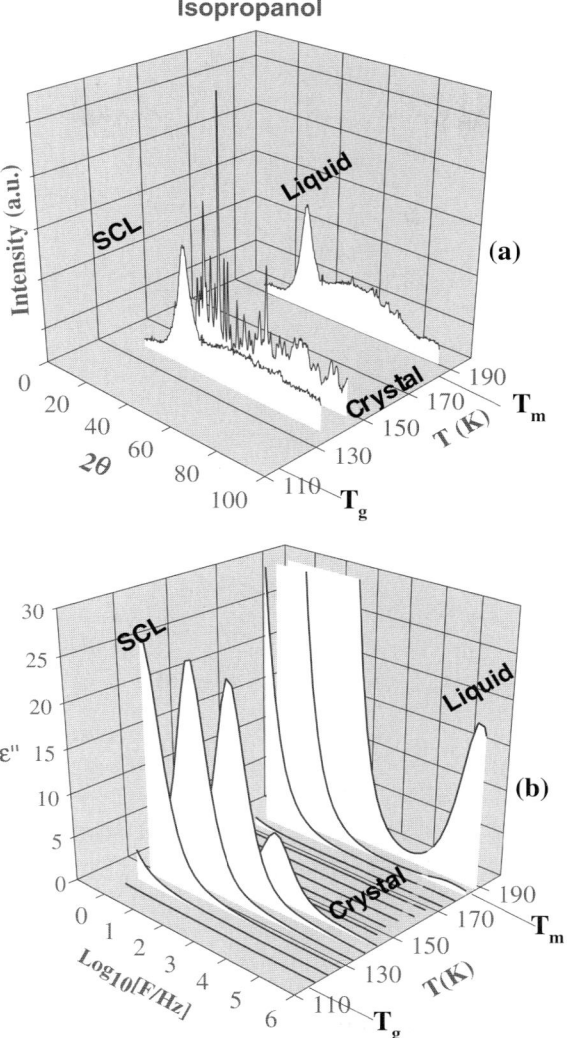

Fig. 21.2. (a)Evolution with temperature of the neutron diffraction patterns of isopropanol. Both in the liquid and in the super cooled liquid state(SCL) the diffraction patterns are characteristic of an amorphous material. The crystalline state presents characteristic narrow Bragg peaks of a crystalline phase. (b) Evolution with temperature of the main relaxation of isopropanol. Molecular mobility at $T > T_g$ is revealed by the main relaxation of isopropanol as reflected by the maximum in frequency of the imaginary part, ϵ'', of the complex dielectric permittivity. The main relaxation appears in both the super-cooled and the liquid state. However, in the crystalline state, where no significant molecular mobility is expected, the relaxation vanishes [6,7]. Isopropanol was quenched from the liquid to the glassy state

dielectric permittivity [12]. Regarding the amorphous phase, it has been shown that, upon crystallization, the α relaxation, which is related to the segmental dynamics, can be used as a probe for crystallization because it is strongly affected by the progressive development of the crystalline phase [13–17]. As an example, Fig. 21.2b shows the evolution with temperature of the main relaxation of isopropanol. This maximum observed in both the supercooled and the liquid state appears as a consequence of the molecular mobility about T_g. Accordingly, in the crystalline state, where no significant molecular mobility is expected, the relaxation vanishes [6, 7].

Therefore by monitoring simultaneously in real time the crystal development, by means of diffraction techniques, and the dynamic changes occurring in the amorphous phase, by means of dielectric spectroscopy, a complete picture of the crystallization process could be obtained. Among the glass forming systems, high molecular weight polymers, tend to develop a characteristic folded chain crystalline lamellar morphology at the nanometer level upon thermal treatment within the temperature range defined between T_g and T_m^0 [3]. The lamellar morphology consists of stacks of laminar crystals and amorphous regions intercalated between them. Although extended chain crystals are thermodynamically more stable, a kinetic factor induces that a polymer chain folds several times, building up thin crystal lamellae. For semicrystalline polymers, this characteristic crystalline nanostructure acts as an internal backbone in the polymer controlling the final mechanical properties of the material.

The objective of this contribution is to review the improvements in the combined use of real time X-ray scattering and dielectric relaxation techniques experiments for a better understanding of polymer crystallization.

21.2 Description of the Experimental Set-up for Simultaneous Small and Wide Angle X-ray Scattering and Dielectric Spectroscopy (SWD)

Dielectric spectroscopy is a technique which allows one to evaluate the complex dielectric permittivity $\epsilon^* = \epsilon' - i\epsilon''$ as a function of frequency and temperature, where ϵ' is the dielectric constant and ϵ'' is the dielectric loss [3, 12]. A schematic view of a dielectric spectroscopy experiment is shown in Fig. 21.3. A dielectric sample of thickness d and area A is subjected to an alternating electric field of angular frequency ω. Through measurements of the complex impedance of the sample it is possible to experimentally determine ϵ^* [12, 18–20]. Dielectric spectroscopy is a very suitable method to study molecular dynamics in polymers above T_g. In this case, segmental motions of the polymeric chains give rise to the so called α-relaxation process, which can be observed as a maximum in ϵ'' and a step-like behavior in ϵ' as a function of frequency. Both, the intensity of the α relaxation, $\epsilon''_{\max}$, and the frequency of maximum loss, $F_{\max}$, are very sensitive to crystallinity, which produces a decrease in $\epsilon''_{\max}$ and a shift of $F_{\max}$ towards lower values when crystallization proceeds [13–17].

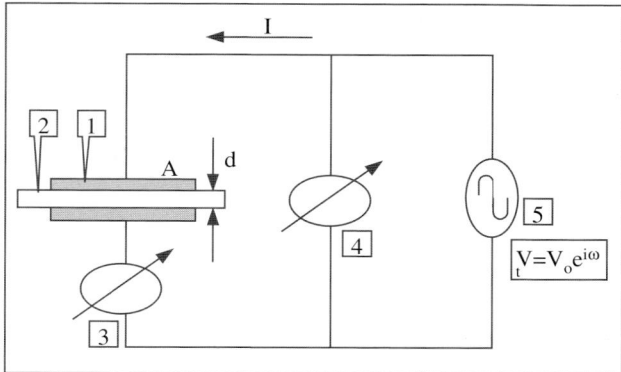

Fig. 21.3. Schematic view of a typical dielectric spectroscopy experiment. (1) Electrodes of area A, (2) Sample film of thickness d, (3) Current Analyzer, (4) Voltage Analyzer, (5) Alternating Voltage generator

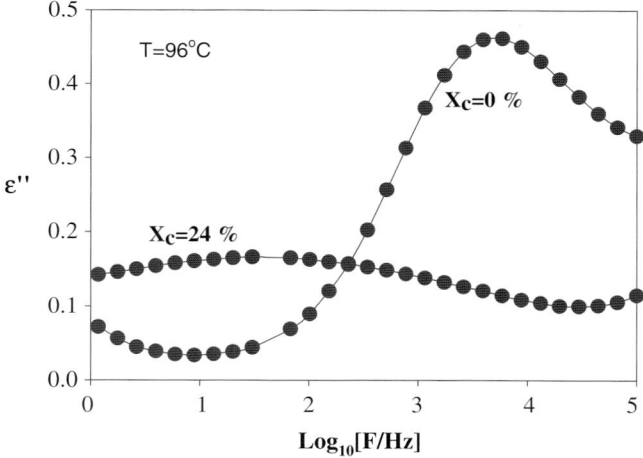

Fig. 21.4. Dielectric loss, ϵ'', as a function of frequency, $F = \omega/(2\pi)$, for amorphous PET (crystallinity $X_c = 0$) and semicrystalline ($X_c = 24\%$)PET at $T = 96°$C.

This behavior is illustrated in Fig. 21.4 for poly(ethylene terephthalate)(PET). Crystallisable polymers tend to develop a certain level of crystallinity provided they are heated at temperatures above the glass transition temperature. The microstructure of semicrystalline polymers typically shows a distinct lamellar morphology consisting of stacks of laminar crystals intercalated by amorphous less ordered regions. The lamellar stacks are characterized by the thickness of the crystals (l_c) and that of the amorphous layers (l_a). Both characteristic lengths define the long period as $L = l_a + l_c$ which can be experimentally determined through small angle X-ray scattering experiments (SAXS) [3, 21].

The overall crystalline fraction X_c can be estimated experimentally from Wide Angle X-ray Scattering measurements (WAXS).

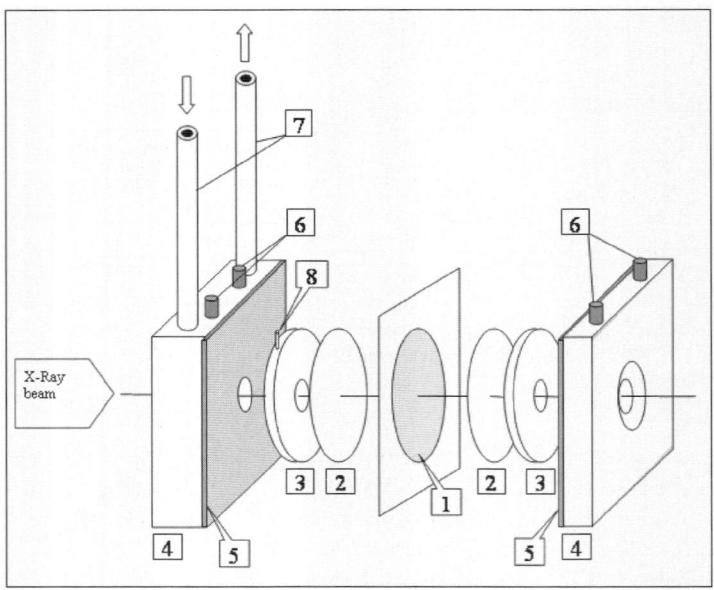

Fig. 21.5. Scheme of the SAXS-WAXS-DS cell. (1) Sample, (2) Aluminum disks, (3) Electrodes, (4) Heating blocks, (5) Insulating polyamide film, (6) heating elements, (7) cooling pipes, (8) PT-100 thermometer

In recent times there have been different works reporting dielectric environments useful to accommodate simultaneous X-ray experiments [10, 22, 23]. A typical scheme of a SAXS-WAXS-DS sample holder (SWD), is illustrated in Fig. 21.5. The sample (1) is placed between two metallic disks (3) acting as electrodes. These are electrically insulated by polyamide films (5) from the heating blocks (4). Heating power is provided by some heating-elements (6) embedded in the sample-cell heating blocks (4). In order to allow the passage of the X-ray beam through the sample, central holes were machined in both, the electrodes and in the heating blocks. The sample, prepared in the form of a film, can be provided with thin circular gold electrodes by sputtering the metal in both free surfaces. The sample film was sandwiched between two thin aluminium disks (2) in order to provide homogeneous heating for the whole sample surface. The sandwich is placed in between the two metallic electrodes (3). Cooling of the device can be obtained by compressed air circulating through a metallic pipe (7) embedded in one of the sample-cell heating blocks. A thermometer (8) is located in one of the metallic electrodes. Electrodes can be connected to a suitable dielectric spectrometer to measure dielectric com-

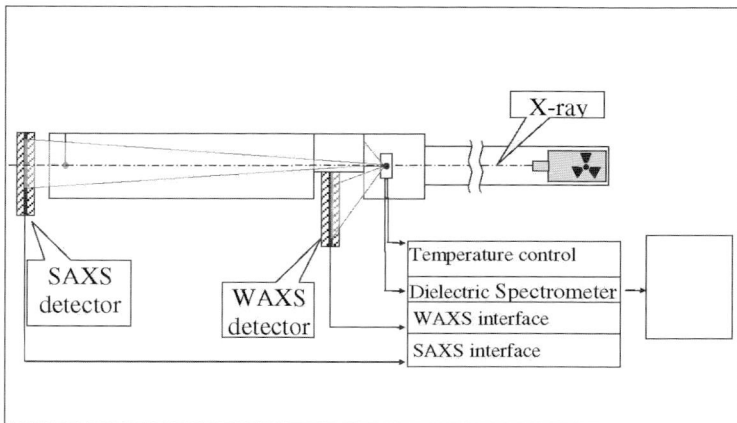

Fig. 21.6. Scheme of a typical experimental set-up for simultaneous SAXS-WAXS-DS experiments at a beam-line

plex permittivity (ϵ^*) in a convenient frequency range [10]. These kind of cells, where simultaneous WAXS and SAXS measurements are performed, can be easily incorporated into a typical synchrotron beam line [10, 23] using two position sensitive detectors as presented in Fig. 21.6. Closer to the sample, the WAXS detector is positioned off the primary beam allowing the SAXS intensity to pass above and to be measured by the SAXS detector located at a larger distance.

21.3 Dielectric Relaxation of Amorphous Polymers: Poly(ethylene terephthalate)

Poly(ethylene terephthalate)(PET) is one of the most common polymers provided by the polymer industry for fiber and packaging purposes [24]. As far as polymer crystallization is concerned, PET can be considered as a paradigm of a crystallisable polymer due to the fact that PET can be obtained either in the amorphous state or with a controlled amount of crystallinity. Therefore, PET has been used to study the influence of crystallinity in a great variety of physical properties including thermal behavior [16, 25–29], structure development [30–33] and mechanical and dielectric behavior among others [13,14,34,35]. Figure 21.7 presents the dielectric loss, ϵ'', and dielectric constant, ϵ', for amorphous PET at $T > T_g$ as a function of frequency for different temperatures.

Amorphous PET (Rhodia S80 from RhodiaSter, $M_\nu = 45000$ g/mol) was prepared by quenching from the molten state as described elsewhere [35]. Broad-band dielectric spectroscopy measurements of the complex dielectric permittivity were performed from 10^{-1} Hz to 10^6 Hz by using a BDS-40

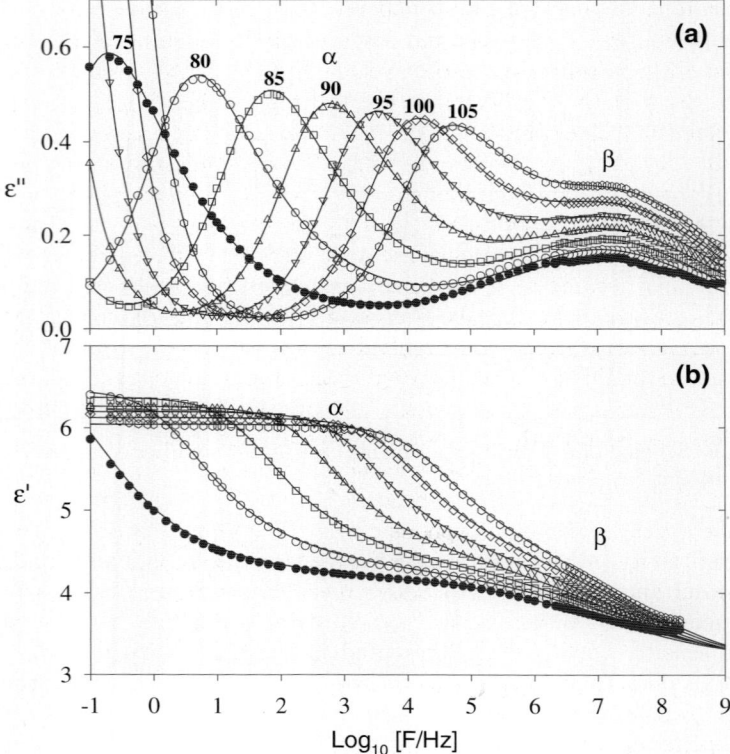

Fig. 21.7. Amorphous PET: (**a**) dielectric loss, ϵ'' and (**b**) dielectric constant, ϵ', as a function of frequency for different temperatures labelled in °C.

Novocontrol system and from 10^6 Hz to 10^9 Hz by means of a Novocontrol BDS-60 coaxial line reflectometer. The broad-band data show the α-relaxation process, at lower frequencies, and the subglass β-relaxation process at higher frequencies. As extensively reported [12], the relaxations manifest themselves as maxima in ϵ'' and concurrent steps in ϵ'. As the temperature increases the frequencies of maximum loss, $F_{\max}$, shift towards higher values. At low frequencies the relaxations are accompanied by a strong increase of ϵ'' corresponding to a dc-conductivity contribution. Isothermal ϵ'' and ϵ' data can be phenomenologically described according to the Havriliak-Negami equation [12, 14, 36] given by:

$$\epsilon^* = \sum_{x=\alpha,\beta} \frac{(\epsilon_0 - \epsilon_\infty)_x}{[1 + (i\omega\tau_x)^{b_x}]^{c_x}} + (\epsilon_\infty)_\beta - i\frac{\sigma}{\epsilon_{vac}\omega^s} \quad (21.1)$$

where $\omega = 2\pi F$, ϵ_0 and ϵ_∞ are the relaxed ($\omega = 0$) and unrelaxed ($\omega = \infty$) dielectric constant values, τ is the central relaxation time of the relaxation time distribution function and b and c ($0 < b, c < 1$) are shape parameters which describe the symmetric and asymmetric broadening of the relaxation time distribution function, respectively [36]. The subscript makes reference either to the α or the β relaxation. The last term of Eq. (21.1) corresponds to the conductivity contribution. Here, σ is related to the direct current electrical conductivity, ϵ_{vac} is the vacuum dielectric constant and s depends on the nature of the conduction mechanism. The results from this phenomenological data analysis are presented in Fig. 21.7 by the continuous curves. Amorphous PET presents an strongly asymmetric α-relaxation as denoted by an asymmetric broadening parameter $c = 0.4$ nearly constant in the studied temperature range [35]. On the contrary, the β-relaxation is symmetric, $c = 1$, in the presented temperature range. The central relaxation time, τ, for the β process follows an Arrhenius behavior characteristic of a non-cooperative process [14] with an activation energy of about 13 Kcal/mol. The origin of the subglass relaxation has been traditionally associated to the local motion of the ester group [14] although recent dielectric measurements indicate a more complex molecular origin [37, 38]. The α-relaxation appears as a consequence of the segmental motions of the amorphous phase above the glass transition temperature and the temperature dependence of its relaxation time can be described by means of the Vogel-Fulcher-Tamann (VFT) [3, 12].

21.4 Time Resolved Cold Crystallization by SWD

21.4.1 Poly(ethylene terephthalate)

Figure 21.8 shows an isothermal cold crystallization experiment for PET SAXS, WAXS and DS data have been simultaneously collected during a cold crystallization experiment at $T_c = 96°C$ and are shown for three different crystallization times during the crystallization process. Each set of measurements was collected during 60 s. The experiments were performed in the Soft Condensed Matter beam-line A2 at HASYLAB in the synchrotron facility DESY in Hamburg, Germany. Both, WAXS and Lorentz corrected SAXS intensities [21] are represented as a function of the scattering vector $s = (2/\lambda)\sin(\theta)$ being 2θ the scattering angle and $\lambda = 0.15$ nm the wavelength of the X-ray used. Complex dielectric permittivity measurements were performed in the frequency range of 10^1 Hz $<$ F $< 10^5$ Hz, using a Novocontrol system integrating a SR 830 Lock-in amplifier with a BDC-L dielectric interface. The dielectric loss data, ϵ'' are given as a function of frequency. The initial amorphous state is characterized by a broad halo in the WAXS pattern, a continuous scattering in the SAXS pattern and a relaxation process characterized as a maximum in ϵ'' centered around a $F_{\max}$ value of $\approx 4 \times 10^3$ Hz. The observed relaxation can be identified with the α process. As time increases, the onset of crystallization manifests itself by the appearance of the characteristic Bragg peaks of the triclinic unit cell of PET in the WAXS patterns. The weight fraction

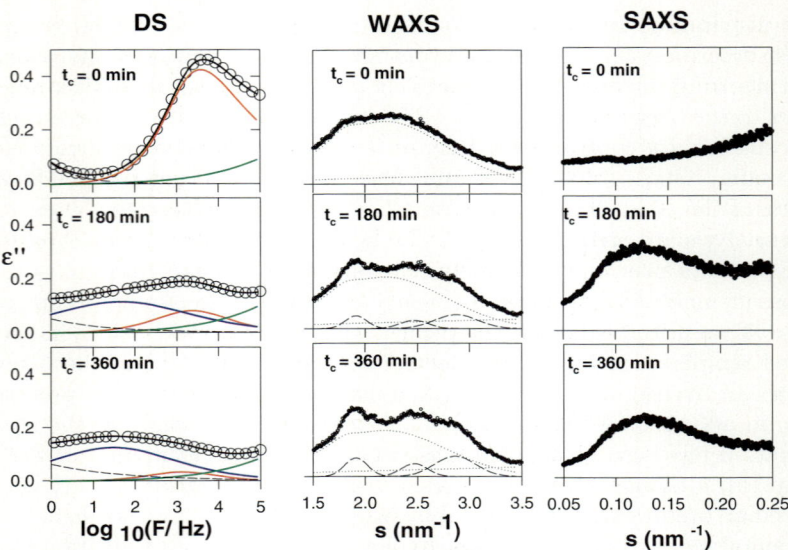

Fig. 21.8. Simultaneous dielectric loss, ϵ'', WAXS and, SAXS experiments during crystallization of initially amorphous PET at $T_c = 96°C$ for three different crystallization times covering the crystallization process. Continuous lines in the DS-data indicate the separate contribution of the primary and secondary α-relaxations, the β-relaxation tail appearing at higher frequencies and the conductivity tail appearing at lower frequencies. *Dotted* and *dashed lines* in the WAXS-data illustrate the peak deconvolution procedure

index of crystallinity (X_c) can be estimated from the ratio between the area below the crystalline Bragg peaks to the total scattered area after appropriate subtraction of a flat background [39]. Dashed lines in Fig. 21.8 illustrate the peak deconvolution procedure. In the SAXS patterns an increase of the scattered intensity at lower s-values that develops into a well defined peak centered around a value of $s = 0.125$ nm^{-1} is observed. This fact indicates that lamellar crystals organize themselves forming lamellar stacks with an average distance between gravity centers of consecutive lamellar crystals of $L = 1/s_{max} \approx 8$ nm. The structural features are accompanied by changes in the dynamics of the amorphous phase as revealed by the simultaneous DS experiment. The α-relaxation exhibits, at the end of crystallization, a decrease in its intensity and a shift towards lower values of its F_{max}. At intermediate crystallization times a significant broadening in the low frequency side of the relaxation is detected. This effect can be described as an additional α'-process appearing as crystallinity develops [22, 35]. The dielectric data can be analyzed in terms of the HN-equation considering the contribution of i) the initial α-process; ii) the second α'-process appearing during crystallization iii) the β-relaxation process which contributes in the higher frequency range of the spectrum and iv) the conductivity which influences the lower frequency part of the spectrum [35]. The separate contribution of every process as well as the total fittings are represented in Fig. 21.8 by the continuous lines. A visualization of the changes in the characteristic parameters simultaneously measured is presented in Fig. 21.9. In this figure we have represented as a function of the crystallization time for both the initial α and the secondary α'-relaxation processes: (a) The dielectric strength, (b) the broadening parameter (c) the asymmetry parameter (d) the central relaxation time. Additionally values for the weight fraction index of crystallinity (X_c) are included (Fig. 21.9 e) calculated from the WAXS data.

From the simultaneous SWD-experiments the following attempt to relate structure and dynamics can be made. For times shorter than a characteristic one ($t \approx 90$ min) the decrease of $(\Delta\epsilon)_\alpha$, Fig. 21.9a, indicates a significant reduction of the mobile material which follows the increase of crystallinity (Fig. 21.9e). However, the reduction of $(\Delta\epsilon)_\alpha$ is stronger than the increment in crystallized material as determined by the increase of X_c. This effect, observed in different polymers [13,17], can be attributed to the formation of an immobilized amorphous phase frequently referred to as rigid amorphous phase (RAP) [40].

During this initial period of crystallization, in spite of the strong reduction in $(\Delta\epsilon)_\alpha$ of about 50%, the remaining mobile material, in the amorphous phase, only slightly change the average segmental mobility in the amorphous phase as reflected by the moderate variation observed in τ_α (Fig. 21.9 d). During this initial period the relaxation tends to become symmetric and increasingly broader, as denoted by the increase of the c_α-parameter and the decrease of the b_α-parameter respectively (Fig. 21.9 b and c). As crystallization proceeds above the characteristic time, a secondary relaxation, α', appears at lower frequencies. The fittings indicate that the α'-relaxation can be treated in a first approach as a symmetric process ($c_{\alpha'} = 1$). As crystallization time increases the dielectric strength of the α'-relaxation, $(\Delta\epsilon)_{\alpha'}$, increases at expenses of $(\Delta\epsilon)_\alpha$ (Fig. 21.9 a). At the end of the crystallization process $(\Delta\epsilon)_\alpha$ tends to vanish and α' becomes the characteristic α-relaxation of the semi-crystalline material.

21.4.2 Poly(ethylene terephthalate)/Poly(ethylene naphthalene 2,6-dicarboxilate) Blends

Blending of poly(ethylene terephthalate) (PET) and poly (ethylene naphthalene-2,6-dicarboxylate) (PEN) has been shown to be an attractive possibility to combine the inherent economics of PET with the superior mechanical, thermal and barrier properties of PEN [24]. The molecular structure of PEN is stiffer than that of PET due to the presence in its main chain of naphthalene instead of benzene rings. The glass-transition temperature, T_g, of PEN is about 50°C higher than that of PET contributing to a better performance in terms of thermal, mechanical, and gas barrier properties [17,24]. PET and

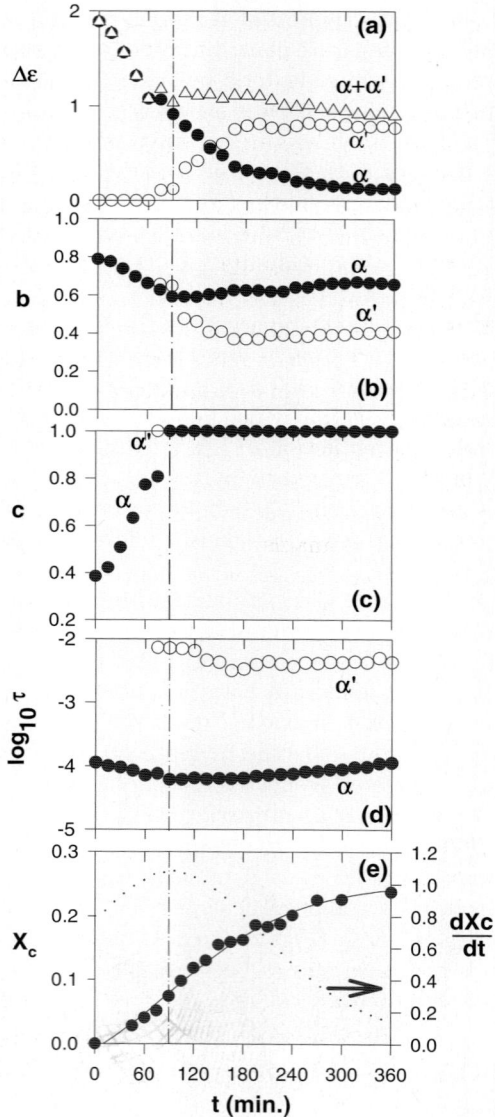

Fig. 21.9. Summary of physical parameters obtained from the SWD experiment for PET at $T = 96°C$ as a function of the crystallization time: (**a**) (●)$\Delta\epsilon$ values for the α relaxation, (○)α' relaxation and ($\triangle$)$(\Delta\epsilon)_\alpha + (\Delta\epsilon)_{\alpha'}$. (**b**) broadening parameter for the (●)α relaxation and (○)α' relaxation (**c**) Asymmetry parameter for the (●)α relaxation and (○)α' relaxation (**d**) Central relaxation times for the (●)α relaxation, (○)α' relaxation (**e**) Crystallinity index (*left*) and slope of the crystallinity curve (*right*)

PEN are immiscible polymers that tend to form separated phases upon blending [41]. However, at temperatures above 270° C certain amounts of PET-PEN block copolymers develop due to transesterification reactions [41,42]. By cryogenic grinding, melt pressing at 300°C and subsequent quenching amorphous films of PET/PEN blends with various degrees of transesterification can be prepared [42]. For low levels of transesterification two glass-transition steps, two peaks of crystallization and melting and two α-relaxation processes are observed indicating the existence of a phase separated system consisting on different PET-rich and PEN-rich phases [42]. Figure 21.10 displays the imaginary part ϵ'' of the dielectric permittivity ϵ^* as a function of frequency and temperature for a PET/PEN (56:44 molar ratio, corresponding to 1/1 in weight) pressed for 3 min at 300°C. The degree of transesterification (f_{TEN}), as estimated, from ^{1}H NMR, by the fraction of terephthalate-ethylene-naphthalene-2,6-dicarboxylate triads (T-E-N) is for this sample of $f_{TEN} = 11\%$ [42,43]. In Fig. 21.10 one can distinguish two α-processes associated with the glass transitions in PET-rich and PEN-rich regions, respectively [42]. Atomic Force Microscope (AFM) studies in samples with low levels of transesterification reveal the presence of PET domains typically smaller than 50 nm well dispersed without significant clustering [43]. The confinement of PET within

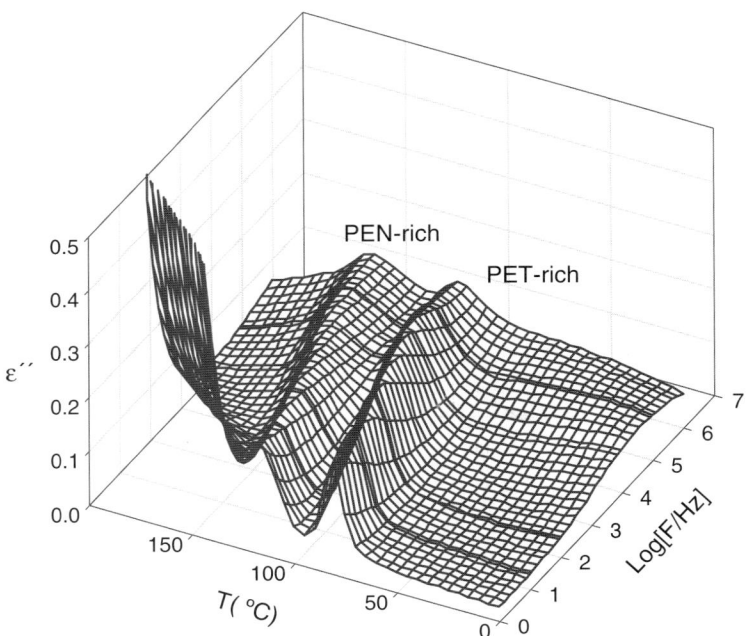

Fig. 21.10. Frequency and temperature dependence of ϵ'' for a 56:44 mol% PET/PEN blend prepared by cryogenic grinding, pressed at 300°C for 3 min and quenched on ice-water. Fraction of T-E-N triads: 11%

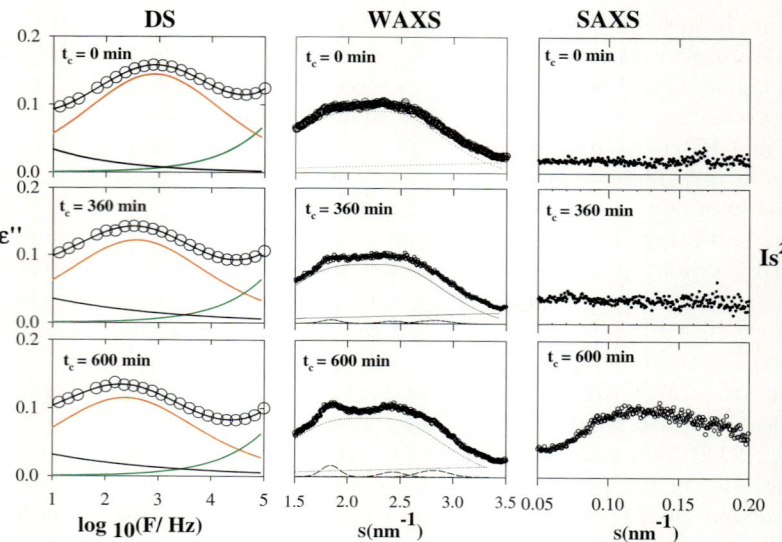

Fig. 21.11. Simultaneous dielectric loss, ϵ'', WAXS and, SAXS experiments during crystallization of initially amorphous PET/PEN blends (1:1 by weight) with a $f_{TEN} = 11\%$ at $T_c = 96°C$ for three selected crystallization times. Continuous lines in the DS-data indicate the separate contribution of the α-relaxation, the β-relaxation tail appearing at higher frequencies and the conductivity tail appearing at lower frequencies. *Dotted* and *dashed lines* in the WAXS-data illustrate the peak deconvolution procedure.

these phase segregated domains is expected to have a strong influence on its crystallization behaviour. The crystallization of polymers in confined environments is a topic of permanent interest because it can be useful to better understand the transition from random coil chains to ordered lamellae which takes place during crystallization [44]. Figure 21.11 shows the time-resolved SAXS-WAXS-DS data simultaneously collected during a cold crystallization experiment of a 1:1 by weight PET/PEN blend with $f_{TEN} = 11\%$ at $T_c = 96°C$ for three selected crystallization times. The PEN is from Eastman with $M_\nu = 25\,000$ g/mol. By comparison with the crystallization of pure PET (Fig. 21.8), here the crystallization kinetics is significantly slower. However, similarly to what it was previously shown for PET, for this PET/PEN sample lamellar crystals organize themselves forming a nanostructure of lamellar stacks with an average distance between gravity centres of consecutive lamellar crystals of about 8 nm. The α-relaxation exhibits, at the end of crystallization, a decrease in its intensity and a shift towards lower values of F_{max}. In contrast with the case for pure PET, now the dielectric data can be described by means of a superposition of a single α-process in addition to the β-relaxation process which contributes in the higher frequency range of the spectrum and the con-

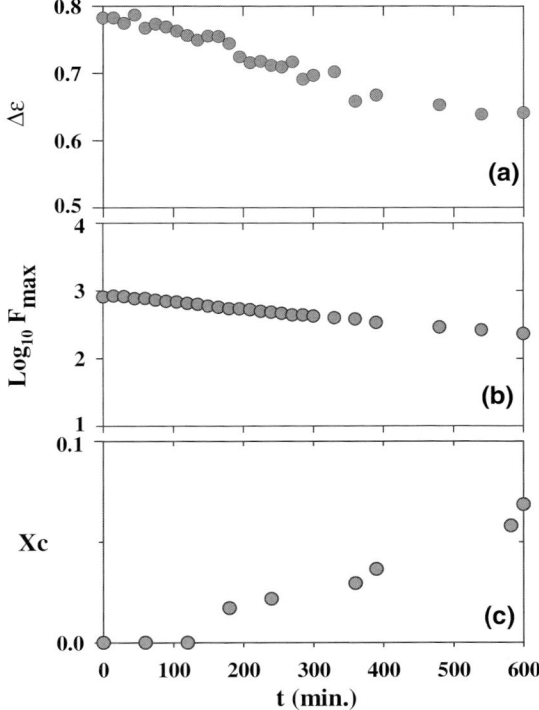

Fig. 21.12. Summary of parameters from the SWD experiment for PET/PEN blends (1:1 by weight) with $f_{TEN} = 11\%$ at $T_c = 96°C$ as a function of the crystallization time: (**a**) $\Delta\epsilon$ values for the α-relaxation, (**b**) Frequency of maximum loss, (**c**) Crystallinity index, X_c

ductivity which influences the lower frequency part of the spectrum [45]. The separate contributions of every process as well as the total fittings are represented in Fig. 21.11 by the continuous lines. The selected crystallization temperature is well below the calorimetric T_g of PEN. Therefore the influence of the α-relaxation of PEN [46] on the low frequency contribution of the DS experiments can be discarded. Figure 21.12 presents the characteristic parameters of the α-relaxation as a function of the crystallization time including the dielectric strength and the frequency of maximum loss which in this case corresponds to $(2\pi\tau_\alpha)^{-1}$. Additionally, values for the weight fraction index of crystallinity (X_c) are included calculated from the WAXS data. In this case ($f_{TEN} = 11\%$), after considering the PET/PEN weight ratio, the final crystallinity of PET is comparable with that reached in pure PET. However a significant slowing down of the PET crystallization with transesterification, as compared with that of pure PET, is evident. During crystallization at $T = 96°C$, there is reduction of the mobile material, reflected by the decrease of $\Delta\epsilon$ (Fig. 21.12a) which is accompanied by an increase of crystallinity (Fig. 21.12c). During crystallization, the remaining mobile material reduces its average segmental mobility in the amorphous phase as reflected by the decrease of $F_{\max}$ with crystallization time (Fig. 21.12b) and the shape of the α-relaxation remains essentially constant. This behaviour is in contrast with that observed under similar conditions for pure PET.

21.5 Development of the Rigid Amorphous Phase (RAP) as Revealed by SWD

21.5.1 Aromatic Polyesters: Poly(ethylene terephthalate), Poly(butylene isophthalate)

The above features, which emerge directly from the simultaneous SAXS, WAXS and DS experiments enable us to propose the following explanation for the cold crystallization of PET. During the initial stages of the process, before the characteristic time, lamellar crystals develop involving a strong formation of rigid amorphous phase as revealed by the increase observed in X_c and the decrease of $(\Delta\epsilon)_\alpha$. In this regime, the average mobility of the remaining mobile amorphous phase is slightly affected, as revealed by the small variation with time of τ_α. One possibility to explain these features is that the amorphous regions located between consecutive crystals within the lamellar stack become immobilized as soon as the lamellar stack is formed during the initial stages of crystallization before the characteristic time. This could explain the strong reduction of $\Delta\epsilon$ for moderate increase of X_c. A similar view was recently proposed to explain oxygen transport properties of PET [47]. In fact, oxygen permeation measurements indicate that the amorphous region within the lamellar stacks can be associated with the rigid amorphous phase (RAP). Moreover, recent AFM observations of the inter-lamellar amorphous phase PET indicate that molecular mobility in these regions should be strongly inhibited [48]. Accordingly, in this initial stage, the α-relaxation should predominantly originate in the inter-lamellar stacks amorphous regions. Around the characteristic time a significantly slower process appears, the α'-relaxation. Due to the fact that in the present case cold crystallization takes place relatively close to T_g the crystallinity evolution spreads in time. Thus, the transition from primary to secondary crystallization regimes is not so well defined as for crystallization at higher temperatures. By calculating the slope of X_c with time, continuous line referred to the right y-axis in Fig. 21.9e, it is observed that the characteristic time at which the α'-relaxation starts to appear is close to the inflexion point of the crystallinity curve. This inflexion point can be associated to the moment in which significant impingement of lamellar stacks may locally occur during primary crystallization. After impingement, secondary crystals are likely to develop. This suggests that secondary crystals growing in the inter-lamellar stacks amorphous phase may act as physical cross-links tending to slow-down

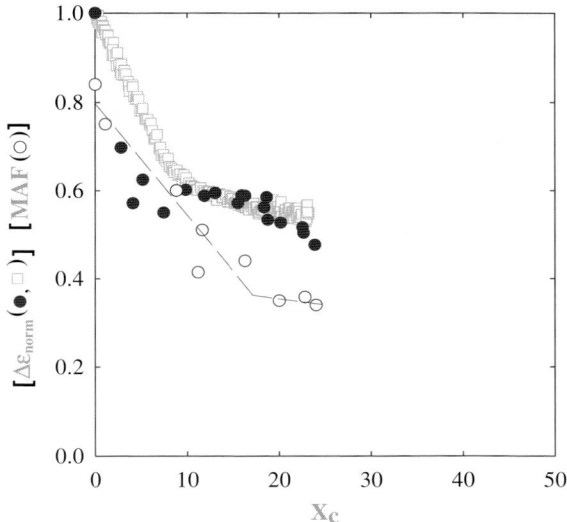

Fig. 21.13. Normalized dielectric strength $(\Delta \epsilon)_{norm}$ as a function of the crystallinity X_c for: (●)PET crystalized at $T = 96°C$ (SWD-data). (□)Poly(butylene isophthalate)(PBI)crystallized at $T = 60°C$(SWD-data [49]). (○)Mobile Amorphous Fraction (MAF)(calorimetry data) for PET crystallized at $T = 117°C$ (data extracted from Fig. 3 of [29]

segmental dynamics and giving rise to the secondary α'-relaxation. The influence of different crystallization regimes on the dynamics is further emphasized in Fig. 21.13. In that plot, $\Delta \epsilon$ values for PET, normalized to its initial value, are presented versus crystallinity. $\Delta \epsilon_{norm}$ can be considered as a measure of the fraction of relaxing species and therefore representative of the mobile amorphous fraction (MAF). As observed in Fig. 21.13, $\Delta \epsilon_{norm}$ decreases with X_c exhibiting two clear tendencies. Initially, $\Delta \epsilon_{norm}$ presents a slope far away from -1, indicating that during initial crystallization the immobilized segments are not only those included in the crystals but also a certain portion of the non-crystallized segments corresponding to the rigid amorphous phase (RAP). For X_c higher than ≈10% the tendency of $\Delta \epsilon_{norm}$ changes drastically and a second slope closer to -1 is observed. This effect can be interpreted assuming that for $X_c > 10\%$ the immobilization of material by RAP formation is not as effective as in the previous period being the amount of immobilized material closer to the amount of material incorporated to the crystals. A rather similar effect has been recently found by calorimetry for PET crystallized at $117°C$ [29]. Here, estimates of the mobile amorphous fraction (MAF) could be calculated by measuring the heat capacity increment at the T_g. In Fig. 21.13 data corresponding to these MAF values have been included for comparison. The existence of two crystallization regimes with two different ratios of RAP formation is also evident. Similar data have been found by means of

SWD for other polymer, namely poly(butylene isophthalate)(PBI) [49], and the corresponding $(\Delta\epsilon)_{norm}$ data are also included in Fig. 21.13 for comparison. One may propose the idea that in the second regime, which we can associate with the secondary crystallization, crystallization takes place essentially in the inter lamellar stacks amorphous phase. These secondary crystals should be arranged either as independent lamellae or as very defective stacks. This mechanism should not produce significant amounts of RAP because, as previously discussed, the RAP can be assigned to an intra lamellar stacks amorphous phase. Additional support for this model on the basis of structural experiments has been discussed extensively for PBI [49]. A similar view has been recently proposed to explain secondary crystallization in poly(ethylene isophthalate-co-terephtalate) copolymers crystallized from the melt [50].

21.5.2 Confined Crystallization of PET in PET/PEN Blends

As mention previously, cryogenic mechanical alloying of PET and PEN, 1:1 by weight, and the subsequent heat treatment followed to obtain films produce a phase separated morphology of PET and PEN rich domains for low levels of transesterification ($f_{TEN} = 11\%$) [42,43]. In this case, significant crystallization of PET within the PET domains is possible as shown in Fig. 21.11. In principle it is difficult to separate the effect of sequence length from those of confinement. However, for low levels of transesterification, the existence of two well defined calorimetric glass transition temperatures [42], and two α-relaxations, as revealed by dielectric spectroscopy (Fig. 21.10) and by dynamic mechanical analysis [43], indicate that T-E-N linkages are more likely to be located within the interface among PET and PEN phase separated domains. In this case ($f_{TEN} = 11\%$), after consideration of the PET/PEN weight ratio the final crystallinity reached by PET in the PET/PEN blends (Fig. 21.11), is comparable with that reached in pure PET (Fig. 21.8). However a significant slowing down of the PET crystallization is observed. From the SWD-experiments a relationship between structure and dynamics for PET crystallization in the PET/PEN blends with low transesterification levels can be attempted. During crystallization at $T = 96°C$, there is a reduction of the mobile material, reflected by the decrease of $\Delta\epsilon$ (Fig. 21.12a) which, similarly as in the neat PET case (Fig. 21.8), parallels the increase of crystallinity (Fig. 21.12c). During crystallization, the remaining mobile material reduces its average segmental mobility in the amorphous phase as reflected by the decrease of F_{max} with crystallization time (Fig. 21.12b) while the shape of the α-relaxation remains essentially constant. This behaviour is in contrast with that observed under similar conditions for pure PET (Fig. 21.8). In order to emphasize these differences, we have represented in Fig. 21.14 the evolution of $\Delta\epsilon$ and F_{max} with crystallinity during the SWD isothermal crystallization experiment at $T = 96°C$ for both PET and PET/PEN 1:1 by weight with $f_{TEN} = 11\%$. The crystallinity values for the PET/PEN blend have been corrected by the weight concentration. Both $\Delta\epsilon$ and F_{max} have been normalized

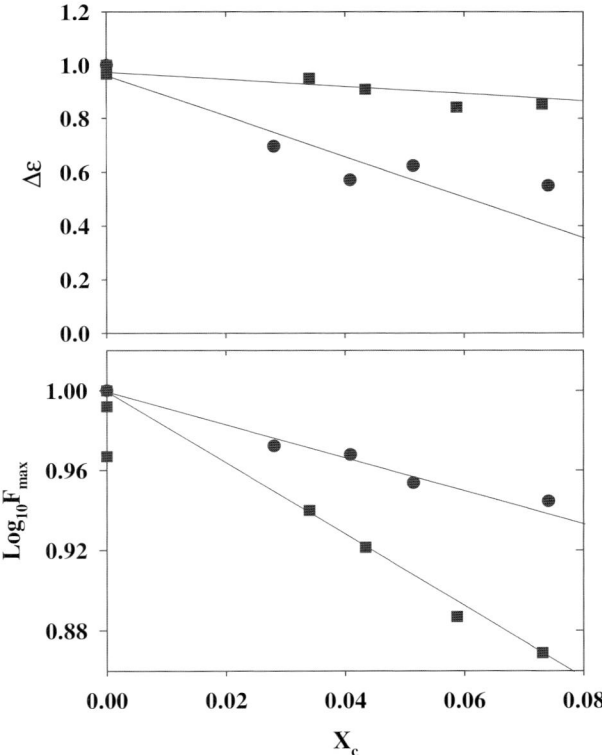

Fig. 21.14. Evolution of normalized $\Delta\epsilon$ and $F_{\max}$ values with crystallinity during an SWD isothermal crystallization experiment at $T = 96°C$ for both PET (●) and PET/PEN 1:1 by weight with $f_{TEN} = 11\%$ (■). The crystallinity values for the PET/PEN blend have been corrected by the weight concentration

to their initial values. It is worth to mention that we are dealing here with the region of low level of crystallinity, $X_c < 10\%$, where PET still exhibits a single α-relaxation (Fig. 21.8). In this crystallization regime the segmental dynamics of the PET/PEN blend seems to be more affected by the crystal development as reflected by the stronger reduction of the $F_{\max}$ values as compared with those of pure PET. In contrast, the blend shows a weaker decrease of $\Delta\epsilon$ with crystallinity than pure PET. Considering that the PET domains in the blend are embedded within the glassy PEN matrix then the observed differences in the evolution of the dynamics with the crystallinity can be attributed to a confinement effect. Firstly, the formation of T-E-N linkages located in the interface among PET and PEN domains in the PET/PEN blend with $f_{TEN} = 11\%$ provokes a slowing down the PET segmental dynamics, as compared to that of pure PET, due to a pinning effect. Secondly, in pure PET the observed strong reduction of the mobile material characterized by the decrease of $\Delta\epsilon$ has been attributed to the formation of an immobilized amorphous phase (RAP) additionally to the crystalline phase. The RAP in PET has been suggested to be assigned to the intra-lamellar amorphous phase where molecular mobility is strongly inhibited [35]. The weaker reduction of $\Delta\epsilon$ of the PET/PEN blend with $f_{TEN} = 11\%$ can be interpreted as due to a lower probability of lamellar stack formation due to the fact that lamellar crystals are forced to grow within a confined space. A smaller population of lamellar stacks in the blend should imply a smaller amount of RAP, as compared with pure PET, and therefore a weaker reduction of $\Delta\epsilon$. For example, in the restricted geometry imposed in thin films of the order of ≈102 nm in thickness, it was shown that PET presents isolated lamellae because spherulitic growth is severely limited [51]. Thirdly, pure PET, during its initial period of crystallization, $X_c < 10\%$, slightly changes the average segmental mobility in the amorphous phase as reflected by the moderate variation observed in $F_{\max}$. This is proposed to be so because, as previously discussed, the α-relaxation mainly originates in the inter-lamellar stacks of amorphous regions. The stronger decrease of $F_{\max}$ for the PET/PEN blend with $f_{TEN} = 11\%$ can be interpreted by considering that lamellar crystals forced to grow within a confined space are more likely to affect the dynamics of the remaining amorphous phase due to a more effective filling of the available space.

21.6 Conclusions

The simultaneous combination of techniques probing structure (WAXS and SAXS) with relaxation methods, detecting modifications of the amorphous phase dynamics (dielectric spectroscopy), can be helpful in order to obtain complementary information about cold crystallization processes in aromatic polyesters. In this review we have presented results which are consistent with a heterogeneous morphology in which lamellar stacks are separated by broad amorphous regions. There are two clearly differentiated regimes of crystallization. Firstly, a primary regime where lamellar stacks are formed. The amorphous phase within the stacks (inter lamellar amorphous phase) seems to be highly constrained and can be assigned to the rigid amorphous phase (RAP). Secondly, a secondary regime in which either isolated lamella or defective lamellar stacks grow in broad amorphous regions located between the stacks. During the second regime of crystallization the rate of RAP formation is reduced. The extension of this kind of experiments to a greater variety of polymers seems to be highly desirable in order to improve our knowledge about polymer crystallization.

Acknowledgements

We would like to express our gratitude to our colleagues C. Alvarez, Z. Denchev, D.R. Rueda, S.S. Funari, M. Jiménez-Ruiz and F.J. Baltá-Calleja whose support contributed to many of the results presented here. Also we are indebted to MCYT from Spain (grant MAT2005-01768) and to the European Community (EC) (MERG-CT-2004-511908 and MERG-CT-2004-505674) for generous support of this investigation. The experiments at HASYLAB (Hamburg, Germany) were supported by the European Community – Research Infrastructure Action under the FP6 "Structuring the European Research Area' Programme (through the Integrated Infrastructure Initiative "Integrating Activity on Synchrotron and Free Electron Laser Science").

References

[1] E.J. Donth, *Relaxation and Thermodynamics in Polymers.* (Akademie Verlag Berlin, 1992).
[2] K.F. Kelton, *Crystal Nucleation in Liquids and Glasses. Solid State Physics 45.* (Academic Press New York, 1991).
[3] G. Strobl, *The Physics of Polymers* (Springer Berlin, 1996).
[4] C. Talon, F.J. Bermejo, C. Cabrillo, G.J. Cuello, M.A. Gonzalez, J.W. Richardson, A. Criado, M.A. Ramos, S. Vieira, F.L. Cumbrera, L.M. Gonzalez: Phys. Rev. Letters 88, 115506 (2002)
[5] H.C. Semmelhack, P. Esquinazi: Physica B, 254, 14 (1998).
[6] M. Jimenez-Ruiz, T.A. Ezquerra, I. Sics, M.T. Fernandez-Diaz: Applied Phys. A-Mat. Sci. & Process., 74, S543 (2002).
[7] A. Sanz, M. Jimenez-Ruiz, A. Nogales, D.M.Y. Marero, T.A. Ezquerra: Phys. Rev. Letters 93, 015503 (2004).
[8] M. Bark and H.G. Zachmann: Acta Polymerica 44, 18 (1993).
[9] E. Eckstein, J. Qian, T. Thurn-Albrecht, W. Steffen and E.W. Fischer: J. of Chem. Phys. **113**, 4751 (2000).
[10] I. Šics, A. Nogales, T.A. Ezquerra, Z. Denchev, F.J. Baltá-Calleja, A. Meyer, R. Döhrmann: Rev. of Sci. Instrum., 71, 1733 (2000).
[11] M. Jimenez-Ruiz, A. Sanz, A. Nogales, T.A. Ezquerra: Rev. of Sci. Instrum., 76, 043901 (2005).
[12] F. Kremer, A. Schönhals (Eds.) *Broadband Dielectric Spectroscopy*, (Springer Berlin, 2003).
[13] G. Williams: G. Adv. Polym. Sci. 33, 59 (1979).
[14] J.C. Coburn, R.H. Boyd: Macromolecules 19, 2238 (1986).
[15] T.A. Ezquerra, J. Majszczyk, F.J. Baltá-Calleja, E. López-Cabarcos, K.H. Gardner, B.S. Hsiao, Phys. Rev. B 50, 6023 (1994).
[16] J. Dobbertin, A. Hensel, C.J. Shick: Thermal Anal. 47, 1027 (1996).
[17] T.A. Ezquerra and A. Nogales in *Lecture Notes in Physics: Polymer Crystallization: Observations, Concepts and Interpretation* ed. by J.-U. Sommer, G. Reiter (Springer Berlin, 2003) pp. 275.
[18] P. Hedvig: *Dielectric Spectroscopy of Polymers* (Bristol: Adam Hilger Ltd, 1997).
[19] N.G. McCrum, B.E. Read, G. Williams: *Anelastic and Dielectric Effects in Polymeric Solids* (Dover, New York, 1991, Original Issue John Wiley, London 1967).

[20] A.R. Blythe: *Electrical Properties of Polymers* (Cambridge: Cambridge University Press, 1979).
[21] F.J. Baltá-Calleja and C.G. Vonk, *X-ray Scattering of Synthetic Polymers* (Elsevier, Amsterdam, 1989).
[22] K. Fukao, Y. Miyamoto: Phys. Rev. Lett. 79, 4613 (1997).
[23] A. Wurm, R. Soliman, J.G.P. Goossens, W. Bras, C. Schick: J. of Non-Crystalline Solids 351(33-36), 2773 (2005).
[24] *Handbook of Thermoplastic Polyesters* ed. by S. Fakirov (Wiley-VCH: Weinheim, 2002).
[25] G. Groeninckx, H. Reynaers, H. Berghmans, G. Smets: J. Polym. Sci. Polym. Phys. 18, 3111 (1980).
[26] S. Monserrat, P. Cortés: J. of Mater. Sci. 30, 1790 (1995).
[27] J.F. Medellin-Rodriguez, P.J. Phillips, J.S. Lin: Macromolecules. 29, 7491 (1996).
[28] N.M. Alves, J. F. Mano, E. Balaguer, J.M. Meseguer Dueñas, J.L. Gómez Ribelles: Macromolecules. 43, 4111 (2002).
[29] R. Androsch, B. Wunderlich: Polymer. 46, 12556 (2005).
[30] A.M. Jonas, T.P. Russell, D.Y. Yoon: Colloid & Polymer Sci. 272, 1344 (1994).
[31] C. Santa Cruz, N. Stribeck, H.G. Zachmann, F.J. Baltá-Calleja: Macromolecules. 24, 5980 (1991).
[32] Z.G. Wang, B.S. Hsiao, B.X. Fu, L. Liu, F. Yeh, B.B. Sauer, H. Chang, J.M. Schultz: Polymer. 41, 1791 (2000)
[33] D.A. Ivanov, Z. Amalou, S.N. Magonov: Macromolecules. 34, 8944 (2001).
[34] C. Schick, E. Donth: Physica Scripta 43, 423 (1991).
[35] C. Alvarez, I. Sics, A. Nogales, Z. Denchev, S.S. Funari, T.A. Ezquerra: Polymer. 45, 3953 (2004).
[36] S. Havriliak and S. Negami: Polymer 8, 161 (1967).
[37] S.P. Bravard, R.H. Boyd: Macromolecules. 36, 741 (2003).
[38] A. Sanz, A. Nogales, T.A. Ezquerra, N. Lotti, L. Finelli. Phys. Rev. E: 70, 021502 (2004).
[39] D.J. Blundell and B.N. Osborn: Polymer 24, 953 (1983).
[40] S.Z.D. Cheng, Z.Q. Wu, B. Wunderlich: Macromolecules. 20, 2802 (1987).
[41] M. Okamoto, T. Kotaka: Polymer: 38, 1357 (1997).
[42] Z. Denchev, T.A. Ezquerra, A. Nogales, I. Šics, C. Alvarez, G. Broza, K. Schulte: J. Polym. Sci.: Part B: Polym. Phys. 40, 2570 (2002).
[43] J.F. Mano, Z. Denchev, A. Nogales, M. Bruix, T.A. Ezquerra: Macromol. Mater. Eng. 288, 778 (2003).
[44] G. Reiter, G. Catelein, J-U. Sommer: Phys. Rev. Letters. 87, 226101 (2001)
[45] C. Alvarez, A. Nogales, M.C. García-Gutiérrez, A. Sanz, Z. Denchev, S.S. Funari, M. Bruix, T.A. Ezquerra: Eur. Phys. J. E. 18, 459 (2005).
[46] A. Nogales, Z. Denchev, I. Šics, T. A. Ezquerra: Macromolecules. 33(25), 9367 (2000).
[47] J. Lin, S. Shenogin, S. Nazarenko: Polymer. 43, 4733 (2002).
[48] D.A. Ivanov, T. Pop, D.Y. Yoon, A.M. Jonas: Macromolecules. 35, 9813 (2002).
[49] A. Sanz, A. Nogales, T.A. Ezquerra, N. Lotti, A. Munari, S.S. Funari: Polymer. 47, 1281 (2006).
[50] B. Lee, T.J. Shin, S.W. Lee, J. Yoon, J. Kim, M. Ree: Macromolecules. 37, 4174 (2004).
[51] H.G. Haubruge, R. Daussin, A.M. Jonas, R. Legras: Polymer. 44, 4733 (2003).

22

Atomistic Simulation of Polymer Melt Crystallization by Molecular Dynamics

Numan Waheed[1], Min Jae Ko[2], and Gregory C. Rutledge[3]

[1] Maurice Morton Institute of Polymer Science, University of Akron, Akron, OH 44325-3909
   nw11@uakron.edu
[2] Department of Chemical Engineering, Massachusetts Institute of Technology, Cambridge, MA 02139
   mjko@mit.edu (Current: Samsung Electronics, HD Display Center, Asan, Korea, minjae.ko@samsung.com)
[3] Department of Chemical Engineering, Massachusetts Institute of Technology, Cambridge, MA 02139
   rutledge@mit.edu

Abstract. We review the use of molecular dynamics (MD) simulation at MIT to gain insight into the molecular level mechanisms of polymer crystallization from the melt. Simulations are constructed to observe nucleation and growth processes separately. For nucleation, we induce elongation under constant load before quenching, to accelerate the process in accord with experimental observations. We observe multiple nucleation events, increased perfection of crystalline lamella on the 5 nm length scale, and lamellar thickening. Nucleation is characterized by a competition of rates. The rate of spontaneous ordering competes with the rate of conformational relaxation to determine the number, size and chain tilt of the crystallites. To measure lamellar growth, we perform simulations for $C_{20}H_{42}$, $C_{50}H_{102}$ and $C_{100}H_{202}$ melts. From these simulations, we obtain data for the growth rate of n-alkane crystals over a range of temperatures and molecular weights. We construct a general crystal growth model that can be parameterized in terms of chemical properties of polymer chains and constants derived from polymer physics. Analysis reveals that the crystal growth rates of alkanes and polyethylene can both be described by the same relationship when the appropriate relaxation time is used to describe the transport barrier to crystallization. For chains shorter than the entanglement length, this is the Rouse time. For chains longer than the entanglement molecular weight, transport limitations are modeled by the local relaxation of an entangled segment at the interface.

22.1 Introduction

The kinetics of polymer crystallization and the structure of a forming crystal are very sensitive to molecular level details. For this reason, different macroscopic representations of crystallization kinetics can reproduce empirical growth rates but cannot discriminate the underlying mechanisms of melt crystallization. The assumptions of these melt crystallization models are often borrowed from knowledge of single chain crystallization from dilute solution. For polymer melts, these assumptions are more difficult to prove experimentally. Atomistic simulation, however, provides an alternative to experimental techniques that can give insight to the molecular level mechanisms occurring and, thus, the correct assumptions on which to base kinetic models.

Experimental methods generally have yielded data at the coarsest level of description: the rate of change in the degree of crystallinity [1–3]. However, homogeneous crystallization is comprised of many competing processes that complicate the interpretation of this quantity. Classical nucleation theory is typically invoked to analyze the process in terms of formation rates and growth rates of nuclei [4]. Each of these rates is the result of several competing processes, including chain ordering, densification, and relaxation, which are difficult to study experimentally. X-ray diffraction techniques [5–7] and high-speed optical measurements [8–11] have provided considerable information on the rates of nucleus formation and growth, respectively, but have not been successful in resolving recurring questions about the mechanisms and intermediate structures of the growing crystal. Some X-ray studies have suggested that a thermodynamic instability resulting from supercooling leads to density fluctuations in the liquid phase, such that the formation of nuclei is preceded by spinodal decomposition and facilitated by the presence of a dense, ordered liquid phase [12]. Other experimental techniques such as TEM [13] and FTIR [14] have suggested a precursor hexagonal phase during crystallization, though different ideas exist about the significance of this phase [15–17]. In addition, while the molecular level behavior is still being debated for crystallization under quiescent conditions [18–23], most industrial processes involve crystallization under conditions of high stress and deformation that accelerate crystallization significantly [2, 24, 25].

Atomistic simulation techniques, although currently unable to provide the complete macroscopic rate data that experiments can provide, are well-suited to enhance our understanding by offering molecular level resolution of systems undergoing crystallization. Simulation techniques such as lattice dynamics, Monte Carlo, and molecular dynamics provide detailed information that is experimentally inaccessible, due to the temporal and spatial resolution of the experimental techniques and the difficulty in analyzing the complex morphologies of crystallizing polymer systems. Through atomistic simulations, however, one can independently observe nucleation [26–30] and growth [31–35] during melt crystallization.

Molecular dynamics (MD) is a staple among simulation techniques for obtaining dynamic data for systems at equilibrium, but only recently has the availability of increased computing power permitted simulation of the large scale reorganization that occurs during polymer melt crystallization. Early MD simulations by Kavassalis and Sundararajan [36, 37] revealed a "global collapse" mechanism for the folding of a single, isolated polyethylene chain. Liu and Muthukumar studied a similar collapse in solution, using Langevin dynamics to represent the surrounding solvent [38]. They observed the spontaneous formation of initial "baby nuclei", their subsequent growth by addition of single chains, and then a merging the baby nuclei to form larger structures reminiscent of lamellae. The theory they developed explains this process in terms of the configurational entropy of the chain [39]. Fujiwara and Sato also found that, through stepwise cooling, a random coil could be transformed into an oriented globular structure [26]. All of these works, however, were conducted at polymer densities well below that of the melt. For concentrated systems of supercooled melts, MD studies of short n-alkanes below the entanglement molecular weight have also been reported [40]. Meyer and Müller-Plathe [27, 41] used a coarse-grained bead-spring model to simulate the crystallization of polyvinyl alcohol. They showed that chain stiffness alone, without an attractive potential, is sufficient to induce the formation of chain-folded lamella in the melt. An annealing strategy similar to the one described here has also been reported by Koyama and co-workers [28, 42], where an initial oriented amorphous configuration was first generated by stretching in the glassy state at lower temperature, 100 K, then allowing the system to evolve at higher temperatures.

Polymer crystal growth has also been studied in several simulations by focusing on secondary nucleation as the governing process. Monte Carlo simulations have suggested folding and thickening mechanisms [43, 44]. Kinetic Monte Carlo has been particularly useful for studying the consequences of certain kinetic assumptions and for understanding secondary nucleation rates [22, 45, 46]. However, kinetic Monte Carlo requires, as input data, parameters that are dependent upon the very mechanisms in question and not directly available from experimental data or physical models. Molecular dynamics has been used before to study nucleation at a surface, usually with simplified force fields and in dilute solution [47, 48].

Our approach decomposes polymer crystallization into the two separate processes of nucleation and growth under processing conditions. This is consistent with recent macroscopic models that seek to predict domain size [49] and the asymmetry of semi-crystalline domains, for example, fibrils, shish kabobs, and lamellae [50]. It allows us to study each process separately, under conditions optimal for that process, so that the massive computing resource requirements necessary for a brute force simulation of polymer melt crystallization are avoided. The rest of this paper is organized as follows. In the Methods section, we discuss non-equilibrium molecular dynamics techniques that can reveal the molecular level phenomena in both nucleation and growth processes. We briefly present the non-equilibrium molecular dynamics technique (NEMD) and describe the spatial and temporal analyses that are required to recognize phase change on an atomic level. In the Results section, we discuss separately the simulation of nucleation and of growth. Finally, we discuss our use of MD results to parameterize a phenomenological model of polymer crystal growth [35,51].

22.2 Methods

The molecular dynamics (MD) technique is based on the numerical solution of Newton's classical equations of motion for many-particle systems using a given interaction potential. The interplay of the non-bonded van der Waals potential well and the barriers between torsional states is essential in capturing the balance between the orienting and densification processes. The interaction potential used in this work is of the united atom type for alkanes and polyethylene and has been verified by simulation of the melt dynamics [52], persistence length [53], and properties of the crystal/amorphous interphase in solid polyethylene [54]. Using this force field, the crystal phase exhibits hexagonal packing lateral to the c-axis. United atoms interact through harmonic bond stretching forces, harmonic angle bending forces, torsional forces, and non-bonded forces. The details of the interaction potential have been given previously [30, 35].

In non-equilibrium molecular dynamics (NEMD), a driving force is introduced which maintains the system out of equilibrium at steady state, or else a perturbation is introduced and the system studied as it relaxes towards equilibrium. Our simulations are of the latter type. Generation of the initial cells is described elsewhere [30, 35].

For nucleation, the perturbation is a temporary extensional deformation that is applied in the melt state, before crystallization is initiated. This deformation produces an initial state of orientation that (i) can be correlated with processing conditions [55,56] and (ii) accelerates homogeneous nucleation into the MD-accessible time scale. After the initial state of orientation is created, the sample is quenched below the melting temperature to observe the formation of nuclei. The simulations were performed in the $N\sigma T$ ensemble, and periodic boundary conditions were imposed along x, y, and z directions in order to eliminate boundary effects. The stress and temperature were controlled using the method of Berendsen et al. [57], while integrating the equations of motion using the velocity Verlet algorithm, as detailed previously [30].

Nucleation studies were conducted on systems of 20 C400 chains. To orient the melts, we applied uniaxial elongational stress to the melt in the z direction, using $\sigma_{xx} = \sigma_{yy} = -0.1$ MPa and $\sigma_{zz} = 100$ MPa, and allowed the system to evolve for 1.6 ns. The temperature was 425 K during the elongation; the melting temperature T_m of C400 is approximately 410 K [58]. Replicas of the elongated system were used as the starting conditions for several thermal quenches; for each replica, we reduced σ_{zz} to -0.1 MPa and quenched the system to one of the following temperatures: 375, 350, 325, 300 or 250 K. The subsequent development of crystallinity was simulated for 30 ns.

To study the growth of crystallites along prescribed crystallographic directions, a second set of simulations was constructed in which crystal lattice planes were simulated at the x-y plane boundaries of the simulation cell. A Steele potential was used to recreate a (110)-like surface, with x and y corrugation, with the y-direction corresponding to the c-axis of the crystal, as detailed previously [34]. The simulations consisted of 100 C20 chains, 42 C50 chains, or 40 C100 chains, in boxes of fixed x- and y-dimensions. Periodic boundary conditions were imposed in the x and y directions to simulate infinite crystal-melt interfaces. In the z-direction, the box dimension (the distance between the surfaces) was held at a constant stress of 0.1 MPa, in order to accommodate the volume change associated with thermal contraction and crystallization. Although full chain extension during crystallization has also been observed for these alkanes from solution at small undercoolings [59], fully extended crystallization of C100 was not expected at large undercoolings. Therefore, in the interest of shorter simulation time, the y-dimension of our C100 simulation is sufficient to allow for the case of once-folded chains; the effects of the finite size have been discussed previously [35]. Systems were equilibrated at 400 K for C20 and C50 and 500 K for C100, above the melting point in each case. Once equilibrated, replicas of the system were quenched to temperatures in the range of the maximum crystallization rate. Growth was typically simulated for 90 ns. Stress and temperature were regulated in the same method mentioned above for the nucleation studies, using the same integration algorithm [35].

The determination of phase change is primarily based on changes in the orientational order parameter $P_2(t)$. This parameter measures the degree of alignment of the chains, for all united atoms in the system. To measure the size or transformation of the crystal or amorphous phase, the order parameter is computed locally, such that orientational order can be measured as a function of position, $P_2(\boldsymbol{r},t)$, or as a function of one dimension (i.e. within planar slices), such as $P_2(x,t)$ or $P_2(z,t)$. The method of calculation of the orientational order parameter, as well as its comparison to order parameters based on density and energy, has been previously discussed [30, 35].

22.3 Results and Discussion

22.3.1 Nucleation

When we refer to nucleation, we mean only the formation of a new phase where one did not exist previously, without assuming a priori the characteristics of the process. Neither classical nucleation theory nor spinodal decomposition is assumed.

After deformation at 425 K, the stress was removed and the systems were quenched to below the melting temperature. We first note that there are several combinations of initial orientation and quench temperature that do not lead to an observable phase change within the 30 ns of MD simulation. Simulations were conducted on systems where the initial amorphous orientation P_2 after elongation was less that 0.3 (0.03 to 0.22); upon quenching the systems to 325 K or 375 K, they did not show any indication of phase change. Instead, the initial orientation was dissipated through conformational relaxations. This result is similar to the behavior of systems that are given high degrees of amorphous orientation, but are not quenched when the stress is removed. Simulations that release the stress but hold the temperature at 425 K also lose their initial stress-induced ordering. The remainder of this section will discuss conditions for which phase change was observed.

For systems with initial amorphous orientation of $P_2 = 0.34$, a transition was observed for quench temperatures 375 K and below. Snapshots of each replica after 30 ns of simulation at different temperatures are shown in Fig. 22.1. The high level of initial orientation facilitates nucleation, such that we can observe the initial nucleation process of polymers within the simulation time scale. Regions that appear crystalline can be identified visually in the snapshots. Crystalline regions display mainly *trans* torsional conformations and the intermolecular packing associated with the crystal phase. At higher quench temperatures (350 and 375 K), two or three thick crystalline domains are seen, while at lower quench temperatures (250, 300 and 325 K), there are more crystalline regions that are thinner and have not developed all the characteristics of the crystalline phase. Crystal domains formed at 350 and 375 K contain chains in the hexagonal packing mode expected for a united atom

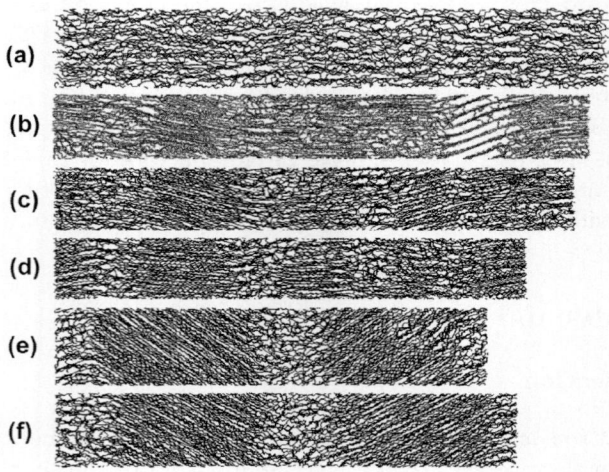

Fig. 22.1. Snapshots from the simulations of 20 C400 chains (**a**) before quench, and 30 ns after quenching from 425 K to (**b**) 250 K, (**c**) 300 K, (**d**) 325 K, (**e**) 350 K, and (**f**) 375 K. Reprinted with permission from [30]. Copyright 2004, American Institute of Physics

model of PE [36, 37, 60, 61]. While the lamellar normals are parallel to the elongation direction, the chains within the crystal at 350 and 375 K are tilted about 41 and 35 degrees with respect to the elongation direction, respectively. For crystallization at 250, 300 and 325 K, the chains within the crystal-like domains are more aligned along the original direction of stretching, but the hexagonal packing pattern is not well resolved.

A temporal and spatial analysis is needed to study the development of the crystal domains. Local orientational order can be calculated from $P_2(z,t)$ at each quench temperature, as a function of time and z coordinate. Figure 22.2 shows the $P_2(z,t)$ profile for the isothermal crystallization simulations at $T_c = 375$, 325, and 250 K. The order profile at 350 K is similar to that at 375 K in Fig. 22.2a, while the profile at 300 K is similar to that shown at 250 K in Fig. 22.2c. At 375 K, this data reveals two distinct nucleation events, which form mature lamella that thicken to almost 6 nm. However, at 250 K we identify a greater number of nucleation events which resemble the initial location of peaks of amorphous orientation and do not exhibit thickening or growth. Thus, even though the overall orientational order parameter and the fraction of *trans* states increases at these lower temperatures, the number and size of ordered domains is determined largely by the initial quenched melt structure, not by the process of crystallization. At the higher temperature, however, the increase in order parameter and *trans* fraction are clearly associated with the emergence of a new phase, and clear interfaces between domains of greater and lesser order can be identified. This evolution of the order profile to one with distinct regions of high and low order is an essential observation that confirms the early stages of phase transition and emergence of a new crystalline phase in these simulations, not simply the gradual relaxation to a new, single phase, equilibrium state.

These results at higher temperature (325, 350 and 375 K) can be understood in terms of three processes that take place during isothermal crystallization. First, the order parameter increases with time at distinct locations, indicative of the nucleation and subsequent perfection of crystalline domains. Second, the width of these ordered domains increases with time, indicative of lamellar thickening. Finally, the decrease of $P_2(z,t)$ with time at other locations corresponds to relaxation of uncrystallized components, removing the initial amorphous orientation. The number of domains and domain size are determined by the rates of these competing processes at each quench temperature.

The formation of tilted chain lamellae at higher temperature is qualitatively consistent with the experiments of el Maaty and Bassett. They reported that the morphology of polyethylene crystals changed when the crystallization occurred above a certain temperature, 400 K in their case [62]. At high temperature, flat lamellae with (201) interfaces (i.e. 34 degree tilted chains) formed directly from the melt, whereas at low temperatures, the lamellae consisted initially of untilted chains with (001) interfaces, which only later relaxed to form tilted-chain and curved lamellae. The (201) interface is thermodynamically

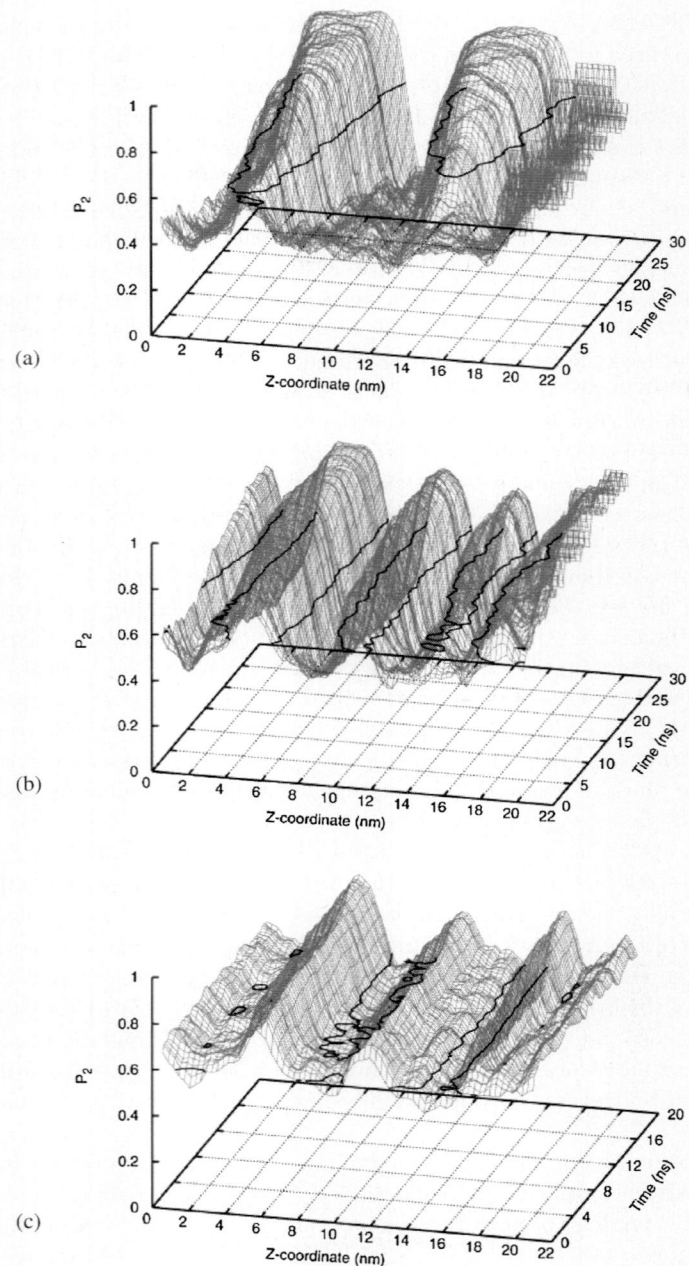

Fig. 22.2. Plot of the orientation order parameter $P_2(z,t)$ as a function of time and position in the simulation cell, for isothermal crystallization at $T_c =$ (**a**) 375 K; (**b**) 325 K; (**c**) 250 K. Reprinted with permission from [30]. Copyright 2004, American Institute of Physics

favored, according to previous calculations for interfacial energy of lamellae as a function of chain tilt [63]. Thus, both simulation and experimental observations can be rationalized as a consequence of competition between the rate of formation of the crystal phase and the mobility of the chains; at high temperatures, the latter is faster and nucleation proceeds immediately via the tilted chain morphology, whereas at lower temperatures, the former occurs first and the nuclei consist of untilted chains. Only in the experiments, where a longer period of observation is possible, are these untilted chain crystallites observed to transform to tilted chain crystallites (believed to occur by an inter-chain sliding mechanism that leads to curvature of the lamellae).

For the highly-ordered domains that are formed at higher temperature, we can quantify the rate of lamellar thickening. The one-dimensional profiles of $P_2(z)$ were fit using a hyperbolic tangent function to describe each interface, and the lamellar thickness obtained as the distance between the midpoints of the two interfaces bounding each crystal domain. A plot of lamellar thickness versus time suggests the existence of a limiting initial lamellar thickness that can be determined from these simulations. The characteristic time required to obtain this limiting thickness is on the order of 5 ns at 325 K, but increases to 30–50 ns at 375 K, beyond the duration of the MD simulation itself. The limiting lamellar thickness was then obtained by extrapolation using a function of the form $l(u) = (l_0 + l_\infty u)/(1 + u)$, where $u = t/\tau_l$, τ_l being the characteristic thickening time. These limiting thicknesses are plotted as a function of inverse undercooling in Fig. 22.3, as suggested originally by Barham et al. [64].

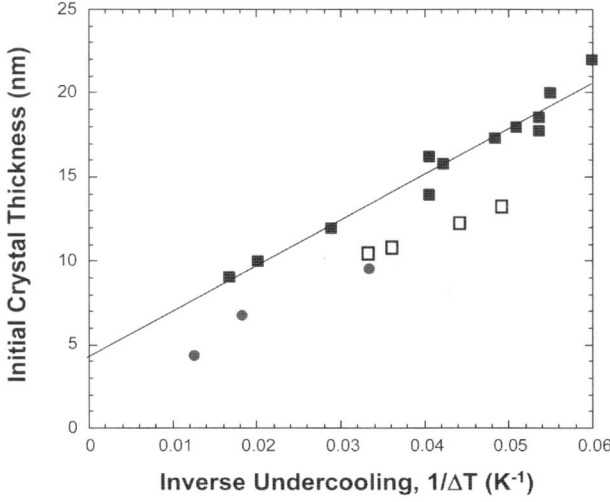

Fig. 22.3. Temperature dependence of the limiting lamellar thickness: simulation data (•); experimental data of Barham et al. [64] (■); experimental data from Hocquet et al. [65] (□)

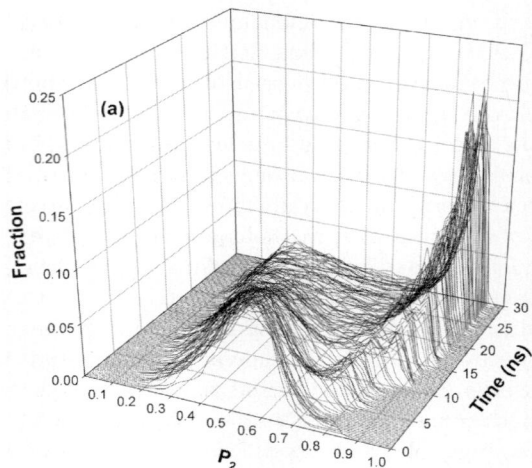

Fig. 22.4. The distribution of the orientation order parameter $P_2(\boldsymbol{r},t)$ for simulation at 375 K. Reprinted with permission from [30]. Copyright 2004, American Institute of Physics

along with subsequent data by Hocquet et al. [65]. According to the secondary nucleation theory of polymer crystallization [66], the lamellar thickness is a linear function of the inverse undercooling, with a slope of $(2\sigma_e T_m/\Delta H_f)$, where σ_e is the interfacial energy, T_m is the melting temperature and ΔH_f is the heat of fusion. We have already seen that for well-formed crystallites, the interfacial energy is well reproduced by the force field used in this work [63]; the good agreement between experiment and simulation exhibited by the slopes in Fig. 22.4 (taking a common value of $T_m = 407$ K for plotting purposes) provides supporting evidence that the ratio $T_m/\Delta H_f$ is also reasonably well described by this force field. The downward shift in the simulation data relative to that of Barham may be due to differences in molecular weight (hence the better agreement with the data of Hocquet et al.) or to the extrapolation required to estimate the limiting lamellar thickness.

The population distribution of the orientation order parameter $P_2(\boldsymbol{r},t)$ over the entire cell at 375 K is shown in Fig. 22.4. $P_2(\boldsymbol{r},t)$ has a unimodal distribution at $t = 0$ ns. Once quenched to 375 K, the distribution shifts towards lower P_2 values, becoming shorter and broader as the chains relax from the initial imposed amorphous orientation. After an induction period of about 3 ns, a peak develops around $P_2 = 0.8$, as the structural heterogeneity associated with the crystalline phase develops. The new peak grows as nucleation and lamellar thickening convert chain segments from amorphous conformations into members of the high order domains. The orientational order parameter distribution was fit to the sum of two Gaussian curves; Takeuchi et al. performed a similar analysis of the energy distribution during crystallization of

n-eicosane [67]. From the time dependence of the position of the amorphous peak, the relaxation time of the oriented amorphous regions was calculated, assuming the time decay is well described by the Kohlraush-Williams-Watt (KWW) stretched exponential equation [68]. The observed relaxation time, τ_{KWW}, decreases by an order of magnitude, from 400 ns to 40 ns, between 325 and 375 K. Comparing the relaxation time to the characteristic thickening time, we observe that $\tau_{\text{KWW}} \gg \tau_l$ at 325 K, but the two become comparable at 375 K, in accord with a shift in the competition between two processes as a function of temperature. At the higher temperature, the relaxation process approaches Arrhenius behavior, as β nears one.

In closing this section, we suggest an analogy between our MD results and the experimental study of Mahendrasingam et al. [69]. In that work, the authors studied the influence of temperature and molecular orientation on crystallization in fibers of polyethylene terephthalate (PET) during drawing. They interpreted their observations as evidence for three regimes or zones of crystallization behavior (we prefer the use of "zone" in this context, to avoid potential confusion with Hoffman's "regime" theory of crystallization [66]), depending on the rate of drawing relative to two material relaxation rates, "retraction" and "reptation". Retraction refers to relaxation of the molecule within a "tube" formed by entanglements, to recover the equilibrium chain length between entanglements, and is proportional to the inverse of temperature and the square of molecular weight. Reptation refers to renewal of the tube and is proportional to the inverse of temperature and the cube of molecular weight. In general, retraction is faster than reptation. In Zone I, the draw rate is faster than the retraction rate. In this Zone, crystallization was observed to be delayed until the end of drawing, and the crystal chain axes were well aligned with the draw direction. In Zone II, the draw rate is intermediate to the retraction and reptation rates; crystallization occurred during deformation and the crystal chain axes were tilted with respect to the draw direction. In Zone III, the draw rate was slower than reptation; no acceleration of crystallization associated with orientation was observed in this zone. In our simulations, there are numerous important differences: the material is C400 rather than PET, P_2 refers to bond-level orientation rather than X-ray scattering at a reciprocal space vector of 0.28 Å^{-1}, we study crystallization as a function of P_2 and temperature rather than draw rate and temperature, and we are limited to observations on very short time scales, such that only oriented crystallization is observed. Nevertheless, if we hypothesize that an appropriate 1:1 mapping exists between strain rate in [69] and initial bond level orientation in this work, then we do indeed observe features reminiscent of Zones I, II and III. For $P_2 = 0.34$ and $T < 325$ K, the rate of nucleation is high and the thickening time is short, while the rate of chain relaxation is low. This creates a large number of thin nuclei with chains well-aligned in the initial orientation direction; these nuclei do not develop further into crystallites on the time scale of our simulations. These observations are consistent with Zone I. For $P_2 = 0.34$ and $T > 325$ K, relaxation occurs commensurate with nucleation, and the thickening time is longer. This produces a smaller number of thicker nuclei with chains tilted relative to the original orientation direction, consistent with Zone II. For a low value of $P_2 = 0.03$, the rate of nucleation is low compared to amorphous phase relaxation, leading to Zone III behavior, where oriented crystallization is not observed. On the time scale of our simulations, no nucleation is observed in this case. Clearly, further simulations are required before the validity of this comparison is fully evident, but we find the similarity of observations encouraging.

22.3.2 Growth

"Growth" refers to the process by which chains from the melt join the pre-existing crystalline nuclei. Once again, by using molecular dynamics, we do not assume any particular mechanism for growth a priori.

Isothermal crystal growth on pre-existing crystal surfaces was simulated for $C_{20}H_{42}$, $C_{50}H_{102}$ and $C_{100}H_{202}$, denoted C20, C50 and C100, for the purpose of studying the molecular weight dependence of the crystal growth rate and parameterizing a phenomenological rate equation. Furthermore, we investigate the properties of growth, including surface nucleation and the growth front characteristics.

Identical samples were quenched to intermediate temperatures between the glass transition temperature and the melt temperature, where growth progressed fast enough to be observed with MD. This approach results in crystal growth data for C20 at 240, 250, 260, 265, 275, 285, 290, and 295 K. C50 samples yielded growth data at 290, 300, 315, 330, 345 and 360 K. C100 samples were tested at 350, 375 and 400 K. (Although the experimental melting point of C100 is 388 K [70], crystallization at 400 K was attempted after successful crystallization was observed for 350 K and 375 K for the simulated polymer.)

Figure 22.5 shows representative snapshots of C20 crystallizing at 285 K. As the simulation progresses, chains located near the growth front become fully extended and arranged in a hexagonal packed structure to become part of the crystal phase. The lamellar growth front thus progresses toward the centerline of the simulation from both x-y boundaries, until the melt is completely converted to crystal. C50 crystallizes similarly and to completion. In the case of C100, the growth process still proceeds through a sequence of layer ordering stages starting at the x-y boundaries of the cell, but the progression is much slower and less complete. Crystal growth for C100 does not completely fill the simulation cell, and defects persist within all layers, including those nearest the surface. Rather than fully extended chains as seen in C20 and C50 or once-folded chains as seen experimentally near T_m, in C100 at high undercoolings [59], only sections of chains are extended and packed in crystallographic registry in the simulated C100.

However, all three systems share some characteristics of the growth front that can be best viewed through sections or layers near the growth front.

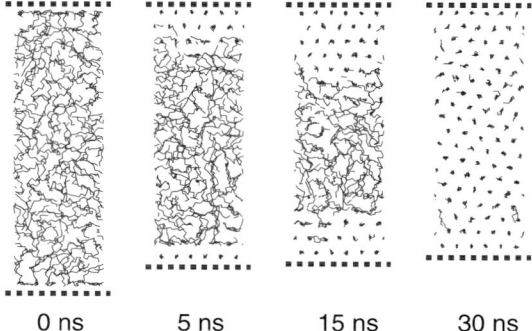

Fig. 22.5. Snapshots from a simulation of 102 *n*-eicosane chains between two surface potentials quenched from 400 K to 285 K at $t = 0$ ns, and then allowed to evolve dynamically at 285 K. *Thick dashed lines* are representative of location of surface potentials. Reprinted with permission from [34]. Copyright 2002, American Institute of Physics

Figure 22.6 shows the first three layers closest to one of the initial surfaces of the C50 system at times $t = 0, 30, 60$ and 90 ns after quenching to 330 K. We interpret these snapshots as indicative of two processes, the first involving ordering and extension of chains within a layer, and the second involving propagation of that order from one layer to the next. Chain extension occurs as the crystalline stems grow through a process of pulling segments in from the amorphous phase. The new surface layer is stabilized when a band of partially extended, aligned, all-*trans* chain segments, approximately 20 CH_2 segments long and a few chains wide, has formed. Once this step is complete, order can begin to propagate to the next layer, by repetition of this process. Meanwhile, order within the stabilized layer continues to improve, as new stems are pulled in and slide into registry. Similar nuclei consisting of several stems 20 CH_2 segments long were also observed in C100. In C20, such nuclei were observed simply as fully extended chains.

The two different processes seen in Fig. 22.6, ordering within a layer, and propagation of that order to the next layer, can be quantified in a three dimensional plot of $P_2(z, t)$ versus z coordinate. Figure 22.7 shows $P_2(z)$ profiles at different times for the C20 system quenched to 285 K. These one-dimensional order profiles were fit to hyperbolic tangent curves to identify the midpoint of the curve as the z-location of the order front. The movement of the growth front is linear in time, in accord with experiments on spherulites (except at the initial and final stages of simulation). From each simulation, we extract two estimates of the linear growth rate based on the orientational order fronts growing at the two surfaces. The analysis described above was repeated for several isothermal crystallization temperatures ranging from 225 to 300 K. From these, we obtained the temperature dependence of linear growth rate

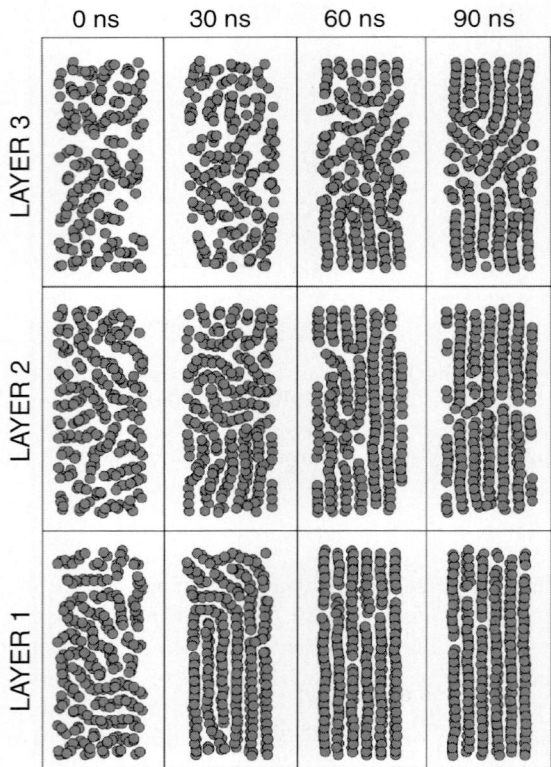

Fig. 22.6. The three layers closest to $z = 0$ plane for 42 C50 chains after quench to 330 K at $t = 0$ ns. Reprinted from [35]. Copyright 2005, with permission from Elvesier

The analysis conducted for C20, C50 and C100 yields the temperature dependence of growth rates at each molecular weight. The growth rate data for these samples is shown in Fig. 22.8a. At each temperature, the error bar is indicative of the high and low estimate obtained from each of the two fronts. For C20, growth is seen between 240 K and 295 K, with a maximum growth rate near 265 K. Above 295 K, crystal is forming and re-melting throughout the simulation, and a growth rate can no longer be calculated. Below 240 K, mobility of the molecules is too slow to obtain a growth rate in the simulation time scale. This trend is also apparent for C50, which reveals a maximum in the growth rate near 347 K. Limited data exists for C100, where three temperatures were simulated, yielding a maximum growth rate occurring near 395 K.

Based on observations from our simulations, we can make several conclusions about the nature of the surface nucleus during crystal growth. We

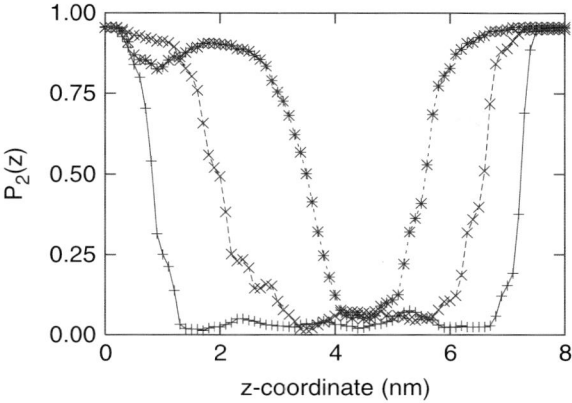

Fig. 22.7. Profile of the orientational order $P_2(z)$ growth front for 102 C20 chains at 285 K at 5 ns (+), 15 ns (×), and 25 ns (⋆).

observe multiple chains adsorbing and desorbing stochastically on the crystal surface. If enough chains absorb in the same region at the same time, a critical nucleus may be formed. Although this surface nucleus is difficult to characterize precisely, it typically consists of a group of chains that are 20-24 beads long and 4-5 chains across (based on observations at $T = 330$ K for C50 and $T = 375$ K for C100), in close agreement with the estimate of 3-4 stems at $T = 392$ K for the critical nucleus size for polyethylene by Wagner and Phillips [11].

From these observations, we can begin to challenge some of the long-standing interpretations of the classical crystal growth theories. For example, it has long been attributed to Lauritzen and Hoffman that the critical rate-limiting step in polymer crystal growth was a stem segment extending to a length matching the underlying lamellar thickness, or in later work, an adsorbed but unattached stem [66,71]. Therefore, they parameterized the thermodynamics of crystal growth in terms of the surface energy of the lateral surfaces and fold surfaces of the final crystal lamella. In our simulations, we do not see nuclei consisting of full-length extended stems, except in the case of C20; instead, we observe the initial formation of surface nuclei of stems of length 20 for all three alkanes studied. Similarly, we do not observe anything that we would interpret as a "monomolecular" surface nucleus [21]; instead we observe segments of several chains coming into registry with the underlying surface to stabilize a crystallographic layer. We believe these molecular level observations of growth could have implications for the estimation of lamellar surface energies from crystal growth kinetics. We cannot conclude whether the surface nucleation produces a mesomorphic or metastable phase, as has been postulated by Strobl and Keller, or whether the nucleus is temperature-dependent. However, compared to our observations, Binsbergen's model of

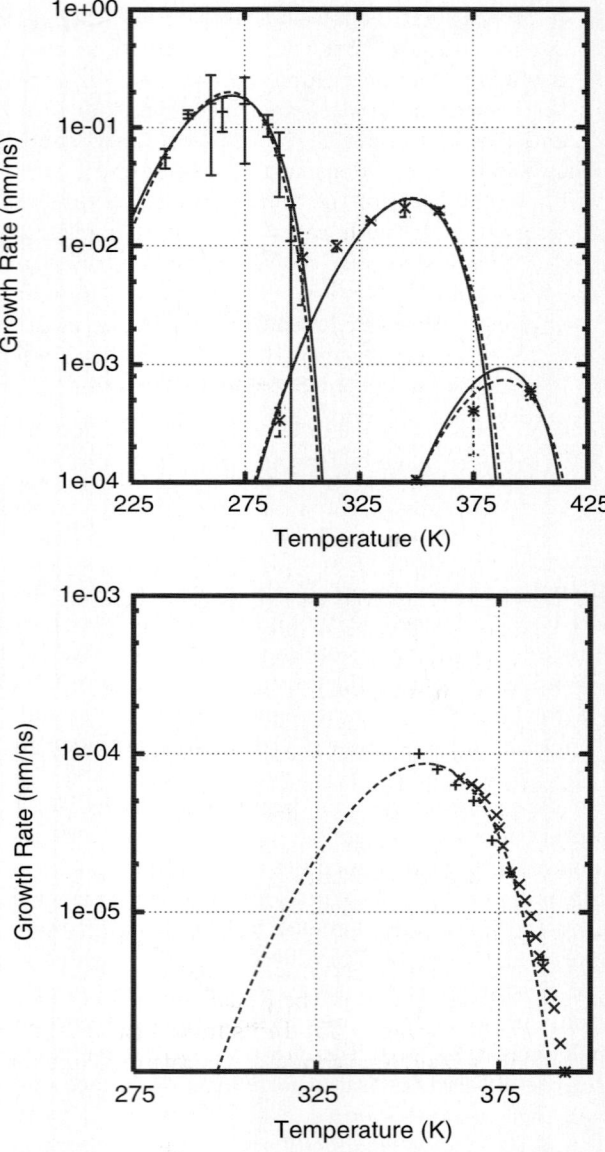

Fig. 22.8. Fit of the model equation to simulation (*solid*) and a combination of simulation and experimental data (*dashed*). (**a**) Simulated growth rates of alkanes fit to Eq. (22.1): C20 (+), C50 (×), and C100 (⋆). (**b**) Experimental growth rates of polyethylene samples fit to the modified version of Eq. (22.1) for entanglements in conjunction with fits to part (a): data of Ratajski et al. [10] (+), and Wagner et al. [11] (×). Part b reprinted with permission from [51]. Copyright 2005, Wiley

stochastic crystal growth, with several adsorption and desorption steps [72], seems most consistent, not only for the stem addition and removal that creates a critical nucleus but also for the random longitudinal diffusion of the chains that continues to thicken the critical nucleus to its final form. In addition, because the boundaries of the surface nucleus are not defined by folds but instead by a change in order and density, the surface energ extracted from the growth rate will not be related to the fold energy as is commonly assumed.

In polymer processing, growth rates (as well as nucleation rates) as a function of temperature are usually described by an empirical equation due to Ziabicki, which describes growth rates in terms of the maximum growth rate for a given polymer. However, despite its success in capturing the temperature dependence of the crystal growth rate, it does not explicitly account for the effects of molecular weight or provide a connection to the underlying phenomena. To accomplish this, we instead adapted a general form of the equation for nucleation derived by Turnbull and Fisher [73], whereby the energy barrier can be decomposed into a thermodynamic part for the formation of a critical nucleus and a diffusive or transport part for activated transport to the phase boundary. The thermodynamic term has been modeled from general nucleation theory and depends on the properties of the monomer only. This result has been theoretically derived for a general description of crystal growth in polymers by Binsbergen [72], and has been applied to specific models of a surface nucleus by Hoffman and Weeks [8] and Mandelkern [21]. For the transport term, however, it was difficult to find a relation that applies over a large temperature range for polymers. Because of the speed of alkane crystal growth compared to experimental techniques, this term has not been parameterized for alkanes at all. Unlike existing data for high molecular weight polymers [71], we observe a strong molecular weight dependence for crystal growth in alkanes. This molecular weight dependence provides insight into the nature of the conformational barrier not only for alkanes, but also for polymers crystallizing at large undercooling, which is usually modeled by an Arrhenius-type free energy barrier [74]. From our molecular dynamics data for alkanes of different molecular weights, the molecular weight dependence was found to be well-described by that of the Rouse relaxation time. Full details of the model can be found elsewhere [35]. In its final form, the model for crystal growth rate of n-alkanes, as a function of temperature T and chain length N, is:

$$G(T,N) = G_0\left(\frac{N_0}{N}\right)^{2n} \exp\left[\frac{2.303\,n\,c_1\,(T-T_{\mathrm{g}}(N))}{c_2 + (T-T_{\mathrm{g}}(N))}\right] \exp\left[-\frac{C}{T}\frac{T_{\mathrm{m}}(N)}{(T_{\mathrm{m}}(N)-T)}\right] \tag{22.1}$$

where G_0 is a rate pre-factor for a reference chain of length N_0 at its glass transition point. T_{m} is the equilibrium melting temperature, and C is a characteristic thermodynamic constant for the polymer, quantifying the ratio of the surface energy of a nucleus to the lattice energy gained by crystallization. The first three factors of (22.1) capture the transport barrier for crystallization. The molecular weight dependence is determined from the ratio of Rouse relaxation times, raised to the power of n. The temperature dependence arises from the monomeric friction coefficient, which can be described by the Williams-Landel-Ferry (WLF) relation, modified by n, using the WLF constants c_1 and c_2 [75]. The glass transition temperature is molecular weight dependent, which is described by the equation of Fox and Loshaek [76], for low molecular weights:

$$\frac{1}{T_g(N)} = \frac{1}{T_g^\infty} + \frac{C_g}{(T_g^\infty)^2} \frac{1}{N} \tag{22.2}$$

where T_g^∞ is the asymptotic value of T_g at infinite molecular weight, and C_g is a constant. Estimates for the glass transition temperature of alkanes, based on kinematic viscosities, suggest that this equation fits well [77]. An analogous molecular weight dependence of the melting temperature which fits well to tabulated experimental data is given by

$$\frac{1}{T_m(N)} = \frac{1}{T_m^\infty} + \frac{C_m}{(T_m^\infty)^2} \frac{1}{N} \tag{22.3}$$

where T_m^∞ is the asymptotic value of T_m at infinite molecular weight and C_m is a constant.

In order to parameterize the crystal growth rate model, 7 parameters need to be determined. There are 3 parameters in (22.1), related to the kinetics, G_0, C, and n. There are also 4 parameters in (3.2, 3.3) that capture the molecular weight dependence of the glass transition and melting temperatures: T_g^∞, C_g, T_m^∞ and C_m; these may often be estimated independently from experimental data or other simulations, but this was not done here. We use C20 as our reference, which defines G_0 as the value for the transport-limited rate prefactor for C20 at its T_g, which is a very small value. A seven parameter fit of the complete alkane data to (22.1–22.3) is conducted using a weighted Levenberg-Marquardt nonlinear least squares algorithm. The resulting parameters are given in Table 22.1 ("Simulation-based" parameters), and the best fit curves are plotted in Fig. 22.8a. Since G_0 changes orders of magnitude during the fitting process, it is fitted as an exponential. In the vicinity of the maximum crystal growth rate, the fitted curve describes the data best.

Although our processing model for crystal growth rates seems to duplicate the behavior adequately, it would be more correct to calculate the phase transition parameters from a series of equilibrium Monte Carlo simulations that could yield the molecular weight dependence of the phase transitions; both melting points [78] and glass transition points [79] have been estimated using molecular dynamics simulations. (However, a great number of simulations would be required to capture the molecular weight dependence effectively.) The model currently overpredicts the T_g and T_m of the n-alkanes. We attribute this mostly to the extrapolation required to determine T_g and T_m as the temperatures at which the growth rate reaches zero. We have discussed elsewhere

Table 22.1. Calculated parameters for the crystal growth model. "Simulation-based" fit uses our simulation data for n-alkanes and the model given by Eqs. (22.1–22.3). "Simulation / Experiment-based" fit additionally models polyethylene data of Wagner et al. [11] and Ratajski et al. [10] using the modified version of Eq. (22.1) for entanglements

Parameter	Simulation-based	Simulation/Experiment-based
$\ln(G_0\ [\text{m/s}])$	-37.9	-56.9
$C\ [\text{K}]$	439	341
n	1.49	2.04
$T_\text{m}^\infty\ [\text{K}]$	500	496
$T_\text{g}^\infty\ [\text{K}]$	346	304
$C_\text{m}\ [\text{CH}_2 \times \text{K}]$	4.78×10^3	5.08×10^3
$C_\text{g}\ [\text{CH}_2 \times \text{K}]$	1.24×10^4	1.17×10^4

the subtle effects of the forcefield in predicting melting temperature [34]. The discrepancy in the simulated melting and glass transition temperatures from experimental data is most likely a consequence of the fact that the force fields do not recreate all aspects of the chain behavior for both phases, for example the hexagonal symmetry of the simulated alkane crystals.

We attempted alternative equations to model the simulation data as well. From models of diffusion-limited reactions, the parameterization of rates can be done from self-diffusion coefficients. We attempted this type of modeling but found the molecular weight dependence could not be captured by the diffusion coefficient. Perhaps this is because of the unique nature of polymer crystallization, where the barrier to crystallization is not diffusion, but rather the time required to yield extension of chains and locking into crystallographic registry. In addition, other forms of the growth rate equation were also considered, such as those discussed by Hoffman [71,74], van Krevelen [24], Strobl [80] and Umemoto [81], but these were generally insufficient to capture both the temperature dependence and molecular weight dependence of the growth rate.

Recent high speed crystallization experiments have measured crystal growth rates for polyethylene over a wide range of temperatures, from the melting temperature to near the temperature where the maximum growth rate occurs [10, 11]. It is worth noting that these recent experiments indicate a maximum crystal growth rate for polyethylene on the of order $10^{-4}\,\text{m/s}$ which coincides with the value predicted by our alkane model for chains of length 150–200, similar to the entanglement length for polyethylene. Thus replacing the Rouse time for the transport barrier to crystal growth in our model for n-alkanes with the time for Rouse relaxation of segments between entanglements provides a remarkably good description of crystal growth kinetics for polyethylene at high undercooling, in the vicinity of the maximum growth rate. This is illustrated in Fig. 22.8, where (3.1) has been applied.

with a few modifications, to model the polyethylene data of Ratajski and Janeschitz-Kriegl [10], and of Wagner and Phillips [11] in addition to our simulation data for n-alkanes. For $N > N_e$, the entanglement length of polyethylene, we replace N_0/N by N_0/N_e in the second factor on the right hand side of (3.1). To describe both our simulated alkane results and the experimental PE data, the equation parameters were then re-determined, this time using the experimentally determined values of $T_g = 190$ K and $T_m = 416$ K for polyethylene, reported by Wagner and Phillips [11]. A common value of the thermodynamic constant C for both alkanes and polymers was assumed, which implies a single common mechanism for nucleation for highly undercooled n-alkanes and polyethylene. The details of the fitting procedure have been reported elsewhere; the best fit parameters are shown in Table 22.1 ("Simulation/Experiment-based" parameters). Figure 22.8a shows the fit to the alkane simulation data. Figure 22.8b shows the fit to the experimental PE data. Reasonable fits are obtained to all 5 data sets, lending support to this simple explanation based on segmental relaxation times. When determining the relevant relaxation time for crystal growth in entangled systems, we discount reptation as the rate limiting step in polymer crystallization at high undercooling on the grounds that it would imply a much stronger molecular weight dependence than is experimentally observed [82]. In our model, the only temperature dependence that persists for $N > N_e$ is that due to the molecular weight dependence of the transition temperatures, T_g and T_m. Furthermore, our previously mentioned observations of surface nuclei, consisting of approximately 4-5 segments, each 20 carbons long, is consistent with a local relaxation mechanism such as that described by the Rouse mode for an entangled segment. Over 20 years ago, Hoffman suggested that the lack of molecular weight dependence of the transport factor in highly undercooled systems could be explained if the transport barrier referred to "the retardation associated with slack portions of the pendant chains" [82]. Our simulation observations and the parametric growth rate equation derived therefrom are consistent with this suggestion, and identify the relaxation of a segment on the order of the length between entanglements as the origin of the retardation associated with crystallization of high molecular weight polymers in highly undercooled systems. Only near the melting temperature, where crystallization is largely nucleation-limited, does reptation offer a credible mechanism for chain transport to the growing crystal.

22.4 Conclusions

We have presented a non-equilibrium molecular dynamics (MD) framework for studying crystallization of polymer melts. By using cleverly constructed simulations, we have independently observed the two phenomena responsible for melt crystallization: nucleation and growth.

Nucleation can be observed on the molecular dynamics time scale using realistic potentials for flexible chains if the initial amorphous orientation is high enough. For crystallization temperatures ranging from 325 to 375 K, these simulations clearly showed the hallmarks of crystal nucleation. We can identify multiple nucleation events, lamellar growth up to the limit imposed by periodic boundaries of the simulation cell, and lamellar thickening. We observed a competition between the rate of nucleation, which results in multiple crystallites, the rate of chain extension, which results in thicker lamellae, and the rate of chain conformational relaxation, which is manifested in lower degrees of residual order in the noncrystalline portion of the simulation. The temperature dependence of lamellar thickness was found to accord with experimental data. At the higher temperatures, tilted chain lamellae were observed to form with lamellar interfaces corresponding approximately to the (201) facet, indicative of the influence of interfacial energy. The different crystallization morphologies observed are analogous to those reported by Mahindrasingam et al. [69], and may be interpreted as consistent with their three zones of growth, after some reinterpretation to equate high process rate with high initial orientation.

We have also observed the characteristics of growth of n-alkane crystals over a range of temperatures and molecular weights. Qualitatively, we see frequent adsorption and desorption of chain segments on the surface for all systems. For C50 and C100, we find evidence for a surface nucleus involving 4-5 chain segments, from multiple chains, that are approximately 20 beads long, shorter than the ultimate thickness of the chain stem in the crystal. We have constructed a general crystal growth model that can be parameterized entirely in terms of universal properties of polymer chains, described by polymer physics and chemically specific quantities that can be estimated polymer by polymer using molecular dynamics simulations. It accounts for the thermodynamic driving force, using classical nucleation theory, and melt relaxation time, using WLF theory. The appropriate relaxation time should used to describe the tranport barrier to crystallization; for chains shorter than the entanglement length, this is the Rouse time. Past the entanglement molecular weight, the analysis reveals that the growth rate of alkanes and polyethylene can both be described by the same relationship. For chains longer than the entanglement molecular weight, transport limitations are modeled by the local relaxation of an entangled segment at the interface.

Acknowledgment

This work was supported by the Center for Advanced Engineering Fibers and Films (CAEFF) of the Engineering Research Centers Program of the National Science Foundation, under NSF Award Number EEC-9731680. This material is also based upon work supported by the National Science Foundation under Grant No. 0079734.

References

[1] M. Avrami: J. Chem. Phys. **8**, 212 (1940)
[2] A. Ziabicki: *Fundamentals of Fibre Formation* (John Wiley, London, 1976)
[3] K. Nakamura, K. Katayama, T. Amano: J. Appl. Polym. Sci. **17**, 1031 (1973)
[4] P.G. Debenedetti: *Metastable Liquids* (Princeton University Press, Princeton, 1996)
[5] R. Qian, J. Shen, L. Zhu: Makromol. Chemie-Rapid Commun. **2**, 499 (1981)
[6] R.H. Somani, L. Yang, I. Sics, B.S. Hsiao, N.V. Pogodina, H.H. Winter, P. Agarwal, H. Fruitwala, A. Tsou: Macromol. Symp. **185**, 105 (2002)
[7] J.A. Kornfield, G. Kumaraswamy, A.M. Issaian: Ind. Eng. Chem. Res. **41**, 6383 (2002)
[8] J.D. Hoffman, J.J. Weeks: J. Chem. Phys. **37**, 1723 (1962)
[9] L. Mandelkern, N.L. Jain, H. Kim: J. Polym. Sci. Polym. Phys. **6**, 165 (1968)
[10] E. Ratajski, H. Janeschitz-Kriegl: Colloid Polym. Sci. **274**, 938 (1996)
[11] J. Wagner, P.J. Phillips: Polymer **42**, 8999 (2001)
[12] P.D. Olmsted, W.C.K. Poon, T.C.B. McLeish, N.J. Terrill, A.J. Ryan: Phys. Rev. Lett. **81**, 373 (1998)
[13] G. Kanig: Colloid Polym. Sci. **269**, 1118 (1991)
[14] K. Tashiro, S. Sasaki, N. Gose, M. Kobayashi: Polymer J. **30**, 485 (1998)
[15] A. Keller, M. Hikosaka, S. Rastogi, A. Toda, P.J. Barham, G. Goldbeck-Wood: J. Mater. Sci. **29**, 2579 (1994)
[16] G. Strobl: Europ. Phys. J. E **18**, 295 (2005)
[17] B. Lotz: Europ. Phys. J. E **3**, 185 (2000)
[18] M. Muthukumar: Europ. Phys. J. E **3**, 199 (2000)
[19] G. Allegra, S.V. Meille: Phys. Chem. Chem. Phys. **1**, 5179 (1999)
[20] K. Armistead, G. Goldbeck-Wood, A. Keller: Adv. Polym. Sci. **100**, 221 (1992)
[21] L. Mandelkern: *Crystallization of Polymers* (McGraw-Hill, New York, 1964)
[22] D.M. Sadler, G.H. Gilmer: Phys. Rev. Lett. **56**, 2708 (1986)
[23] J.J. Point: Macromolecules **12**, 770 (1979)
[24] D.W. van Krevelen: Chimia **32**, 279 (1978)
[25] A.K. Doufas, A.J. McHugh, C. Miller: J. Non-Newtonian Fluid Mech. **92**, 27 (2000)
[26] S. Fujiwara, T. Sato: J. Chem. Phys. **110**, 9757 (1999)
[27] H. Meyer, F. Müller-Plathe: Macromolecules **35**, 1241 (2002)
[28] A. Koyama, T. Yamamoto, K. Fukao, Y. Miyamoto: Phys. Rev. E **65**, 050801 (2002).
[29] M.S. Lavine, N. Waheed, G.C. Rutledge: Polymer **44**, 1771 (2003)
[30] M.J. Ko, N. Waheed, M.S. Lavine, G.C. Rutledge: J. Chem. Phys. **121**, 2823 (2004)
[31] W.B. Hu, D. Frenkel, V.B.F. Mathot: Macromolecules **36**, 549 (2003)
[32] T. Shimizu, T. Yamamoto: J. Chem. Phys. **113**, 3351 (2000)
[33] T. Yamamoto: Polymer **45**, 1357 (2004)
[34] N. Waheed, M.S. Lavine, G.C. Rutledge: J. Chem. Phys. **116**, 2301 (2002)
[35] N. Waheed, M.J. Ko, G.C. Rutledge: Polymer **46**, 8689 (2005)
[36] T.A. Kavassalis, P.R. Sundararajan: Macromolecules **26**, 4144 (1993)
[37] P.R. Sundararajan, T.A. Kavassalis: J. Chem. Soc. Faraday Trans. **91**, 2541 (1995)
[38] C. Liu, M. Muthukumar: J. Chem. Phys. **109**, 2536 (1998)

[39] M. Muthukumar: Philos. Trans. Royal Soc. London Series A – Math. Phys Eng. Sci. **361**, 539 (2003)
[40] K. Esselink, P.A.J. Hilbers, B.W.H. van Beest: J. Chem. Phys. **101**, 9033 (1994)
[41] H. Meyer, F. Müller-Plathe: J. Chem. Phys. **115**, 7807 (2001)
[42] A. Koyama, T. Yamamoto, K. Fukao, Y. Miyamoto: J. Chem. Phys. **115**, 560 (2001)
[43] P.G. Higgs, G. Ungar: J. Chem. Phys. **100**, 640 (1994)
[44] C.M. Chen, P.G. Higgs: J. Chem. Phys. **108**, 4305 (1998)
[45] G. Goldbeck-Wood: Polymer **31**, 586 (1990)
[46] J.P.K. Doye, D. Frenkel: J. Chem. Phys. **110**, 2692 (1999)
[47] T. Yamamoto: J. Chem. Phys. **109**, 4638 (1998)
[48] H.X. Guo, X.Z. Yang, T. Li: Phys. Rev. E **61**, 4185 (2000)
[49] W. Schneider, A. Köppl, J. Berger: Int. Polym. Processing **2**, 151 (1988)
[50] M. Hütter, G.C. Rutledge, R.C. Armstrong: Phys. Fluids **17**, 014107 (2005)
[51] N. Waheed, G.C. Rutledge: J. Polym. Sci. B Polym. Phys. **43**, 2468 (2005)
[52] W. Paul, D.Y. Yoon, G.D. Smith: J. Chem. Phys. **103**, 1702 (1995)
[53] J.C. Horton, G.L. Squires, A.T. Boothroyd, L.J. Fetters, A.R. Rennie, C.J. Glinka, R.A. Robinson: Macromolecules **22**, 681 (1989)
[54] P.J. In't Veld, G.C. Rutledge: Macromolecules **36**, 7358 (2003)
[55] D.J. Blundell, D.H. MacKerron, W. Fuller, A. Mahendrasingam, C. Martin, R.J. Oldman, R.J. Rule, C. Riekel: Polymer **37**, 3303 (1996)
[56] R.Y. Qian: J. Macromol. Sci. Phys. **B40**, 1131 (2001)
[57] H.J.C. Berendsen, J.P.M. Postma, W.F. Van Gunsteren, A. Dinola, J.R. Haak: J. Chem. Phys. **81**, 3684 (1984)
[58] G.M. Stack, L. Mandelkern, I.G. Voigt-Martin: Macromolecules **17**, 321 (1984)
[59] G. Ungar, J. Stejny, A. Keller, I. Bidd, M.C. Whiting: Science **229**(4711), 386 (1985)
[60] S. Fujiwara, T. Sato: J. Chem. Phys. **107**, 613 (1997)
[61] A. Koyama, T. Yamamoto, K. Fukao, Y. Miyamoto: J. Macromol. Sci. Phys. **B42**, 821 (2003)
[62] M.I.A. el Maaty, D.C. Bassett: Polymer **42**, 4957 (2001)
[63] S. Gautam, S. Balijepalli, G.C. Rutledge: Macromolecules **33**, 9136 (2000)
[64] P.J. Barham, R.A. Chivers, A. Keller, J. Martinez-Salazar, S.J. Organ: J. Mater. Sci. **20**, 1625 (1985)
[65] S. Hocquet, M. Dosiere, Y. Tanzawa, M.H.J. Koch: Macromolecules **35**, 5025 (2002)
[66] J.D. Hoffman, R.L. Miller: Polymer **38**, 3151 (1997)
[67] H. Takeuchi: J. Chem. Phys. **109**, 5614 (1998)
[68] G. Williams, D.C. Watts: Trans. Faraday Soc. **66**, 80 (1970)
[69] A. Mahendrasingam, D.J. Blundell, C. Martin, W. Fuller, D.H. MacKerron, J.L. Harvie, R.J. Oldman, C. Riekel: Polymer **41**, 7803 (2000)
[70] D.M. Small: *The Physical Chemistry of Lipids: From Alkanes to Phospholipids* (Plenum Press, New York, 1986)
[71] J.I. Lauritzen, J.D. Hoffman: J. Appl. Phys. **44**, 4340 (1973)
[72] F.L. Binsbergen: Kolloid Zeitschrift **237**, 389 (1970)
[73] D. Turnbull, J.C. Fisher: J. Chem. Phys. **17**, 71 (1949)
[74] J.P. Armistead, J.D. Hoffman: Macromolecules **35**, 3895 (2002)
[75] M.L. Williams, R.F. Landel, J.D. Ferry: J. Amer. Chem. Soc. **77**, 3701 (1955)
[76] T.G. Fox, S. Loshaek: J. Polym. Sci. **15**, 371 (1955)

[77] A.A. Miller: J. Polym. Sci. Polym. Phys. **6**, 249 (1968)
[78] Y. Tsuchiya, H. Hasegawa, T. Iwatsubo: J. Chem. Phys. **114**, 2484 (2001)
[79] R.H. Boyd, R.H. Gee, J. Han, Y. Jin: J. Chem. Phys. **101**, 788 (1994)
[80] G. Strobl: Europ. Phys. J. E **3**, 165 (2000)
[81] S. Umemoto, N. Kobayashi, N. Okui: J. Macromol. Sci. Phys. **B41**, 923 (2002)
[82] J.D. Hoffman: Polymer **23**, 656 (1982)

A Multiphase Model Describing Polymer Crystallization and Melting

Gert Strobl

Physikalisches Institut, Albert-Ludwigs-Universität Freiburg, 79104 Freiburg, Germany
strobl@uni-freiburg.de

Abstract. The results of temperature dependent small angle X-ray scattering experiments on a variety of crystallizing polymers contradict conventional wisdom and suggest that polymer crystallization generally uses a route which includes a passage via a mesomorphic phase. We construct a thermodynamic scheme dealing with the transitions between melt, mesomorphic layers and lamellar crystallites, assuming for the latter ones that they exist both in an initial 'native' and a final 'stabilized' form. Application of the scheme in a quantitative evaluation of small angle X-ray scattering and calorimetric results yields the equilibrium transition temperatures between the various phases, latent heats of transition and surface free energies. As an example, the data obtained for s-polypropylene are given. Here, the mesomorphic phase has thermodynamic properties which place this state intermediate between melt and crystals.

23.1 Introduction

When the fundamentals of the structure of semi-crystalline polymers – with stacks of layer-like crystallites with thicknesses in the nm-range being embedded in an amorphous matrix – were revealed in the Fifties, considerations about the mechanism of the formation of these structures started immediately. In the Sixties and Seventies, they became a major field of research and a focus of interest, discussed as a central topic in all structure oriented polymer conferences (see, for example, [1] with the lectures at the Faraday Discussion in Cambridge 1979). In the years which followed, one approach gained superiority – the one put forward by Hoffman, Lauritzen and their co-workers [2]. It was accepted and used in data evaluations by more and more workers, due to some appealing features:

- The picture envisaged by the treatment – a crystalline lamella with an ordered fold surface and smooth lateral faces, growing layer by layer with a secondary nucleation as rate determining step – is clear and easy to grasp.

- The theory developed by Hoffman and Lauritzen yields a simple equation for the growth rate.
- Growth rates can easily be measured, either in an optical microscope or globally with various techniques which probe the temporal development of the crystallinity.

The impression of many in the community that the mechanism of polymer crystallization is principally understood and the issue essentially settled was, however, wrong. With the Nineties a renewed thinking set in, triggered by new experimental observations. In fact, the experimental basis of the Hoffman-Lauritzen theory had always been rather narrow. Putting the focus on growth rates only, the basis of validation were growth rate measurements exclusively. The Hoffman-Lauritzen treatment includes several implications. In particular, it assumes that

- lamellae grow by a direct attachment of chain sequences from the melt onto essentially smooth lateral faces,
- the lamellar thickness is determined by the supercooling below the equilibrium melting point, being given by the Gibbs-Thompson equation, apart from a minor correction which is necessary to provide a thermodynamic driving force.

These assumptions looked quite natural, and nobody would have questioned them without very good reasons. Such reasons, however, now came up:

- Keller and his co-workers, when crystallizing polyethylene at elevated pressures, observed the formation of crystals out of the hexagonal phase and speculated that this may also happen under normal pressure conditions [3,4].
- Kaji and co-workers interpreted scattering which arose before the appearance of the crystallites as indicating the buildup of a precursor phase in the first step of polymer crystallization [5], and Olmsted constructed a corresponding theory [6].
- Time- and temperature-dependent small angle X-ray scattering (SAXS) experiments, at first carried out by us for syndiotactic polypropylene and related copolymers, contradicted the basic assumption of a control of the lamellar thickness by the supercooling below the equilibrium melting point [7]. As it turned out, lamellar thicknesses are determined by the supercooling below another temperature which is always located above the equilibrium melting point. In addition, the thicknesses are not affected by the presence of co-units.

With these new observations the fundamental question about the mechanism of polymer crystallization was reopened.

This article describes our own approach in the search for a new understanding. It begins in the next section with the reproduction of some selected experimental results which are typical. They established the basis of our considerations and led us to a model which assumes for the formation of polymer crystallites a passage through a transient mesomorphic phase [8]. The model was first introduced in purely qualitative manner. Then, in a next step, a thermodynamic multiphase scheme was set up [9]. It is explained in Sect. 22.3. On the basis of this scheme the results of SAXS experiments – crystal thicknesses as a function of the crystallization temperature, their variation during heating and their values at melting points – can be evaluated. This is exemplified in section 4 with results for s-polypropylene and related octene-copolymers.

23.2 Experimental Findings

Considerations about mechanisms of crystallization and melting in polymers require as some basic ingredients

- a knowledge of the variation of the crystal thickness, d_c, with the crystallization temperature, T_c,
- a monitoring of possible structure changes during a heating to the melting point, and
- a knowledge of the variation of the final melting temperature, T_f, with the final crystal thickness.

Time and temperature dependent small angle X-ray scattering (SAXS) experiments made it possible to determine these properties. They were carried out for several polymer systems, including syndiotactic polypropylene (sPP) with copolymers [7], poly(ethylene-co-octene)s (PEcO) [10, 11], isotactic polypropylene [12], isotactic polystyrene (iPS) [13], poly(1-butene) [14] and poly(ϵ-caprolactone) (PϵCL) [10].

23.2.1 Crystallization Line and Melting Line

Figures 23.1, 23.2, 23.3 present as three selected examples the results obtained for an octene copolymer of sPP (sPPcO15 with 15% per weight of octene units), an octene copolymer of polyethylene (PEcO14 with 14% per weight of octene units) and PϵCL. The Gibbs-Thompson equation describes the melting point T_f of a crystallite with thickness d_c (heat of fusion: Δh_f, surface free energy: σ_{ac}) as

$$T_f(d_c) = T_f^\infty - T_f^\infty \frac{2\sigma_{ac}}{\Delta h_f} \frac{1}{d_c}. \tag{23.1}$$

The equation suggests plotting the melting points as a function of the inverse crystal thickness d_c^{-1}, and the same representation is used here also for the relation between T_c and d_c. The appearance of the plots is typical for all investigated samples: Two straight lines are found which cross each other. The 'melting line', giving the dependence between T_f and d_c^{-1}, confirms the Gibbs-Thompson equation. This allows a determination of the equilibrium melting point T_f^∞ by a linear extrapolation to $d_c^{-1} = 0$. The 'crystallization line' gives

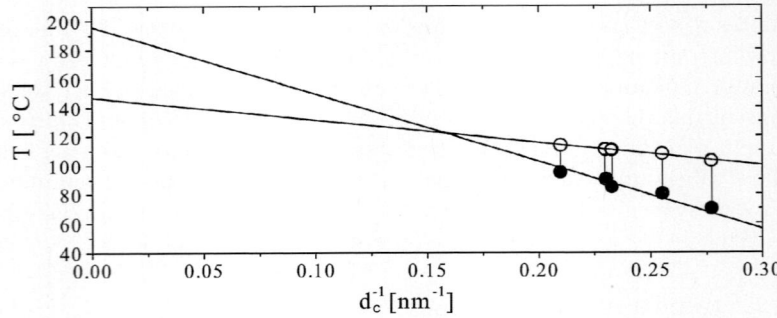

Fig. 23.1. sPPcO15: Crystallization line T_c versus d_c^{-1} (*filled symbols*) and melting line T_f versus d_c^{-1} (*open symbols*) [7]

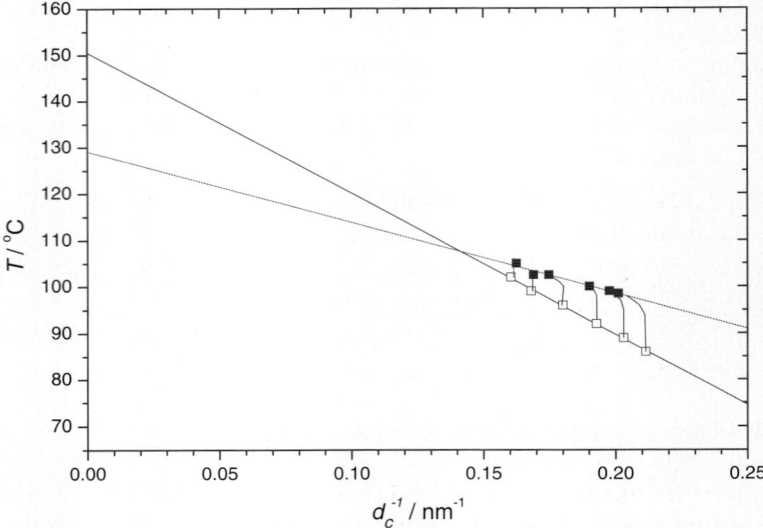

Fig. 23.2. PEcO14: Crystallization line and melting line. The connecting lines show the temperature dependence of d_c^{-1} during the heating [10]

the relationship between T_c and d_c^{-1}. It has a higher slope than the melting line, intersects the latter at a finite value of d_c^{-1} and has a limiting temperature for $d_c^{-1} \to 0$, denoted T_c^∞, which differs from T_f^∞. The crossing implies that there is $T_c^\infty > T_f^\infty$. The results of the temperature dependent measurements during heating are given by the thin lines which connect respective points on the crystallization and the melting line. The lines are vertical when the thickness remains constant and are curved when the thickness increases during heating.

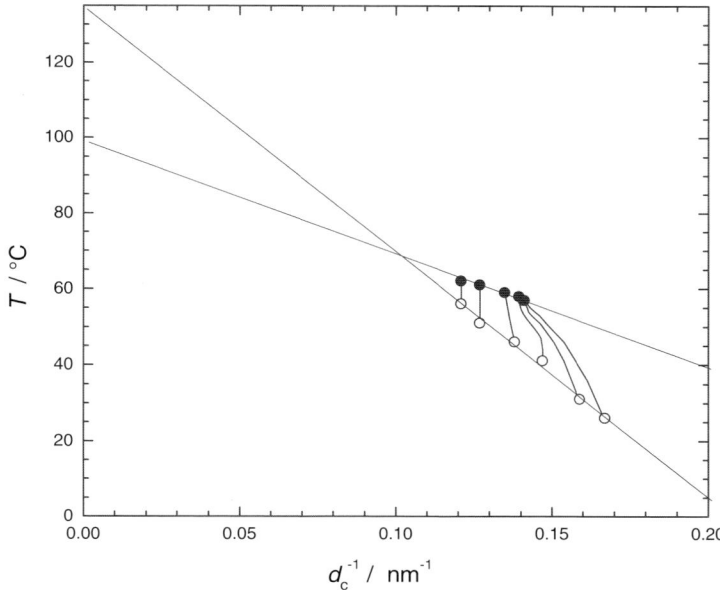

Fig. 23.3. PϵCL: Crystallization line and melting line [10]

The existence of straight crystallization lines in all investigated systems expresses a simple law: Crystal thicknesses are inversely proportional to the distance from a certain characteristic temperature, T_c^∞, different from the equilibrium melting point T_f^∞. In the examples given here, T_c^∞ is 20 to 50°C above T_f^∞.

Lamellar crystallites principally exist at temperatures below the melting line only. Therefore, crystals with thicknesses as given by the crystallization line cannot be formed any longer when the temperature of the intersection point is approached. This is indeed experimentally confirmed [15]. SAXS results in the interesting temperature range were obtained for sPPcO20 and they are shown in Fig. 23.4. Points deviate from the crystallization line already before reaching the point of intersection. In the experiment crystallization was conducted using the self-seeding procedure, i.e., the sample was just shortly heated above its melting point and then crystallized again. Memory effects then reduce the crystallization time.

23.2.2 Effects of Counits and Diluents

The presence of co-units or stereo defects in a chain, which cannot be included in the crystal lattice, and of low molar mass diluents in the melt modifies the crystallization and melting properties. Time- and temperature dependen

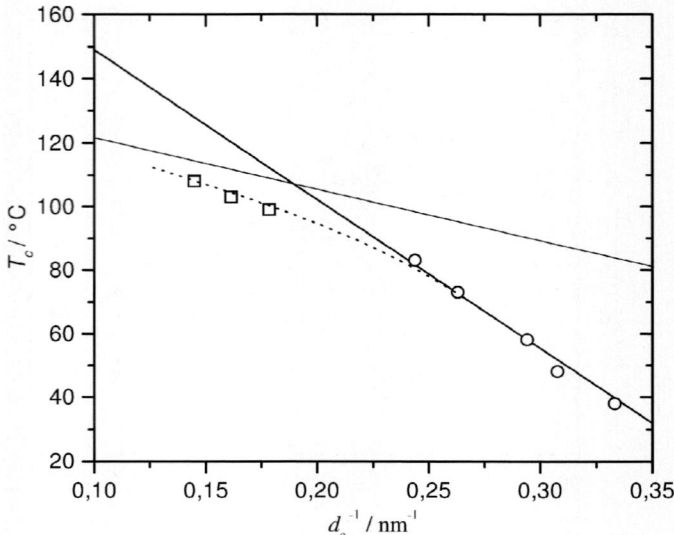

Fig. 23.4. sPPcO20: Relationship between crystallization temperature and crystal thickness in the range around the point of intersection between the melting- and the crystallization line. Isothermal crystallizations were carried out with the aid of the self seeding technique [8, 15]

SAXS and wide angle X-ray scattering (WAXS) studies were carried out to see these effects.

The findings for sPP and a variety of related octene-copolymers [7] are depicted in Fig. 23.5. Contrasting the normal behavior of the melting lines, which shift to lower temperatures when the co-unit content increases, the crystallization line is invariant within this set of samples. One observes a unique T_c vs d_c^{-1} relationship common to all of them, which determines d_c as being inversely proportional to the supercooling below $T_c^\infty = 195°C$.

In a recent experiment the same invariance was also found for the lateral size of the blocks in the lamellar crystals [16]. Crystal block diameters can generally be derived from the linewidth of Bragg reflections in WAXS patterns, by application of the Scherrer equation

$$D_{hkl} = \frac{1}{\Delta s_{hkl}} \quad \text{with} \quad s = \frac{2\sin\theta}{\lambda}, \qquad (23.2)$$

where Δs_{hkl} denotes the integral linewidth of the hkl-reflection (θ: Bragg angle, λ: X-ray wave length). Figure 23.6 presents the diameter thus obtained for a set of different sPPs, with a sample of high syndiotacticity (sPP), a commercial sample with lower tacticity (Fina sPP), and two octene copolymers (sPPcOx). The given lengths were derived from the linewidth of the 200 reflection, i.e., they refer to the direction perpendicular to the 200 lattice planes.

23 A Multiphase Model Describing Polymer Crystallization and Melting

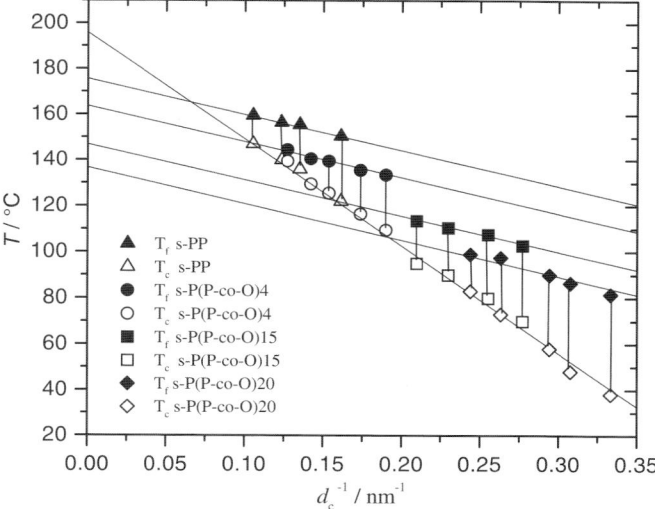

Fig. 23.5. sPP and sPPcOx (x:% per weight): Unique crystallization line (*open symbols*) and series of melting lines (*filled symbols*) [7]

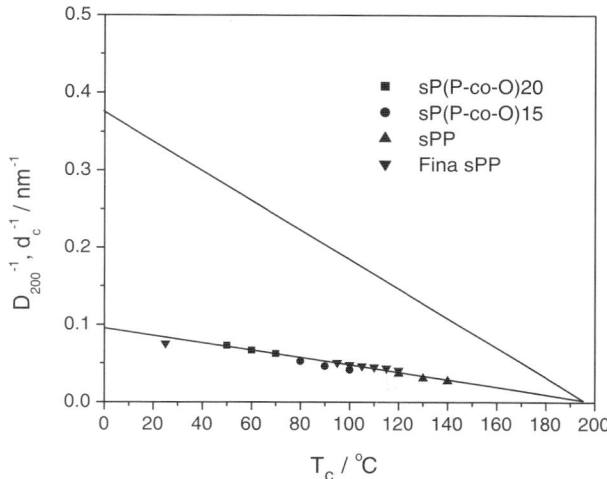

Fig. 23.6. Samples of sPP and sPPcOx crystallized at various T_cs: Lateral block lengths D_{200} derived from the linewidth of the 200-reflection (*filled symbols*) and crystallization line from Fig. 23.5 [16]

As can be seen, all points $D_{200}^{-1}(T_c)$ are located on one common line. When continued, this line ends again at $T_c^\infty = 195°C$, like the crystallization line of sPP which is also included in the figure. Analogous results were obtained for PE and copolymers with butene- and octene units [16].

In order to see the effect of diluents on the crystallization and melting behavior, SAXS experiments were carried out for mixtures of PEcO14 with two diluents, namely n-hexadecane (n-$C_{16}H_{34}$) and methylanthracene ($C_{15}H_{12}$) [17]. Results are reproduced in Fig. 23.7, and they show the following:

- The presence of $C_{15}H_{12}$ leaves the crystallization line unaffected, opposite to $C_{16}H_{34}$, which results in shifts even larger than those of the melting lines.
- Both diluents equally suppress the melting points.
- The shift of the melting line increases linearly with the co-unit content x, in agreement with Raoult's law

$$T_f^\infty(x) = T_f^\infty - \frac{R(T_f^\infty)^2}{\Delta h_f} x \qquad (23.3)$$

Surveying the observations a most remarkable constancy of the crystal layer thickness at a given crystallization temperature is seen, the only exception being the effect arising from an addition of n-$C_{16}H_{34}$ to PE. We understand these findings as evidence for an interference of a third, mesomorphic phase along the crystallization pathway. If co-units or diluents are already rejected when the mesomorphic phase forms, they have indeed no influence on the crystal formation. On the other hand, if the diluent is included in the mesomorphic phase (n-$C_{16}H_{24}$), the crystallization line shifts to lower temperatures.

23.2.3 Recrystallization Processes

Heating an isothermally crystallized polymer is not always accompanied just by a melting of the crystallites according to their stability. In many cases the melting is immediately followed by the formation of a new crystal. These 'recrystallization processes' can have different characteristics, depending on the crystallization temperature.

Easy to interpret are the results obtained when samples are crystallized at high temperatures. Figure 23.8 presents as a typical example the behavior of an sPP with high tacticity (91% of syndiotactic pentades) during heating scans after a crystallization at 115°C [18]. The interface distribution functions derived from measured SAXS curves indicate a continuous slow decrease of the crystallinity without crystal thickness changes, and a final melting at 145°C. Subsequently new crystallites form, with a step-like increased thickness corresponding to the temperature of their new formation. On further heating, to 153°C, these melt again. The DSC scans presented on the right show this

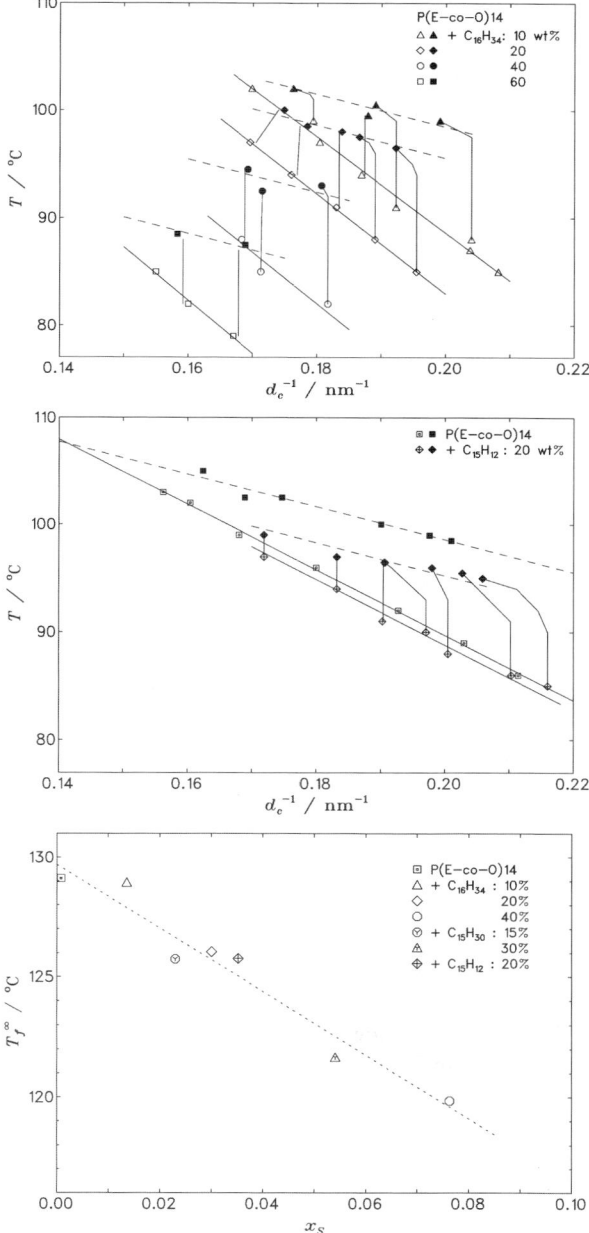

Fig. 23.7. (*Top*) Polymer-diluent mixtures PEcO14/$C_{16}H_{34}$ (90/10; 20/80; 40/60; 40/60): Crystallization lines and melting lines. (*Center*) Mixtures PEcO14/$C_{15}H_{12}$ (100/0; 80/20): Crystallization lines and melting lines. (*Bottom*) Equilibrium melting points T_f^∞ in dependence on the mole fraction of the diluent, determined by linear extrapolations of the respective melting lines [17]

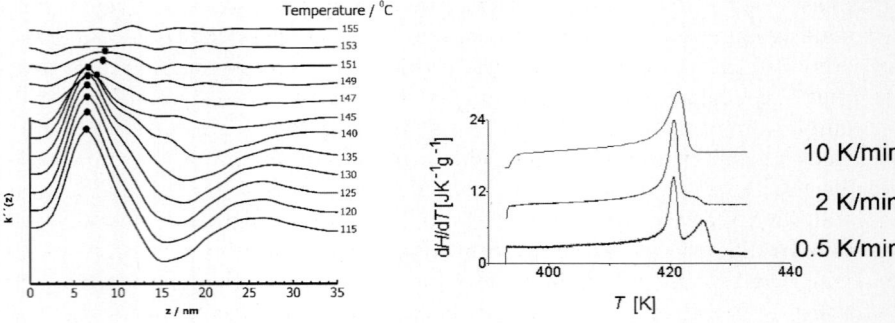

Fig. 23.8. sPP crystallized at 115°C: (*Left*) Variation of the interface distribution function during a heating to the melt. The peak location gives the crystal thickness [18]. (*Right*) DSC curves measured with different heating rates after the crystallization

melting-recrystallization-melting process also, and indicate that it occurs only if sufficient time is provided; for the higher heating rates this was not the case.

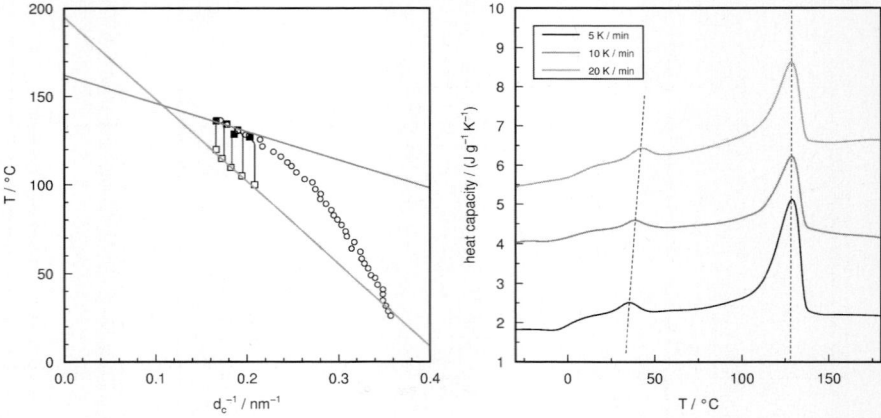

Fig. 23.9. sPP-Mitsui, quenched to the glassy state and then crystallized at 25°C and some temperatures in the range 100°C–120°C: (*Left*) Variation of the crystal thickness during subsequent heating processes. Crystallization line from Fig. 23.5 and melting line. (*Right*) DSC curves measured with three heating rates [9, 19]

Figure 23.9 presents as a second example SAXS and DSC results which were obtained for an sPP sample with lower tacticity [9, 19]. The vertical connecting lines between the points of crystallization (open circle) and melting (filled circles) at the highest T_cs indicate the same properties as in the first case – a melting possibly followed by a recrystallization – however, for a crystallization at 25°C, conducted coming from the glassy state, a completely different behavior is found. The crystal thickness increases immediately when the heating begins, and the reorganization processes steadily continue up to a complete melting at 130°C. The final melting shows up also in the DSC thermograms shown on the right. The ongoing reorganization gives no signal in the thermogram, which means, that it proceeds at a practically constant crystallinity. Only at the onset of the recrystallization processes several degrees above T_c a weak signal appears. Shifting to lower temperatures on decreasing the heating rate is indicative for the nature of this low temperature endotherm: It reflects a competition between crystal disaggregation and reformation processes. Properties intermediate between the two limiting cases following for $T_c = 15°C$ and $T_c = 25°C$ are found for $T_c = 100°C$. Here, crystals are at first stable, i.e., keep their thickness constant, but when a certain limiting temperature is reached reorganization processes set in. The final melting point is again 130°C, as for $T_c = 25°C$.

Figure 23.10 reproduces SAXS and DSC results which were obtained for iPS, and they show the same scenario [9, 20]. All samples crystallized at temperatures below 220°C experience a continuous recrystallization during heating, and melt at a constant temperature of 230°C. Then, for $T_c > 220°C$, the melting temperature begins to vary, shifting up to higher values with rising T_c.

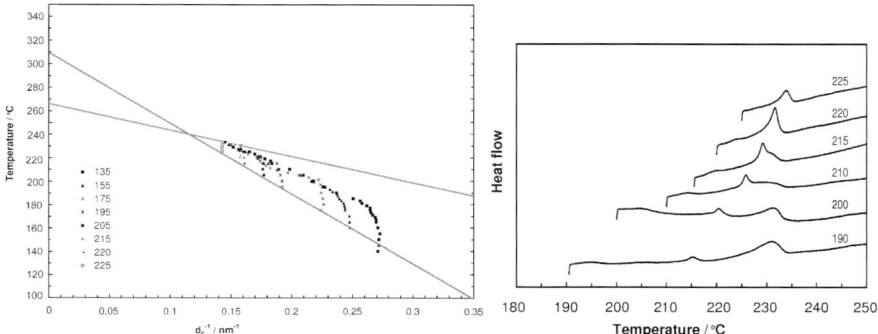

Fig. 23.10. iPS, quenched to the glassy state and then crystallized at various temperatures: (*Left*) Variation of d_c^{-1} during a subsequent heating, obtained by SAXS experiments. Crystallization line and melting line. (*Right*) DSC thermograms of samples, measured after isothermal crystallization processes with a heating rate of 0.5 K min^{-1} [9, 20]

The structural reorganization setting in immediately for low T_cs is an extremely rapid process, much faster than the recrystallization process of Fig. 23.8 where the initial crystallization was conducted at a high T_c. Whereas the latter no longer occurs when choosing a heating rate of 10 K min^{-1}, suppression of the first one requires heating rates which are 4 orders of magnitude higher. This was shown by Schick et al. [21] in studies of the melting of cold crystallized PET ($T_c = 130°C$) with the aid of a 'chip calorimeter' allowing for thin film heating rates up to 10^5 K min^{-1}.

Hence, two different scenarios for the structural reorganization during heating scans subsequent to isothermal crystallizations are found,

- a low T_c pathway associated with a continuous crystal thickening up to a fixed melting point, and
- a high T_c pathway with a constant crystal thickness and a melting point which rises together with T_c.

There are good reasons to invoke for the fast reorganization a passage through a mesomorphic phase, rather than the melt. They are presented in the next section.

23.3 A Multiphase Model of Polymer Crystallization and Melting

In the beginning of the 1990ies, Keller, Hikosaka, Kawabata, and Rastogi carried out crystallization experiments for polyethylene at elevated pressures using a polarizing optical microscope [3]. They observed a crystal formation via the hexagonal phase. Crystals nucleate into the hexagonal phase, then grow to sizes in the micrometer range before they transform into the orthorhombic phase after a statistically initiated, second nucleation step. Authors interpreted their observations as a new example for Ostwald's rule of stages. This rule, formulated about 100 years earlier, states that crystals always nucleate into that mesomorphic or crystalline structure which is the most stable one for nm-sized crystals. Due to differences in the surface free energy this state may differ from the crystal modification which is macroscopically stable.

Ostwald's rule of stages might also provide the clue for an understanding of polymer crystallization at normal pressures. The observed controlling temperature for d_c, which is T_c^∞ and not T_f^∞, indeed indicates the interference of a transient mesomorphic phase. Different from the statistically induced mesomorphic-crystalline transformation process observed for PE at elevated pressures, crystal thicknesses are now sharply selected. Figure 23.11 displays a sketch of a qualitative model which can explain the basic observations [8]. The model is meant to describe different stages which are passed through when a lamellar crystallite is growing. The process starts with an attachment of chain sequences from the melt onto a growth face of a mesomorphic layer with minimum thickness, which then spontaneously thickens. When a critical thickness is reached, the layer solidifies immediately under the formation of block-like crystallites. A next, equally important step in the crystal development is a stabilization of the crystallites in time, leading to a further decay in the Gibbs free energy. In the sketch this last step is addressed as a merging of the blocks, but this represents only one possibility.

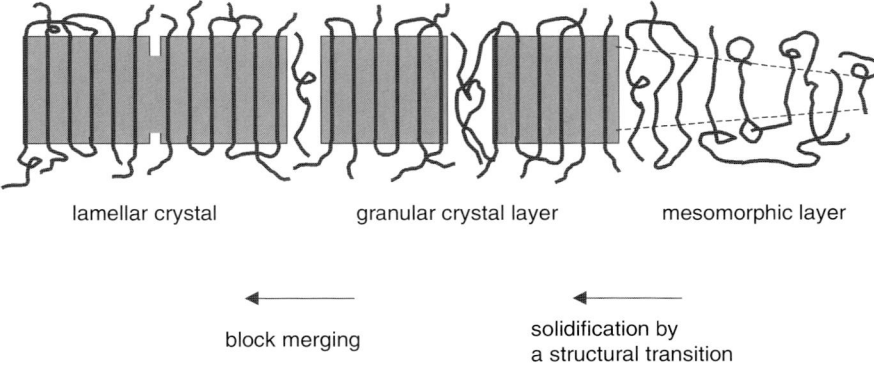

Fig. 23.11. Sketch of the pathway followed in the growth of polymer crystallites

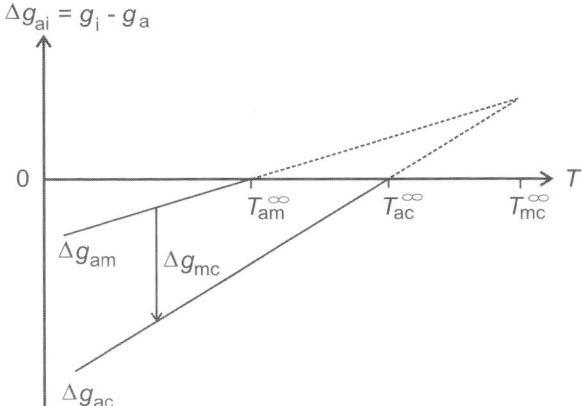

Fig. 23.12. Thermodynamic conditions assumed for crystallizing polymers: Temperature dependencies of the bulk chemical potentials of a mesomorphic and the crystalline phase. The potentials are referred to the chemical potential of the melt and denoted Δg_{am} and Δg_{ac} respectively

The basic thermodynamic conditions under which a mesomorphic phase can interfere and thus affect the crystallization process are described in the drawing of Fig. 23.12. The schematic plot shows for both the crystalline phase (c) and the mesomorphic phase (m) the difference of the bulk chemical potential to that of the melt (a):

$$\Delta g_{ac} = g_c - g_a \;,$$
$$\Delta g_{am} = g_m - g_a \;. \qquad (23.4)$$

Coming from high temperatures the chemical potential of the crystalline phase drops below the value of the melt when crossing the equilibrium melting point T_{ac}^∞. The mesomorphic phase requires a lower temperature to fall with its chemical potential below that of the melt, here at T_{am}^∞. The plot includes also a temperature T_{mc}^∞. It represents the temperature of a virtual transition, namely that between the mesomorphic and the crystalline phase. The transition temperatures have the order $T_{mc}^\infty > T_{ac}^\infty > T_{am}^\infty$. Since the bulk chemical potential of the crystal is always below that of the mesomorphic phase, the mesomorphic phase is only metastable for macroscopic systems. However, for small objects, with sizes in the nm range, stabilities can be inverted. Due to a usually lower surface free energy, thin mesomorphic layers can have a lower Gibbs free energy than a crystallite with the same thickness. Then Ostwald's rule of stages applies.

23.3.1 Thermodynamic Scheme

The model can be associated with a thermodynamic scheme [9]. It includes four different phases:

- the amorphous melt
- mesomorphic layers (label 'm')

and, in order to account for the stabilization processes, two limiting forms of the crystallites, namely

- native crystals (labeled 'c_n') and
- stabilized crystals (with label 'c_s').

The scheme, being displayed in Fig. 23.13, delineates the stability ranges and transition lines for these phases. The variables in this phase diagram are the temperature and the crystal size, whereby the inverse crystal thickness serves as size parameter. The thickness is given by the number n of structure units in a stem, i.e., $n = d_c/\Delta z$ with Δz denoting the stem length increment per structure unit. The transition lines are denoted $T_{mc_n}, T_{ac_n}, T_{mc_s}, T_{ac_s}, T_{am}$, all to be understood as functions of n^{-1}.

Of particular importance are the 'crossing points' X_n and X_s. At X_n both mesomorphic layers and native-crystalline layers have the same Gibbs free energy as the melt, at X_s this equality holds for the stabilized crystallites. The positions of X_n and X_s control what happens during an isothermal crystallization followed by heating. There are two different scenarios, exemplified by the pathways A and B in the figure; in experiments they are realized by crystallizations at low or high temperatures respectively. Pathway B: At the point of entry, labeled '1', chains are attached from the melt onto the lateral growth face of a mesomorphic layer with the minimum thickness. The layer

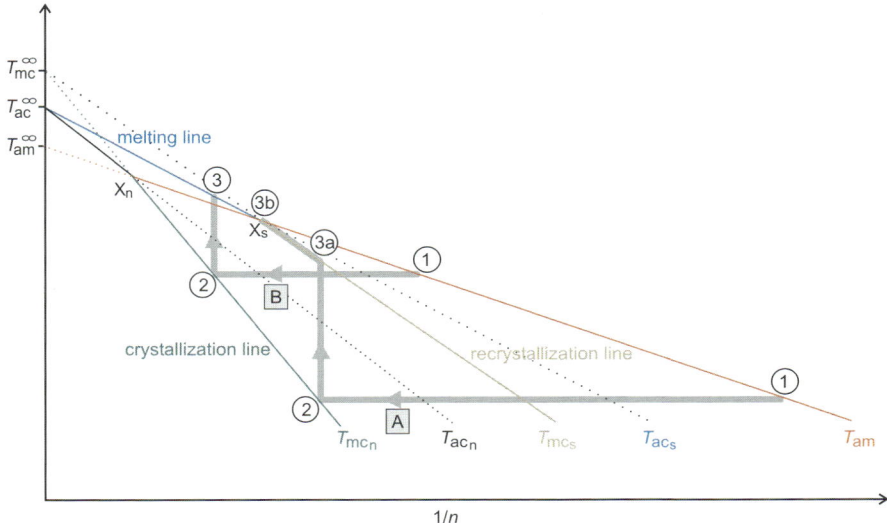

Fig. 23.13. (T/n^{-1})-phase diagram for polymer layers in a melt ("a") dealing with three phases: mesomorphic "m", native crystalline "c_n" and stabilized crystalline "c_s". Two pathways for an isothermal crystallization followed by heating, A (low crystallization temperatures) and B (high crystallization temperatures). The experimental 'crystallization line' is identical with T_{mc_n}, the 'melting line' is identical with T_{ac_s}, the 'recrystallization line' is to be identified with T_{mc_s}

spontaneously thickens until the transition line T_{mc_n} is reached at point '2', where native crystals form immediately. The subsequently following stabilization transforms them into a lower surface free energy state. The consequence of the stabilization shows up during a subsequent heating. Without stabilization a heating would immediately transform the native crystals back into the mesomorphic state, but after the stabilization the situation has changed: Since the crossing point is shifted to location X_s, the crystallites remain stable upon heating until the next transition line is reached. As shown by the scheme, this transition is now a direct melting without the interference of a mesomorphic phase. Pathway A: The beginning is the same – starting at point 1 with an attachment of chain sequences onto a spontaneously thickening mesomorphic layer, then, on reaching T_{mc_n}, the formation of native crystals followed by a stabilization. Heating the stabilized crystals they at first retain their structure. However, as shown in the scheme, at first the transition line T_{mc_s} is reached which relates to a transformation into the mesomorphic state instead of a crystal melting. The consequences which follow are obvious ((3a) to (3b)): The same two steps are repeated again and again, first a transition into the mesomorphic phase and then a thickening until crystals form. The end of this multi-sequence is reached at the crossing point X_s where the crystal melts.

Thermodynamics determines the different transition lines. T_{ac_s} relates to the equilibrium between stabilized crystals and the melt where

$$g_c + \frac{2\sigma_{ac_s}}{n} = g_a . \quad (23.5)$$

σ_{ac_s} describes the surface free energy per crystal stem end for a stabilized layer in the melt. With

$$g_a - g_c \approx \frac{\Delta h_{ac}}{T_{ac}^\infty}(T_{ac}^\infty - T) \quad (23.6)$$

one obtains

$$T_{ac}^\infty - T \approx \frac{2\sigma_{ac_s} T_{ac}^\infty}{\Delta h_{ac}} \frac{1}{n} . \quad (23.7)$$

In experiments this line is addressed as the 'melting line'. Proceeding in analogous manner one obtains for the 'crystallization line' the equation

$$T_{mc}^\infty - T \approx \frac{(2\sigma_{ac_n} - 2\sigma_{am})T_{mc}^\infty}{\Delta h_{mc}} \frac{1}{n} \quad (23.8)$$

and for the 'recrystallization line' the equation

$$T_{mc}^\infty - T \approx \frac{(2\sigma_{ac_s} - 2\sigma_{am})T_{mc}^\infty}{\Delta h_{mc}} \frac{1}{n} \quad (23.9)$$

(σ_{am} and σ_{ac_n} denote respective surface free energies). The transition between the melt and the amorphous state, described by the line T_{am}, occurs for

$$T_{am}^\infty - T \approx \frac{2\sigma_{am} T_{am}^\infty}{\Delta h_{am}} \frac{1}{n} . \quad (23.10)$$

The function $T_{am}(n^{-1})$ begins at T_{am}^∞ and then passes through the two crossing points X_n and X_s. A knowledge of two of these three points is required in order to fix the a⇔m transition line.

23.4 Examples of Application

Figure 23.5 demonstrated for sPP with related copolymers the independence of the crystal thickness from the co-unit content. Figure 23.7 showed for PEcO14 the effect of two different diluents, namely of n-hexadecane and of methylanthracene [17]. The results demonstrated that the effect of diluents can be different: A dissolution of methylanthracene leaves the crystallization line unchanged, producing only a shift in the melting line, but the dissolution of n-hexadecane results in shifts of both the melting- as well as the crystallization line. The thermodynamic scheme provides an understanding, and the two different situations are dealt with in Fig. 23.14. Effects depend on whether or

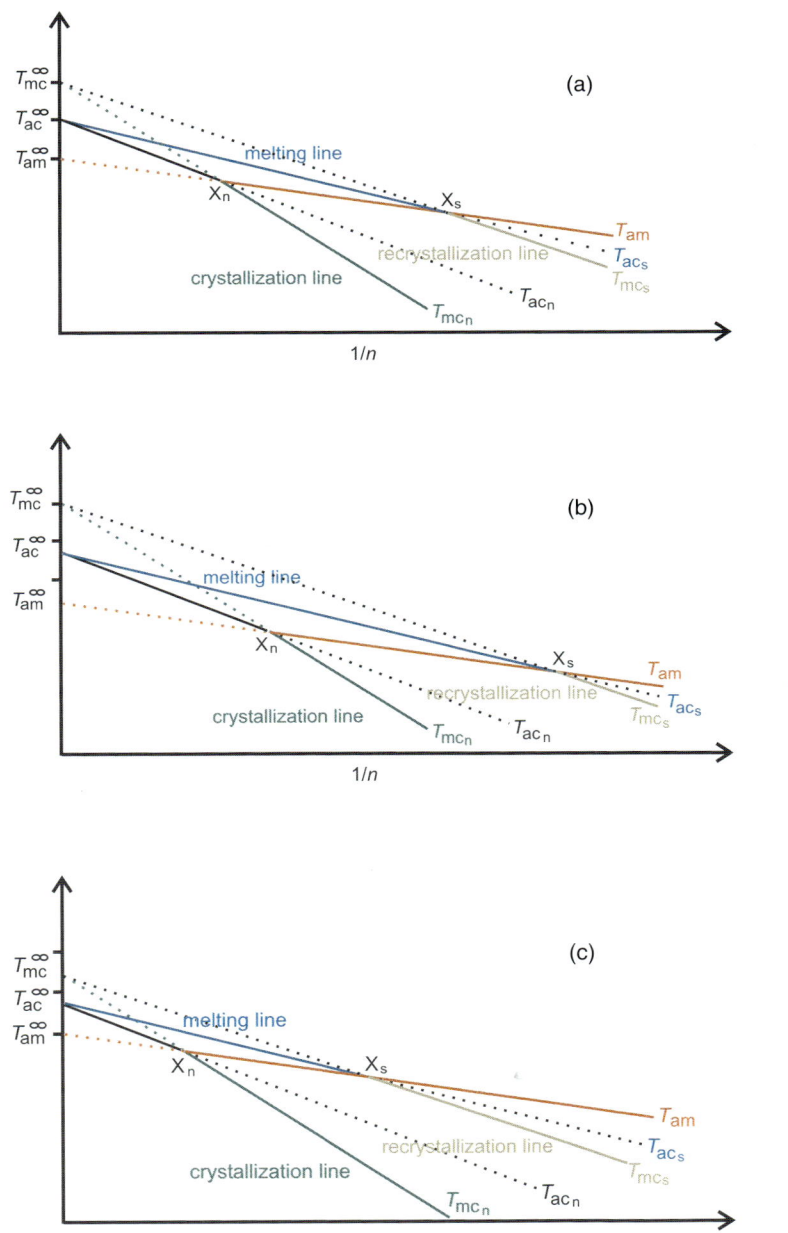

Fig. 23.14. Variations in the (T/n^{-1})-phase diagram introduced by co-units and diluents: (**a**) Homopolymer crystallization (**b**) Effect of co-units or a diluent which remains in the melt: Shift of the melting line but invariant crystallization line; (**c**) Effect of a diluent which enters the mesomorphic phase: Shifts of both the melting line and the crystallization line

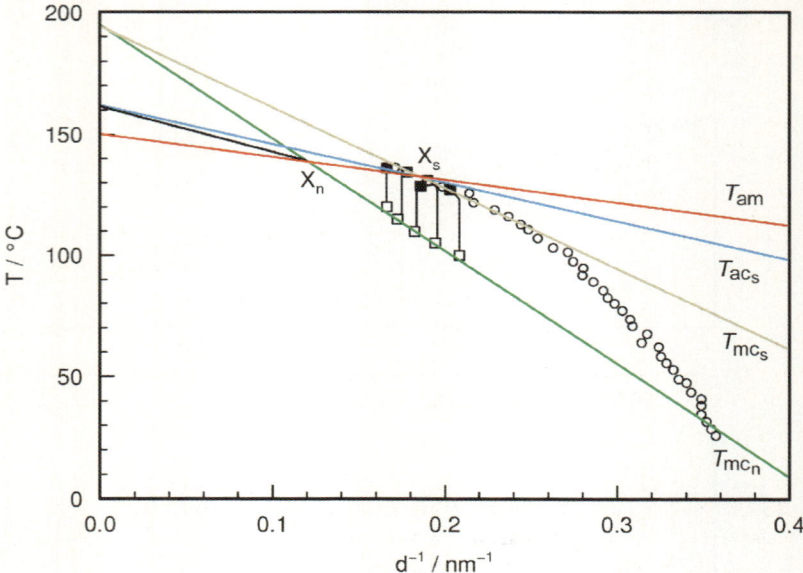

Fig. 23.15. sPP-Mitsui: SAXS data from Fig. 23.9 represented on the basis of the multiphase scheme. In addition to the crystallization line (T_{mc_n}) and the melting line (T_{ac_s}), the figure includes now the recrystallization line (T_{mc_s}), the a⇔m transition line and the crossing points X_n and X_s [9]

not the diluent molecules or the co-units can enter the mesomorphic phase. If they are rejected those transformation lines which include the melt, i.e., T_{ac_n}, T_{ac_s} and T_{am}, are shifted to lower temperatures, but the line T_{mc_n} remains unaffected. This is the situation sketched in part (b). The other situation is encountered if the diluent becomes incorporated into the mesomorphic phase and is only rejected subsequently when the crystals form. Under these conditions (part (c)) all transitions which include the crystalline state are shifted while the transition between the melt and the mesomorphic phase, T_{am}, remains on its place. Such a situation is obviously met if n-hexadecane is used as a diluent for PE, which leads to a shifting of both the crystallization line T_{mc_n} and the melting line T_{ac_s}.

Next examples refer to the results of SAXS and DSC studies on sPP and its copolymers which were reproduced in Figs. 23.4, 23.5 and 23.9. The SAXS data in Fig. 23.9 obtained for the commercial sPP-Mitsui are shown again in Fig. 23.15, now with additional features referring to the scheme. The sample was both cold crystallized from the glassy state at 25°C and crystallized from the melt at several temperatures between 100°C and 120°C. The thicknesses for the various crystallization processes are all located on the crystallization line. As mentioned previously, the changes of the thickness with temperature observed during heating greatly differ. For the three highest temperatures

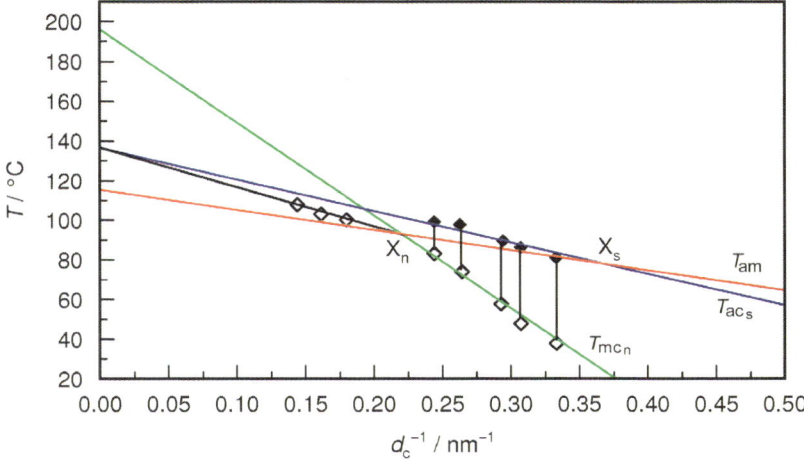

Fig. 23.16. SAXS data of sP(PcO20) from Figs. 23.4 and 23.5: Representation on the basis of the multiphase scheme, with crystallization line (T_{mc_n}), melting line (T_{ac_s}), crossing points X_n and X_s and the a⇔m transition line [9]

thicknesses remain constant up to the melting points. For the cold crystallized sample changes set in immediately when the heating starts. d_c^{-1} approaches and then follows the recrystallization line, until melting occurs near to or at the crossing point X_s. As is obvious, with a crystallization at the three highest temperatures one enters pathway B of the scheme, for the lower temperatures the structure changes during heating are those of pathway A. The temperature at the crossing point X_s appears also in the DSC scans on the right of Fig. 23.9. After an extended range of continuous reorganization, which extends up to 110°C, crystals melt at about 130°C, in agreement with the location of X_s found in the SAXS experiments. The melting line has to pass through the measured melting points.

Figure 23.16 collects SAXS data obtained for sPPcO20 and already displayed in Figs. 23.4 and 23.5. The data well fix the crystallization and the melting line, and the line plotted through the three high temperature points – it represents T_{ac_n} – determines X_n. An additional DSC scan carried out for a cold-crystallized sample yielded the temperature location of X_s, which turned out to be 80°C. The now known locations of the two crossing points X_n and X_s allow to draw the a⇔m transition line. With this, the scheme is fully established.

Having adjusted the scheme for both samples, all relevant thermodynamic data can be derived, and they are collected in Table 23.1.

The heat of fusion $\Delta h_{ac} = 7.7$ kJ/mol C_3H_6 is taken from the literature. The heat of transition $\Delta h_{am} = 5.8$ kJ/mol C_3H_6 follows from a simple consideration based on Fig. 23.12. Thermodynamics relates the three transition

Table 23.1. s-Polypropylene and s-poly(propylene-co-octene): Thermodynamic data following from the experiments

	T_{mc}^∞ °C	T_{ac}^∞ °C	T_{am}^∞ °C	$T(X_n)$ °C	$T(X_s)$ °C	Δh_{ac} kJ/mol C$_3$H$_6$	Δh_{am} kJ/mol C$_3$H$_6$	σ_{ac_n} kJ/mol	σ_{ac_s} kJ/mol	σ_{am} kJ/mol
sPP-Mitsui	195	162	150	139	132	7.7	5.8	9.0	7.5	3.4
sP(PcO20)	195	137	113	92	80	7.7	5.8	9.0	7.5	3.4

temperatures T_{am}^∞, T_{ac}^∞, T_{mc}^∞ with the heats of transition Δh_{ac} and Δh_{am}. Since the slopes of Δg_{am} and Δg_{ac} are given by the entropy changes Δs_{am} and Δs_{ac} respectively, one can write

$$(T_{mc}^\infty - T_{ac}^\infty)\Delta s_{ac} = (T_{mc}^\infty - T_{am}^\infty)\Delta s_{am} , \qquad (23.11)$$

and therefore obtains

$$\frac{\Delta h_{am}}{\Delta h_{ac}} \approx \frac{\Delta s_{am}}{\Delta s_{ac}} = \frac{T_{mc}^\infty - T_{ac}^\infty}{T_{mc}^\infty - T_{am}^\infty} . \qquad (23.12)$$

The three surface free energies were derived from the slopes of the respective transition lines.

23.5 Conclusion

The consistency of the data representation within the framework corroborates the validity of the proposed thermodynamic multiphase scheme. In particular, the correct description of the two modes of structural changes upon heating after crystallizations at low and high temperatures looks convincing. It may be surprising that a mesomorphic phase with properties in between the crystal and the melt should exist for all polymer systems, but the experiments indicate it clearly. Indeed, it is the interference of this mesomorphic phase which generally controls polymer crystallization via the selection of the crystal thickness. Of equal importance for the non-reciprocity of crystallization and melting in polymer systems is the stabilization process which transfers the initial native crystallites into their final stabilized form.

Although not being particularly simple with its multitude of transition lines the proposed scheme still refers to an ideal case in the sense that it addresses the melting behavior of stabilized crystals only. As is demonstrated by the finding of straight Gibbs-Thompson melting lines, these possess definite properties, i.e., constant values of the heat of fusion and the surface free energy. In fact, as is known from various observations, not all the crystals experience a stabilization. In particular those which develop at later times often remain in the native or in an only partially stabilized state. Their presence shows up, for example, in the thickness change during heating of the cold-crystallized sPP shown in Fig. 23.9. Stabilized crystals would not change their thickness up to the temperature at which the recrystallization line is reached, as it is observed for a crystallization at 100°C. In case of the sample crystallized at room temperature thickness changes set in immediately, which indicates that at least a part of the lamellar crystallites has remained in the initial native state. Even a small temperature increase then brings them back into the mesomorphic state, from where they immediately recrystallize after some thickening.

Choosing straight lines for all the transitions in the phase diagram is of course an approximation, the same one which leads to the corresponding equations in Sect. 23.3.1. There are cases, where the data justify this linearization approximation, as for example the s-PP crystallization line in Fig. 23.5 which straightly extends over a range of 100°C. Existing curvatures would modify the three bulk transition temperatures, but it is difficult to estimate the amount of these changes and thus the accuracy of the values given in the tables.

Acknowledging the importance of the mesomorphic phase for the crystal formation in polymers a legitimate question comes up: Why has its occurrence not been reported so far, or only in special cases like the polyethylene crystallization at high pressure? In particular, there exist now many AFM observations with high resolution but so far no images which would have the character of the sketch of Fig. 23.5. The answer could be that the mesomorphic phase is passed through very rapidly, maybe even in the manner that it exists as a transient state during the formation of a block only. The block formation would then resemble the formation of a nucleus, and the building of a crystal lamella consequently a repeated self-supported and guided nucleation. That crystal nucleation can be facilitated by a passage through an intermediate phase is known since Ostwald's time, and it is corroborated by convincing experiments, for example, by the nucleation studies on n-alkanes carried out by Sirota et al [22].

There exist certainly more questions and additional observations. Hopefully, the proposed scheme can serve as a sound basis to discuss them. It appears that it takes up main properties of polymer crystallization and melting in a correct manner.

Acknowledgements

The presented experiments were carried out by Mahmoud Al-Hussein, Michael Grasruck, Georg Hauser, Barbara Heck, Thomas Hippler, Torsten Hugel and Jürgen Schmidtke, and I would like to express my appreciation for their engaged work. Gratefully acknowledged is the support provided by the Deutsche Forschungsgemeinschaft.

References

[1] Faraday Discussion. *Faraday Disc. Chem. Soc.*, 68, 1979.
[2] J.D. Hoffman, G.T. Davis, and J.I. Lauritzen. In *Treatise on Solid State Chemistry* Vol. 3, N.B. Hannay Ed., p. 497. Plenum, 1976.
[3] S. Rastogi, M. Hikosaka, H. Kawabata, and A. Keller. *Macromolecules*, 24:6384, 1991.
[4] A. Keller, M. Hikosaka, S. Rastogi, A. Toda, P.J. Barham, and G. Goldbeck-Wood. *J. Mater. Sci.*, 29:2579, 1994.
[5] M. Imai, K. Kaji, T. Kanaya, and Y. Sakai. *Phys. Rev. B*, 52:12696, 1995.
[6] P.D. Olmsted, W.C.K. Poon, T.C.B. McLeish, T.C.B. Terrill, and A.J. Ryan. *Phys. Rev. Lett.*, 81:373–376, 1998.
[7] G. Hauser, J. Schmidtke, and G. Strobl. *Macromolecules*, 31:6250, 1998.
[8] G. Strobl. *Eur. Phys. J. E*, 3:165, 2000.
[9] G. Strobl. *Eur. Phys. J. E*, 18:295, 2005.
[10] B. Heck, T. Hugel, M. Iijima, E. Sadiku, and G. Strobl. *New J. Physics*, 1:17, 1999.
[11] T.Y. Cho, B. Heck, and G. Strobl. *Colloid Polym Sci*, 282:825, 2004.
[12] M. Iijima and G. Strobl. *Macromolecules*, 33:5204, 2000.
[13] M. Al-Hussein and G. Strobl. *Macromolecules*, 35:8515, 2002.
[14] Q. Fu, B. Heck, G. Strobl, and Y. Thomann. *Macromolecules*, 34:2502, 2001.
[15] T. Hugel. *Diplomarbeit*. Fakultät für Physik, Universität Freiburg, 1999.
[16] T. Hippler, S. Jiang, and G. Strobl. *Macromolecules*, 38:9396, 2005.
[17] B. Heck, G. Strobl, and M. Grasruck. *Eur. Phys. J. E*, 11:117, 2003.
[18] M. Al-Hussein and G. Strobl. *Eur. Phys. J. E*, 6:305, 2001.
[19] M. Grasruck and G. Strobl. *Macromolecules*, 36:86, 2003.
[20] M. Al-Hussein and G. Strobl. *Macromolecules*, 35:1672, 2002.
[21] A.A. Minakov, D.A. Mordvintsev, and C. Schick. *Polymer*, 45:3755, 2004.
[22] E.B. Sirota and A.B. Herhold. *Science*, 283:529, 1999.

Index

ABC triblock terpolymers 230, 255
activation energy 106, 395, 411
adjacent re-entry 286
alkane 457
alkyl nanodomains 204
α'-process 98, 444
α relaxation 97, 436
amphiphilic character 231
analytical shish-kebab model 142
annealing experiments 318
antinucleation effect 255
Arrhenius law 106
athermal nucleation 65, 82
atomic force microscopy (AFM) 182
attachment probability 180
average loop length 23
Avrami constant 392, 419
Avrami law 108
axial flow deformations 67, 69

baby nuclei 7
background over-subtraction 176
bending rigidity 30, 34
β-relaxation process 107
bimodal blend 139
block copolymers 201
boundary layer 12
branched n-alkanes 303
breakout behavior 212
brush-decorated crystal 24

cavitation 348
chain entropy 15
chain folding 48, 288

chain microstructure 349
chain sliding diffusion 297
chain tilt 287, 304
characteristic time scales 146
classical nucleation theory 47
co-unit content 488
coil explosion 316, 322
coil-stretch transition 135, 137
coincident crystallization 234
confined crystallization 201
conformational order parameter 98
contraction during thickening 299
cooperative growth 11
critical cluster volume 82
critical nucleus 462, 471, 473
critical stem length 49
critical strain 347
cross-β orientation 362
{201} crystal surface 262, 269
crystal growth 181, 468
crystal growth front 49
crystal growth rate 391, 393, 419
crystal plasticity 348
crystal structure 163
crystal thickening 303
crystal-melt interface 189, 262, 263, 269, 275, 279
crystalline stems 23
crystalline-crystalline block copolymers 230
crystallization conditions 286
crystallization during processing 329
crystallization kinetics 5

crystallization line   486, 496
crystallization morphology   463
crystallization of polyamide 6   151
crystallization precursor scaffold   134
crystallization temperature depression   157
crystallization theory   3

deformation at break   353
deformed entanglement network   145
dendritic growth   377
denritic square-shaped single-crystal   184
depletion zone   181, 184, 379
detection sensitivity of SANS   173
diblock copolymer   230
dielectric complex permittivity   440
dielectric spectroscopy   436
differential scanning calorimetry   154, 166
diffusion field   379
diffusion limited aggregation (DLA)   180
disentangled crystallizable interphase   285
dissolution of polyamide 6   151
dominant lamellae   51
double concentric spherulites   234

early stage crystallization   119
edge-on lamellae   189, 196
effect of diluents   488
energy barrier for secondary nucleation   244
energy barriers   1
entanglement density   347
entanglements   286
enthalpic elasticity   367, 369
enthalpy of crystallization   154
entropic barrier   12
entropic surface tension   29
epitaxial crystallization   246
equilibrium lamellar thickness   10
equilibrium shape of polymer crystals   20
equilibrium state of polymer crystals   19
ethylene sequence length distributions   308

excess surface free energy   21
extended chain crystals   20, 50
extended-chain conformation   137

fiber morphology   362
finite chain extensibility   67, 69
flat-on lamellae   189, 195
flow potential   79, 81, 83
flow-induced crystallization   134
fold surface   261, 262, 279, 411
fold-end surface free energy   48
folded chain crystals   19
fractal dimension   129
fractal objects   125
fractionated crystallization   250
free energy landscape   2, 8, 15, 47
frustrated alkyl groups   203

Gaussian chain model   34
Gibbs dividing surface   261, 276
Gibbs-Thompson relation   16, 22, 208, 482, 483
Green function   27
growth front nucleation probability   185
growth kinetics   471, 475
growth morphology   375
growth rate measurements   482
growth sector   288

heat capacity   261, 271, 278, 280
Herring equation   261, 269, 279, 281
heterogeneous nucleation   202, 399, 400, 404, 421
hierarchies of molecular organization   1
hierarchy of length scales   202, 212
hierarchy of side-branches   184
high cooling rates   332
Hoffman-Lauritzen treatment   482
homogeneous nucleation   202, 399, 402, 421, 423
hydrolysis of polyamide 6   156

immobilized amorphous phase   445
induction period of crystallization   169
industrial processing   329
infilling growth   377
influence of confinement   202

influence of dissolution 158
instabilities of the growth front 180
integer fraction 298
inter-lamellar amorphous phase 450
interface distribution function 488
interface internal energy 261, 269, 278
interface stresses 261, 262, 278, 279, 281
interlamellar domain 261, 263, 267, 271, 279–281
interphase 268
interphase of intermediate order 286
intra-lamellar slip processes 346
intra-molecular nucleation 418
intramolecular crystal nucleation 51
isotactic polypropylene (iPP) 349
isothermal cold crystallization 443

kinetic barrier 20
kinetically controlled structures 181

lamellar doubling 296
lamellar structure 181
lamellar thickening 463, 465
lamellar thickness 3
localization parameter 28
long oriented structures 89
long range elasticity 348
loop reentry 270, 280
loop-train adsorption 411, 412
loops and tails 23
low molecular weight polyethylene 428

many chain crystals 32
mechanisms of plastic deformation 346
melting line 486, 496
melting process 386
melting temperature depression 157
memory effect 399, 404
mesomorphic form 363
mesomorphic precursor phase 6
metallocene catalysts 349
metastability of morphological entities 330
metastable hexagonal phase 294
metastable states 2, 20
microphase-separation 203
molar mass distribution 158

molar mass fractionation 51
molecular dynamics 457
molecular simulation 457
molecular transport term 392, 394, 397, 402, 403, 413, 417
molecular weight dependence 392, 409, 410, 414, 416, 419
Monte Carlo simulations 261, 262, 268, 276
morphology development 181
Mullins-Sekerka instability 377

nascent morphology 294
negative spherulites 240
nematic structure 104
non-Gaussian chain 72
non-integer fraction 303
non-reversing process 433
nucleating effect 377
nucleation 461
nucleation and growth 1, 118, 169
nucleation exclusion zone 401
nucleation rate 4, 391, 399, 402, 404, 416
nucleus density 400
number of folds per chain 32

order disorder transition 230
oriented nucleation 65
Ostwald's rule of stages 492
overall crystalline fraction 440
overall crystallization rate 393, 419

phase diagram 362
phase segregated microdomains 230
plastic deformation 348, 364
poly(ethylene oxide)-b-poly(ε-caprolactone) 230, 232
polyamide 4.6 165
polyamide 6 153, 158, 161
polyethylene 261, 263, 268, 457
polyethylene-like glass transition 204
polymer processing 460, 473
polymorphic transformations 348
pre-crystalline structure 102
pre-ordering before crystallization 118
precursor phase 482
precursor of shish-kebab structure 87
primary crystal nucleation 47

primary crystallization 208
process of "reeling-in" 289
protein molecules 20
pulse shear 89

rapid scanning AFM (VideoAFM™) 373
recrystallization line 496
recrystallization processes 488
regime theory 396
regiodefects 347
relaxation time distribution function 442
2D rheo-SAXS patterns 139
rigid amorphous fraction 313
ripples 186, 189
rr triads 349

2D scattering pattern 91
screw dislocations 185, 196
secondary crystal nucleation 49
secondary crystallization 208
segmental motion 97
segregation of impurities 51
sheared polymer melts 137
shish-kebab crystalline structures 134
shish-kebab structure 87
side chain crystallization 205
side chain polymers 202
single chain crystal 26, 29
size exclusion chromatography 156, 158
sliding entropy 24
slip-loop model 26
small angle light scattering 122
smectic structure 99
spatial restriction 248
spinodal decomposition 169
square-like spirals 188
stable orthorhombic phase 294
step-scan alternating DSC 428
stereodefects 347
stereoregularity 349, 356
stiffness matrix 272, 274

strain hardening 292, 354
stress-induced crystallization 462
stress-induced melting 346
stress-induced phase transitions 349
stress-strain curves 347, 354, 364
stretching direction 362
subsidiary lamellae 51
supercooled liquid state 435
surface crystal nucleation 49
surface nucleation 471
synchrotron radiation 441, 443
syndiotactic polystyrene (sPS) 332

temperature modulated DSC 99
terrace length 397
thermal expansion coefficients 275, 280
thermodynamic multiphase scheme 483
thermodynamic stability 20
thermoplastic elastomers 356
thickening process 301
tight loops 30
tilting of stems 37
time scale of structure formation 98
tip-splitting 378
transesterification 447
transient mesomorphic phase 483, 501
transient orientation distribution of chain segments 65, 85
type of chain folding 246

ultra-long monodisperse alkanes 298
ultra-small-angle X-ray scattering 117

vapor pressure of water and ethanol 154

welding behavior 296
wide-angle X-ray diffraction 155
worm-like chain model 35
Wulff's construction 22

X-ray scattering patterns 99

Lecture Notes in Physics

For information about earlier volumes
please contact your bookseller or Springer
LNP Online archive: springerlink.com

Vol.668: H. Ocampo, S. Paycha, A. Vargas (Eds.), Geometric and Topological Methods for Quantum Field Theory

Vol.669: G. Amelino-Camelia, J. Kowalski-Glikman (Eds.), Planck Scale Effects in Astrophysics and Cosmology

Vol.670: A. Dinklage, G. Marx, T. Klinger, L. Schweikhard (Eds.), Plasma Physics

Vol.671: J.-R. Chazottes, B. Fernandez (Eds.), Dynamics of Coupled Map Lattices and of Related Spatially Extended Systems

Vol.672: R. Kh. Zeytounian, Topics in Hyposonic Flow Theory

Vol.673: C. Bona, C. Palenzula-Luque, Elements of Numerical Relativity

Vol.674: A. G. Hunt, Percolation Theory for Flow in Porous Media

Vol.675: M. Kröger, Models for Polymeric and Anisotropic Liquids

Vol.676: I. Galanakis, P. H. Dederichs (Eds.), Half-metallic Alloys

Vol.677: A. Loiseau, P. Launois, P. Petit, S. Roche, J.-P. Salvetat (Eds.), Understanding Carbon Nanotubes

Vol.678: M. Donath, W. Nolting (Eds.), Local-Moment Ferromagnets

Vol.679: A. Das, B. K. Chakrabarti (Eds.), Quantum Annealing and Related Optimization Methods

Vol.680: G. Cuniberti, G. Fagas, K. Richter (Eds.), Introducing Molecular Electronics

Vol.681: A. Llor, Statistical Hydrodynamic Models for Developed Mixing Instability Flows

Vol.682: J. Souchay (Ed.), Dynamics of Extended Celestial Bodies and Rings

Vol.683: R. Dvorak, F. Freistetter, J. Kurths (Eds.), Chaos and Stability in Planetary Systems

Vol.684: J. Dolinšek, M. Vilfan, S. Žumer (Eds.), Novel NMR and EPR Techniques

Vol.685: C. Klein, O. Richter, Ernst Equation and Riemann Surfaces

Vol.686: A. D. Yaghjian, Relativistic Dynamics of a Charged Sphere

Vol.687: J. W. LaBelle, R. A. Treumann (Eds.), Geospace Electromagnetic Waves and Radiation

Vol.688: M. C. Miguel, J. M. Rubi (Eds.), Jamming, Yielding, and Irreversible Deformation in Condensed Matter

Vol.689: W. Pötz, J. Fabian, U. Hohenester (Eds.), Quantum Coherence

Vol.690: J. Asch, A. Joye (Eds.), Mathematical Physics of Quantum Mechanics

Vol.691: S. S. Abdullaev, Construction of Mappings for Hamiltonian Systems and Their Applications

Vol.692: J. Frauendiener, D. J. W. Giulini, V. Perlick (Eds.), Analytical and Numerical Approaches to Mathematical Relativity

Vol.693: D. Alloin, R. Johnson, P. Lira (Eds.), Physics of Active Galactic Nuclei at all Scales

Vol.694: H. Schwoerer, J. Magill, B. Beleites (Eds.), Lasers and Nuclei

Vol.695: J. Dereziński, H. Siedentop (Eds.), Large Coulomb Systems

Vol.696: K.-S. Choi, J. E. Kim, Quarks and Leptons From Orbifolded Superstring

Vol.697: E. Beaurepaire, H. Bulou, F. Scheurer, J.-P. Kappler (Eds.), Magnetism: A Synchrotron Radiation Approach

Vol.698: S. Bellucci (Ed.), Supersymmetric Mechanics – Vol. 1

Vol.699: J.-P. Rozelot (Ed.), Solar and Heliospheric Origins of Space Weather Phenomena

Vol.700: J. Al-Khalili, E. Roeckl (Eds.), The Euroschool Lectures on Physics with Exotic Beams, Vol. II

Vol.701: S. Bellucci, S. Ferrara, A. Marrani, Supersymmetric Mechanics – Vol. 2

Vol.702: J. Ehlers, C. Lämmerzahl, Special Relativity

Vol.703: M. Ferrario, G. Ciccotti, K. Binder (Eds.), Computer Simulations in Condensed Matter Systems: From Materials to Chemical Biology Volume 1

Vol.704: M. Ferrario, G. Ciccotti, K. Binder (Eds.), Computer Simulations in Condensed Matter Systems: From Materials to Chemical Biology Volume 2

Vol.705: P. Bhattacharyya, B.K. Chakrabarti (Eds.), Modelling Critical and Catastrophic Phenomena in Geoscience

Vol.706: M.A.L. Marques, C.A. Ullrich, F. Nogueira, A. Rubio, K. Burke, E.K.U. Gross (Eds.), Time-Dependent Density Functional Theory

Vol.707: A.V. Shchepetilov, Calculus and Mechanics on Two-Point Homogenous Riemannian Spaces

Vol.708: F. Iachello, Lie Algebras and Applications

Vol.709: H.-J. Borchers and R.N. Sen, Mathematical Implications of Einstein-Weyl Causality

Vol.710: K. Hutter, A.A.F. van de Ven, A. Ursescu, Electromagnetic Field Matter Interactions in Thermoelastic Solids and Viscous Fluids

Vol.711: H. Linke, A. Månsson (Eds.), Controlled Nanoscale Motion

Vol.712: W. Pötz, J. Fabian, U. Hohenester (Eds.), Modern Aspects of Spin Physics

Vol.713: L. Diósi, A Short Course in Quantum Information Theory

Vol.714: Günter Reiter and Gert R. Strobl (Eds.), Progress in Understanding of Polymer Crystallization

Printing: Krips bv, Meppel
Binding: Stürtz, Würzburg